COASTAL UPWELLING

Its Sediment Record

Part B: Sedimentary Records of Ancient Coastal Upwelling

NATO CONFERENCE SERIES

I Ecology
II Systems Science
III Human Factors
IV Marine Sciences
V Air–Sea Interactions
VI Materials Science

IV MARINE SCIENCES

Volume 1 Marine Natural Products Chemistry
edited by D. J. Faulkner and W. H. Fenical

Volume 2 Marine Organisms: Genetics, Ecology, and Evolution
edited by Bruno Battaglia and John A. Beardmore

Volume 3 Spatial Pattern in Plankton Communities
edited by John H. Steele

Volume 4 Fjord Oceanography
edited by Howard J. Freeland, David M. Farmer, and Colin D. Levings

Volume 5 Bottom-interacting Ocean Acoustics
edited by William A. Kuperman and Finn B. Jensen

Volume 6 Marine Slides and Other Mass Movements
edited by Svend Saxov and J. K. Nieuwenhuis

Volume 7 The Role of Solar Ultraviolet Radiation in Marine Ecosystems
edited by John Calkins

Volume 8 Structure and Development of the Greenland–Scotland Ridge
edited by Martin H. P. Bott, Svend Saxov, Manik Talwani, and Jörn Thiede

Volume 9 Trace Metals in Sea Water
edited by C. S. Wong, Edward Boyle, Kenneth W. Bruland, J. D. Burton, and Edward D. Goldberg

Volume 10A Coastal Upwelling: Its Sediment Record
Responses of the Sedimentary Regime to Present Coastal Upwelling
edited by Erwin Suess and Jörn Thiede

Volume 10B Coastal Upwelling: Its Sediment Record
Sedimentary Records of Ancient Coastal Upwelling
edited by Jörn Thiede and Erwin Suess

COASTAL UPWELLING

Its Sediment Record

Part B: Sedimentary Records of Ancient Coastal Upwelling

Edited by

Jörn Thiede

Department of Geology
University of Kiel
Kiel, Federal Republic of Germany

and

Erwin Suess

School of Oceanography
Oregon State University
Corvallis, Oregon

Published in cooperation with NATO Scientific Affairs Division

PLENUM PRESS · NEW YORK AND LONDON

Library of Congress Cataloging in Publication Data

NATO Advanced Research Institute on Coastal Upwelling and Its Sediment Record (1981: Vila Moura, Portugal)
Sedimentary records of ancient coastal upwelling.

(Coastal upwelling, its sediment record; pt. B) (NATO conference series. IV, Marine sciences; 10B)
"Published in cooperation with NATO Scientific Affairs Division."
Includes bibliographical references and index.
1. Sedimentation and deposition—Congresses. 2. Upwelling (Oceanography)—Congresses. I. Thiede, Jörn, 1941– . II. Suess, Erwin. III. North Atlantic Treaty Organization. Scientific Affairs Division. IV. Title. V. Series: NATO Advanced Research Institute on Coastal Upwelling and Its Sediment Record (1981: Vila Moura, Portugal). Coastal upwelling, its sediment record; pt. 2. VI. Series: NATO conference series. IV, Marine sciences; 10B.
QE471.N39 1981 pt. 2 551.3′6s [551.3′6] 83-8011
[QE571]
ISBN 0-306-41352-3

Second half of the proceedings of a NATO Advanced Research Institute on Coastal Upwelling and Its Sediment Record, held September 1–4, 1981, at Vilamoura, Portugal

PREFACE

NATO Advanced Research Institutes are designed to explore unresolved problems. By focusing complementary expertise from various disciplines onto one unifying theme, they approach old problems in new ways. In line with this goal of the NATO Science Committee, and with substantial support from the U.S. Office of Naval Research and the Seabed Assessment Program of the U.S. National Science Foundation, such a Research Institute on the theme of Coastal Upwelling and Its Sediment Record was held September 1-4, 1981, in Vilamoura, Portugal.

The theme implies a modification of uniformitarian thinking in earth science. Expectations were directed not so much towards finding the key to the past as towards exploring the limits of interpreting the past based on present upwelling oceanography. Coastal upwelling and its imprint on sediments are particularly well-suited for such a scientific inquiry. The oceanic processes and conditions characteristic of upwelling are well understood and are a well-packaged representation of ocean science that are familiar to geologists, just as the magnitude of bioproduction and sedimentation in upwelling regimes --among other biological and geological processes--have made oceanographers realize that the bottom has a feedback role for their models.

The organization of these two volumes of proceedings reflects much of the initial intentions of this conference. This is the first time that the sedimentary response to coastal upwelling has been examined exclusively and in a joint effort by oceanographers and geologists. Experts in upwelling oceanography have kept sedimentologic, geochemical and paleoceanographic implications in mind and geologists relate findings and interpretations to their colleagues in terms of known oceanographic processes. The first volume examines those physical, chemical and biological phenomena which are unmistakably linked to the oceanography of upwelling and which have the potential of leaving an imprint in the sediment record. The goal was to identify and evaluate processes whose very existence, magnitude and products can be interpreted from the sediment record so that ultimately the evolution of upwelling regimes might be traced through time. The success of such attempts depends on finding appropriate answers to the following questions:

What controls the distribution and duration of upwelling centers?

What characteristic terrigenous, biogenous and authigenic constituents directly reflect temperature, salinity, nutrient make-up and current distributions of upwelling systems?

What is the relationship between bioproduction, vertical transfer, recycling and burial of biogenic components, particularly of organic matter?

What is the role of the well-developed oxygen minimum layer that impinges onto the sea floor in upwelling regimes and which imprints and modifies organic geochemical and inorganic geochemical signals?

The second volume explores the ancient sediment record for indications of upwelling as we know it today. Here the focus is on questions such as:

Are there sufficient independent micropaleontological, geochemical and sedimentological criteria for most of the large upwelling zones in today's ocean which could be identified and recognized in the fossil record?

How well do different time- and space-scales permit detailed interpretations of ancient upwelling records?

Is there solid evidence from the sedimentary regimes of fossil upwelling zones for which there is no present analogous oceanographic situation?

Each of the conference topics is introduced by a review or conceptual article designed to summarize up-to-date information within oceanography and geology. These are followed by articles which treat in detail circulation patterns of coastal upwelling; particulate organic matter production, transfer and preservation; patterns of dissolved nutrients; geochemistries of organic and inorganic geochemical constituents; and regional patterns of upwelling facies. The histories of Holocene upwelling regimes are examined as well as the upwelling records from the Pleistocene and Tertiary, the Mesozoic and Paleozoic. As with most conference proceedings, the theme is covered heterogenously. Some topics enjoy multiple contributions, others are less well covered, a few are not represented at all. The same is true for documentation; some contributions contain new and original data published here for the first time, others contain selected material previously published in different contexts. Still others summarize earlier and ongoing research under the conference's theme. Such variety, paired with the personal flavor of individual contributors, represent the true state-of-the-science in this field and the spirit of Vilamoura. This format should allow the reader rapid exposure to the theme of the conference as it has contributed to intensive interdisciplinary exchange among the conference participants. All participants and contributors gave generously and enthusiastically of their wisdom and knowledge and we extend to them, also in behalf of the conference sponsors, our gratitude.

Special thanks are due to our colleagues from the organizing committee, R.T. Barber (Beaufort), S.E. Calvert (Vancouver), J.H. Monteiro (Lisbon), E. Seibold (Bonn) and R.L. Smith (Corvallis), and the conference session chairmen without whose dedication the initial plan of the conference could not have been brought to a successful completion. Many thanks are also due numerous colleagues who diligently reviewed the individual contributions in preparation for this publication.

We gratefully acknowledge the superb clerical assistance of J.L. Dickson and M.J. Armbrust and the editorial advice and help of Z. Suess (all of Corvallis); their continued enthusiasm was invaluable in the task of editing the proceedings. We want to thank L. Schmidt of Plenum Publishing Corporation (New York) for continued and efficient work on the technical editing.

Finally, we acknowledge the financial contributions towards printing costs of these volumes by Exxon Research Corporation (Houston), British Petroleum Corporation (London), Deutsche Texaco A.G. (Wietze) and the OSU Foundation, Research Office and School of Oceanography (Oregon State University, Corvallis).

Erwin Suess Jörn Thiede*

Corvallis and Oslo
January, 1983

* Present address:
Geologisch-Paläontologisches Institut
der Universität
Kiel, Olshausenstrasse
Federal Republic of Germany

CONTENTS OF PART B

High Resolution Holocene Time Scales

Pleistocene Time Scales

Pre-Pleistocene Phanerozoic Time Scales

CONTENTS OF PART A

SEDIMENTARY RECORDS OF ANCIENT COASTAL UPWELLING

INTRODUCTION

Jörn Thiede

Institute for Geology
University of Oslo
Blindern, Oslo 3, Norway

Erwin Suess

School of Oceanography
Oregon State University
Corvallis, Oregon 97331, U.S.A.

Coastal upwelling is a large-scale perturbation of the oceanic surface and subsurface circulation which develops along continental margins in response to specific elements of the tropospheric circulation. These elements, originating in the atmosphere, are predominantly equatorward winds along the coast resulting from quasi-stationary mid-ocean high pressure systems. The configuration of the modern oceans and the present climatic zonation have led to the development of four major areas of year-round coastal upwelling as part of the eastern boundary currents of the Atlantic and Pacific oceans: off northwest Africa, southwest Africa, western North America, and northwestern South America (Figure 1). A major seasonal upwelling area exists along the northern rim of the Indian Ocean basin in response to the monsoonal wind regimes. In addition, locally persistent coastal upwelling may develop anywhere in surface waters when segments of coast lines, or even ice margins, parallel regional surface wind regimes.

It has been known for some time that the sediments under such coastal upwelling regimes are different from those of the adjacent shelf, slope and deep-sea floors and that they record some complex signal of upwelling oceanography (Brongersma-Sanders, 1948). However, the details had remained vague as to how these signals were to be understood in oceanographic terms and which specific parameters were actually recorded in the sediments. Such details have now been reviewed and documented in the first part of this series of articles (e.g., Jones et al.; Smith; Traganza et al.). The second part

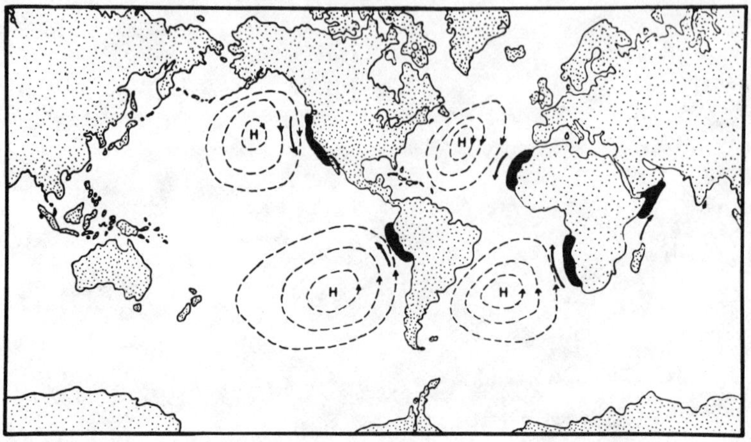

Fig. 1. The five major coastal upwelling regions of the world, the sea level atmospheric pressure systems and major currents that influence them. The dashed circles represent mean idealized positions of isobars during the season of strongest upwelling, from Hartline (1980). By permission from Science, Copyright (c) 1980, AAAS Washington.

addresses the "Sedimentary Records of Ancient Coastal Upwelling". Here the regional distributions of specific upwelling sediment facies are described which illustrate vividly how fluxes of biogenic, clastic and authigenic components combine to form consecutive facies belts and which, in turn, are related in very specific ways to the upwelling process. Distinctive upwelling sediments are deposited in well-defined regions; they have unique chemical and physical properties, textural characteristics and sedimentary structures. However, in most cases, and particularly in the fossil record, it is a combination of these properties which makes it possible to identify upwelling facies, since individual parameters are often insufficient or ambiguous. Micropaleontological and organic geochemical methods have probably come furthest within the range of presently available observational capabilities in identifying specific upwelling regions (Brassel and Eglinton; Dunbar; Schrader and Baumgartner; Wefer et al.; all this volume).

In trying to describe upwelling regimes of the geologic past and understand the geologic observations in terms of oceanographic processes one immediately faces sizable problems. The incomplete geological record, both in space and time, often precludes establishing the diverse data base needed to arrive at a unique interpretation. Likewise, limited time-resolution curtails quasi-synoptic, map-like reconstructions of well-defined stratigraphic horizons deposited under upwelling conditions and it affects time-series measurements of "indicator-parameters" from selected locations. The nature of geological samples in general does not allow the identification of events

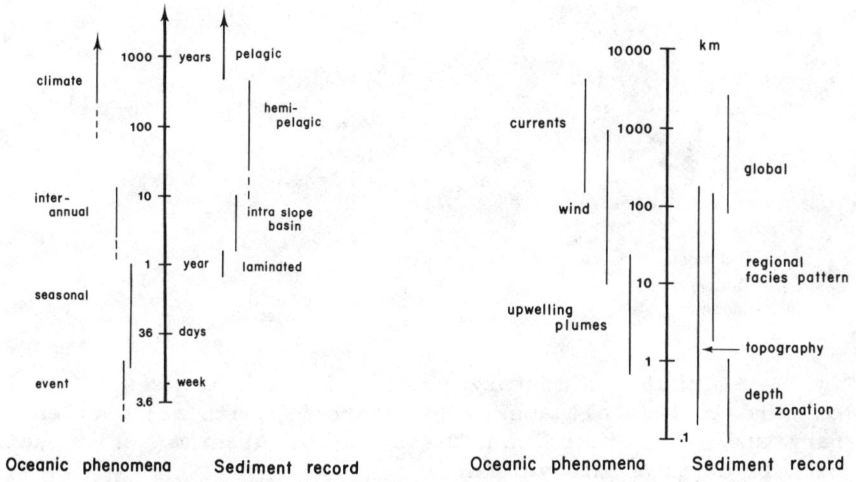

Fig. 2. Schematic illustration of (a) time- and (b) length-scales of oceanic phenomena and of the resolvable geologic record preserved in the sediments.

less than one year in duration, as is possible in laminated sediments or in growth rhythms documented from marine shell material. Such records are available only from very few localities, often not in an optimal geographic position for prolonged upwelling to have occurred and, they usually cover only the youngest geologic past, the Tertiary and Quaternary periods. Therefore, in most cases the upwelling signals found in the geological record average oceanographic processes over a much longer time scale than modern oceanographic observations do (Fig. 2). It is not clear how this "smoothing function" changes the sediment properties; rapid changes in the hydrography of upwelling regimes might prevent the geologic record from accumulating enough detail. For example, the controversy over whether or not Quaternary phosphorites formed as a result of coastal upwelling in western and eastern boundary current regimes (Fig. 3) appears to be in large part such a "smoothing" problem (Burnett, Roe and Piper; O'Brien and Veeh; this volume). It is possible that the slow-growing Holocene phosphorite nodules at western boundary current regimes began forming in pre-Holocene times when these regimes were experiencing coastal upwelling, just as the eastern boundary current regimes do today, thereby obscuring the geologic record.

So far we have attempted to understand geological records in terms of the dynamics of modern upwelling regimes as we know them from today's oceanographic conditions. However, when looking back into the geologic past we encounter situations without modern analogs, and it remains to be seen if we can use today's available insights to adequately describe ancient oceans and their upwelling re-

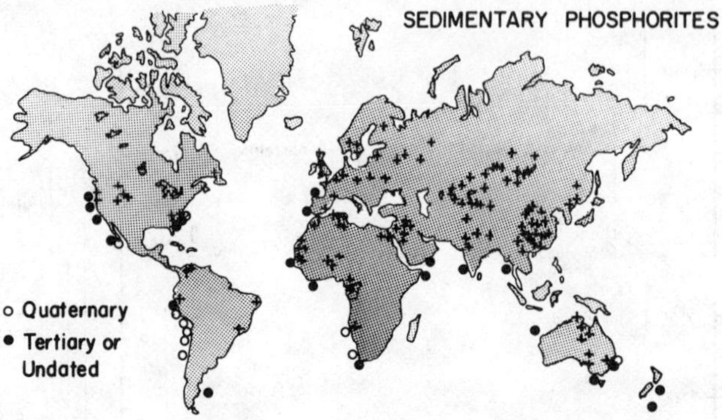

Fig. 3. Global sedimentary phosphorite occurrences. Deposits shown on land include all ages. Submarine deposits are divided into Quaternary and "all others" on the basis of uranium series dating (from Burnett et al., this volume).

gimes. The situations for which we do not have modern analogs result from dynamics of the lithosphere, atmosphere, hydrosphere and biosphere which have changed through time and which jointly control the evolution of the depositional environments in the oceans and along continental margins. A change in one of the elements of the global paleoenvironment will often find a response in another, but the nature of the geological record prevents us from establishing the proper causal relationships between these changes.

Studies of modern upwelling regimes have taught us their dimensions and geographic locations (Figs. 1 and 2). However, in the geologic past when the physiography of the global surface was considerably different from today and when coastal upwelling took place with other dimensions, in other geographic locations and in different climatological settings we have no observable situation to guide our interpretations. Short- and long-term sea level changes due to climatic and tectonic processes, in their broadest sense throughout the entire Phanerozoic, have resulted in large displacements of the coast lines (Vail et al., 1977). During times of low sea level stands the coast lines were located close to the shelf edge, during times of high stands entire continental platforms were flooded. What were then the dimensions of coastal upwelling regimes with coastlines located close to the shelf edge? Where did upwelling develop in wide shelf seas? How does sea level respond to changes of the geoid, how fast can such changes occur, and how do these changes affect coastal upwelling?

Plate tectonic history and the resulting paleogeography of our globe have been reconstructed in detail for the past 150-180 million years. Reconstructions of older intervals of the Phanerozoic (Fig.

4) are considerably more ambiguous, but they do allow the identification of situations where orientation and location of continental margins, mountain chains and oceans were such that upwelling regions could develop in different settings from those of today. The long zonal coastlines developed during mid-Paleozoic times (Fig. 4); the asymmetric distributions of continent and ocean during the existence of Pangaea and Panthalassa are cases with no present-day analog.

Today, rather peculiar flow patterns of the atmosphere and of oceanic water masses are observed because of the development of an extreme climate in the late Cenozoic which resulted in the cooling of the polar atmosphere and hydrosphere and in the growth of large ice sheets. The steep temperature gradients between the poles and the equator are important forcing functions of the atmospheric circulation and of the oceanic current regimes. Only 60-120 million years ago the polar oceans were relatively warm and the oceanic bottom-water current regimes were relatively sluggish. We observe black-shale deposition over wide areas of the Cretaceous oceans and shelf seas. Was coastal upwelling during these times an important process in the formation of vast volumes of marine hydrocarbon source rocks? Can we use our knowledge about the effectiveness and location of modern upwelling regimes to predict where large quantities of organic carbon might have accumulated during the geologic past? Do we have enough data to model flow patterns of the atmosphere and of the oceanic water masses in sufficient detail to locate former upwelling regimes?

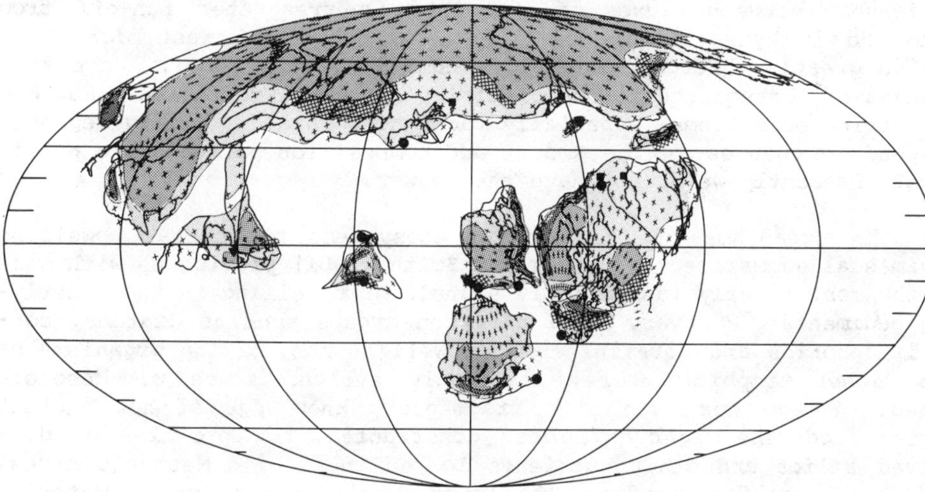

Fig. 4. Silurian phosphorites (triangles), cherts (dots) and organic-carbon-rich rocks (squares). Paleogeography: Light shading= continental shelf; Medium shading=low lands; Dark shading=high lands; White=oceans (from Parrish et al., this volume).

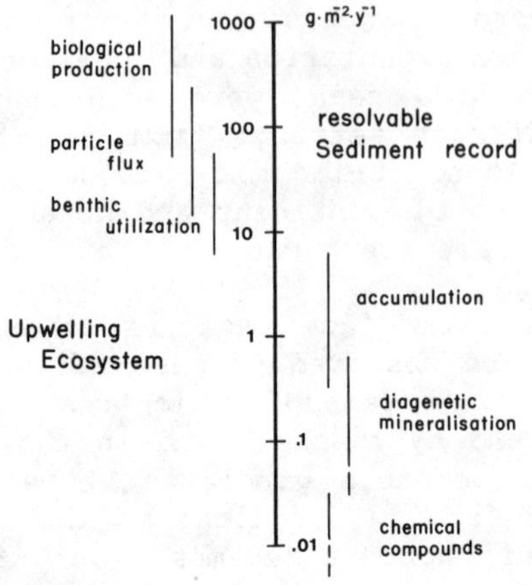

Fig. 5. Changes in flux rates of organic matter in an upwelling ecosystem between initial production in surface waters to eventual burial in sediments.

We know little about the possibilities of when and how the chemistry of ocean water masses and of the atmosphere might have changed through geologic time. Recent studies suggest, for example, the relationship between fluxes of nutrients in freshwater run-off from land and in hydrothermal solutions from the ocean crust might have varied greatly through time (e.g., Edmond et al., 1982). Were such changes of consequence to the chemistry of upwelling facies? How did the marine ecosystems, especially those of coastal upwelling regimes, respond to changes in the amount and composition of available nutrients? Presently we do not have the answer.

The organisms of the present ecosystems of coastal upwelling regimes as summarized by Barber and Smith (1981) provide us with some of the most readily identifiable signals of upwelling in the underlying sediments. However, major plankton groups such as diatoms, coccolithophorids and foraminifers, as well as many of the organisms of the higher trophic levels, have only evolved since mid-Mesozoic times. We do not even have fragmentary knowledge of how ancient marine food-webs might have been constructed, how profiles of dissolved silica and other nutrients looked in pre-mid Mesozoic times, and which organisms produced the large quantities of organic material preserved in the early Phanerozoic upwelling records. These concerns explain why ancient upwelling records are so difficult to interpret.

Studies of the sediments under the modern upwelling regimes have clearly demonstrated how the various signals generated by coastal upwelling are selectively preserved and thereby partially reduced and obliterated. The fate of one group of upwelling signals contained in organic matter is illustrated here in terms of flux rate changes from initial bioproduction to eventual burial (Fig. 5). The significance here is to realize that organic matter is largely lost in this process and that the ultimate signal in the sediment represents the minutest fraction of that which was initially present and which we wish to reconstruct. Sediment components as signal carriers continue to change through time even after burial due to the interaction with the fluids of the sedimentary pore space and under the influence of changing temperatures and pressures. Siliceous components may be altered to chert and organic carbon to hydrocarbons and/or carbonates, to mention the fate of two of the more important components of upwelling facies which we observe in the geologic record (Garrison, 1981). The following chapters in this volume clearly demonstrate how difficult it is with increasing geologic time to achieve high time resolution in a series of samples, but also how remarkably consistent the facies associations remain which are believed to have formed during Upper and Middle Cretaceous, Ordovician and Cambrian upwelling.

REFERENCES

Barber, R.T. and Smith, R.L., 1981, Coastal upwelling ecosystems, in: "Analysis of Marine Ecosystems," A.R. Longhurst, ed., Academic Press, London, 31-68.

Brongersma-Sanders, M., 1948, The importance of upwelling water to vertebrate paleontology and oil geology, Verhandelingen der Koninklijke Nederlandsche Akademie van Wetenschappen, Afd. Naturkunde, Sec. 2, 45(4):1-112.

Edmond, J.M., von Damm, K.L., McDuff, R.E., and Measures, C.J., 1982, Chemistry of hot springs on the East Pacific rise and their effluent dispersal, Nature, 297:187-192.

Garrison, R.E., ed., 1981, "The Monterey Formation and Related Siliceous Rocks of California," Society of Economic Paleontologists and Mineralogists, Pacific Section, Bakersfield, 327 pp.

Hartline, B.K., 1980, Coastal upwelling: physical factors feed fish, Science, 208:38-40.

Vail, P.R., Mitchum, R.M. and Thompson, S., 1977, Seismic stratigraphy and global changes of sea level, American Association of Petroleum Geologists, Memoir, 26:63-97.

SURFACE SEDIMENT DISTRIBUTIONS

SOME UNIQUE SEDIMENTOLOGICAL AND GEOCHEMICAL FEATURES OF DEPOSITS IN COASTAL UPWELLING REGIONS

Gleb N. Baturin

Institute of Oceanology
Academy of Sciences of the U.S.S.R.
Moscow, U.S.S.R.

ABSTRACT

Coastal upwelling which brings to ocean shelves about 200,000 km^3/year of water enriched in nutrients is influencing the shelf sedimentation possibly more than the world's river runoff. Sediments beneath the upwelling regions, in particular the Namibian and Peru-Chile shelves, may be greatly enriched in opaline silica, organic matter, phosphorus and associated elements, including Ni, Zn, Mo, Cd, U. The same pertains to the interstitial waters of these sediments. Specific indicators of upwelling in the sedimentary record include bone breccias, coprogenic material, phosphorites, and pyrite. Besides being highly concentrated, the organic matter bears a number of specific features in its composition. The stable isotopes of carbon, oxygen and sulphur in various sedimentary materials from these regions may also be indicative of upwelling, as can the ratios of some elements including rare earth elements (REE) and transition metals.

A GLOBAL ASSESSMENT OF UPWELLING CIRCULATION, NUTRIENTS, BIOTA AND SEDIMENTATION

Coastal upwelling is a remarkable feature of circulation in the world's ocean. It is the vertical upward movement of water from a subsurface layer caused by the interaction of coastal currents, winds, and the Coriolis force (Gunther, 1936; Hart and Currie, 1960; Wooster and Reid, 1963; Smith, 1968; and many others). Evidently all three forces have operated in the ocean over a considerable period of time in the past; hence the phenomenon of coastal upwelling may be as old as the ocean itself. Because water masses are generally stratified in their physical and chemical properties, the upwelled waters, by locally disturbing these properties, inevitably influence the climate, biological productivity and sedimentation in the area.

The combined length of the coasts influenced by permanent or seasonal upwelling in the Pacific, Atlantic and Indian Oceans is approximately 10,000 km with an average width of about 50 km. Thus, the total area presently affected by coastal upwelling is estimated to be $0.5 \cdot 10^6$ km^2, or only ∿0.14% of the ocean's surface. With an average upwelling rate of 0.5-1 m/day the total mass of upwelled water may reach 200,000 km^3/year (Wooster and Reid, 1963; Ryther, 1969; Burkov, 1972). This exceeds global river runoff by a factor of 3 to 5 (Table 1).

When compared to open ocean surface waters, the upwelled waters are enriched in all nutrients, particularly in phosphate and silica (Bogoyavlensky and Shishkina, 1971; Calvert and Price, 1971). Coastal upwelling may carry about $(10\text{-}20) \cdot 10^6$ tons of dissolved phosphate and $(200\text{-}400) \cdot 10^6$ tons of dissolved silica annually to the near-shore surface waters, enhancing biological productivity by 2 to 3 orders of magnitude above that of pelagic waters (Ryther et al., 1971). The total masses of nutrients utilized by biota in these zones are comparable to those delivered by upwelling or are even greater due to repeated utilizations in the biologic food web.

The high biological productivity of upwelling zones is accompanied by a number of unique phenomena: episodic outbursts of plankton blooms ("red tides"), mass mortality of fishes and other animals, H_2S-invasions of the water column, and accumulations of guano on nearby coasts or islands (Hutchinson, 1950; Copenhagen, 1953; Brongersma-Sanders, 1957). The high biological productivity further causes rapid sedimentation of organic detritus, thereby favoring the accumulation of sediments enriched in elements from the biogenic cycle. The preferential enrichment of biogenic elements is further favored by an arid climate on the nearby land, absence of rivers, shallow water depth, low temperatures of surface waters, limited or entirely absent bottom fauna, and low oxygen conditions in the bottom waters.

Sediments beneath upwelling zones may be enriched in amorphous silica, organic matter, phosphorus and elements associated with these carriers. An example of one such recent accumulation is a lens of diatomaceous ooze on the Namibian shelf at 18-24°S and a water depth of 60-130 m which contains averages of 40 wt-% amorphous silica, 5 wt-% organic carbon and 0.2-0.3 wt-% phosphorus, excluding phosphorus bound to bones, coprolites and phosphatic nodules (Baturin, 1970; Bliskovsky, Baturin and Kuzmina, 1975; Burnett, 1977; Bremner; Calvert and Price; and Brongersma-Sanders, all this volume). The linear rate of sedimentation of these oozes, having a dry density of ∿0.5 g/cm^3 wet volume, is approximately 0.5 mm/year (Baturin, 1969; Veeh, Calvert and Price, 1974). Diatoms in this area produce about

Table 1. Some general features of recent coastal upwelling

Characteristic	Zone of Coastal Upwelling	Pelagic Ocean
Total area, $10^6 km^2$	0.5	361
Depth, m	100-300	3800 (mean)
Flux of water, km^3/year	$(100-200)\cdot10^3$	$38\cdot10^3$ (rivers)
Transport of nutrients,		
10^6 tons/year: Phosphorus	10-20	1.5
Silica	100-400	230
Primary productivity of		
C-org, g/m^2/day	2-11	0.05-0.5
Absolute masses of elements taking part in primary production, 10^6 t/year:		
Organic carbon	0.5-1.5	$(35-70)\cdot10^3$
Silica	500 (?)	$(80-160)\cdot10^3$
Phosphorus	up to 20	$(1.5-3)\cdot10^3$
Rate of biogenic sedimentation,		
mm/1000 years	up to 500	1-5
Concentration of biogenic elements in sediments:		
Organic carbon, wt-%	2-20	<0.5
Silica, wt-%	5-70	traces-80
Phosphorus, wt-%	0.2- >1	0.01-0.07
Nickel, ppm	35-455	97 (mean)
Zinc, ppm	18-337	130 (mean)
Molybdenum, ppm	30-500	4 (mean)
Cadmium, ppm	20-30	0.08-0.7
Uranium, ppm	5-60	0.1-3
Concentration of elements dissolved in interstitial waters:		
Alk, mg-eq/ℓ	5-46	1-7
NH_4, mg/ℓ	4-66	0.2-8
Si, mg/ℓ	13-42	4-24
P, mg/ℓ	0.1-8.7	0.01-0.9
U, μg/ℓ	3-650	1-60

Compiled using data from: Baturin, 1969, 1975, 1978; Baturin and Oreshkin, 1981; Bogoyavlensky and Shishkina, 1971; Calvert and Price, 1970; Emelyanov, 1973; Gunther, 1936; Hart and Currie, 1960; Ryther et al., 1971; Wooster and Reid, 1963.

1-2 kg of carbon and 4.5 kg $SiO_2/m^2/yr$, utilizing 40-80% of the phosphorus input each year. Thus as much as 2% of the organic carbon, 2% of the phosphorus, and 8% of the silica involved in primary production may be preserved in the sediments.

The highly reducing chemical environment produced in these sediments by an abundance of reactive organic matter enhances a number of early diagenetic processes. As a result, interstitial waters show sharp concentration differences for a number of nutrients and microelements when compared to seawater or to interstitial waters of pelagic sediments (Table 1). Phosphatization, pyritization and redistribution of several micro-elements in sediments of an upwelling area are among the results of these early diagenetic processes (Baturin, 1969; 1982).

Optimal conditions favoring the long-term accumulation of biogenic sediments beneath upwelling zones do not always occur, however. Unfavorable conditions include steep shelf platform morphologies, high-energy wave climates, strong bottom currents, high rates of terrigenous sedimentation, bioturbation by bottom fauna, and finally, on a geologic time scale, erosion during repeated transgressive-regressive cycles which periodically remove the sediment record. Recent regional examples where unfavorable conditions prevail and no upwelling sediment signals accumulate include the Moroccan shelf, the eastern Arabian margin, and the Oregon margin. Here biogenic material is swept off the shelf by onshore, offshore and alongshore flows or is not deposited in the first place due to steep bottom reliefs.

Thus, one is obliged to turn to the classic idea that there are only a few favorable areas where biogenic element assemblages of amorphous SiO_2, organic carbon, and phosphorus accumulate, which are the most reliable indicators of upwelling. From this point of view the wide-spread association of diatomites, carbonaceous shales and phosphorites in the geologic record is quite reliable proof of active and persistent ancient upwelling.

Sedimentological and Geochemical Criteria

Fecal matter. Specific indicators of upwelling include bone breccias and coprogenic material. Bone breccias are formed during mass mortality events which occasionally occur in upwelling zones (Brongersma-Sanders, 1957; Baturin, 1974). Similar breccias are found in a number of ancient sediment accumulations, in particular the Maikop Formation which also contains rare metal mineralizations (Mstislavsky and Kochenov, 1960). Sedimentation of coprogenic material, equally widespread in recent and ancient marine and oceanic shelf sediments, enhances transport of organic matter and phosphorus to the sediment-water interface (Bushinsky, 1966; Baturin, 1969; 1982; Mikhailov et al., 1972). If the characteristic rod or spindle shapes of fecal matter are preserved, they are easy to recognize ir-

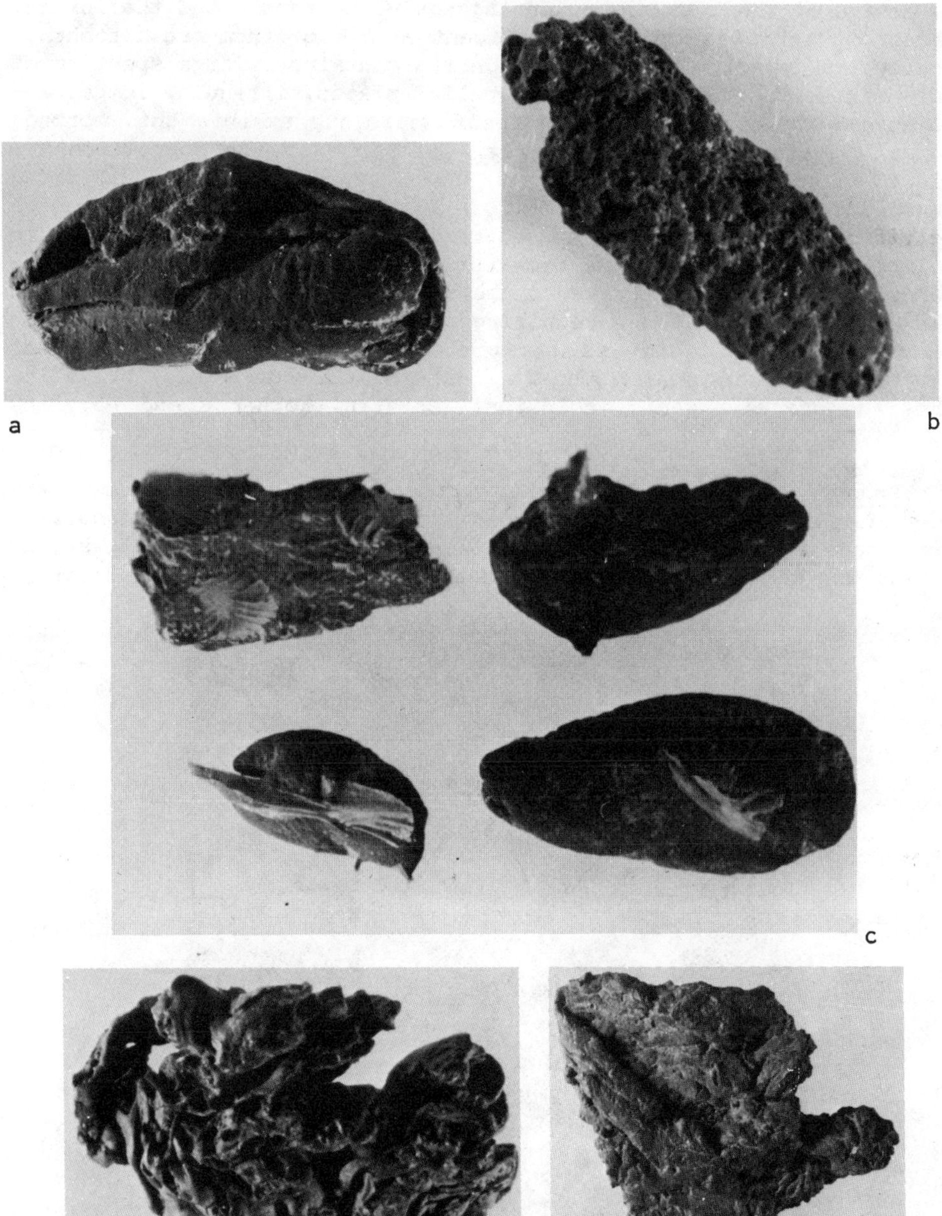

Fig. 1. Coprogenic material from Recent diatomaceous oozes of Namibian shelf. (a) Folded structure of sea-lion coprolite, x2; (b) coprolite with corrosion cavities, x4; (c) debris of coprolites with inclusions of fish bones and scales, x2; (d),(e) some odd forms of coprolite debris, x3.

respective of their size, but during early diagenesis and the initial stages of lithification, a significant number of them are autochthonically fractured. The resulting debris contains a large spectrum of grain sizes and shapes including pellet plates, irregular lumps, aggregates and grains. Namibian shelf sediments exhibit this morphological breakdown quite well (Fig. 1). Some of the coprolites are made of several concentric layers which can be unfolded and fragmented into separate plates (Fig. 1a). The uncertainty of coprolite identification, even in recent sediments, suggests that their role in shelf sedimentation is still underestimated (Scheidegger and Krissek; Staresinic et al., both this volume). At present it seems that the sole formal feature for recognizing coprolites in thin sections (although not too reliable) is the presence of characteristic fluidal structures together with abundant (5-15% of the view field) branching tubes and opaque organic inclusions (Baturin, Kochenov and Petelin, 1970).

Phosphorites. The biogenic sediments of upwelling zones and some of their components, including diagenetic phosphatic nodules, carry several characteristic mineralogical and geochemical signatures. The most common of these is the occurrence of biogenic struc-

Fig. 2. Recent diatom *Rhaphoneis* aff. Wetzelli Mertz included in phosphorite from Peru shelf. Scanning electron microscope, x4000 (Baturin and Dubinchuk, 1979).

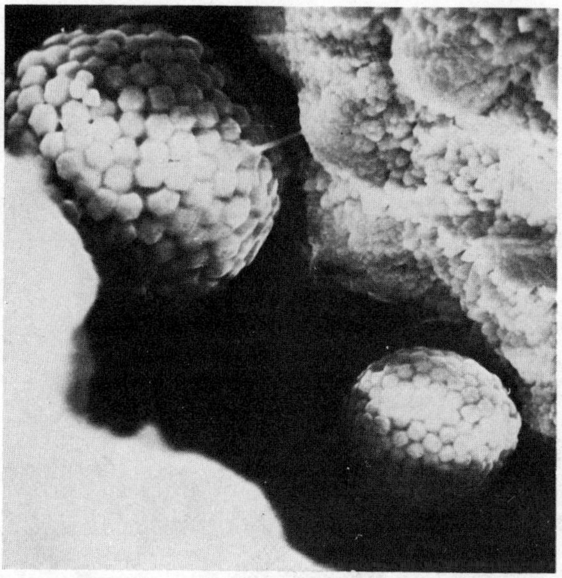

Fig. 3. Globular pyrite from a Recent phosphorite nodule of Namibian shelf. Scanning electron microscope, x12000.

tures in phosphorites. The youngest phosphorites from the Namibian shelf retain the structure of numerous diatom frustules clearly enough for exact micropaleontological identification (Fig. 2). In Miocene phosphatic nodules from the same region, diatom structures can be recognized but not identified (Baturin and Dubinchuk, 1979). Cretaceous phosphorites from north Africa, which are associated with diatomites, often contain dispersed opal spherules comparable in size to diatoms but wholly devoid of diatom-like structures (Mikhailov et al., 1972).

A typical feature of sediments and phosphorites of upwelling zones is the frequent occurrence of framboidal pyrite impregnating biogenic debris and forming microglobules and irregular aggregates (Fig. 3) (Logvinenko, Nikolaeva and Romankevich, 1973; Lein et al., 1978; Baturin and Dubinchuk, 1979). A unique microstructural feature of young phosphorites formed in the sediments of upwelling zones is the extensive distribution of spindle-shaped apatite crystals 1-2 μm in length, occurring separately or in x-shaped pairs, triplets, rosettes, balls or irregular aggregates of various dimensions (Fig. 4) (Burnett and Veeh, 1977; Baturin and Dubinchuk, 1979; O'Brien and Veeh, this volume) (Fig. 4).

Populations of sulphate-reducing bacteria in the biogenic sediments of upwelling zones may reach 10^7 cells per gram of wet ooze (Novozhilova and Baturin, 1973). Recent soft phosphorites also con-

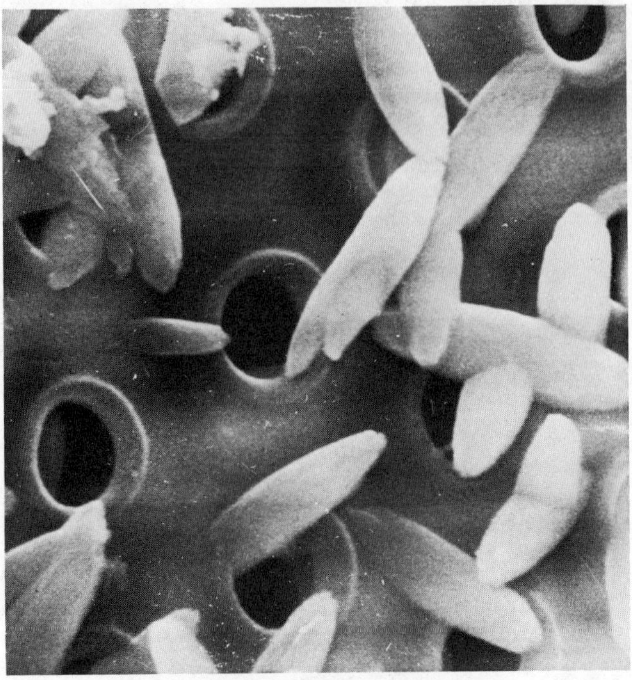

Fig. 4. Spindle-shaped apatite crystals in Recent phosphorite from the Namibian shelf. Scanning electron microscope, x24000.

tain viable cells (Mishustina, 1973). Cayeux (1936) reported the occurrence of abundant bacterial cell molds in phosphorites of all ages. A detailed electron-microscopic study of Recent phosphorites from upwelling zones, however, found only a few identifiable bacterial cells in several of the samples (Fig. 5) (Baturin and Dubinchuk, 1979). Future progress in sample preparation techniques may change this situation and allow bacterial remains in phosphorites to act as a distinctive upwelling signal (O'Brien and Veeh, this volume).

The biogenic sediments of upwelling zones and their associated phosphatic nodules, phosphatized coprolites and bone detritus are enriched in several trace metals, particularly U, Mo, Cd, Zn, Ni (Table 1). For example U-concentrations average 6-8 (rarely 30-60) ppm in sediments, 80-120 ppm in phosphatic nodules, 100-700 ppm in fish remains and 30-70 ppm in phosphatized coprolites (Baturin, 1975). These values are from 2 to 200 times higher than the U content of average pelagic sediments. The highly reducing chemical environment beneath upwelling zones promotes the reduction of uranium and the prevalence of U^{4+} over U^{6+}. Uranium in phosphorites associated with recent upwelling is largely of the U^{4+}-form (Burnett and Veeh, 1977) and is commonly found in well-developed uraninite crys-

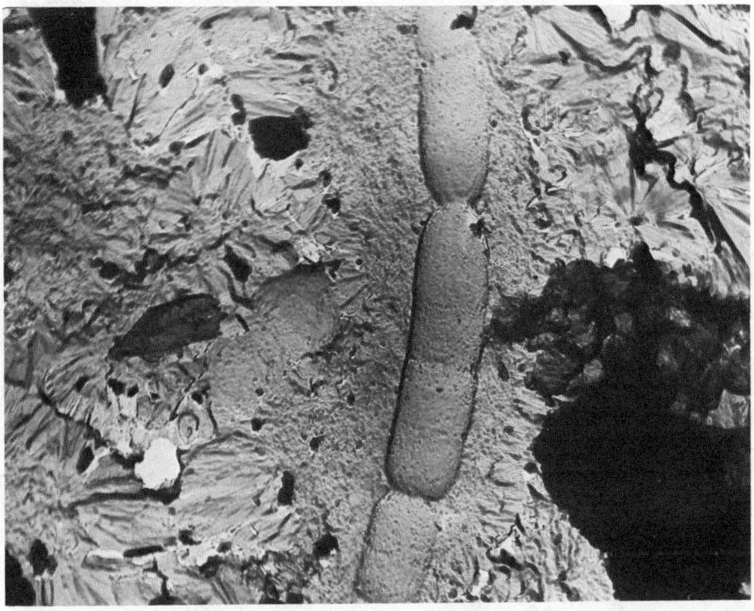

Fig. 5. Phosphatized microbial molds in Recent phosphorite from the Namibian shelf. Electron microscope, x26000 (Baturin and Dubinchuk, 1979).

Fig. 6. Uraninite crystals in Recent phosphorite from Namibian shelf. Electron microscope, x20000 (Baturin and Dubinchuk, 1979).

tals (Baturin and Dubinchuk, 1979) (Fig. 6). The uranium contained in young phosphorite nodules (of absolute age <10^6 years is characterized by a $^{234}U/^{238}U$ activity ratio higher than unity. The ratio may be as high as 1,16-1,17, similar to that found in ocean waters (Baturin, Merkulova and Chalov, 1972; Burnett and Veeh, 1977).

Organic matter. While the total amount of organic matter of sediments and phosphatic materials from upwelling zones generally reflects the enhanced productivity of the surface waters, its composition may be even more informative for environment reconstructions. A thorough study of the organic matter component of diatomaceous oozes from a core taken on the Namibian shelf showed that a large number of organic compounds, including lipids, amino acids, fatty acids, alcohols, carbohydrates and sterols, might provide useful information (Morris and Calvert, 1977; Brassel and Eglinton; Gagosian et al.; and Simoneit, all this volume).

Calcareous and diatomaceous surface sediments from the west African shelf contain organic matter enriched in bitumen amounting to 10.9 and 17.8% of the total organic matter, respectively, relative to terrigenous sediments, which have only 3.6% bitumen. The organic matter of diatomaceous oozes is also enriched in hydrolizable organic matter, 17% of the total; depleted in humic acids, 41% of the total; and depleted in insoluble residue, 24% of the total (Romankevich and Baturin, 1974; Reimers and Suess, this volume). The organic matter content in Recent phosphorites from the Namibian shelf decreases by a factor of 4-6 during lithification, but the degree of preservation of the various compounds increases in the following order: carbohydrates, free lipids, nitrogeneous substances, bound lipids (Romankevich and Baturin, 1972). Organic matter contained in bone phosphate and coprolites undergoes similar transformations. Our new data indicate that the stability of organic matter in these materials increases from fish bones to mammal bones to phosphatized coprolites to phosphatic nodules.

It is significant to note that the high relative content of extractable organic fractions and the composition of kerogen in many ancient phosphorites indicate that their organic matter originated as phytoplankton remains accumulating in a reducing environment. These characteristics may indicate a genetic relationship between phosphorite occurrence and oil formation (Powell, Cook and McKirdy, 1975).

Stable isotopes. Particular attention must be paid to isotopic investigations of carbon, oxygen and sulphur in sediment components from beneath upwelling zones. The carbon isotopic composition of the bulk organic matter in diatomaceous oozes from the Namibian shelf is fairly constant with $\delta^{13}C$ values ranging from -19.5 to -20.5°/₀₀ PDB (Table 2). Terrigenous-diatomaceous oozes of the Peru-Chile shelf have similar values. The carbon isotopic composition of the bulk organic matter in Recent oceanic phosphorites has not been deter-

Table 2. Stable carbon and oxygen isotopes in sedimentary materials from upwelling zones

Source Material	Sample and Location	Age	$\delta^{13}C$ ‰PDB	$\delta^{18}O$ ‰SMOW
Total organic matter	Diatomaceous ooze, Namibian shelf(5)*	Recent	-19.5	--
Total organic matter	Terrigenous-diatomaceous sediments, Peru-Chile shelf(6)	Recent	-20.5 -21.3	--
Total organic matter	Phosphorites of Russian platform and North Africa(2)	Cretaceous-Paleogene	-22.2 -23.5	--
Lipids	Sediments of upwelling zones(6)	Recent	-21.0 -24.1	--
Lipids	Phosphorites of upwelling zones(7)	Recent	-20.7 -24.2	--
Lipids	Fish bones(7)	Recent	-22.1	--
Lipids	Mammal bones(7)	Pleistocene	-21.6	--
CO_2 of phosphate	Phosphorite from Namibian shelf(2)	Recent	- 5.7	--
CO_2 of phosphate	Phosphorites of California basins(3)	Miocene Pleistocene	- 1.2 - 2.2	-0.5 +0.4
CO_2 of phosphate	Nodular phosphorites of Russian Platform(2)	Cretaceous	- 2.1 - 3.1	--
CO_2 of phosphate	Oolite-microgranular phosphorites of Karatau(2)	Cambrian	- 5.1 - 6.1	--
PO_4 of phosphorites	Guano from Peru shelf (1)	Recent	--	18.2
PO_4 of phosphorites	Fish bones from Namibian, Peru and Chile shelves(1)	Recent	--	18.2 19.9
PO_4 of phosphorites	Mammal bones from Namibian shelf(1)	Recent-Late Quaternary	--	12.2 14.5
PO_4 of phosphorites	Phosphorites from Namibian shelf(1)	Recent	--	20.0 21.1
PO_4 of phosphorites	Phosphorites from Morocco and Peru shelves and Agulhas Bank(1)	Eocene-Late Quaternary	--	15.3 17.4
PO_4 of phosphorites	Phosphorites of Eurasia and America(4)	Mesozoic-Cenozoic	--	15.6 23.2
PO_4 of phosphorites	Phosphorites of Eurasia and America(4)	Pre-Cambrian-Paleozoic	--	8.6 15.5

* References: (1) Baturin, Strizhov and Isaeva, 1980 with additions; (2) Bliskovsky, Bondar and Lein, 1981; (3) Kolodny and Kaplan, 1970; (4) Longinelli and Nuti, 1968; (5) Morris and Calvert, 1977; (6) Shadsky et al., 1979; (7) Shadsky et al., 1980, with additions.

mined, but values from Cretaceous-Paleogene phosphorites of the Russian Platform and northern Africa are slightly less than the organic matter of Recent sediments.

The isotopic composition of carbon in lipids of Recent sediments ranges from -21.0 to -24.1°/₀₀ (average $\delta^{13}C$=-22.0°/₀₀), heavier than the lipid composition in terrigenous shelf sediments (average $\delta^{13}C$= -23.5°/₀₀) and pelagic muds (average $\delta^{13}C \sim$-28.0°/₀₀). The carbon of free lipids extracted from Recent phosphatic nodules of the Namibian shelf appears to become lighter during phosphorite lithification (from -20.7 to -24.2°/₀₀). Carbon isotopes in bound lipids, however, become heavier during the same process (from -24.0 to -23.0°/₀₀) (Shadsky, Baturin and Grinchenko, 1979; Shadsky, Romankevich, and Baturin, 1980).

The carbon isotopic composition of CO_2 incorporated in the fluorcarbonate-apatite phase of phosphorites is variable. Miocene-Pleistocene phosphorites from California basins show nearly the same $\delta^{13}C$ range as Cretaceous phosphorites from the Russian Platform. Recent nodules from the Namibian shelf (-5.7°/₀₀) and Cambrian phosphorites from Karatau have similar compositions (Table 2). Further investigation of this indicator may allow its use in the interpretation of phosphorite formation, diagenesis, and lithification.

The oxygen isotopic composition of PO_4-groups in young biogenic and diagenetic phosphates from upwelling zones is approximately the same as in phosphate dissolved in ocean water ($\delta^{18}O$ ranging from 18.2 to 21.2°/₀₀) (Table 2). Cetacean bones are an exception with substantially lighter oxygen isotopes. As recent phosphorites become lithified and bones fossilized, the oxygen isotopic composition of the PO_4-groups tends to become lighter. Thus, pre-Quaternary phosphorites usually have lower $\delta^{18}O$ values than Recent ones. A similar tendency was observed in a large set of phosphorites of various ages from land deposits (Longinelli and Nuti, 1968).

The sulfur isotopic composition of SO_3-groups which form isomorphic admixtures with phosphate has been determined so far for only one Recent phosphorite sample from the Namibian shelf. Its composition was the same as the sulphate of ocean water ($\delta^{34}S$=20.6°/₀₀). Although similar results have been reported for a number of pre-Quaternary phosphorites, the range of $\delta^{34}S$ in ancient phosphorites is fairly large (from 6.3 to 32.1°₀₀) depending on their age and the extent of post depositional alteration (Bliskovsky et al., 1977).

Rare earth elements. Because phosphorites are typically formed beneath upwelling zones and contain associated rare earth elements, it is informative to follow the REE depositional and diagenetic pathways during the initial phase of phosphorite formation. Recent phosphorites from oceanic shelves are exceedingly poor in REE (<0.005%), even poorer than the enclosing biogenic sediments. Most of the REE in these phosphorites are bound not to phosphate but to organic matter and terrigenous components. Late Quaternary phosphorites are slightly enriched in REE, but it is only in Pre-Quaternary deposits that the REE concentration levels rise to values typical of ancient phosphorites presently found on land. These data suggest that REE enrichment in marine phosphorites begins only in the late stages of diagenesis and continues during subsequent processes (Baturin, Bliskovsky and Mineev, 1972).

The transition elements, Fe, Mn, Ti, V, Cr, Co, Ni, Cu, Sc, also show interesting behavior during Recent phosphorite formation. During progressive lithification of phosphatic nodules the abundance of these elements drops considerably due to recrystallization of phosphate and degradation of organic matter. This abundance decrease is not uniform, however, as elements with odd atomic numbers migrate from phosphate much more efficiently than even-numbered elements. This difference may be due to differences in their outer electronic shell configuration (Tambiev and Baturin, 1981).

At this time it is difficult to predict which of the above mineralogical and geochemical peculiarities of sediments and sedimentary materials will prove to be most effective as indicators of coastal upwelling in the past, but there is no doubt that a set of appropriate regional sedimentary criteria for most of the large upwelling zones in today's ocean could be identified and recognized in the fossil record. A tentative list of specific features of Recent upwelling zones must evidently include such criteria as:

- Lithology: accumulation of organic-rich sediments of variable composition --calcareous, siliceous, clayey--, with occasional phosphatic debris --pellets, nodules, coprolites, bones.
- Mineralogy: occurrence of pyrite-glauconite-phosphate associations among purely biogenic minerals (calcium carbonate and silica).
- Chemistry: high contents of organic matter, as compared to pelagic sediments, with specific components (bitumen, hydrolizable organic matter, humic acids, high P, U and Cd contents).
- Isotopes: excess ^{234}U in sediments, phosphorites and bone phosphates; relatively low $\delta^{13}C$ values in lipids of biogenic sediments; values of $\delta^{18}O$ and $\delta^{34}S$ in phosphorites which are approximately the same as in seawater.

Some of these features change with geologic time leading to the loss of clear evidence, for example, of biogenic origin of silica and coprolites. Such losses occur through alteration of original pyrite-glauconite-phosphate relations, of phosphate structures, from changes in minor element concentrations and isotopic ratios, particularly of $^{234}U/^{238}U$, and from oxidation of organic matter. But the most general lithologic and geochemical trends are recognizable throughout geologic time unless, of course, the entire sedimentary column is eroded.

REFERENCES

Baturin, G.N., 1969, Authigenic phosphorite concretions in Recent sediments of the South-West African shelf, Doklady Academii Nauk S.S.S.R., 189(6):1359-1362 (in Russian).

Baturin, G.N., 1970, Recent authigenic phosphorite formation on the South-West African shelf, in: "The Geology of the East Atlantic Continental Margin," F.M. Delaney, ed., Institute of Geological Science, Cambridge, Report 70/13, 173 pp.

Baturin, G.N., 1974, About geologic consequences of mass mortality of fish in the ocean, Okeanologia, 14(1):101-105 (in Russian).

Baturin, G.N., 1975, "Uranium in Recent Marine Sedimentation," Atomizdat, Moscow, 152 pp. (in Russian).

Baturin, G.N., 1978, "Phosphorite on the Sea Floor," Nauka, Moscow, 230 pp. (in Russian).

Baturin, G.N., 1982, "Phosphorite on the Sea Floor: Distribution, Composition and Origin," Elsevier Scientific Publishing Company, Amsterdam, 330 pp.

Baturin, G.N. and Dubinchuk, V.T., 1979, "Microstructures of oceanic phosphorites," Nauka, Moscow, 200 pp. (in Russian).

Baturin, G.N. and Pokryshkin, V.I., 1980, Upwelling and phosphorite formation, Okeanologia, 20(1):87-96 (in Russian).

Baturin, G.N. and Oreshkin, V.N., 1981, Cadmium in process of Recent oceanic phosphorite formation, Geochimia, 11:1727-1732 (in Russian).

Baturin, G.N., Kochenov, A.I. and Petelin, V.P., 1970, Phosphorite formation on the South-West African shelf, Lithology and Mineral Resources, 3:15-26 (in Russian).

Baturin, G.N., Bliskovsky, V.Z. and Mineev, D.A., 1972, Rare-earth elements in phosphorite from the sea-floor, Doklady Academii Nauk S.S.S.R., 207(4):954-957 (in Russian).

Baturin, G.N., Merculova, K.I. and Chalov, P.I., 1972, Radiometric evidence for recent formation of phosphatic nodules in marine shelf sediments, Marine Geology, 13:M37-M42.

Baturin, G.N., Strizhov, V.P. and Isaeva, A.B., 1980, The isotopic composition of oxygen in bone phosphate and phosphorite nodules from the sea floor, Preprints, 8th All-Union Symposium on Stable Isotopes in Geology, Academii Nauk S.S.S.R., Moscow, 11-14 November, 215-217 (in Russian).

Bliskovsky, V.Z., Baturin, G.N. and Kuzmina, T.S., 1975, About phosphatic matter of some phosphorites from the sea floor, Lithology and Mineral Resources, 4:3-9 (in Russian).

Bliskovsky, V.Z., Bondar, B.A. and Lein, A.Y., 1981, About isotopic composition of carbon in minerals of phosphorite ores, Geochimia, 4:616-618 (in Russian).

Bliskovsky, V.Z., Grinenko, V.A., Migdisov, A.A., and Savina, L.I., 1977, Isotopic composition of sulphur in minerals of phosphorite ores, Geochimia, 8:1208-1218 (in Russian).

Bogoyavlensky, A.N. and Shishkina, O.V., 1971, Major features of the hydrochemistry of the Peru-Chile region, Trudy Institute Oceanology, Academy of Sciences U.S.S.R., 89:96-105 (in Russian).

Brongersma-Sanders, M., 1957, Mass mortality in the sea, in: "Treatise on Marine Ecology and Paleoecology," J.W. Hedgpeth, ed., Geological Society of America, Memoir, 67:941-1010.

Burkov, V.A., 1972, "General circulation of waters of the Pacific," Nauka, Moscow, 196 pp. (in Russian).

Burnett, W.C., 1977, Geochemistry and origin of phosphorite deposits from off Peru and Chile, Geological Society of America, Bulletin, 8(6):813-823.

Burnett, W.C. and Veeh, H.H., 1977, Uranium series disequilibrium studies in phosphorite nodules from the West coast of South America, Geochimica et Cosmochimica Acta, 41(6):755-764.

Bushinsky, G.I., 1966, "Ancient Phosphorites of Asia and their Genesis," Nauka, Moscow, 192 pp. (in Russian).

Butlin, K.P., 1949, Some maladorous activities of sulfate-reducing bacteria, Proceedings, Society of Applied Bacteriology, 12(1): 39-42.

Calvert, S.E. and Price, N.B., 1970, Minor metal contents of recent organic-rich sediments off South-West Africa, Nature, 227:593-595.

Calvert, S.E. and Price, N.B., 1971, Upwelling and nutrient regeneration in the Benguela current, Deep-Sea Research, 18:505-523.

Cayeux, L., 1936, Existence de nombreuses bactéries dans les phosphates sédimentaires de tout âge, Comptes Rendue de l'Academie des Sciences de Paris, 203:1198-1200.

Copenhagen, W.J., 1953, The periodic mortality of fish in Walvis region, Dept. of Commerce and Industry, Union of South Africa, Division of Fisheries, Investigation Report No. 14, Cape Town, 35 pp.

Emelyanov, E.M., 1973, Distribution and composition of muds on the South West African shelf, Trudy Institute Oceanology, Academy of Sciences U.S.S.R., 95:211-238 (in Russian).

Gunther, E.R., 1936, A report on oceanographical investigations in the Peru coastal current, Discovery Report, 13:107-276.

Hart, T.Y. and Currie, R.J., 1960, The Benguela current, Discovery Report, 31:123-298.

Hutchinson, G.E., 1950, Survey of contemporary knowledge of biogeochemistry. 3. The biogeochemistry of vertebrate excretion, Bulletin, American Museum of Natural History, 96:554 pp.

Kolodny, Y. and Kaplan, I.R., 1970, Carbon and oxygen isotopes in apatite CO_2 and co-existent calcite from sedimentary phosphorite, Journal of Sedimentary Petrology, 40(3):954-959.

Lein, A.Y., Sidorenko, G.A., Volkov, I.I., and Shevchenko, A.Y., 1978, The diagenetic makinavite, melnikovite (greigite) and pyrite in sediments on the profile across the Pacific ocean and in sediments of the Gulf of California, Doklady Academii Nauk S.S.S.R., 223(3):698-700 (in Russian).

Logvinenko, H.V., Nikolaeva and Romankevich, E.A., 1973, Authigenic minerals of Recent sediments of the South-Eastern part of the Pacific ocean, Lithology and Mineral Resources, 4:15-27 (in Russian).

Longinelli, A. and Nuti, S., 1968, Oxygen isotopic composition of phosphorites from marine formations, Earth and Planetary Science Letters, 15(1):13-16.

Mikhailov, I.A., Baturin, G.N., Kochenov, A.V., Mirtov, Y.V. and Razvalyaev, A.V., 1972, Environmental conditions of phosphorite accumulation in the Nile syncline and on the South-West African shelf, Lithology and Mineral Resources, 5:3-13 (in Russian).

Mishustina, I.E., 1973, Ultramicroscopic bodies in diatomaceous oozes of the South-West African shelf, Izvestiya Academii Nauk S.S.S.R., Series Biology, 3:139-140 (in Russian).

Morris, R.J. and Calvert, S.E., 1977, Geochemical studies of organic-rich sediments from the Namibian shelf. 1. The organic fractions, in: "A Voyage of Discovery: George Deacon 70th Anniversary Volume," M. Angel, ed., Pergamon Press, Oxford, 647-665.

Mstislavsky, M.M. and Kochenov, A.V., 1960, Bone breccia of Maikop and fish mortality during red tides, Doklady Akademii Nauk S.S.S.R., 134(5):1169-1172 (in Russian).

Novozhilova, M.I. and Baturin, G.N., 1973, Some data on bacterial microflora in sediments of South-Eastern part of the Atlantic ocean, Trudy Institute Oceanology, Academy of Sciences U.S.S.R., 95:268-273 (in Russian).

Powell, T.G., Cook, P.Y. and McKirdy, D.M., 1975, Organic geochemistry of petroleum genesis, American Association of Petroleum Geologists, Bulletin, 59(4):618-632.

Romankevich, E.A. and Baturin, G.N., 1972, On the composition of organic matter of phosphorites from the South-West African shelf, Geochimia, 6:719-726 (in Russian).

Romankevich, E.A. and Baturin, G.N., 1974, Biochemical composition of sediments of the South-West African shelf (5-23°S), Okeanologiya, 14(4):660-664 (in Russian).

Ryther, J.H., 1969, Photosynthesis and fish production in the sea, Science, 166:72-72.

Ryther, J.H., Menzel, D.W., Hulburt, E.M., Lorenzen, C.J., and Corwin, N., 1971, The production and utilisation of organic matter in the Peru coastal current, Investigation Pesquera, 35(1): 43-59.

Shadsky, I.P., Baturin, G.N. and Grinchenko, Y.I., 1979, Isotopic composition of carbon in lipids in bottom sediments of the Atlantic ocean, Preprints, 2nd Symposium on Isotopes in Nature, Academy of Sciences of the German Democratic Republic, Leipzig, 5-9 November, p. 98-99.

Shadsky, I.P., Romankevich, E.A. and Baturin, G.N., 1980, Isotopic composition of carbon in Recent phosphatic nodules and the problem of their formation, Doklady Academii Nauk S.S.S.R., 254(3):743-744 (in Russian).

Smith, R.L., 1968, Upwelling oceanography, Marine Biology, Annual Review, 6: 11-46.

Tambiev, S.B. and Baturin, G.N., 1981, Elements of the iron family in the process of Recent phosphorite formation, Doklady Academii Nauk S.S.S.R., 258(1):201-207 (in Russian).

Veeh, H.H., Calvert, S.E. and Price, N.B., 1974, Accumulation of uranium in sediments and phosphorites on the South West African shelf, Marine Chemistry, 2(2):189-202.

Wooster, W.S. and Reid, J.L., 1963, Eastern boundary currents, in: "The Sea, 2," M.N. Hill, ed., John Wiley and Sons, New York, 253-280.

SEDIMENTATION OF ORGANIC MATTER IN UPWELLING REGIMES

Colin P. Summerhayes

Exxon Production Research Company
P.O. Box 2189
Houston, Texas 77001, U.S.A.
Present address:
BP Research Centre
Sunbury-on-Thames
Middlesex TW16 7LN
United Kingdom

ABSTRACT

This is a review of the nature and processes of organic sedimentation in upwelling regimes, with special reference to the continental margins of northwest and southwest Africa, the California Borderland, and the Gulf of California. Pelagic and hemipelagic sediments predominate in these areas, with turbidites locally on basin floors. Most of the organic matter is amorphous, bacterially degraded, marine material; minor terrestrial components occur near humid coasts, and are significant in turbidites off those coasts. Arid coasts may have some aeolian terrestrial organic matter. Massive, short-term productivity is a first order control on the accumulation of organic matter. High productivity rather than oxygen depletion in bottom waters accounts for much organic enrichment (e.g., continental slope: southwest Africa). Organic matter decomposing at the sea floor strips oxygen from bottom waters making them locally anoxic and contributing to the overall oxygen depletion of the oxygen minimum zone (e.g., southwest African shelf; Santa Barbara Basin). Lateral movement of an anoxic oxygen minimum zone into a basin favors preservation of organic matter locally (e.g., Gulf of California). Anoxic conditions in bottom waters preserve laminations (varves), especially where there are seasonal changes in the sediment supply (e.g., in terrigenous clastics - Gulf of California, and Santa Barbara Basin). Diagenetic loss of organic matter by oxidation destroys the record of organic sedimentation where bottom waters are oxidizing and either rates of sedimentation are very slow, or productivity is low (e.g., deep Gulf of California). There is no consistent relation between organic carbon and rate of sedimentation in the four areas examined. Absence of organic enrichment off northwest Africa today reflects primarily low productivity, and a high degree of turbulence that prevents the settling of fecal pellets; the moderately high oxygen content of bottom waters there helps to prevent organic enrichment. During glacials productivity was higher there than it is today, and organic rich sediments accumulated. High productivity seems to be spasmodic in the geological record. Organic-rich or highly phosphatic sediments possibly associated with upwelling formed mainly in the Mid Cretaceous (northwest and southwest Africa), Miocene (those areas and California Borderland), and Quaternary (all four areas). This periodicity suggests major periodic worldwide redistribution of oceanic nutrients.

INTRODUCTION

The objective of this review is to determine how the sedimentation and preservation of organic matter takes place in upwelling regimes. It is well known that there are large accumulations of organic matter on the seafloor beneath highly productive upwelled surface waters; many petroleum source beds may be ancient analogs of these modern deposits (Dow, 1978; Demaison and Moore, 1980). But, close inspection of the available evidence shows that while some upwelling centers have rich deposits of organic matter (Peru, Chile, California, southwest Africa), others do not (central Brazil, northwest Africa). In order to understand how organic matter accumulates in upwelling regimes, we need to study examples of both of these types. For this review, I have chosen the California Borderland, the Gulf of California, southwest Africa, and northwest Africa, because these are the upwelling regimes with which I am most familiar. Hopefully, we shall gain some insight into the controls on organic matter sedimentation in upwelling regimes by comparing and contrasting the oceanographic and sedimentary processes of these different regions. In this review I address the following questions:

- What is the source of the organic matter (marine versus terrestrial)?
- How was it deposited (hemipelagic versus turbidite)?
- How was it preserved (oxic versus anoxic environments)?
- How important is productivity (spasmodic versus steady upwelling)?
- What is the effect of topography (deep versus shallow water)?
- How is the oceanographic signal modified by diagenesis after deposition?

Implied in these questions is the recognition that sedimentary organic matter may contain the oceanographic signal of upwelling and can reflect the productivity of surface waters and degree of oxygenation of bottom waters in upwelling regimes. In any given area, the answers to these questions may be used to resolve important paleoenvironmental problems concerning paleocirculation and the relative intensity of upwelling.

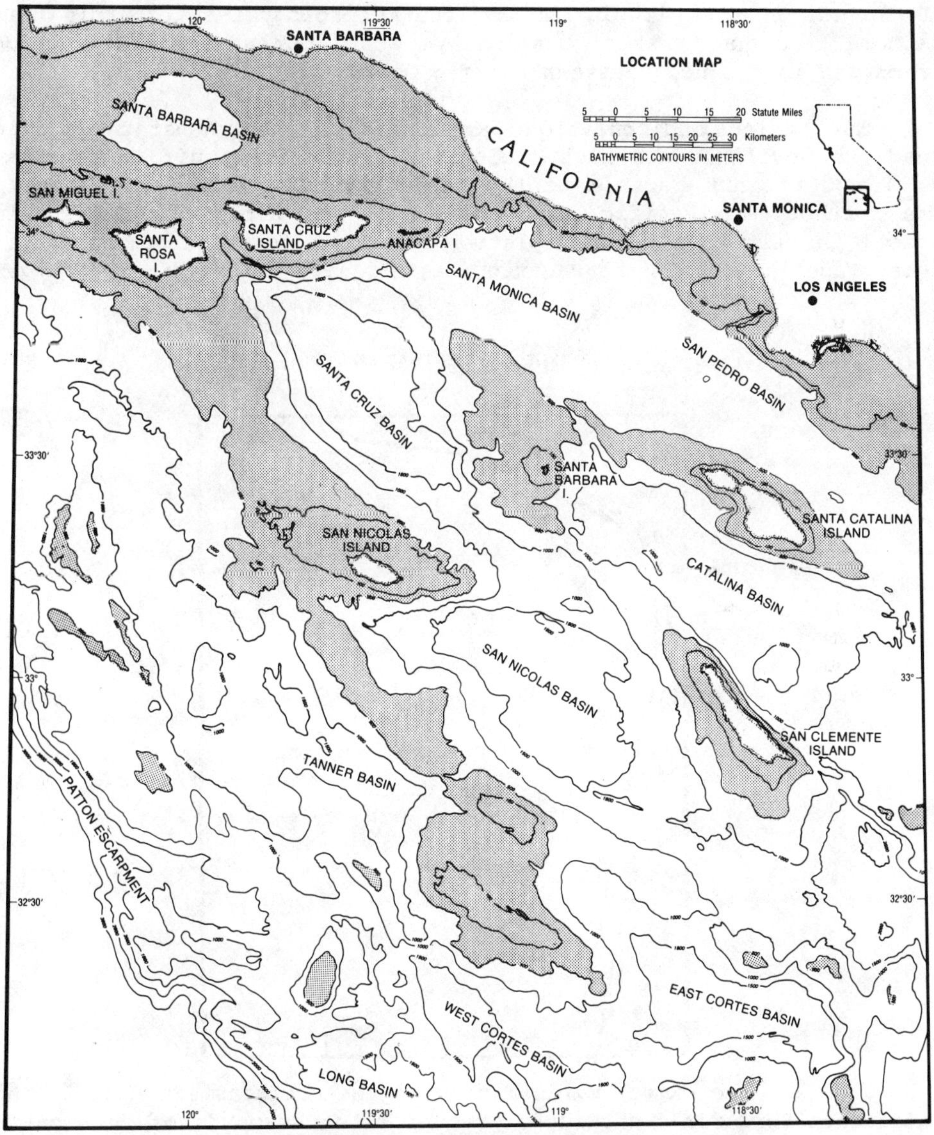

Fig. 1. Bathymetry of California Borderland showing major basins (drawn by Lisa Pratt, from Coast and Geodetic Survey Maps).

DISTRIBUTION OF ORGANIC MATTER IN SELECTED UPWELLING AREAS

California Borderland

General oceanographic setting. Within the California Borderland are several basins that are separated by sills from each other and from the open Pacific (Fig. 1). TOC (total organic carbon) is high (3-5%) in the basin deeps, and low on the intervening ridges (Emery, 1960). Is it the circulation, or the morphology that controls this pattern? To answer this question, we must look at the distribution of oxygen in the northeastern Pacific Ocean.

The regional circulation pattern of the northeastern Pacific leads to upwelling and high biological productivity off California. At least in part because of the productivity, subsurface waters in the northeastern Pacific are poor in oxygen along the California coast; in part the oxygen depletion is also a function of the age of subsurface waters in the Pacific (Kester, 1975). Within the oxygen

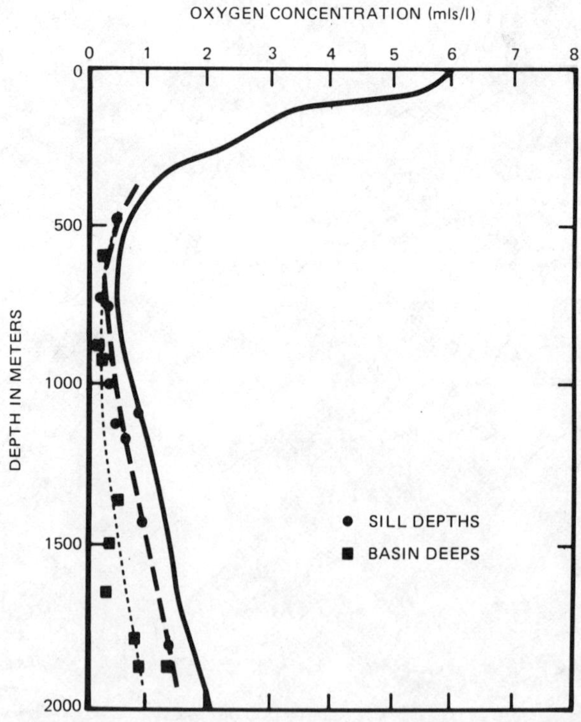

Fig. 2. The oxygen minimum zone in the northeastern Pacific and California Borderland showing oxygen profiles seawards of the continental slope, at the entrances to silled basins, and at the bottoms of those basins (adapted from Emery, 1960).

minimum zone, oxygen depletion is severe, dropping to less than 1.0 mℓ/ℓ O_2 at depths of 400-1400 m (Fig. 2). This water enters the Borderland basins across sills that are somewhat shallower than the basin floors. At the sill depth there is less oxygen than in the open northeastern Pacific (Fig. 2). The landward decrease in oxygen may be a response to upwelling and heightened productivity nearshore (Brongersma-Sanders, 1971). In the basin deeps there is slightly less oxygen than at sill depth, because of use of oxygen by decomposition of organic matter at the bottom (Fig. 2) (Emery and Hülsemann, 1962). Clearly, the main control on oxygen depletion off this coast is the regional circulation, rather than the restricted circulation in basin deeps. Nevertheless, it is probably the restricted circulation that accentuates the regional depletion in oxygen, causing bottom water to become locally anoxic. Nitrate concentrations are 30-40 µg atoms/ℓ (Rittenberg, Emery and Orr, 1955).

Controls on organic matter. Accumulation of organic matter here most probably occurs as follows:

1. Upwelling causes high productivity, leading to massive production of fecal pellets and other organic rich particles;

2. Organic matter is rapidly transported to the bottom, for example in fecal pellets, sinking at rates of as much as 1700 m/day (Dunbar and Berger, 1981), or as marine snow, sinking at rates of as much as 95 m/day (Shanks and Trent, 1980).

3. Low oxygen levels in the oxygen minimum zone prevent rapid decomposition of organic matter. As implied above, there is a feedback mechanism involved here, with the decomposition that takes place on and in the surface sediments maintaining the low levels of oxygen (Emery and Hülsemann, 1962).

Total organic carbon values (TOCs) are high not only in the anoxic basins close to shore, like Santa Barbara (O_2 = 0.1-0.2 mℓ/ℓ; Emery, 1960), but also in dysaerobic offshore basins, like Tanner (Fig. 1) (O_2 = 0.8 mℓ/ℓ; Emery, 1960). Thus conditions do not have to be completely anoxic to encourage the accumulation of organic matter, as pointed out also by Rhoads and Morse (1971) and Demaison and Moore (1980).

That the presence of silled basins is not a prerequisite for the accumulation of organic matter here is shown by the accumulation of richly organic Miocene sediments on the open continental slope at DSDP Sites 467 and 468 at the seaward edge of the Borderland (Fig. 3) (Gilbert and Summerhayes, 1981; Simoneit, Meyers and Summerhayes, 1981; Summerhayes, 1981a). I consider that these sediments were deposited in an oxygen minimum zone that became very well-developed along this coast in response to regional changes in Pacific circulation that were induced by the growth of Antarctic ice sheets (Summerhayes, 1981a). A drop in sea level 10-11 million years ago

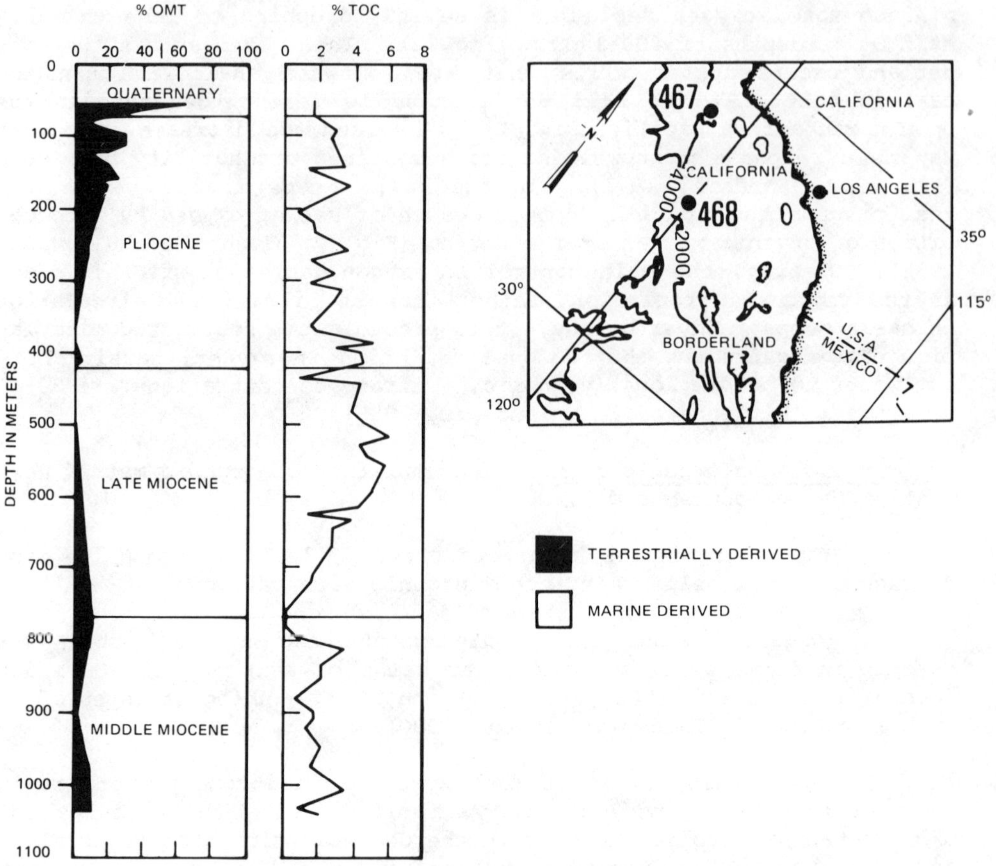

Fig. 3. Downhole distribution of TOC and organic matter (OMT) at DSDP Site 467. Note also locations of DSDP Sites 467 and 468 (from Summerhayes, 1981a).

lowered the base of the oxygen minimum zone to depths of 2500-3000 m favoring the accumulation of organic matter on the open slope in Late Miocene-Early Pliocene times. These oxygen depleted conditions accompanied an intensification of upwelling that gave rise to deposition of the organic-rich Monterey Shales along the California coast (Pisciotto, 1978). Phosphorites, indicative of a high rate of supply of nutrients, formed at the same time. A lessening in upwelling and in oxygen depletion from the Pliocene into the Quaternary then made conditions less favorable for the formation of diatomaceous ooze and phosphorite; these changes are associated with a warming trend and a rise in sea level (Pisciotto, 1978; Summerhayes, 1981a). In the younger sediments, rates of sedimentation may be inversely related to organic content (Calvert, 1976; Malouta, 1978).

Organic facies. The organic matter in the organic-rich Miocene and Pliocene deposits of the Borderland at DSDP Sites 467 and 468 is dominantly marine (Simoneit et al., 1981). Most of it is amorphous, and there are subordinate amounts of structured marine and terrestrial materials (Fig. 3) (Gilbert and Summerhayes, 1981). Pyrolysis shows that this amorphous material can be classified as Type II and is probably marine (Rullkötter, Von der Dick and Welte, 1981). The proportions of terrestrial organic matter increase in younger deposits at these sites, in response to seaward progradation of clastics (Fig. 3) (Gilbert and Summerhayes, 1981; Simoneit et al., 1981; Summerhayes, 1981a). Organic facies in organic-rich surface sediments are either mixed marine and terrestrial, with marine predominating, as in the nearshore Santa Barbara Basin (Sweeney and Kaplan, 1980) or are strongly marine, as in the more offshore Tanner Basin (J. Kulla, unpublished data).

As found in other environments (Summerhayes, 1981b), there is a close relationship between TOC and the type of organic matter in organic rich sediments from offshore California (Figs. 4 and 5). In

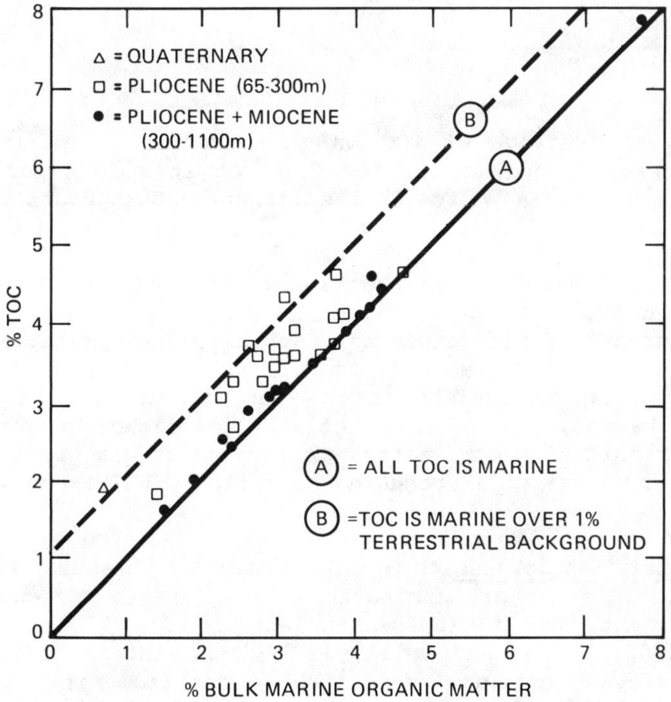

Fig. 4. Dependence of TOC on preservation of marine-derived organic matter in Tertiary and Quaternary sediments from DSDP Site 467. Bulk marine organic matter is TOC x (total amorphous + structured marine organic matter) (from Summerhayes,1981a).

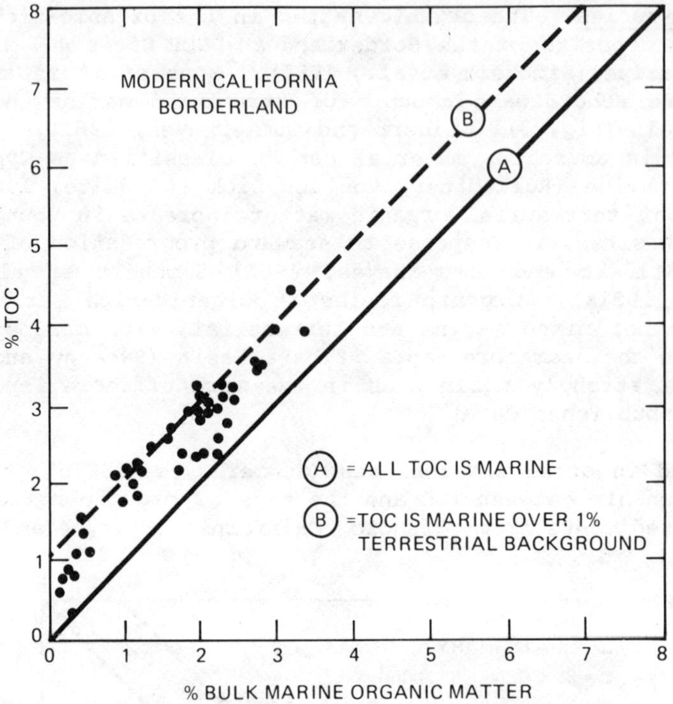

Fig. 5. Dependence of TOC on preservation of marine-derived organic matter (calculated as for Fig. 4) in modern sediments of Santa Barbara and Santa Monica Basins (drawn by Lisa Pratt).

effect, TOC is controlled by the preservation of labile marine and marine-derived amorphous organic matter. This increase takes place above a background of refractory terrestrial organic material. Where the rate of supply of terrestrial organic matter is high, as near the mainland, near canyon outlets in the deep basin, and in turbidite units in the basins, the TOC is relatively low either because of deposition under oxidizing conditions, or because of clastic dilution of marine organic matter under reducing conditions (Hülsemann and Emery, 1961; Malouta, 1978).

Sedimentary structures. In the nearshore basins, like Santa Barbara, the sediments are laminated in response to seasonal changes in the sediment supply (Hülsemann and Emery, 1961; Emery and Hülsemann, 1962; Soutar and Crill, 1977). In winter, runoff is high and productivity is low, so layers rich in detrital minerals are deposited; in spring and summer, when runoff is low and productivity is high, fecal pellets and diatom tests dominate the sediment supply. The resulting laminations are preserved because the bottom water contains too little oxygen to support benthic organisms. However, in the subsurface there are alternations of laminated and homogeneous sediments. The latter are bioturbated and represent times in which the basin was flushed, perhaps by internal waves (Emery and Hülsemann, 1962). Apparently there is no difference in organic content between the laminated and homogeneous layers (Hülsemann and Emery, 1961). In the organic-rich Mio-Pliocene sediments from DSDP Sites 467 and 468, some sediments are laminated and others are homogeneous, indicating that oxygen levels fluctuated between the more or less completely anoxic (less than say, 0.2 mℓ/ℓ), for laminations to be preserved, and the dysaerobic (0.5 to 1.0 mℓ/ℓ; Demaison and Moore, 1980), for sediments to become bioturbated and homogeneous but not to lose much of the organic matter.

Diagenesis. Downcore, TOC decreases by about 30% over the top 1-2 m of the sediment column in the Borderland Basins (Hülsemann and Emery, 1961; Kaplan, Emery and Rittenberg, 1963). This early diagenetic loss is caused by the decomposition of organic matter by sulfate-reducing bacteria (Kaplan et al., 1963; Sholkovitz, 1973). More extensive loss of organic matter (say 75% or more) would be anticipated if the depositional environment were strongly oxidizing (Müller and Mangini, 1980).

Gulf of California

General oceanographic setting. The Gulf of California is an open marine basin with an unrestricted connection to the Pacific (Fig. 6). In the center of the Gulf are several small restricted deeps, like the Guaymas Basin. The same oxygen minimum zone that penetrates the Borderland Basins penetrates the Gulf of California, because subsurface water from the Pacific moves into the Gulf at depths of 300 m or so to compensate for surface outflow (Fig. 6) (Roden, 1964; Van Andel, 1964; Brongersma-Sanders, 1971). Circulation in the Gulf is characterized by upwelling and high productivity and this helps to reinforce oxygen depletion in the oxygen minimum zone. Oxygen values are anoxic (less than 0.2 mℓ/ℓ) at water depths of about 400-1000 m, and dysaerobic to oxidizing (0.8-1.5 mℓ/ℓ) in the deep basins on the floor of the Gulf at 1600-2000 m (Fig. 6) (Roden, 1964).

Controls on organic matter. As in the California Borderland, organic enrichment occurs in response to upwelling, high productivity, fecal pellet transport of organic matter to the bottom, and preservation of organic matter in anoxic or poorly oxygenated dysaerobic bottom water.

TOCs are highest along the continental margins at about the depth where the oxygen minimum impinges on the continental slope (Fig. 7). Many of these sediments are highly diatomaceous (Fig. 7). Here TOCs are usually 3-4%, reaching 5% or more at the mouth of the Gulf (Van Andel, 1964). In deeper water the TOCs drop to 2-3%; TOCs are also low on the shelf and upper continental slope above the oxy-

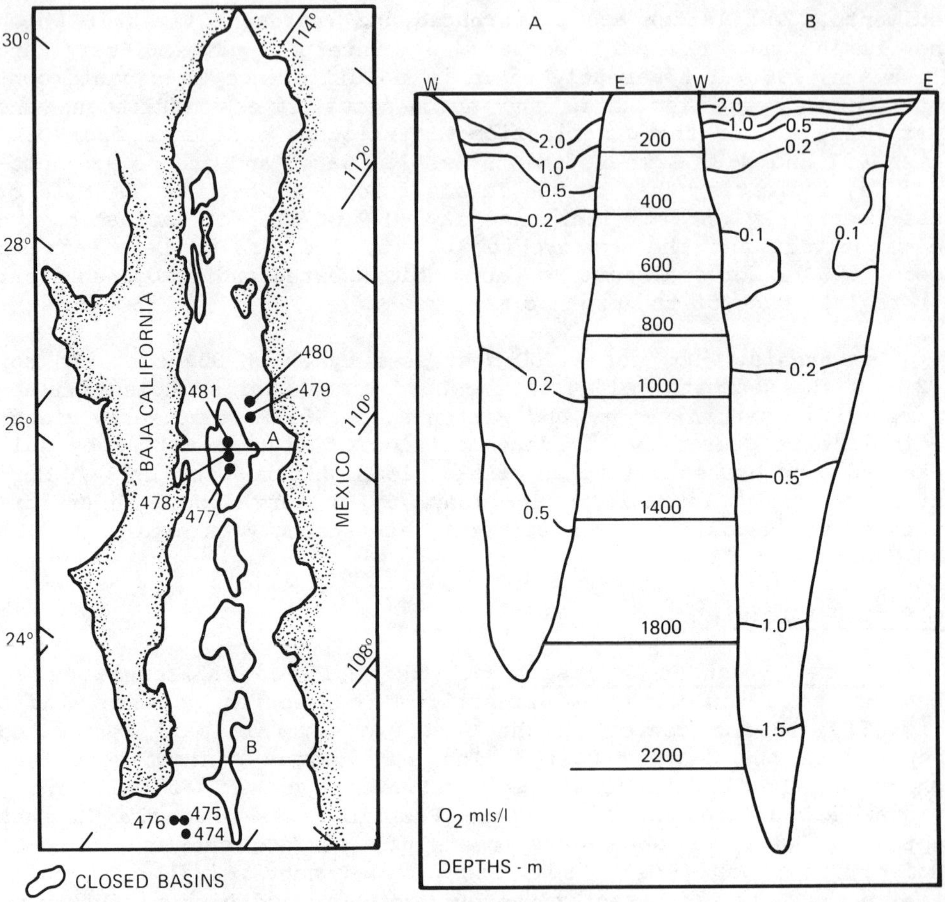

Fig. 6. The oxygen minimum zone in the Gulf of California (based on Roden, 1964), with locations of deep basins from Van Andel (1964). Line A crosses Guaymas Basin. Leg 64 DSDP drill sites are shown. Note excessive oxygen depletion (<0.1 mℓ/ℓ) on upper continental slope is a modification of the oxygen minimum that is probably caused by decomposition of bottom sediments (profile B).

gen minimum zone. Thus, as in the California Borderland, we do not need complete anoxia in order to find large amounts of organic matter accumulating on the sea floor. Again, the presence of sills does not control the distribution of organic matter; in fact, the deep-silled basins have less organic matter than do the open continental slopes around the edges of the Gulf (Fig. 7) (Van Andel, 1964; Gilbert and Summerhayes, 1982; Simoneit, Summerhayes and Meyers, 1982). As in the California Borderland, Calvert (1976) shows there is an inverse relationship between TOC and sedimentation rate. But the data of Gilbert and Summerhayes (1982) show that this trend is locally reversed.

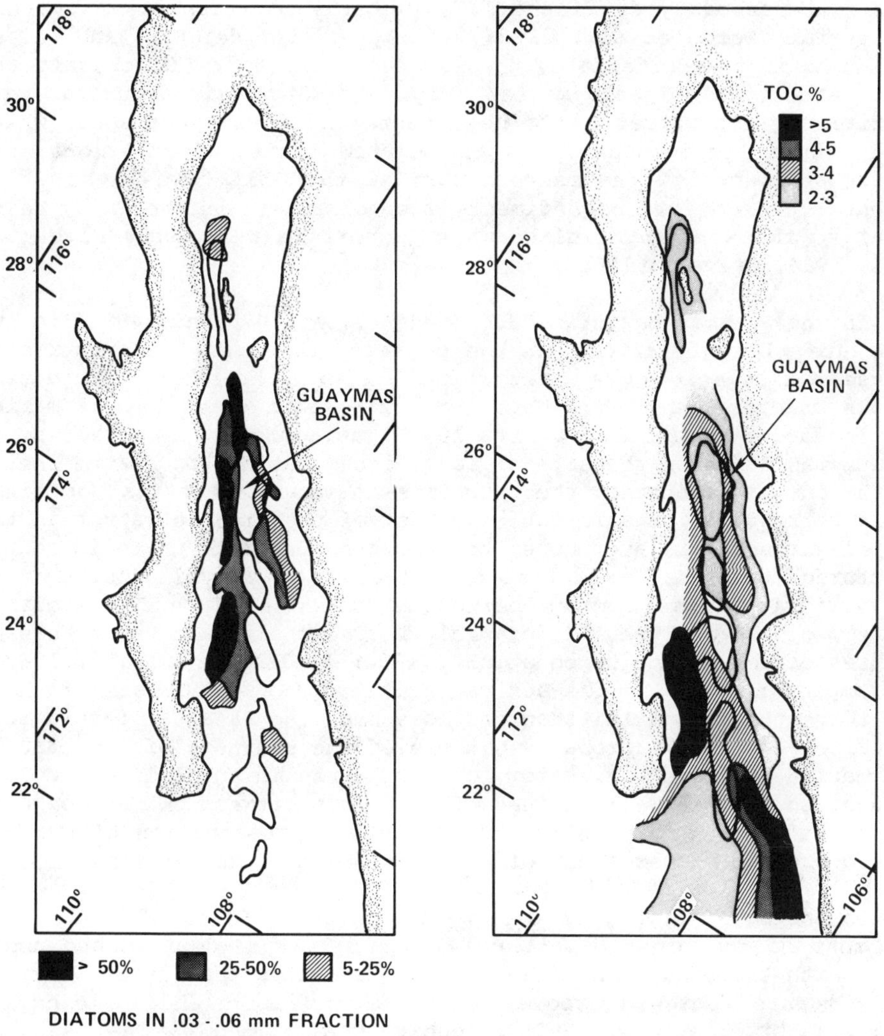

Fig. 7. Distribution of diatoms in the 0.03-0.06 mm size fraction (left), and of TOC in the bulk sediments (right), in the Gulf of California (from Van Andel, 1964); note enrichment of both in the zone where the oxygen minimum zone impinges on the bottom (compare with Fig. 6), not in deep basins.

Organic facies. The organic matter in Quaternary sediments from the zone of organic enrichment on the margins of the Gulf, at DSDP Sites 479 and 481, is dominantly marine (Simoneit et al., 1982). Most of it is amorphous, and there are minor amounts of structured marine and terrestrial components. Pyrolysis shows that the amorphous material can be classified either as Type II, probably marine, or as mixtures of Types II and III, probably mixtures of marine and terrestrial (Deroo et al., 1982; Rullkötter, Von der Dick and Welte, 1982). As in the Borderland, the samples richest in TOC contain the least terrestrial organic matter (Fig. 8). A certain amount of terrestrial organic matter is to be expected in these sediments, which contain both terrigenous clays and diatom tests. Furthermore, the surface sediments of the eastern part of the Gulf have higher clay contents than do surface sediments from elsewhere in the Gulf, being nearer to the Mexican mainland where runoff is relatively high (Van Andel, 1964; Calvert, 1966a).

In the Guaymas Basin at DSDP Sites 477 and 478, and 480 (Fig. 6) there are wide variations in the organic facies in the subsurface, because hemipelagic ooze is interbedded with numerous turbidites. Various chemical data show that the hemipelagic oozes have a marine organic facies (Simoneit et al., 1982), as might be expected, since the Guaymas Basin is the site of large accumulations of diatom skeletal remains; in contrast, the turbidites have a terrestrial or mixed marine-terrestrial organic facies. Most of the organic matter in the hemipelagic units is amorphous; that in the turbidite units is mostly structured terrestrial material or charcoal. Pyrolysis confirms the wide variation, some samples being classified as Type II (probably marine), others as Type III (probably terrestrial), and yet others as mixtures of these two (Deroo et al., 1982; Rullkötter et al., 1982). The samples nearest to the Mexican mainland (Site 480) seem to have more terrestrial material than do those near the Baja California margin (Sites 477, 480), probably because of the nearness of Site 480 to a submarine fan (see Fig. 6 for location). As in the California Borderland, there is a well defined relationship between TOC and organic facies, reflecting the control of TOC by the preservation of marine-derived organic matter (most of which is now amorphous) (Fig. 8).

On the Baja margin, in deep water, at DSDP Sites 474, 475, and 476, most of the organic matter is, again, marine-derived and amorphous, with the exception of some turbidite units that have a terrestrial organic facies (Deroo et al., 1982; Gilbert and Summerhayes, 1982; Rullkötter et al., 1982; Simoneit et al., 1982). The ratio of marine to terrestrial organic matter in the kerogen fraction is much higher here than at other sites in the Gulf, reflecting the small influence of the arid Baja California peninsula on detrital sedimentation (Fig. 8). For the same reason, the surface sediments of this margin have very high carbonate contents compared with the eastern Gulf margin.

The marine character of the organic facies agrees with the mineralogical character of the surface sediments. The main locus of diatom sedimentation is in the central Gulf deeps presumably because this point is furthest from terrigenous clastic sources along the coast. However, examination of the sand fraction shows that diatoms and radiolaria are abundant everywhere, especially along the margins

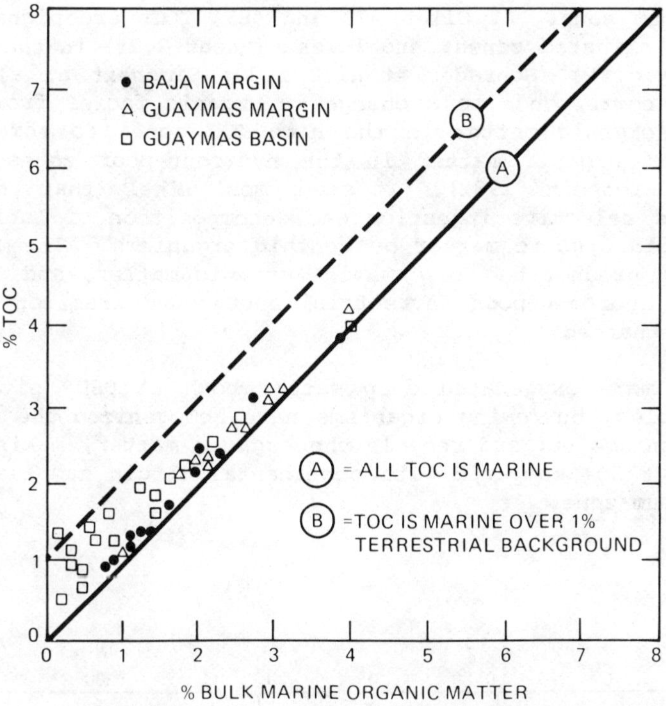

Fig. 8. Dependence of TOC on preservation of marine organic matter in sediments from Guaymas Basin and slope (DSDP Site 477, 478, 479, 481), and off Baja California (Cabo San Lucas slope DSDP Sites 474, 475, 476) (adapted from Simoneit et al., 1982); see Fig. 6 for site locations.

in the oxygen minimum zone; carbonate is abundant, too, especially in the west, and terrigenous clastics are important only on the eastern margin.

Sedimentary structures. Laminations characterize surface sediments collected where the oxygen minimum impinges on the continental slope. Oxygen levels in bottom waters in contact with these laminated sediments are usually less than 0.5 mℓ/ℓ. As in the Santa Barbara Basin, the laminations are varves that formed in response to seasonal changes from high runoff and low productivity in summer and fall, to intensified upwelling and productivity, and lower runoff, in winter and spring. Each varve couplet consists of a layer dominated by diatoms, and one dominated by clay minerals (Calvert, 1964; 1966b; Schrader et al., 1980).

In the subsurface, zones of laminated sediment alternate with zones of homogeneous sediment at DSDP Site 481. The alternations probably represent differing degrees of oxygen depletion within the oxygen minimum zone. At Sites 479 and 481, TOCs are highest (about 3%) in the laminated zones, and lowest (about 2.2%) in the homogeneous zones (Fig. 9) (Schrader et al., 1980; Simoneit et al., 1982). This change corresponds to a change in organic facies from more hydrogen-rich organic matter in the high TOC zones to somewhat less hydrogen-rich organic matter in the hydrogen-poor zones (Fig. 9) (Peters and Simoneit, 1982). It seems most likely that the loss in TOC reflects selective ingestion and decomposition of labile hydrogen-rich lipid organic matter by benthic organisms. The relatively TOC-poor end product has less marine organic matter, and the influence of the hydrogen-poor terrestrial background fraction is correspondingly enhanced.

In the more oxygenated deep water, both at DSDP sites and in surface samples, burrowing organisms have homogenized the sediments. Benthic organisms eat and recycle the organic matter, making the TOC drop by about 50% compared with continental margin samples from the oxygen minimum zone.

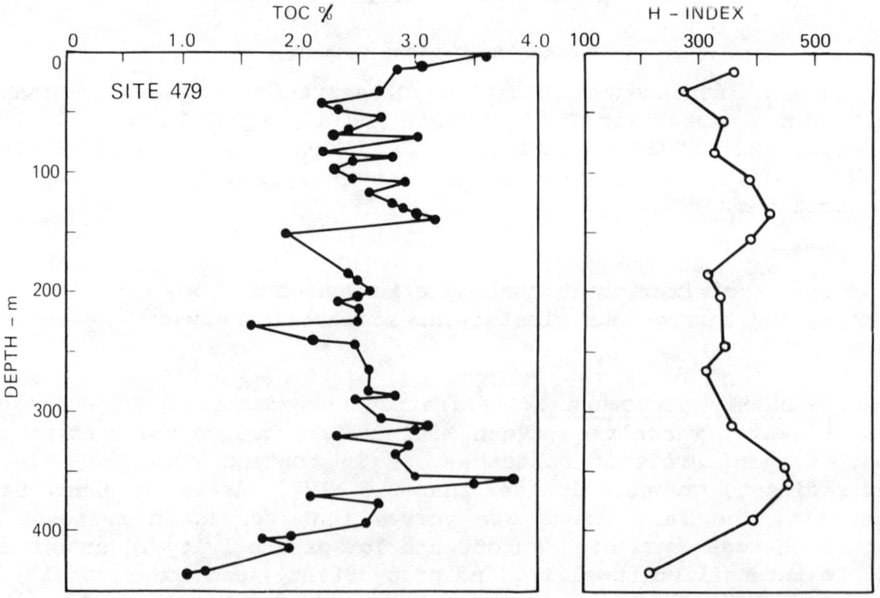

Fig. 9. Downhole variations in TOC, and in hydrogen index as measured by pyrolysis, at DSDP Site 479. High TOC correlates with high hydrogen index, and is typical of laminated zones; low TOC goes with low hydrogen index and correlates with homogeneous zones (adapted from Simoneit et al., 1982).

Diagensis. Off Cabo San Lucas, at the deep water DSDP sites, subsurface sediments tend to contain about 50% less TOC than do surface samples. Most probably this drop is caused by the consumption of organic matter by sulfate-reducing bacteria in the subsurface. The average TOC of deep subsurface samples stabilizes at about 1.5%. This same pattern can be seen in the Guaymas Basin, though the record is complicated by both thermal affects (organic matter is cooked off to hydrocarbons in some places), and by erratic swings in TOC concentration brought about by alternation from pelagic to turbiditic sedimentation. On the continental margin, in the oxygen minimum zone, a downhole decrease in TOC over the top few meters (Fig. 9) coincides with a change from laminated to homogeneous sediments, so may not be diagenetically controlled (Simoneit et al., 1982).

At the deep water DSDP sites there is a ubiquitous TOC maximum at a subsurface depth of about 10-20 m, where the sediments are calculated to be about 10,000 years old. If this age is correct, then these high values may represent a more oxygen depleted environment for Gulf bottom waters at the start of the Holocene transgression. However, the radiocarbon ages disagree with biostratigraphic ages (Spiker and Simoneit, 1982), and this deposit may be older, perhaps having been deposited near the peak of the last glaciation when winds were stronger than today and upwelling may have been more intense. A TOC-rich laminated diatomaceous sediment 300 m deep in DSDP hole 478 represents some time in the past when conditions were temporarily anoxic in the depths of the Guaymas Basin.

Southwest Africa

General oceanographic setting. Upwelling off southwest Africa takes place at the surface in two regions, one along the coast, the other in the Offshore Divergence Belt located over the shelf edge (Fig. 10). The Offshore Divergence Belt is a zone of shear between the coastal Benguela Current and the Trade Wind Drift. It is independent of wind stress, and is characterized by upward movement of cold nutrient-rich water. At the coast, wind-driven upwelling is concentrated in three main centers: (1) between Cape Frio and the Kunene River, in the north; (2) off Walvis Bay, north of Lüderitz, in the central part of the coast; and (3) in the Cape Town area, in the south (Hart and Currie, 1960; Bang, 1971). Upwelling is seasonal along the coast, being at a peak in spring (Hart and Currie, 1960). Productivity is high (Koblentz-Mishke, Volkovinsky and Kabanova, 1970), and biological evidence suggests that both the coastal and shelf edge systems have the highest productivity (Summerhayes, Hofmeyer and Rioux, 1974). Nitrate concentrations are high (NO_3 usually 10-30 μg atom/ℓ; Calvert and Price, 1971a) like those off Peru (Huntsman and Barber, 1977).

Oxygen concentrations are notably higher in the Atlantic than in the Pacific in all subsurface waters, including the oxygen minimum zone (Kester, 1975). On the continental slope of southwest Africa

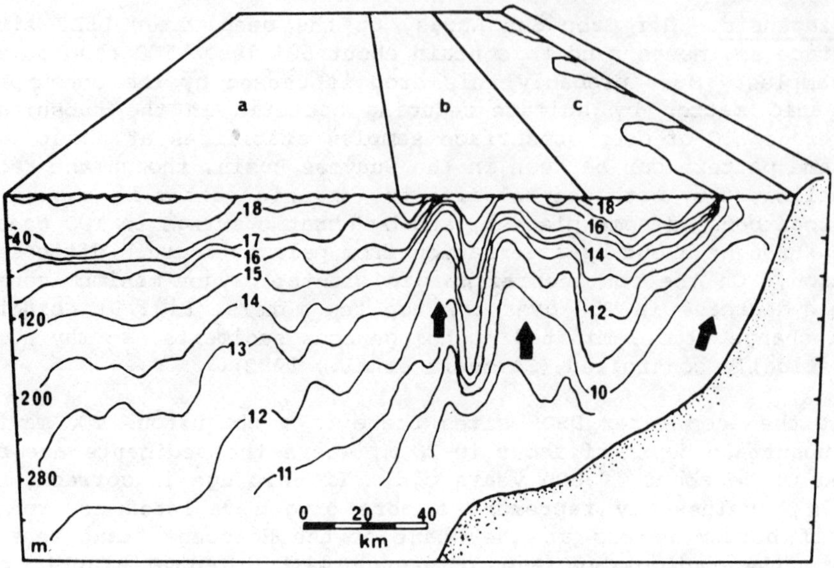

Fig. 10. Mechanism of upwelling at shore and shelf edge off southwest Africa (adapted from Bang, 1971); a = Trade Wind Drift, b = Offshore Divergence Belt, and c = Benguela Current. Upwelling cells are outlined on the surface at the coast. Isotherms in °C; depths in meters; arrows represent upwelling.

there is a well-defined oxygen minimum zone at depths of 400-800 m, as in the northeastern Pacific (Fig. 11). But oxygen levels at these depths off southwest Africa are much higher than in the Pacific, being 2-3 mℓ/ℓ instead of 1 mℓ/ℓ. It is only on the shelf, which is unusually deep off southwest Africa (the shelf edge lies at 300-400 m), that oxygen depletion is severe (Fig. 11). Much of the shelf shallower than 200 m is covered by a thin layer of oxygen-poor water with less than 2 mℓ/ℓ O_2, and locally as little as 0.1 mℓ/ℓ. Low oxygen values are sometimes found also on the outer shelf north of Walvis Bay. In general, oxygen values are lower on the shelf north of Luderitz than south of it.

Controls on organic matter. Upwelling, high productivity, and the low content of oxygen in bottom waters seem to control the distribution of organic matter here, as off California, and fecal pellets play a major role in the sedimentation of organic matter (Bishop, Ketten and Edmond, 1978).

The most spectacular end product of upwelling here is the accumulation of organic-rich diatomaceous ooze in extensive patches on the inner shelf (less than 150 m deep) between Luderitz and the Kunene River (Fig. 12) (Emilianov and Senin, 1969; Calvert and Price, 1971b; Summerhayes, 1972a; Bremner, 1980). This is a desert coast and there is little supply of terrigenous clastics. TOCs in this

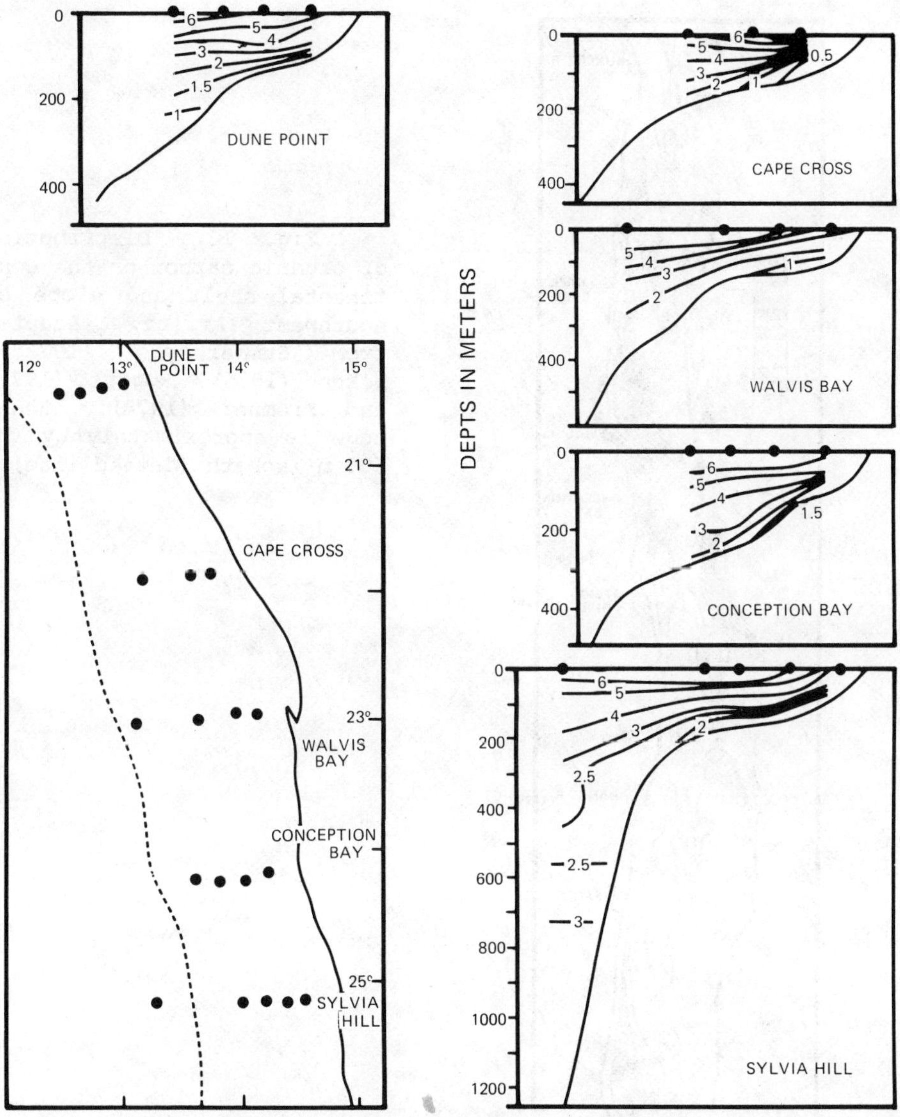

Fig. 11. Cross sections through the oxygen minimum zone off southwest Africa (adapted from Calvert and Price, 1971a); dashed line is shelf edge. Oxygen is in mℓ/ℓ.

sediment are consistently greater than 5-7%, and locally range up to 14% (Summerhayes, 1972a; Bremner, 1974; Rogers, 1977). The landward edge of this deposit is locally as little as about one mile offshore, especially near Walvis Bay. The deposit is mostly located in water 30-130 m deep, but is locally as shallow as 1 m and as deep as 220 m.

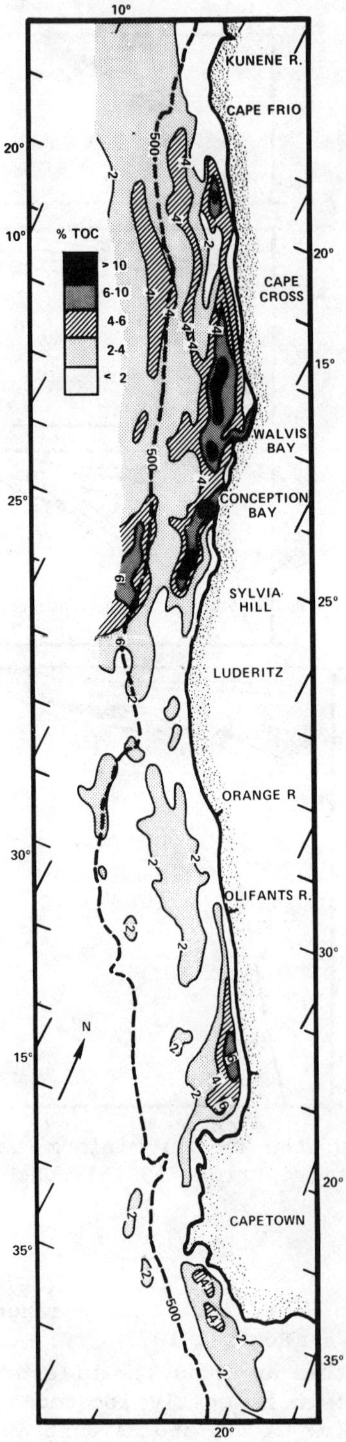

Fig. 12. Distribution of organic carbon on the continental shelf and slope of southwest Africa. Adapted from Summerhayes (1972b), Birch (1975), Rogers (1977) and Bremner (1974). Shelf edge is approximately by the 500 m isobath (dashed line).

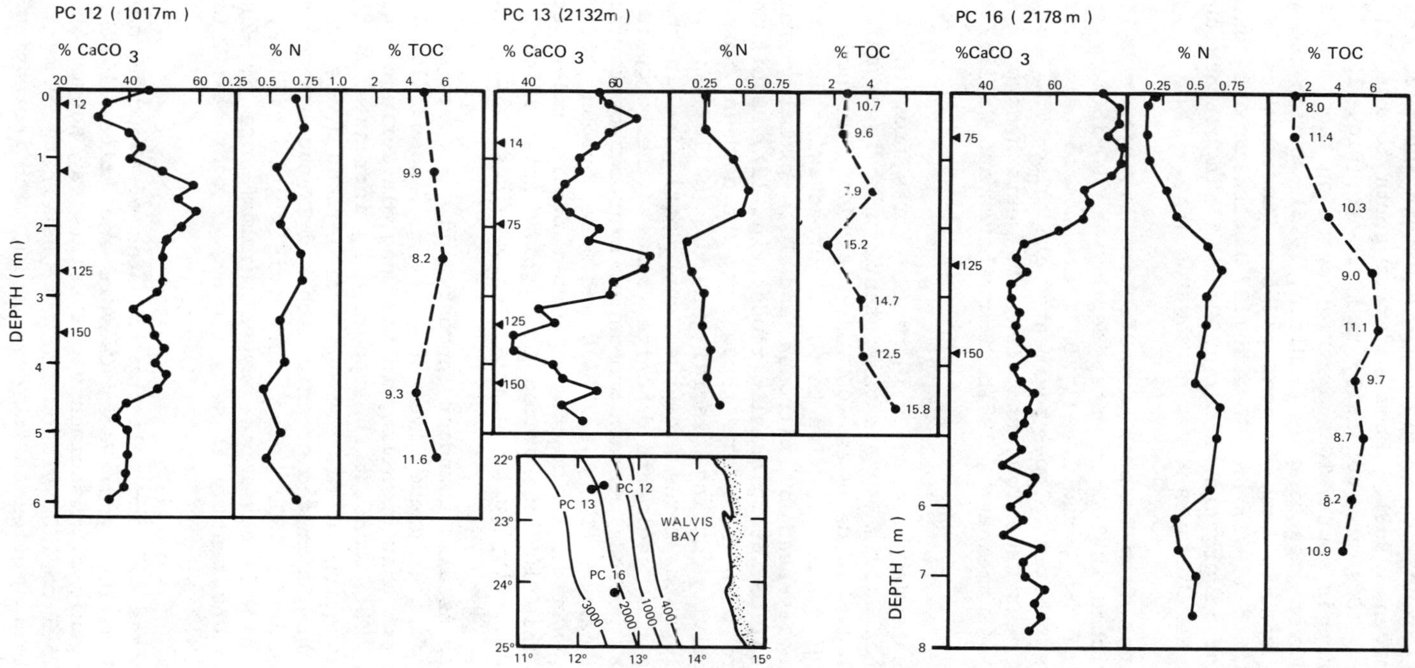

Fig. 13. Downcore variations in TOC, organic (Kjeldahl) nitrogen and $CaCO_3$ on the continental slope near Walvis Bay, southwest Africa. Organic analyses made at Woods Hole Oceanographic Institution. Depths in meters. Ages in thousands of years before present given by triangle symbols. $CaCO_3$ and age data from Summerhayes, Bornhold and Embley (1979).

Diatomaceous muds extend into Walvis Bay, to within about 30 feet of the shore (Copenhagen, 1953). From Lüderitz south to the Olifants River, inner shelf TOCs are generally much lower (Fig. 12) (1-2%; Rogers, 1977). Further south, TOCs are high again on the inner shelf off the Olifants River (reaching 6-7%; Birch, 1975), and on the western edge of the Agulhas Bank (reaching 3-5%; Summerhayes, 1972b). Bremner (1974), Birch (1975) and Rogers (1977) also report zones of moderate enrichment in organic matter (TOCs 2-7%) in patches on the middle shelf from the Olifants River northwards (Fig. 12).

Slope sediments from beneath the Offshore Divergence Belt are significantly enriched in TOC at water depths of 500-1500 m (Fig. 12). TOC values of 4-5%, and locally 5-7% are common here north of Lüderitz. More southerly slope sediments are also TOC enriched (1-3%) especially near Cape Town and along the western edge of Agulhas Bank (Fig. 12) (Summerhayes, 1972b; Summerhayes and Willis, 1972; Birch, 1975).

Subsurface Quaternary sediments in piston cores from the continental slope west of Walvis Bay are consistently rich in organic matter in water at least as deep as 2000 m (Fig. 12). Downcore TOCs average 3-5%, ranging from 1.4% up to 6.2%. Broad swings in the organic content (represented by nitrogen and TOC) probably reflect changes from glacial to interglacial periods (Fig. 12), as off northwest Africa, where high TOCs are thought to have been caused by heightened productivity during glacial periods (Müller and Suess, 1979). Studies of piston cores from the continental rise in the Angola Basin confirm that sediments deposited beneath the Benguela Current during glacial periods contain markedly lower carbonate, higher levels of organic carbon (TOC 1.8-2.3%), and more abundant siliceous skeletal remains than in sediments deposited during interglacials (TOC 0.9-1.2%). This is interpreted to signify intensification of upwelling and productivity during glacial periods (Bornhold, 1973).

Organic carbon is also abundant in post Middle Miocene sediments at DSDP Site 362 off the Kunene River (Fig. 14). Siesser (1980) uses these data to suggest that upwelling in the Benguela Current began in the Late Miocene. This concept is supported by increases in the abundance of preservation of diatoms and radiolarians (Diester-Haass and Schrader, 1979). Laminated organic-rich Mid-Cretaceous sediments (black shales) found in DSDP holes along this margin probably were deposited within a well-developed oxygen minimum zone 500-2500 m deep. The laminations testify to very low oxygen contents; TOCs are up to more than 10%.

The diatom oozes are accumulating at rates of about 56 cm/1000 yrs (range 15-103 cm/1000 yrs; Veeh, Calvert and Price, 1974). On the slope, the organic-rich sediments off Walvis Bay accumulated at rates of about 2.5-3 cm/1000 yrs during the Quaternary (Summerhayes et al., 1979). This implies an increase in TOC with increasing rate

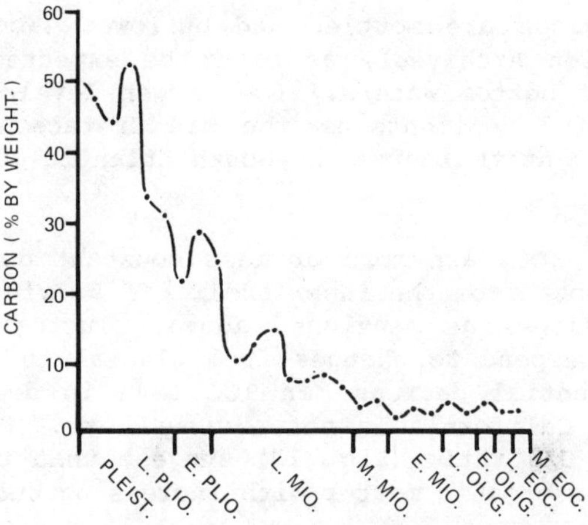

Fig. 14. Variation in organic carbon content (TOC) with age at DSDP Site 362, off southwest Africa (from Siesser, 1980).

of sedimentation, as found elsewhere by Müller and Suess (1979). It is the opposite of what Calvert (1976) records for offshore California.

Organic facies. Almost all (90-95%) of the organic matter in samples from the continental shelf and slope off southwest Africa is amorphous, the rest being structured marine material. Geochemical analyses show that off Walvis Bay the organic matter is derived from algal sources and has been modified by microbial activity (Boon, de Leeuw and Schenk, 1975). Some terrestrial organic matter occurs close inshore off the major river mouths.

The marine character of the organic matter agrees well with the biogenic character of the sediments on the slope and on much of the shelf north of Cape Town (Summerhayes, 1972a; 1972b; Summerhayes and Willis, 1972; Bremner, 1974; Birch, 1975; Rogers, 1977). In the south, the organic-rich muds of the inner shelf off the Olifants River (Birch, 1975), and on the western Agulhas Bank (Rogers, 1971; Summerhayes 1972b; Summerhayes and Willis, 1972) are terrigenous. The terrigenous components of the inner shelf sediments of Agulhas Bank are relict (Rogers, 1971); it seems likely that their high organic content is a reflection of past upwelling and high productivity.

Sedimentary structures. There are no reports of laminations in the gravity cores from the diatomaceous oozes off Walvis Bay, perhaps because of a lack of seasonal terrigenous influx. Fecal pellets, indicative of extensive bioturbation, are abundant in most of the organically enriched shelf and slope surface sediments. Benthic polychaete worms seem to produce most of the pellets. The piston cores from the slope are mottled and burrowed (Woods Hole Oceanographic Institution Archives), as might be expected from the high oxygen content of bottom waters. Low oxygen levels favored accumulation of laminated sediments in the Mid Cretaceous in deep water along this coast. At that time the South Atlantic was a more or less closed basin.

Diagenesis. TOCs are more or less constant down gravity cores of diatomaceous ooze from the inner shelf off Walvis Bay. In piston cores from the slope, as mentioned above, fluctuations in TOC and nitrogen may correspond to changes from glacial to interglacial periods; the exponential decrease in TOC seen in deep water samples from the Gulf of California is not obvious here. However, downcore increases in the C/N ratio (Fig. 13) suggest that there may be some diagenetic loss of organic matter with increasing burial time.

Diagenetic processes in Walvis Bay diatom oozes cause redistribution of phosphate and the formation of phosphorite in the sediments (Veeh et al., 1974). Phosphatic sediments and rocks are widespread on the continental margin of southwest Africa (Summerhayes, 1973; Summerhayes et al., 1973; Parker, 1975; Price and Calvert, 1978; Birch, 1979). Some of the phosphatic rock is Paleocene-Eocene but most of it is Late Miocene-Pliocene (Dingle, 1977; Siesser, 1978). Thus extensive phosphorite formation coincided with the inception of upwelling in the Benguela Current.

Northwest Africa

General oceanographic setting. Coastal upwelling under the influence of the NE Trade Winds is well-developed off the Saharan coast, especially off Cap Blanc (Fig. 15). The upwelling is variable not only in time, being strongest in summer, but also in space, being centered both on the inner shelf and over the shelf edge (Jones, 1972; Hughes and Barton, 1974; Barton, Huyer and Smith, 1977). On the slope the system is further complicated by the presence of a compensatory poleward current beneath the south flowing Canary Current that dominates surface circulation. There are similar subsurface currents in the upwelling systems of California, Peru, and southwest Africa (Hughes and Barton, 1974; Huyer, 1976; Barber and Smith, 1981). Another important feature of northwest African upwelling is that there is strong onshore flow along the bottom over the shelf. An onshore flow beneath the offshore surface flow is common to upwelling systems, but usually occurs at mid-water depths above a near-bottom offshore flow.

Upwelling causes high biological productivity in northwest African surface water although nutrient levels are lower NO_3 < 12 μg atoms/ℓ) than in the more productive upwelling systems like that off Peru (NO_3 10-25 μg atoms/ℓ). The lower productivity off northwest Africa may be a function not only of a low nutrient supply, but also

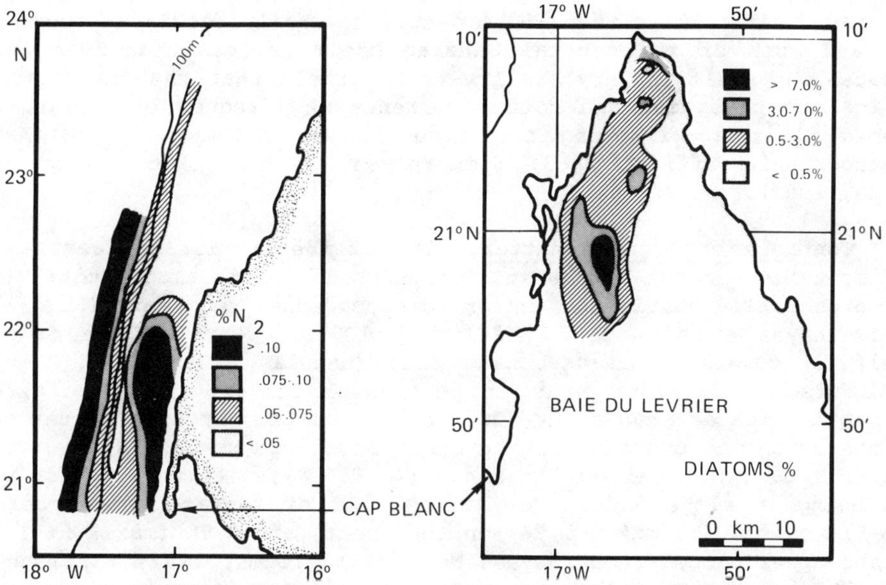

Fig. 15. Distribution of organic matter (as % Kjeldahl nitrogen) on the shelf and upper slope of the southern Sahara (from Milliman, 1977), and of diatoms in sediments in the Baie Du Levrier (from Koopmann, Sarnthein and Schrader, 1978). The Bay is <20 m deep.

of greater turbidity, winds being stronger and less steady, thus causing more turbulence, off northwest Africa than off Peru. Turbidity limits productivity by restricting light penetration. Rapid recycling of organic matter, including diatoms, occurs in the water column (Koblentz-Mishke et al., 1970; Huntsman and Barber, 1977; Friederich and Codispoti, 1979).

In the oxygen minimum zone off northwest Africa, oxygen values fall to 1.5-2.0 mℓ/ℓ near Cap Blanc, and to 1.0-1.5 mℓ/ℓ south between Cap Blanc and Cape Verde, on the continental slope between 300-500 m deep; on the upper slope and outer shelf oxygen is usually 2.5-3.0 mℓ/ℓ near Cap Blanc, but may be as low as 1.5-2.0 mℓ/ℓ at times. Oxygen values of just under 1.0 mℓ/ℓ have been recorded on the outer shelf off Cape Verde. As in the Gulf of California (Fig. 6) and off southwest Africa (Fig. 11), oxygen levels off Cap Blanc are lowest at the seabed, probably because of decomposition of organic matter at the sediment-water interface. Oxygen levels are not as low as they are on the shelf off southwest Africa (Jones and Folkard, 1970; Weichart, 1974; Friebertshäuser et al., 1975; Kester, 1975).

Another important feature of the continental margin of northwest Africa is its exceptionally shallow continental shelf. Most of the

Saharan shelf is less than 100 m deep (the shelf edge is at about 105 m), and much of the central Saharan shelf is less than 50 m deep. Because the shelf is very shallow it is likely that the large swells typical of this area and bottom currents may frequently disturb the seabed, thereby winnowing out fine-grained sediment (U.S. Naval Oceanographic Office, 1963; Summerhayes et al., 1976; Barber and Smith, 1981).

Controls on organic matter. One of the surprising features of sedimentation on the northwest African shelf is the almost total lack of organically enriched and/or diatomaceous sediment (Fig. 15) (Summerhayes et al., 1976; Milliman, 1977). Most of the Saharan shelf is covered with carbonate sand (usually >90% $CaCO_3$). Muddy sediments ($CaCO_3$ <75%) occur in patches on the innermost shelf, and at depths greater than about 500 m on the slope. Organic matter correlates with mud content, so the muddy sediments contain more organic matter (>0.1% N) than do the sands (<0.05% N), with the highest values being in slope muds. The average TOC of surface sediments off Cap Blanc is 0.4% (max = 0.7%) on the shelf, and 0.9% (max 1.5%-1.7%) on the upper slope (Summerhayes, Nutter and Tooms, 1972; Hartmann et al., 1976). Higher values reaching about 3% TOC occur in muddy sediments at 1000-2000 m depth; this material may have been washed off the shelf (Fütterer, this volume). Both the muds and the sands off Cap Blanc contain more organic matter than their equivalents off the northern Sahara or Morocco, reflecting the increase in upwelling and high productivity towards Cap Blanc where upwelling is strong year round. Surface TOCs are also high, reaching 2.6%, south of Cap Blanc, on the slope near Cape Verde. The TOCs in the northwest African cores increase with increasing rate of sedimentation so their variation does not simply reflect changes in productivity. The association of TOC and rate of sedimentation, and the occurrence of the highest TOCs at 1000-2000 m, where oxygen concentrations are 3-5 mℓ/ℓ also shows that oxygen distribution here does not control organic accumulation.

There are accumulations of organic-rich diatomaceous sediments, with up to 10% diatoms and 1.5-2.0% TOC, in the Baie du Levrier in the lee of Cap Blanc (Fig. 15) (Koopman et al., 1978). It is not clear if the diatoms were produced *in situ* or transported into the Bay from the Cap Blanc area, although the latter possibility seems more likely. These muddy sediments are extensively reworked and full of fecal pellets of benthic organisms.

Subsurface sediments in piston cores from deep on the continental slope off the central Saharan coast reveal the history of deposition of organic matter in the Quaternary (Fig. 16). Cores from 1800-2800 m on the lower continental rise within about 100 miles west and 300 miles north of Cap Blanc have much more organic matter in sediments of glacial age (TOC = 2-4%) than in overlying postglacial sediments (0.5-1.7% TOC) (Fig. 16) (Hartmann et al., 1973; 1976; Debyser

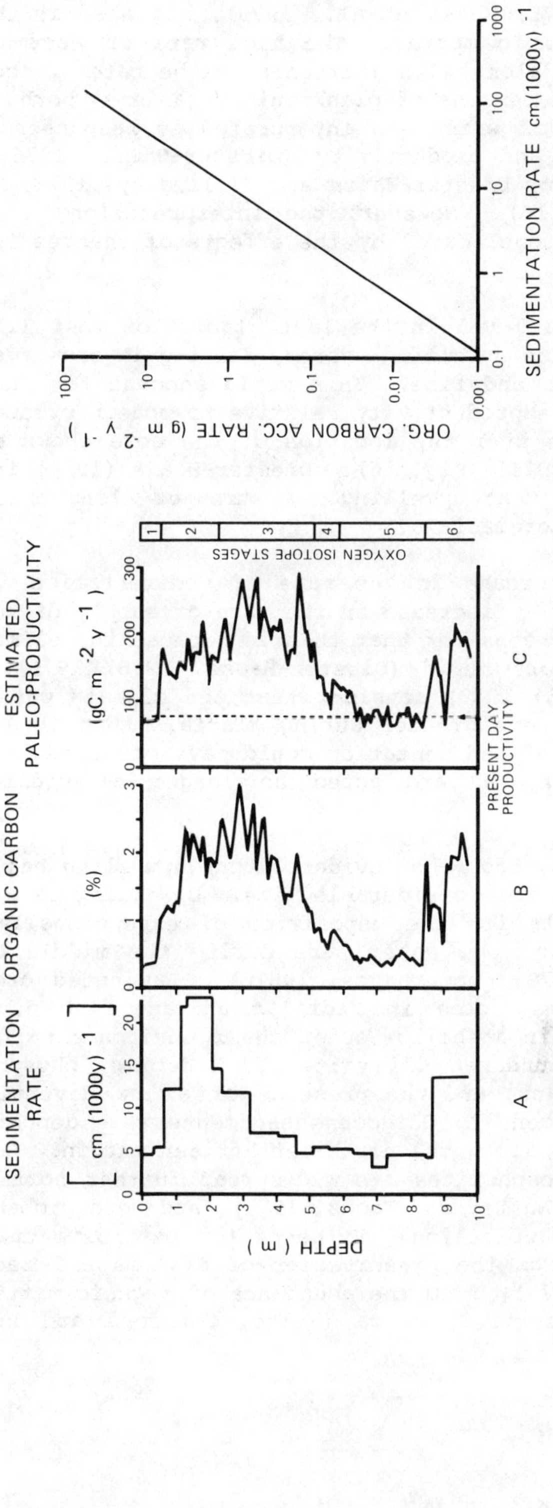

Fig. 16. Downcore variations in TOC, sedimentation rate and estimated paleoproductivity in a core from 2575 m on the Saharan continental slope (left), and relation of organic carbon accumulation to sedimentation rate (right) (from Müller and Suess, 1979).

and Gadel, 1979; Müller and Suess, 1979). This increase is caused by an increase not only in the rate of sedimentation, from 5-8 cm/1000 yrs in the Holocene to 10-20 cm/1000 yrs in the last glaciation (Gaskell et al., 1975; Moyes et al., 1979), but also in the rate of accumulation of organic matter. The high rate of accumulation of organic matter goes along with increases in the rate of accumulation of biogenic skeletal remains of planktonic organisms, both calcareous and siliceous, all of which are interpreted as responses to an increase in upwelling and productivity (Diester-Haass, 1976; Hartmann et al., 1976; Berger, Diester-Haass and Killingley, 1978; Müller and Suess, 1979) (Fig. 16). However, the interpretation is not quite that simple--it is complicated by the effects of changes in both sea level and climate.

The drop in sea level in the last glaciation most likely had a profound impact on the upwelling system, driving it seawards from the shelf over the slope and rise. This would account for the substantial increase in paleoproductivity relative to modern productivity on the continental rise over the location of the core shown in Fig. 16 (Müller and Suess, 1979) (Fig. 16). Diester-Haass (1976) interpreted her data to suggest that upwelling was stronger along this coast in glacials than in interglacials.

Much of the increase in the rate of sedimentation in glacial periods is caused by an increase in the rate of supply of terrigenous mineral grains; some consider that this is because the climate of the hinterland became more humid (Diester-Haass, 1976; 1979). Others, like Sarnthein (1978; 1979) consider that the climate was more arid and that dust loads were greater during glacials than interglacials. An increase in rate of sedimentation could have occurred also because the shelves were exposed and acted no longer as sediment traps (Thiede, 1977).

Older sediments, too, show evidence that upwelling has persisted off northwest Africa for considerable time. Upwelling is believed to have been responsible for the deposition of exceptionally organic-rich black shales in deep water here during the middle Cretaceous (Von Rad et al., 1979; Summerhayes, 1981b). Laminated organic-rich mid-Cretaceous shales, rich in radiolarians and fish debris, were also deposited here in a shallow outer shelf environment in the Tarfaya Basin. The abundance of pyrite, fish debris, phosphate, diatoms, and radiolarians, and the presence of a low diversity fauna, suggest that mid-Eocene to Oligocene sediments were deposited under upwelling conditions along the northwest African margin. Late Cretaceous and Eocene phosphorites are widespread further north, off Morocco (Summerhayes, Nutter and Tooms, 1971), and were probably deposited in response to upwelling. Evidence for extensive upwelling in the Miocene comes from the preservation of diatoms and radiolarians (Von Rad et al., 1979), from the abundance of organic matter at DSDP Site 397, off the central Sahara (Welte, Cornford and Rullkötter,

1979), and from the abundance of phosphorites, especially on the Moroccan shelf and slope. Glauconitic phosphorites from the Moroccan margin range in age from 10.6 ± 0.5 to 14.4 ± 0.5 million years, averaging 12 million years (K-Ar data from Summerhayes, 1970), placing them in the late middle Miocene. Pliocene phosphorites have been dredged from the outer shelf off the Sahara, attesting to the continuation of extensive upwelling.

Organic facies. Most of the organic matter in surface samples from the shelf and slope off the Sahara is amorphous (80-90%), the rest being structured marine material. Caratini, Bellet and Tissot (1979) show that amorphous organic matter completely dominates not only the surface, but also the subsurface sediments of cores from the continental slope off Cap Blanc. The cores contain accessory amounts of structured terrestrial material and charcoal that is probably aeolian. Aeolian dusts from off this coast, and from off the coast of southwest Africa, carry terrestrial organic matter and must be responsible for introducing small amounts of this material into deep-sea sediments locally (Simoneit and Eglinton, 1977). Chemical data from the Cap Blanc cores shows that most of the amorphous organic matter is marine --kerogens are rich in hydrogen and are classified as type II (marine) organic matter (Debyser and Gadel, 1979). In one core isotopic data suggest that the organic-rich surface sediment has a predominantly marine organic facies ($\delta^{13}C$=-21°/$_{oo}$), while the less rich subsurface sediment 1 m down is a mixture of terrestrial and marine material ($\delta^{13}C$=-23°/$_{oo}$) (Gaskell et al., 1975). As discussed below, this change may reflect subsurface diagenesis.

The mid-Cretaceous black shales of the Saharan coast contain mostly amorphous organic matter chemically classified as type II (probably marine) (Tissot et al., 1980; Summerhayes, 1981b). The TOC content of these shales is directly proportional to the preservation of marine organic matter, and there is little or no background of terrestrial organic matter. Within the black shale sequence, organic-rich, black, laminated sediment deposited under reducing conditions alternates with light-colored, TOC-poor sediment that is bioturbated and was deposited under oxidizing conditions. Amorphous and probably marine organic matter also dominates these oxidized layers. Deposition off an arid coast characterized by extensive upwelling and little terrigenous runoff is inferred from this evidence.

Diagenesis. High concentrations of carbon dioxide, phosphate, and ammonia in the interstitial waters of piston cores from the Saharan slope show that organic matter is now being or has recently been oxidized in the subsurface. Diagenesis probably explains the downhole decrease in organic matter in the top 1 m of the Cap Blanc cores of Gaskell et al. (1975) and Debyser and Gadel (1979). Hartmann et al. (1973) calculate that 15-20% of the carbon in their cores was lost to diagenesis in the upper 1-2 m of the sediment column. In Gaskell's core, which does not contain much organic matter

(1.3% at surface; 0.9% at 1 m depth), isotopic data suggest there is a downcore increase in the terrestrial organic component. This increase could occur because bacteria and benthic organisms are selectively recycling the lipid-rich marine organic matter.

DISCUSSION

General Controls on the Distribution of Organic Matter in Upwelling Regimes

The main controls on the distribution of organic matter are (i) its supply, and (ii) its preservation, both of which are functions of the local and global circulation [see general reviews by Dow (1978) and Demaison and Moore (1980) for example]. In upwelling regimes, nutrients are brought to the surface, stimulating high biological productivity. The rate of supply of organic mater to the bottom is correspondingly high, especially on continental shelves and slopes. Undoubtedly the rate of sedimentation plays some part in preserving the organic matter that reaches the bottom. For instance, where the rate is high, in oxidizing environments, more organic matter is preserved than where the rates are low (Fig. 16) (Müller and Suess, 1979). Another important control of preservation is the amount of oxygen in bottom waters (Demaison and Moore, 1980). If we accept that organic matter is rapidly transported to the bottom mainly in fecal pellets (Suess, 1980; Dunbar and Berger, 1981), then it seems unlikely that the oxygenation of the water column has much effect on sinking organic matter. However, the oxygen content of bottom waters will control how much organic matter decomposes at the sediment-water interface. Under anoxic conditions ($O_2 < 0.5$ mℓ/ℓ) there will tend to be more preservation than under either dysaerobic conditions (0.5-1.0 mℓ/ℓ O_2) or truly oxidizing conditions (>1.0 mℓ/ℓ) (Demaison and Moore, 1980).

Throughout the world's oceans subsurface waters at intermediate depths (300-1550 m) tend to be depleted in oxygen, as a response to the decomposition of organic matter (Fig. 17) (Kester, 1975). In the western parts of ocean basins this oxygen minimum zone usually has more than 2.0 mℓ/ℓ O_2. On the eastern sides of the same basins there is usually less than that, and often less than 1.0 mℓ/ℓ O_2 (Fig. 17). The greatest oxygen deficiency occurs in the eastern Pacific Ocean. There are at least two reasons why the eastern margins of the Pacific have less oxygen than their Atlantic counterparts. Firstly, the Pacific Ocean contains less oxygen than the Atlantic (Fig. 18); secondly, Pacific waters contain more nutrients and so are potentially more productive than most Atlantic waters (Redfield, Ketchum and Richards, 1963; Berger, 1970). Where high productivity occurs in the Pacific, it leads to more loss of oxygen from subsurface waters. This pattern of nutrient enrichment and oxygen depletion occurs partly because the circulation of the Pacific is essentially "estuarine,"

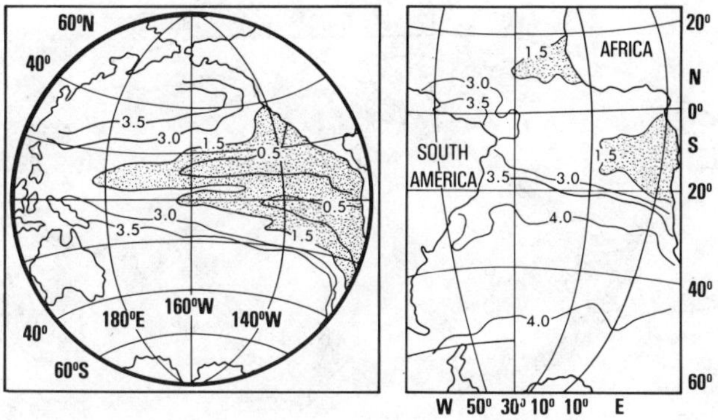

Fig. 17. Distribution of minimal oxygen values in the oxygen minimum zone (at about 400 m depth) in the Atlantic and Pacific Oceans (adapted from Kester, 1975).

while that of the Atlantic is "anti-estuarine (which means that the Pacific acts as a nutrient trap), and partly because much of the world's deep water originates in the North Atlantic. North Atlantic Deep Water is rich in oxygen (Fig. 18) and poor in nutrients. Along its path towards the Indian and Pacific Oceans its oxygen is lost (Fig. 18) by decomposition of sinking organic matter from which, in return, it gains nutrients (Redfield et al., 1963; Berger, 1970).

If we were to obtain all of our insights about organic sedimentation in upwelling centers by examining world productivity maps, we would suspect that northwest and southwest Africa had the most organic matter, in about equal proportions, while there was somewhat less organic matter off Peru and California (Fig. 19) (Koblentz-Mishke et al., 1970). We might also suspect that the Grand Banks and eastern U.S. continental shelf had organic rich sediments, these also being

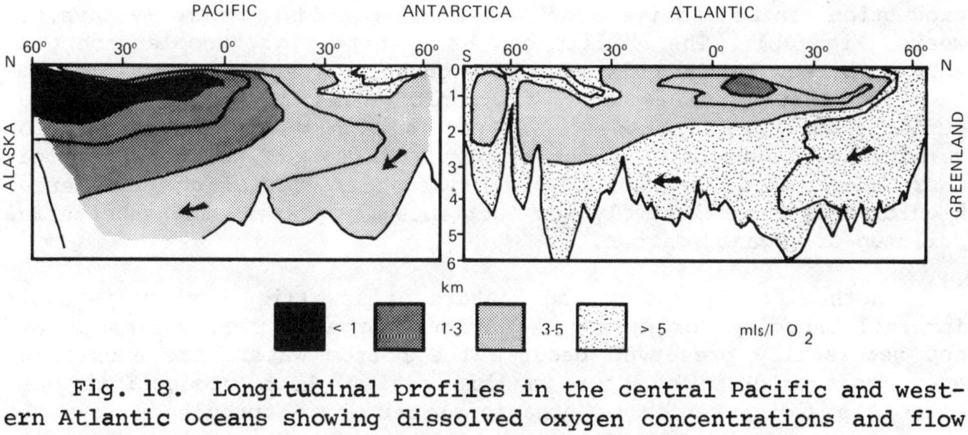

Fig. 18. Longitudinal profiles in the central Pacific and western Atlantic oceans showing dissolved oxygen concentrations and flow of deep water (adapted from Berger, 1970 and Kester, 1975).

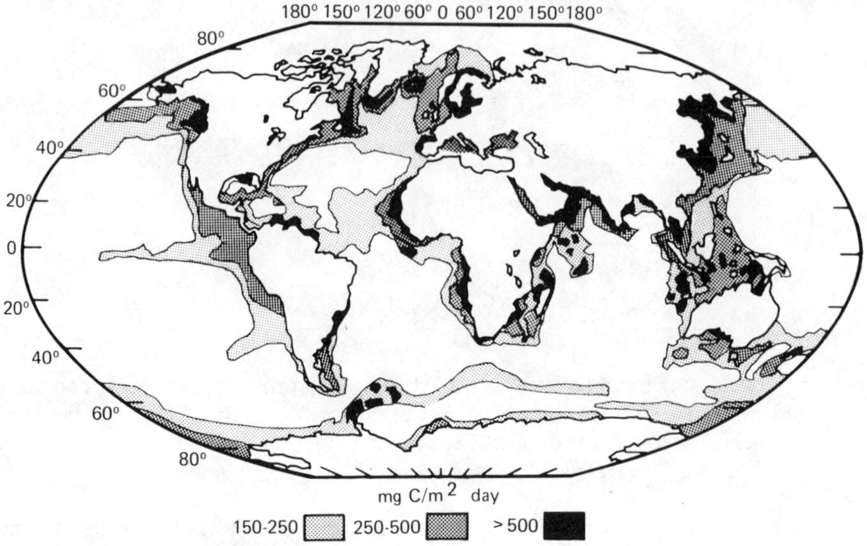

Fig. 19. Distribution of primary productivity in the world's oceans (modified from Koblentz-Mishke et al., 1970).

highly productive areas (Fig. 19). As pointed out by Demaison and Moore (1980), among others, productivity maps are essentially meaningless for sediment studies. In fact, as we know from this and earlier studies, northwest Africa is a poor area for the accumulation of organic matter, there is hardly any organic matter in sediments from the Grand Banks or the eastern U.S. margin, and Peru and California have about as much organic matter as we find off southwest Africa.

One of the significant, and usually overlooked, features of the productivity of upwelling systems (L. Pratt, pers. comm.) is neither the annual production, nor the maximal production, but the fact that production can be massive over very short periods of time -- days, or weeks (Fig. 20). The ability of the system (e.g., zooplankton grazers), to recycle the bulk of this production (phytoplankton) is limited, so that large amounts of organic matter can reach the bottom during short upwelling events. This is true where the circulation permits, but evidently is not happening today off northwest Africa, where the combination of landward cross-shelf flow, high turbulence, high oxygen, and relatively low nutrients may prevent any substantial build up of organic matter.

Another poorly understood feature of upwelling systems is their internal capacity for oxygen depletion. Organic rich sediments are not necessarily preserved because the bottom waters are anoxic (or even dysaerobic). As shown in this review, high productivity can force the sedimentation of organic matter where turbulence does not

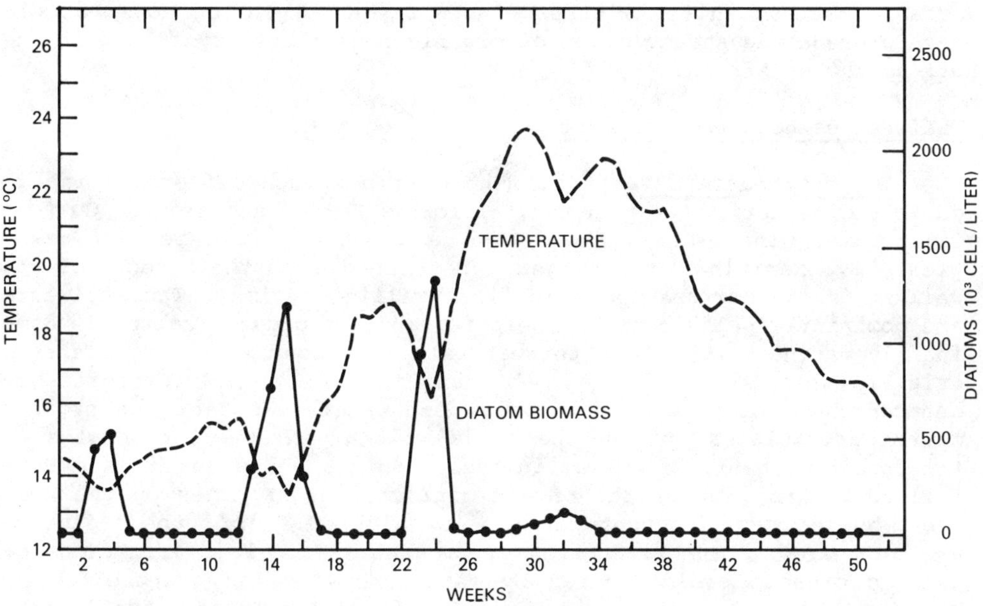

Fig. 20. Effect of upwelling on primary productivity in surface waters off southern California (modified from Tont, 1976 by Lisa Pratt). Most of the year's productivity is focused on massive, short-term blooms coincident with cold water upwelling events.

prevent organic particles from settling. This organic matter will be preserved, with some loss by diagenesis, if it is buried at a reasonable rate, regardless of the oxgyen content of the bottom waters (as on the southwest and northwest African slopes, for instance). Decomposition of the organic matter at the sediment-water interface and in the top few meters of the sediment column will use oxygen not only from the pore waters, but also from the overlying water column, thereby modifying the oxgyen levels in the oxygen minimum zone (Figs. 6 and 11) (Emery, 1960; Emery and Hülsemann, 1962; Calvert and Price, 1971a). Anoxic bottom waters in these areas, even in basins like Santa Barbara, are not anoxic in the sense of being stagnant (i.e., still): they are constantly moving. For example, Santa Barbara Basin's bottom water is replenished at least every 20 years. The oxygen depleted bottom waters from the basins of the California Borderland, or from the shallow shelf off southwest Africa, are constantly moved away from the places where oxygen is being lost to the seafloor, to become part of the overall oxygen minimum zone. Their incorporation into the regional oxygen minimum zone further lowers oxygen values in that zone (Fig. 17). This leads us to a chicken-and-egg situation, in that in some places, like the Gulf of California (Brongersma-Sanders, 1971), "estuarine" circulation moves oxygen depleted water into the basin, thereby allowing the preservation of organic matter, while in other places oxygen depletion does not always precede the accumulation of organic matter but may follow it, as off southwest Africa.

Differences Between Different Upwelling Regimes

The northeastern Pacific. There is not much difference in organic sedimentation between the California Borderland and the Gulf of California. In essence, these may be taken as complementary examples. By examining both we get a better idea of the effects of upwelling in the northeastern Pacific upwelling regime. The sediments are dominantly pelagic or hemipelagic, and the organic matter is dominantly marine, with minor terrestrial plant contributions. Terrestrial organic matter is only abundant in the low TOC turbidite sequences that occur locally in very deep water. Most of the organic matter accumulates not on the shelf, but at intermediate depths on the open continental slope or in basins on the slope (as in the Borderland basins), where the oxygen minimum zone impinges on the seabed. Where oxygen contents fall below about 0.2 mℓ/ℓ the sediments are laminated. The laminations represent seasonal interruptions of pelagic organic sedimentation by terrigenous influx, controlled by runoff. The influx of terrigenous clastics not only adds small background amounts of terrestrial organic matter, but also dilutes the organic carbon content of the sediment. The amount of dilution may be proportional to sedimentation rate (see Calvert, 1976), but is not always so (Gilbert and Summerhayes, 1982). Downcore, laminated sediments tend to alternate with zones of homogenous and bioturbated sediment, suggesting that oxygen depletion is not persistent. More detailed stratigraphic analyses are needed to ascertain what controls these fluctuations in oxygen depletion. Diagenesis appears to be important not in the anoxic waters of the oxygen minimum zone, but in the dysaerobic water below that zone. For instance, in the deeper parts of the Gulf of California as much as 50% of the original organic matter that is found in surface sediments seems to be lost by decomposition within the top few meters of the sediment column. This process reduces considerably the oceanographic signal of upwelling in these sediments.

Southwest Africa. Off southwest Africa, as in the northeastern Pacific, the organic matter is nearly all marine, its sedimentation is pelagic, and fecal pellets are probably the most important transporting agent. Nutrient levels appear roughly similar, but there is much more runoff in the northeast Pacific, and much less oxygen.

There are several important differences in the distribution of organic matter between these two areas. For instance, off southwest Africa, organic matter is enriched not only on the continental slope, but also on the continental shelf (Fig. 12). TOC's are highest in the nearshore upwelling zone, and increase northward partly because upwelling increases in intensity in that direction, and partly because there is more clastic sedimentation on the southern shelf. We see the same regional pattern on the slope, with higher TOC's in the north, though TOC's are always less than on the adjacent inner shelf. Another difference from the northeast Pacific is that TOC is richest in the most rapidly accumulating sediments off southwest Africa; TOC's are much less on the slope, where the rate of sedimentation is slower. Within the shelf muds off Walvis Bay, there is no systematic relation between TOC and sedimentation rate (Veeh et al., 1974). The TOC's of the slope muds, though less than on the shelf, are higher than in comparable sediments from the northeast Pacific, though accumulating at half the rate. Evidently, there is no simple relation between TOC and sedimentation rate in these areas, although such relations do exist in other areas (Fig. 16) (Müller and Suess, 1979).

As off California, there is off southwest Africa a general relation between TOC enrichment and the oxygenation of bottom waters. Like the TOC-rich slope muds of the northeast Pacific, the organic-rich shelf muds of southwest Africa accumulate under bottom waters that have as little as 0.1 mℓ/ℓ O_2 (Fig. 6). But the TOC-rich slope muds of southwest Africa accumulate where oxygen levels are 2-3 mℓ/ℓ or more (Fig. 6). By contrast, at oxygen levels lower than 1 mℓ/ℓ O_2 in the deep Gulf of California, there is less TOC, and the sediments are losing much organic matter through subsurface diagenesis. Since the African margin sediments are accumulating at a slower rate than those of the Guaymas Basin and Baja slope, one cannot call upon sedimentation rates to explain the difference. The difference may reflect neither the oxygenation of bottom waters nor the sedimentation rate, but instead, the magnitude of the productivity and the rate of supply of organic matter to the bottom. As pointed out by Müller and Suess (1979), both productivity and rate of sedimentation can control the TOC in these types of environments. Probably the higher TOC's off southwest Africa reflect both higher productivity, and lower runoff (i.e., less dilution by clastics) compared with the northeast Pacific. The differences in oxygen content may be immaterial.

Subsurface diagenesis, which is controlled by the rate of sedimentation as well as by the oxygenation of bottom water, is an important regulator of organic content. It substantially reduces the organic content of subsurface sediments in the dysaerobic (i.e., moderately oxygenated) waters of the deeper parts of the Gulf of California. As shown by the downhole increase in the C/N ratio, diagenesis is taking place also on the well-oxygenated southwest African slope (Fig. 13). Despite this process, subsurface TOC's are consistently high there. This strongly implies higher productivity off southwest Africa. As a result, the oceanographic signal of upwelling is still clearly preserved on the slope off Walvis Bay, but not deeper than the oxygen minimum zone in the Gulf of California.

Laminations, which are commonly found in sediments where bottom waters are anoxic --as in the northeastern Pacific-- apparently are not preserved off southwest Africa (Bremner, pers. comm.). The main reason seems to be the lack of the seasonally controlled runoff that is a major control on the development of laminations off California and Mexico. On the southwest African slope the high oxygen content of bottom waters militates against the preservation of laminations even if there were seasonal variations in runoff.

The high runoff of the northeast Pacific introduces a background of terrestrial organic matter. This background is virtually non-existent off southwest Africa, except near rivers, where there is some clastic input, or in relict deposits of terrigenous mud on the inner shelf.

Northwest Africa. Off northwest Africa, the organic matter is mostly marine, sedimentation is pelagic, nutrient levels are relatively low, oxygen is relatively abundant, runoff is negligible but there is some aeolian input of terrestrial organic matter. At 105 m deep, the shelf edge is anomalously shallow compared with the worldwide average of 132 m. This may help to explain why there is a high level of turbulence and turbidity in the water column, and why landward cross-shelf flow is well-developed in the bottom waters. Rates of sedimentation and oxygen levels on the slope are about the same as off southwest Africa.

Assuming that fecal pellets are the main transporting agent for organic matter, how can we explain the lack of any significant accumulation of organic matter on the shelf and upper slope? Over the slope, productivity may be sufficiently low, relative to that off southwest Africa, so that much of the organic matter that reaches the bottom is rapidly recycled by benthic organisms and never becomes part of the sediment. The small accumulation of organic matter on the middle slope may consist of material that has been swept off the shelf (Fütterer, this volume). Over the shelf, where productivity is high, turbulence may prevent accumulation of organic matter and favor its recycling in the water column. The predominance of coarse carbonates on both the shelf and upper slope attests to a strong bottom current regime or to well-developed stirring by long period waves. But, even the inner shelf muds are not particularly rich in organic matter, except in the lee of Cap Blanc. Apart from physical processes interfering with the settling of organic matter, it is possible that upwelling is too steady here, so that there is not enough of the massive, short-term productivity needed to overcome the ability of the system to recycle organic matter in the water column. More work is needed to confirm the validity of these suggestions.

The relatively low nutrient concentrations off northwest Africa reflect the general depletion of the North Atlantic in nutrients. The Benguela upwelling system, off southwest Africa, has more nutrients apparently because it derives its upwelled water by the sinking of surface water in subantarctic regions where nutrients are abundant. Thus, it is not only the vertical settling of organic matter that governs oxygen depletion and nutrient enrichment in subsurface waters, but also the lateral movement of subsurface water of differing nutrient content. The source of the subsurface water in upwelling zones plays a vital role in determining the productivity and oxygen depletion of upwelling cells, hence the amount of organic sedimentation.

Paleocirculation

Changes in the pattern of accumulation of organic matter with time reveal the history of upwelling in these different areas. For instance, upwelling and high productivity gave rise to organic-rich sediments and/or phosphatic sediments in the mid-Cretaceous and in the Miocene off northwest and southwest Africa, and in the Miocene-Pliocene of California. During glacial periods in the Quaternary, when sea level exposed the continental shelves, centers of upwelling shifted to the continental slopes. Productivity increased off northwest Africa, and organic-rich sediments accumulated there on the continental slope (Müller and Suess, 1979). Similar changes account for subsurface increases in TOC off southwest Africa, where the lack of laminations shows that conditions remained oxidizing. Downhole fluctuations from laminated to homogeneous sediments occur in both the Gulf of California and Santa Barbara Basin, and reflect climatically controlled changes in the degree of oxygenation of bottom waters. Detailed stratigraphic analyses of changes like these are needed to provide the precise time-stratigraphic framework against which climatic changes in upwelling systems can be measured.

A striking feature of these histories is the tendency for periods of organic enrichment to be worldwide in nature. This applies especially to the Miocene, which represents a major change in ocean circulation, and perhaps a major redistribution of oceanic nutrients.

CONCLUSIONS

The following conclusions are attempts to answer the questions addressed in this review:

1. What is the source of the organic matter? Nearly all of it is the amorphous product of bacterial degradation of marine organic matter, probably mostly phytoplankton. Terrestrial organic matter is present in minor amounts off humid coasts where runoff is appreciable (e.g., California, Mexico), but is less noticeable or absent off arid coasts (e.g., northwest and southwest Africa).

2. How was the organic matter deposited? The organic-rich deposits are either pelagic, especially off arid coasts (e.g., Baja, southwest Africa, northwest Africa), and far from shore off humid coasts (e.g., seaward parts of California Borderland), or hemipelagic, especially close to humid coasts where runoff is high (e.g., Mexico, California). Turbidites are important off humid coasts, especially on basin floors (e.g., Santa Barbara and Guaymas Basins).

3. Was the organic matter deposited in oxic or anoxic environments? In some places the oxygenation of bottom waters appears strongly to influence organic accumulation (e.g., Gulf of California); in other places it does not (southwest African slope, northwest African slope, and Tanner Basin, where TOC and oxygen are both abundant, and Santa Barbara Basin, where laminated and homogeneous sediments have the same TOCs). Oxygenation does help to control, but does not cause, laminations. In oxygenated areas the sediments are bioturbated regardless of TOC (e.g., southwest African slope), while in anoxic areas they are not, so that seasonal changes can be preserved as laminations (e.g., Mexico, Santa Barbara Basin). Where conditions are anoxic and there is no seasonal runoff, there may be no laminations despite complete anoxia (southwest African shelf). Lateral movement of an oxygen depleted oxygen minimum zone into a basin may favor the accumulation of organic matter there (e.g., Gulf of California).

4. How important is productivity? Massive, short-period production, rather than steady state high productivity, controls the accumulation of organic matter in upwelling regimes, by overwhelming the system's ability to recycle organic matter as food in the water column. Productivity rather than the oxygen content of bottom waters accounts for the deposition of organic matter in many environments (e.g., southwest African and northwest African slopes). The presence of organic matter on the bottom can exert a strong control on the oxygenation of bottom waters, and locally affects strongly the oxygen depletion of the oxygen minimum zone (e.g., southwest African shelf, Santa Barbara Basin).

5. What is the effect of topography? Topography by itself may not be important. But the combination of wave energy, topography, productivity, and the depth to the top of the oxygen minimum zone may together account for the absence of organic-rich sediments from the continental shelves of three of the four areas examined (northwest Africa, California, Mexico). In all four regions, the continental slope proved to be a major locus of deposition of organic matter.

6. Is diagenesis important? Diagenetic processes can modify the upwelling signal in the sediments by destroying organic matter where bottom waters are oxygenated; there is much less diagenetic loss of organic matter where bottom waters are anoxic. Under oxidizing conditions, the effects of diagenesis may be limited, and the signal of upwelling preserved where productivity is very high (e.g., southwest Africa), rather than where productivity is weak (e.g., Gulf of California). Relations between TOC and rate of sedimentation do not appear to be consistent in these deposits, although there is a relation between these two parameters in some areas, as shown by Müller and Suess (1979) and Calvert (1976).

ACKNOWLEDGEMENTS

I thank Exxon Production Research Company for permission to publish this review. Stimulating discussion with Erwin Suess, Wolf Berger, Rob Dunbar and Lisa Pratt about sedimentation in upwelling regimes helped catalyze my ideas on this topic.

REFERENCES

Bang, N.D., 1971, The southern Benguela Current region in February 1966: Part II: bathythermography and air-sea interactions, Deep- Sea Research, 18:209-224.

Barber, R.T. and Smith, R.L., 1981, in: "Analysis of Marine Ecosystems," A.R. Longhurst, ed., Academic Press, London, 31-68.

Barton, E.D., Huyer, A. and Smith, R.L., 1977, Temporal variation observed in the hydrographic regime near Cabo Corviero in the northwest African upwelling region, February to April, 1974, Deep-Sea Research, 24:7-24.

Berger, W.H., 1970, Biogenous deep-sea sediments: fractionation by deep-sea circulation, Geological Society of America, Bulletin, 81:1385-1402.

Berger, W.H., Diester-Haass, L. and Killingley, J.S., 1978, Upwelling off northwest Africa: The Holocene decrease as seen in carbon isotopes and sedimentologic indicators, Oceanologica Acta, 1:3-7.

Birch, G.F., 1975, "Sediments on the Continental Margin off the West Coast of South Africa," Ph.D. Thesis, University of Cape Town, Cape Town, 210 pp.

Birch, G.F., 1979, Phosphatic rocks on the western margin of South Africa, Journal of Sedimentary Petrology, 49:93-110.

Bishop, J.K., Ketten, D.R. and Edmond, J.M., 1978, The chemistry, biology, and vertical flux of particulate material from the upper 400 m of the Cape Basin in the southeast Atlantic Ocean, Deep-Sea Research, 25:1121-1161.

Boon, J.J., DeLeeuw, J.W. and Schenck, P.A., 1975, Organic geochemistry of Walvis Bay diatomaceous ooze - 1: occurrence and significant of the fatty acids, Geochimica et Cosmochimica Acta, 39: 1559-1565.

Bornhold, B.D., 1973, Late Quaternary sedimentation in the eastern Angola Basin, Technical Report WHOI-73-80, Woods Hole Oceanographic Institution, 212 pp.

Bremner, J.M., 1974, Texture and composition of surficial continental margin sediments between the Kunene River and Sylvia Hill, South West Africa, Technical Report 6, Geological Survey, University of Cape Town Marine Geology Program, Cape Town, 39-43.

Bremner, J.M., 1980, Physical parameters of the diatomaceous mud belt off South West Africa, Marine Geology, 34:M67-M76.

Brongersma-Sanders, M., 1971, Origin of major cyclicity of evaporites and bituminous rocks: an actualistic model, Marine Geology, 11: 123-144.

Calvert, S.E., 1964, Factors affecting distribution of laminated diatomaceous sediments in Gulf of California, in: "Marine Geology of the Gulf of California," T.J. van Andel and G.G. Shor, eds., American Association of Petroleum Geologists, Memoirs, 3:311-330.

Calvert, S.E., 1966a, Accumulation of diatomaceous silica in the sediments of the Gulf of California, Geological Society of America, Bulletin, 77:569-596.

Calvert, S.E., 1966b, Origin of diatom-rich, varved sediments from the Gulf of California, Journal of Geology, 74:546-565.

Calvert, S.E., 1976, The mineralogy and geochemistry of nearshore sediments, in: "Chemical Oceanography," J.P. Riley and R. Chester, eds., Vol. 6, 2nd edition, Academic Press, New York, 187-280.

Calvert, S.E. and Price, N.B., 1971a, Upwelling and nutrient regeneration in the Benguela Current, October, 1968, Deep-Sea Research, 18:505-523.

Calvert, S.E. and Price, N.B., 1971b, Recent sediments of the South West African shelf, in: "Institute of Geological Sciences Report," F.M. Delany, ed., 70/16, 171-185.

Caratini, C., Bellet, J. and Tissot, C., 1979, Microscopy of the organic matter: palynology and palynofacies, ORGON III - Mauritanie, Sénégal, Îles du Cap Vert, Editions du Centre National de la Recherche Scientifique, Paris, 215-265.

Copenhagen, W.J., 1953, The periodic mortality of fish in the Walvis Bay region: a phenomenon within the Benguela Current, Investigation Report, Division of Fisheries, Union of South Africa, 14: 1-35.

Debyser, Y. and Gadel, F., 1979, Kerogen geochemistry in the sediments, ORGON III - Mauritanie, Sénégal, Îles due Cap Vert, Editions du Centre National de la Recherche Scientifigue, Paris, 375-403.

Demaison, G.J. and Moore, G.T., 1980, Anoxic environments and oil source bed genesis, American Association of Petroleum Geologists, Bulletin, 64:1179-1209.

Deroo, G., Herbin, J.P., Roucaché, J., Boudon, J.P., Rober, P., Jardiné, S., and Marestang, P., 1982, Geochemistry and optical study of organic matter in some Pleistocene and Pliocene sediments from DSDP/IPOD Site 474 to 481 of Leg 64, Gulf of California, in: "Initial Reports DSDP," 64, J.R. Curray, D.S. Moore et al., U.S. Government Printing Office, Washington, 855-864.

Diester-Haass, L., 1976, Quaternary accumulation rates of biogenous and terrigenous components on the east Atlantic slope off northwest Africa, Marine Geology, 21:1-24.

Diester-Haass, L., 1979, Indicators of continental climates in marine sediments, a reply, "Meteor" Forschungs-Ergebnisse, C31:53-58.

Diester-Haass, L. and Schrader, H.-J., 1979, Neogene coastal upwelling history of northwest and southwest Africa, Marine Geology, 29:39-53.,

Dingle, R.V., 1977, Agulhas Bank phosphorites: A review of 100 years of investigation, Transactions of the Geological Society of South Africa, 77:261-264.

Dow, W.G., 1978, Petroleum source beds on continental slopes and rises, American Association of Petroleum Geologists, Bulletin, 62:1584-1606.

Dunbar, R.B. and Berger, W.H., 1981, Fecal pellet flux to modern bottom sediment of Santa Barbara Basin (California) based on sediment trapping, Geological Society of America, Bulletin, 92:212-218.

Emery, K.O., 1960, "The Sea Off Southern California: A Modern Habitat of Petroleum," John Wiley and Sons, New York, 366 pp.

Emery, K.O. and Hülsemann, J., 1962, The relationships of sediments, life and water in a marine basin, Deep-Sea Research, 8:165-180.

Emilianov, E.M. and Senin, I.M., 1969, Composition of the sediments of the South West African shelf, Litologiya i Poleznëe Iskopaemëe, 2:10-25 (in Russian).

Friebertshäuser, M.A., Codispoti, L.A., Bishop, D.D., Friederich, G.E., and Westhagen, A.A., 1975, Joint 1 - Hydrographic station data RV "Atlantis" cruise 82, Data Report 18, Coastal Upwelling Ecosystems Analysis, International Decade of Ocean Exploration, 243 pp.

Friederich, G.E. and Codispoti, L.A., 1979, On some factors influencing dissolved silicon distribution over the northwest African shelf, Journal of Marine Research, 37:337-353.

Gaskell, S.J., Morris, R.J., Eglinton, G. and Calvert, S.E., 1975, The geochemistry of a recent marine sediment off northwest Africa. An assessment of source of input and early diagenesis, Deep-Sea Research, 22:777-789.

Gilbert, D. and Summerhayes, C.P., 1981, Distribution of organic matter in sediments along the California continental margin, in: "Initial Reports DSDP," 63, R.S. Yeats, B.U. Haq et al., U.S. Government Printing Office, Washington, 757-761.

Gilbert, D. and Summerhayes, C.P., 1982, Organic facies and hydrocarbon potential in the Gulf of California, "Initial Reports DSDP," 64, J.R. Curray, D.S. Moore et al., U.S. Government Printing Office, Washington, 865-870.

Hart, T.J. and Currie, R.I., 1960, The Benguela Current, Discovery Report, 31:123-298.

Hartmann, M., Müller, P., Suess, E. and Van Der Weijden, C.H., 1973, Oxidation of organic matter in recent marine sediments, "Meteor" Forschungs-Ergebnisse, C12:74-86.

Hartmann, M., Müller, P.J., Suess, E. and Van Der Weijden, C.H., 1976, Chemistry of Late Quaternary sediments and their interstitial waters from the northwest African continental margin, "Meteor"Forschungs-Ergebnisse, C24:1-67.

Hughes, P. and Barton, E.D., 1974, Stratification of water mass structure in the upwelling area off northwest Africa in April/May 1969, Deep-Sea Research, 21:611-628.

Hülsemann, J. and Emery, K.O., 1961, Stratification in recent sediments of Santa Barbara Basin as controlled by organisms and water character, Journal of Geology, 69:279-290.

Huntsman, S.A. and Barber, R.T., 1977, Primary production off northwest Africa: the relationship to wind and nutrient conditions, Deep-Sea Research, 24:25-33.

Huyer, A., 1976, A comparison of upwelling events in two locations: Oregon and northwest Africa, Journal of Marine Research, 34:531-546.

Jones, P.G.W., 1972, The variability of oceanographic observations off the coast of northwest Africa, Deep-Sea Research, 19:405-431.

Jones, P.G.W. and Folkard, A.R., 1970, Chemical oceanographic observations off the coast of northwest Africa with special reference to process of upwelling, Conseil Permanente International pour l'Exploration de la Mer, Rapport, 159:38-60.

Kaplan, I.R., Emery, K.O. and Rittenberg, S.C., 1963, The distribution and isotopic abundance of sulfur in recent sediments off southern California, Geochimica et Cosmochimica Acta, 27:297-332.

Kester, D.R., 1975, Dissolved gases other than CO_2, in: "Chemical Oceanography," Vol. 1, 2nd edition, J.P. Riley and G. Skirrow, eds., Academic Press, New York, 498-556.

Koblentz-Mishke, O.J., Volkovinsky, V.V. and Kabanova, J.G., 1970, Plankton primary production of the world ocean, in: "Scientific Exploration of the South Pacific," W.S. Wooster, ed., National Academy of Science, Washington, 183-193.

Koopmann, B., Sarnthein, M. and Schrader, H.-J., 1978, Sedimentation influenced by upwelling in the subtropical Baie Du Levrier (West Arica), in: "Upwelling Ecosystems," E. Boje and M. Tomczak, eds., Springer Verlag, New York, 282-288.

Malouta, D.N., 1978, "Holocene Sedimentation in Santa Monica Basin, California," M.S. Thesis, University of Southern California, Los Angeles.

Milliman, J.D., 1977, Effects of arid climate and upwelling upon the sedimentary regime off southern Spanish Sahara, Deep-Sea Research, 24:95-103.

Moyes, J., Duplantier, F., Duprat, J., Faugeres, J.C., Pujol, C., Pujos-Lamay, A., and Tastet, J.P., 1979, Étude stratigraphique et sedimentologique, ORGON III, Mauritanie, Sénégal, Îles du Cap Vert, Editions du Centre National de la Recherche Scientifigue, Paris, 125-213.

Müller, P.J. and Suess, E., 1979, Productivity, sedimentation rate, and sedimentary organic matter in the oceans - 1. Organic carbon preservation, Deep-Sea Research, 26:1347-1362.

Müller, P.J. and Mangini, A., 1980, Organic carbon decomposition rates in sediments of the Pacific manganese nodule belt dated by ^{230}Th and ^{231}Pa, Earth and Planetary Science Letters, 51:94-114.

Parker, R.J., 1975, The petrology and origin of some glauconitic and glauco-conglomeratic phosphorites from the South African continental margin, Journal of Sedimentary Petrology, 45:230-242.

Peters, K.E. and Simoneit, B.R.T., 1982, Rock-eval pyrolysis of Quaternary sediments from DSDP-IPOD Leg 64, Sites 479 and 480, Gulf of California, in: "Initial Reports DSDP," 64, J.R. Curray, D.S. Moore et al., U.S. Government Printing Office, Washington, 925-932.

Pisciotto, K.A., 1978, "Basinal Sedimentary Facies and Diagenetic Aspects of the Monterey Shale, California," Ph.D. Thesis, University of California, Santa Cruz, 450 pp.

Price, N.B. and Calvert, S.E, 1978, The geochemistry of phosphorites from the Namibian Shelf, Chemical Geology, 23:151-170.

Redfield, A.C., Ketchum, B.H. and Richards, F.A., 1963, The influence of organisms on the composition of seawater, in: "The Composition of Seawater, Comparative and Descriptive Oceanography," M.N. Hill, ed., THE SEA, Vol. 2, Wiley and Sons, New York, 26-77.

Rhoads, D.C. and Morse, J.W., 1971, Evolutionary and ecological significance of oxygen-deficient marine basins, Lethaia, 4:413-428.

Rittenberg, S.C., Emery, K.O. and Orr, W.L., 1955, Regeneration of nutrients in sediments of marine basins, Deep-Sea Research, 3: 23-45.

Roden, G.I., 1964, Oceanographic aspects of Gulf of California, in: "Marine Geology of the Gulf of California," T.J. van Andel and G.G. Shor, eds., American Association of Petroleum Geologists, Memoir, 3:30-58.

Rogers, J., 1971, Sedimentology of Quaternary deposits on the Agulhas Bank, Bulletin 1, University of Cape Town Marine Geology Program, Cape Town, 117 pp.

Rogers, J., 1977, Sedimentation on the continental margin off the Orange River and the Namib Desert, Bulletin 7, Geological Survey University Cape Town Marine Geology Program, Cape Town, 162 pp.

Rullkötter, J., Von der Dick, H. and Welte, D.W., 1982, Organic petrography and extractable hydrocarbons of sediments from the eastern North Pacific Ocean, Deep Sea Drilling Project Leg 63, in: "Initial Reports DSDP," 63, R.S. Yeats, B.U. Hag et al., U.S. Government Printing Office, Washington, 819-836.

Rullkötter, J., Von der Dick, H. and Welte, D.H., 1982, Organic petrography and extractable hydrocarbons of sediments from the Gulf of California, Deep Sea Drilling Project Leg 64, in: "Initial Reports DSDP," 64, J.R. Curray, D.S. Moore, et al., U.S. Government Printing Office, Washington, 837-854.

Sarnthein, M., 1978, Sand deserts during glacial maximum and climatic optimum, Nature, 272:43-46.

Sarnthein, M., 1979, Indicators of continental climates in marine sediments, a discussion, "Meteor"Forschungs-Ergebnisse, C31: 49-51.

Schrader, H. and others, 1980, Laminated diatomaceous sediments from the Guaymas Basin Slope (Central Gulf of California): 250,000-year climate record, Science, 207:1207-1209.

Shanks, A.L. and Trent, J.D., 1980, Marine snow: sinking rates and potential role in vertical flux, Deep-Sea Research, 27:137-144.

Sholkovitz, E., 1973, Interstitial water chemistry of the Santa Barbara Basin sediments, Geochimica et Cosmochimica Acta, 37:2043-2073.

Siesser, W.G., 1978, Age of phosphorites on the South African continental margin, Marine Geology, 26:M17-M28.

Siesser, W.G., 1980, Late Miocene origin of the Benguela upwelling system off northern Namibia, Science, 208:283-285.

Simoneit, B.R.T. and Eglinton, G., 1977, Organic matter of eolian dusts and its input to marine sediments, Advanced Organic Geochemistry 1975, Actas 7 Congreso Internacional de Geoquimica Organica, Madrid, 415-430.

Simoneit, B.R.T., Meyers, P.A. and Summerhayes, C.P., 1981, Sources, preservation, maturation and migration of organic matter in Neogene sediments from the continental margin off California and Baja California: a synthesis of organic geochemical studies from DSDP Leg 63, in: "Initial Reports DSDP," 63, J.R. Curray, D.S. Moore et al., U.S. Government Printing Office, Washington, 943-948.

Simoneit, B.R.T., Summerhayes, C.P. and Meyers, P.A., 1982, Sources, preservation and maturation of organic matter in Pliocene and Quaternary sediments of the Gulf of California: a synthesis of organic geochemical studies from DSDP Leg 64, in: "Initial Reports DSDP," 64, J.R. Curray, D.S. Moore et al., U.S. Government Printing Office, Washington, 939-952.

Soutar, A. and Crill, P.A., 1977, Sedimentation and climatic patterns in the Santa Barbara Basin during the 19th and 20th centuries, Geological Society of America, Bulletin, 88:1161-1172.

Spiker, E.C. and Simoneit, B.R.T., 1981, Radiocarbon dating of recent sediments from Leg 64, Gulf of California, in: "Initial Reports DSDP," 64, J.R. Curray, D.S. Moore et al., U.S. Government Printing Office, Washington, 757-758.

Suess, E., 1980, Particulate organic carbon flux in the oceans - Surface productivity and oxygen utilization, Nature, 288:260-263.

Summerhayes, C.P., 1970, "Phosphate Deposits on the Northwest African Continental Shelf and Slope," Ph.D. Thesis, University of London, 282 pp.

Summerhayes, C.P., 1972a, South West African Shelf sediments, Technical Report 4, University of Cape Town Marine Geology Program, Cape Town, 95-102.

Summerhayes, C.P., 1972b, Aspects of the mineralogy and geochemistry of Agulhas Bank sediments, Technical Report 4, University of Cape Town Marine Geology Program, Cape Town, 64-82.

Summerhayes, C.P., 1973, Distribution origin, and economic potential of phosphatic sediments from the Agulhas Bank, South Africa, Transactions of the Geological Society of South Africa, 73:291-277.

Summerhayes, C.P., 1981a, Oceanographic controls on organic matter in the Miocene Monterey Formation, offshore California, in: "The Monterey Formation and Related Siliceous Rocks," R. Garrison and R. Douglas, eds., Special Publication, Society of Economic Paleontologists and Mineralogists, Pacific Section, 213-219.

Summerhayes, C.P., 1981b, Organic facies of Mid-Cretaceous black shales in the deep North Atlantic, American Association of Petroleum Geologists, Bulletin, 65:2364-2380.

Summerhayes, C.P. and Willis, J.P., 1972, Mineral and element geochemistry of some ocean floor sediments from around South Africa, Technical Report 4, University of Cape Town Marine Geology Program, Cape Town, 83-94.

Summerhayes, C.P., Nutter, A.H. and Tooms, J.S., 1971, Geological structure and development of the continental margin of northwest Africa, Marine Geology, 11:1-25.

Summerhayes, C.P., Nutter, A.H. and Tooms, J.S., 1972, The distribution and origin of phosphate in sediments off northwest Africa, Sedimentary Geology, 8:3-28.

Summerhayes, C.P., Birch, G.F., Rogers, J. and Dingle, R.V., 1973, Phosphate in sediments off southwestern Africa, Nature, 243:509-511.

Summerhayes, C.P., Hofmeyer, P.K. and Rioux, R.H., 1974, Seabirds off the southwestern coast of Africa, Ostrich, 45:83-109.

Summerhayes, C.P., Milliman, J.D., Briggs, S.R., Bee, A.G., and Hogan, C., 1976, Northwest African shelf sediments: influence of climate and sedimentary processes, Journal of Geology, 84: 277-300.

Summerhayes, C.P., Bornhold, B.D. and Embley, R.W., 1979, Surficial slides and slumps on the continental slope and rise off South West Africa: A reconnaissance study, Marine Geology, 31:265-277.

Sweeney, R.E. and Kaplan, I.R., 1980, Natural abundances of ^{15}N as a source indicator for nearshore marine sedimentary and dissolved nitrogen, Marine Geology, 9:81-94.

Thiede, J., 1977, Aspects of the variability of the glacial and interglacial North Atlantic eastern boundary current (last 150,000 years), "Meteor"Forschungs-Ergebnisse, C28:1-36.

Thiede, J. and Van Andel, T.H., 1977, The paleoenvironment of anaerobic sediments in the Late Mesozoic South Atlantic Ocean, Earth and Planetary Science Letters, 33:301-309.

Tissot, B., Demaison, G., Masson, P., Delteil, S.R., and Combaz, A., 1980, Paleoenvironment and petroleum potential of middle Cretaceous black shales in Atlantic Basins, American Association of Petroleum Geologists, Bulletin, 64:2051-2063.

Tont, S.A., 1976, Short period climatic fluctuations: effect on diatom biomass, Science, 194:942-944.

U.S. Naval Oceanographic Office, 1963, Oceanographic Atlas of the North Atlantic Ocean, IV, Sea and Swell, Publication 700, U.S. Naval Oceanographic Office, Washington, 227 pp.

Van Andel, T.J., 1964, Recent marine sediments of Gulf of California, in: "Marine Geology of the Gulf of California," T.J. van Andel and G.G. Shor, eds., American Association of Petroleum Geologists, Memoirs, 3:216-310.

Veeh, H.H., Calvert, S.E. and Price, N.B., 1974, Accumulation of uranium in sediments and phosphorites of the South West African shelf, Marine Chemistry, 2:189-202.

Von Rad, U., Cepek, P., Von Stackelberg, U., Wissman, G., and Zobel, B., 1979, Cretaceous and Tertiary sediments from the northwest

African slope (dredges and cores supplementing DSDP results), Marine Geology, 29:213-312.

Weichart, G., 1974, Meereschemische Untersuchungen im nordwestafrikanischen Auftriebsgebiet 1968, "Meteor"Forschungs-Ergebnisse, A14:33-70.

Welte, D.H., Cornford, G. and Rullkötter, J., 1979, Hydrocarbon source rocks in deep sea sediments, Proceedings Offshore Technology Conference, Houston, 457-461.

BIOGENIC SEDIMENTS ON THE SOUTH WEST AFRICAN (NAMIBIAN) CONTINENTAL MARGIN

J. Michael Bremner

Marine Geoscience Unit of the Geological Survey
University of Cape Town
Rondebosch 7700, South Africa

ABSTRACT

Based largely on differences in composition of the surficial sediment, the South West African continental margin is divided into three distinct latitudinal zones, the Kunene, Walvis and Orange margins. Sediment mantling the Kunene margin is practically devoid of biogenic material; the Walvis margin is dominated by it; and the Orange margin is intermediate. The most widespread biogenic component on the Walvis and Orange margins is $CaCO_3$. It derives mainly from planktonic foraminifers, which reach greatest abundance (max. 96.6%) on the outer shelves and upper slopes. Benthic foraminifers are important on the middle and inner shelves, as are patches of relict molluscs--the principle gravel-size constituent of the sediment.

Opal is next in abundance (max. 88%), but confined exclusively to the Walvis inner shelf. It occurs as a coast-parallel elongate body of diatomaceous mud measuring 740 km in length. Budget calculations indicate that approximately 5 x 10^6 metric tons of Si are deposited there annually. Compared with a similar deposit in the Gulf of California, it is very pure, with an average concentration of 54% opal. In keeping with the rich deposits of $CaCO_3$ and opal on the Walvis margin, organic matter is most abundant there as well. Three sub-parallel belts exist, the richest (max. 24.6%) of these being associated with diatomaceous mud on the inner shelf. The other two belts, on the middle shelf and upper slope, are rich in fecal pellets, thus indicating a copious food supply for benthic scavengers. An estimate of the annual organic matter supply rate for the diatomaceous mud belt is 2.3 x 10^6 m^3.

INTRODUCTION

Open ocean surface waters are invariably undersaturated with respect to dissolved silica, yet productivity in the euphotic zone of the Benguela Current system is such that a 740 km-long diatomaceous mud belt has accumulated off Walvis Bay, and extensive deposits of calcareous material (predominantly foraminifers) exist farther offshore. This unusual condition has been brought about by upwelling due to a favorable combination of three factors, namely the orientation of the coastline, the direction and strength of prevailing winds, and the Coriolis effect. In essence, upwelling has left a clear imprint on the character of sediments on the southwest African continental margin.

The first sample of diatomaceous mud was recovered by a Lieutenant Gutsche of the Cape Garrison Artillery in 1900; since this time, scientists of many nationalities have been drawn to the area by intriguing aspects of the environment. First, the water was scrutinized to examine its productivity, chemistry and physics (Hart and Currie, 1960; Calvert and Price, 1971; Carmack and Aagaard, 1977); then the sediments were looked at and described cursorily (Senin, 1968; Bremner, 1978); and finally, some detailed studies have recently been undertaken on the sediment geochemistry, and on specific components of the sediment, e.g., the phosphorite and the organic matter (Baturin, 1971; Brongersma-Sanders et al., 1980). The work described here represents part of a detailed, systematic investigation of the continental margin sediments off South West Africa. Complementing this are two papers currently in preparation, one on the authigenic components of the sediment, and the other on terrigenous minerals.

A division of the margin into three distinct zones seems appropriate (Fig. 1) based to some extent on the morphology of the shelf, but mainly on the composition of the surficial sediments. Each of the zones are explicitly defined in a footnote of Table 1 where, in addition, average concentrations of the various biogenic components in the sediment are listed.

HYDROGRAPHIC SETTING

The southern part of the South West African continental margin is overlain by the Benguela Current. It is a sluggish body of water with maximum velocities near the Orange River of approximately 25 cm/sec (1/2 knot) (Stander, 1964). The main stream of the current flows westward away from the coast at Walvis Bay (23°S) (Fig. 2) but three narrow, ancillary branches parallel the coast in a northerly direction (Moroshkin, Bubnov and Bulatov, 1970). Upwelling is most persistent in spring when the prevailing southerly winds increase in velocity and frequency (Schell, 1968). A stable, active center of upwelling exists south of Lüderitz where the wind reaches greatest

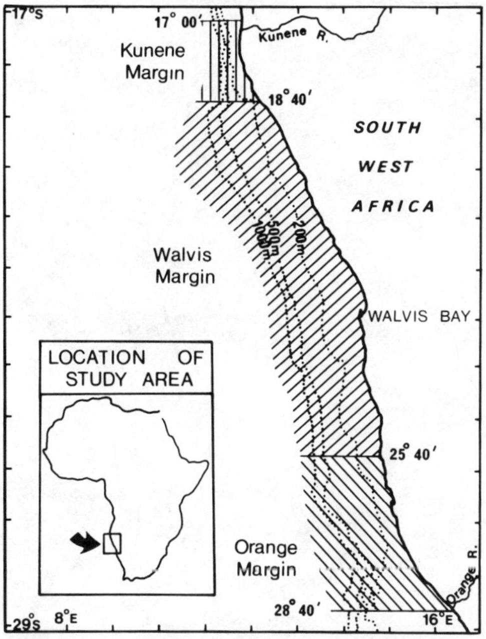

Fig. 1. Division of the South West African continental margin into three zones based largely on the composition of the surficial sediment.

intensity (Bang, 1971). The depth of upwelling shallows from about 350 m in the Orange River-Lüderitz area to about 200 m between Palgrave Point and Rocky Point. The water becomes progressively warmer, more saline, richer in oxygen and poorer in phosphate, nitrate and silica towards the north (Calvert and Price, 1971; Carmack and Aargaard, 1977).

In the northern part of South West Africa, the southward-flowing Angola Current converges with the eastern branch of the Benguela Current, and penetrates south to at least latitude 23°S. The origin of this water is an oxygen-poor layer in the eastern tropical South Atlantic (Visser, 1969). It is brought into the area by a cyclonic gyre off the Angola coast (Fig. 2) and reaches a depth of around 300 m (Moroshkin et al., 1970).

Interaction between the Angola and Benguela Currents has not been rigorously studied for long periods of time. The intrinsic weakness of the two systems where they converge and intermingle over the shelf is apparent from bottom photographs, however, where well-preserved animal trails are seen to criss-cross the muddy sea floor.

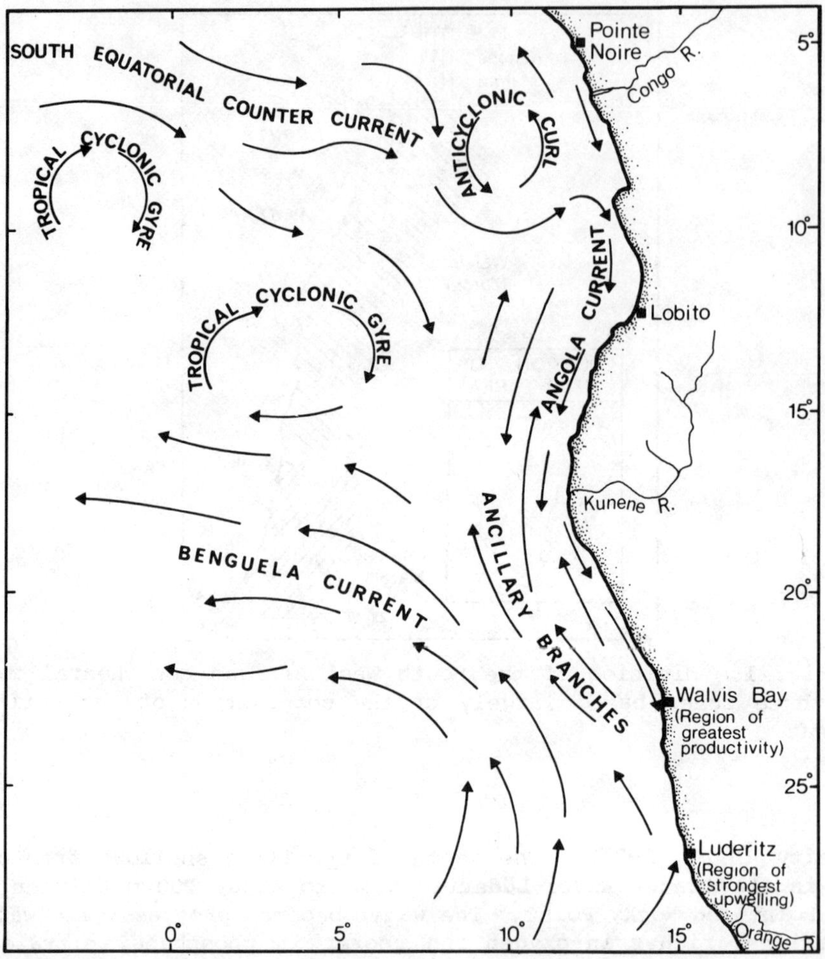

Fig. 2. Geostrophic water circulation in the southeast Atlantic Ocean (after Moroshkin et al., 1970 and Bang, 1971).

METHODS

792 surficial sediment samples were collected between 10 and 1500 m depths with a Van Veen grab (water depth <250 m) and a short gravity corer. Sea salt was effectively removed from the samples by dialysing them in fresh water for 24 hours while on board ship. They were then stored at room temperature in airtight plastic containers for subsequent analysis in the laboratory.

Calcium carbonate was determined gasometrically according to Hülsemann (1966). The procedure involves CO_2 measurement at ambient temperature and atmospheric/hydrostatic pressures by reacting the sample with acid. Accuracy was maintained at <2% by frequent checks against analytical reagent grade $CaCO_3$, and reproducibility was found to be 25% and 4% at the 95% confidence level for $CaCO_3$ concentrations of 5% and 50%, respectively. Correction was made for CO_2 liberated by francolite, the concentration of this mineral having previously been determined for all samples.

Volume percentages of the most commonly occurring benthic and planktonic foraminifers, molluscs and radiolarians were determined by point counting 100 particles in the >63 μm size fraction of 84 widely-separated samples. The relative deviation of a 20% population density was found to be about 10%. In addition, the gravel fractions of all samples (555) were examined for mollusc diversity, the greatest number of left or right valves of bivalves being taken as the population density of the species.

The concentration of opal was determined semiquantitatively following a method devised by van Andel (1964). The technique has been described by Calvert (1966) and Bremner (1980a). Organic carbon in the sediment was determined by wet-oxidation (Morgans, 1956). The samples were analyzed in duplicate, and the reagents were regularly standardized following each batch of 20 samples. The accuracy of the method using several mixtures of quartz and sucrose was found to be ±3% at the 95% confidence level (Summerhayes, 1972), and the maximum relative deviation was less than 5%. An organic matter : organic carbon ratio of 1.8 : 1 was used.

CALCIUM CARBONATE

Regional Distribution

Calcareous sediment, with >50% $CaCO_3$, covers more than 60% of the southwest African continental margin, and for the most part, lies in water depths greater than 150 m (Fig. 3a). In the Sylvia Hill-Lüderitz area, however, these sediments diverge seaward to depths of around 1000 m. Rogers (1977) contends that the divergence may be related to seasonal variations in the intensity of coastal upwelling, and to a strong near-bottom current on the upper slope known as the Good Hope Jet (Bang, 1976). The Kunene margin is another area of $CaCO_3$ impoverishment with a maximum value of only 35.7%. The low concentrations there reflect the nutrient-poor quality of water brought in by the Angola Current.

The richest deposits of $CaCO_3$ (max. 96.6%) lie on the mid-shelf and upper slope of the Walvis margin. The $CaCO_3$ skeletal assemblage is typical of temperate-water biogenic deposits that are dominated by foraminifers and molluscs - an association referred to as 'foramol' (Lees, 1975). Zooplankton graze the seaward edge of the nearshore phytoplankton-rich zone, pelagic fish feed along the common boundary

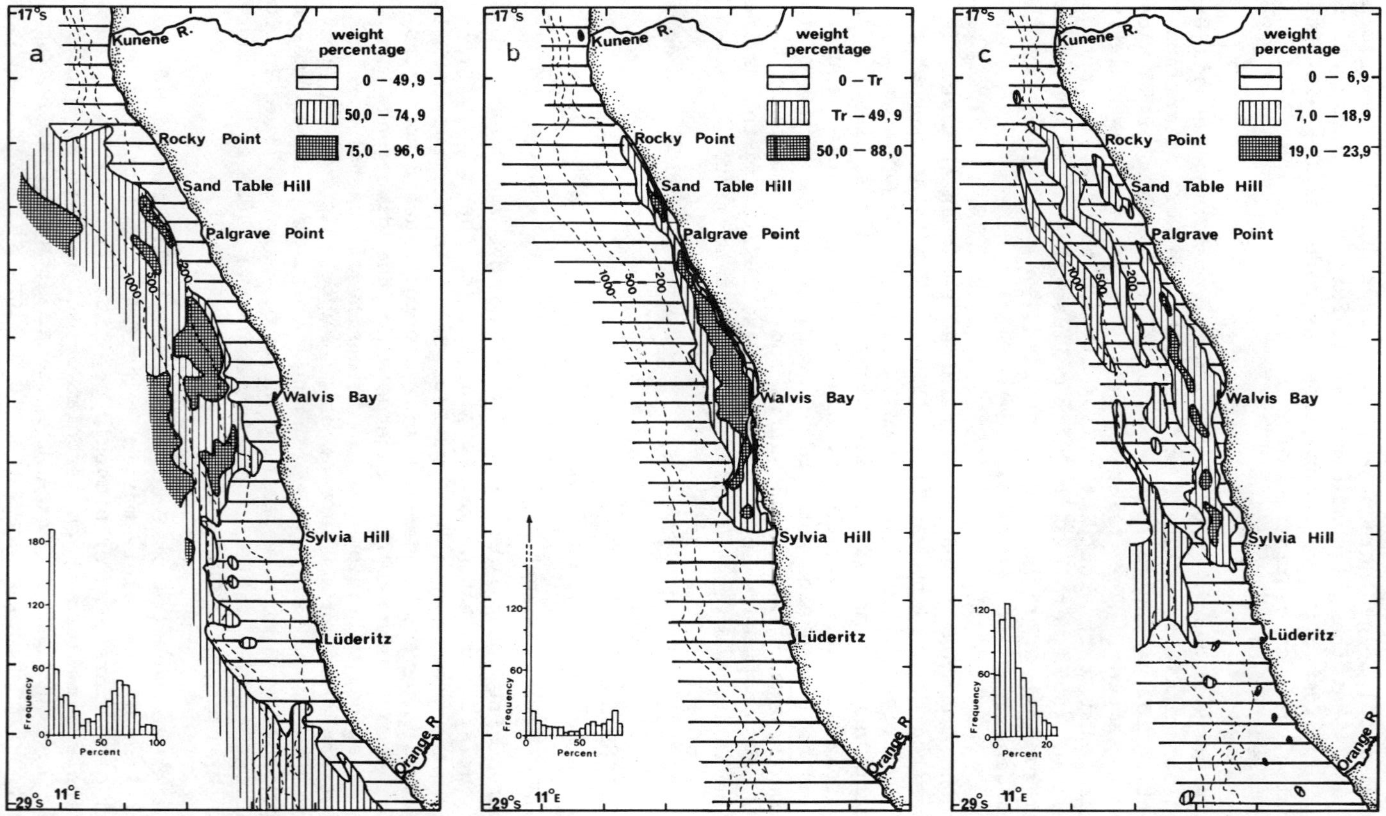

Fig. 3. Distribution of three biogenic components in the surficial sediments of the continental margin derived from analysis of 729 samples (Rogers, 1977; Bremner, 1978). Isobaths are shown in meters. (a) $CaCO_3$; b. Opal; c. Organic matter.

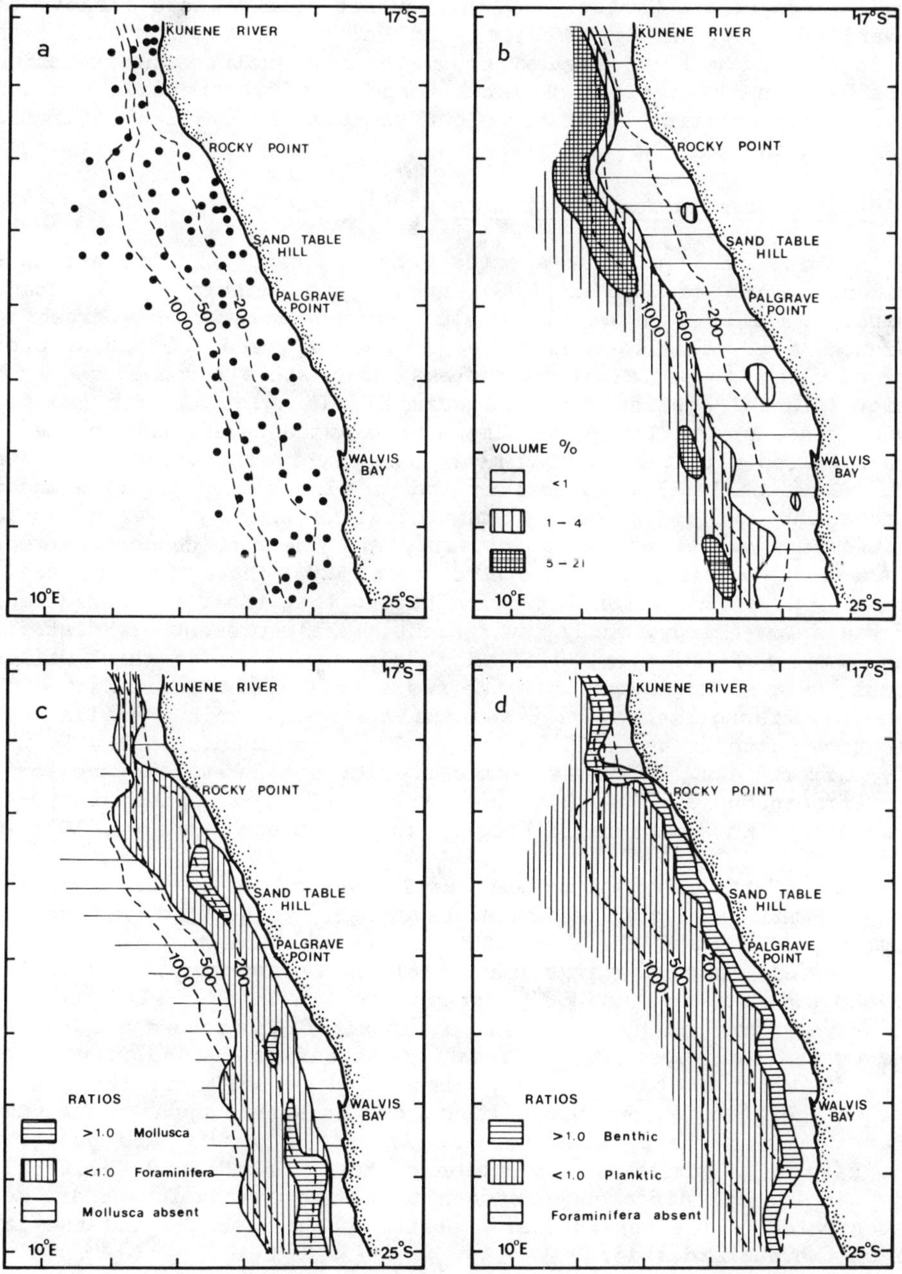

Fig. 4. a. Location of samples used for point counting biogenic components in the sand + gravel fractions (>63 μm)

b. Distribution of radiolarians

c. Mollusc:foraminifer ratios

d. Benthic:planktonic foraminifer ratios

(Visser, Kruger and Coetzee, 1973), benthic foraminifers inhabit the seaward edge of the diatomaceous mud belt on the inner shelf, and relict beds of mollusc shells occupy the mid shelf. Interrelationships between the various skeletal components for the area north of Sylvia Hill (latitude 25°S) are shown pictorially by means of ratios in Fig. 4c and 4d.

Mollusc and Foraminifer Assemblages

A total of 52 mollusc species were identified in the surficial sediments north of Sylvia Hill (Tankard and Kilburn, pers. comm.; Bremner, 1978). Some of the shells possess a remarkably fresh appearance, e.g., the ligament of *Venus chevreuxi* and the glossy aperture of *Sveltia lyrata*. Nevertheless, they are all considered to be relict (probably dating from the Würm II pleniglacial) for the following reasons: no living specimens were encountered; the shells are always filled with ambient sediment; they typically occur associated with relict pelletal phosphorite; and in places they underlie modern diatomaceous mud. The most common bivalvia are *Lucinoma capensis*, *Dosinia lupina* and *Tellina gilchristi*, and the most common gastropod is *Nassarius analogicus*. *Lucinoma* (max. 87%) have fragile shells with a high articulation ratio indicating that their preferred habitat was a low-energy, muddy environment. Their present-day distribution (Fig. 5a), however, differs little from that of thick-walled, robust *Dosinia* suggesting that the two species were mixed during the Flandrian transgression. *Tellina* and *Nassarius* exhibit similar distribution patterns as well, although their concentration levels usually differ. None of these common species are endemic to the southwest African shelf. They have been recovered from depths of ∿50-400 m, and occur in sediments of Neogene to Recent age (Kensley, 1973).

Some of the most common and easily recognized benthic foraminifers in the surficial sediment have been identified and counted (Bremner, 1978) and the distribution of two having greatest abundance, *Brizalina spathulatus* (max. 76%) and *Bolivina* sp. (max. 71%), are shown in Fig. 5c and 5d. Except for the inner shelf, which is extensively covered by a deposit of anoxic diatomaceous mud, *Brizalina* occurs over most of the remaining shelf area. *Bolivina*, however, is confined to a narrow, discontinuous strip on the seaward side of the diatomaceous mud belt and on the Kunene upper slope. Two other common benthic foraminifers, *Ammonia beccarii* and *Uvigerina* sp., have very similar distribution patterns on the mid shelf. The only significant difference between them is that the former are more concentrated on the landward side of the 200 m isobath, and the latter on the seaward side.

The foraminiferal assemblage, except for a narrow strip on the inner shelf, is overwhelmingly dominated by planktonics (Fig. 4d), but only one species, *Orbulina universa* (max. 21%), has been systematically counted because of its ease of recognition (Fig. 5b). It is

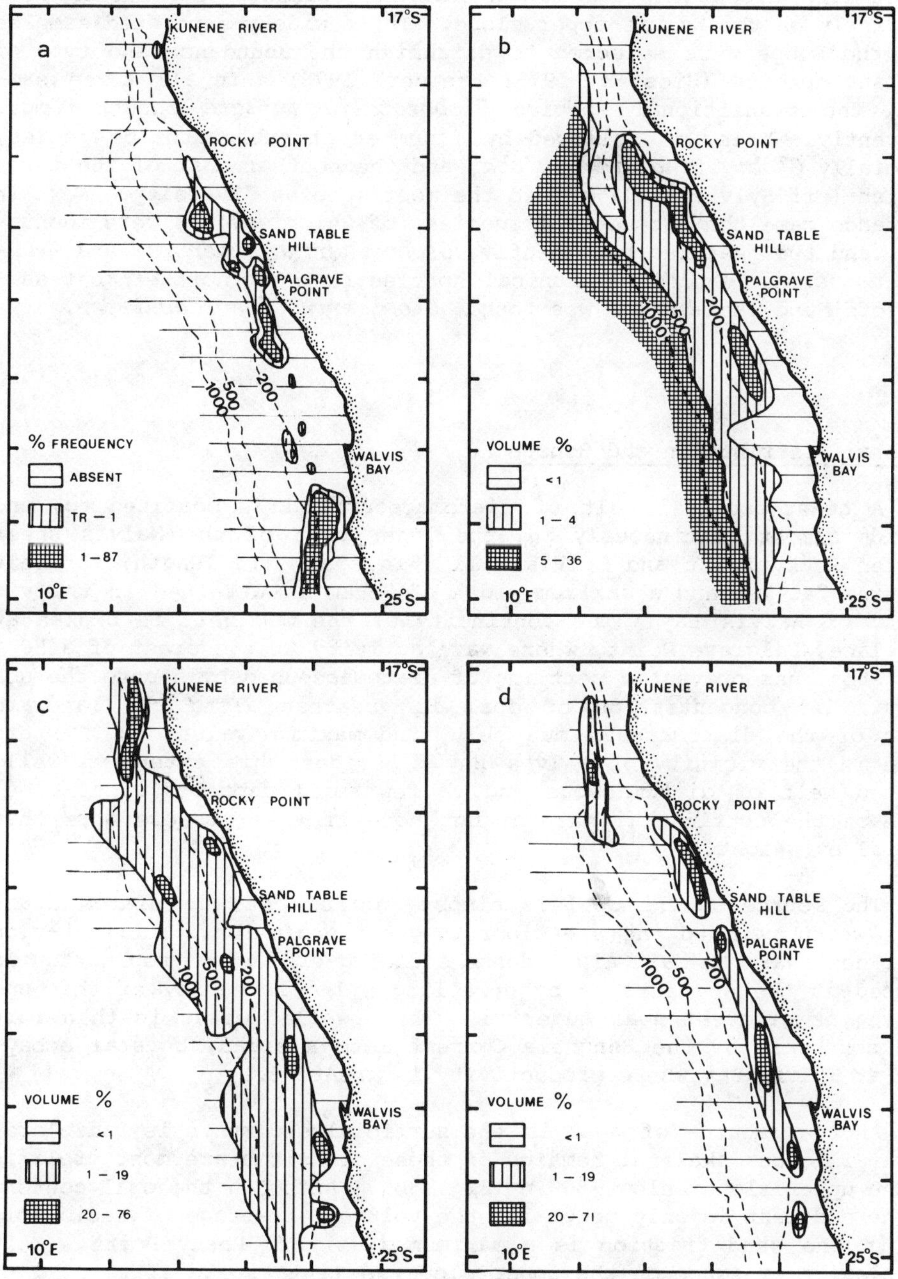

Fig. 5. a. Distribution of the mollusc *Lucinoma capensis*
b. Distribution of the planktonic foraminifer *Orbulina universa*
c. Distribution of the benthic foraminifer *Brizalina spathulata*
d. Distribution of the benthic foraminifer *Bolivina* sp.

essentially absent in the Kunene margin sediments and occurs most abundantly on the Walvis upper slope. Three widely-separated samples from the slope were selected to establish the abundance and type of dominant species (Siesser, 1975; Bremner, 1978). In all three samples, the transitional species *Globorotalia inflata* occurred most frequently. This was followed by a number of subantarctic species, especially *Globigerina quinqueloba*, and the southernmost of the three samples (off Sylvia Hill) showed the most species diversity. Next in abundance came the subtropical species, of which several were identified, and two featured prominently, *Globoquadrina dutertrei* and *Orbulina universa*. Only one tropical species from the northernmost sample (off Sand Table Hill) was found, *Globigerinoides sacculifer*.

OPAL

Regional Distribution and Source

A coast-parallel belt of diatomaceous mud is confined to, and extends almost continuously on, the inner part of the Walvis shelf between Rocky Point and Sylvia Hill (Fig. 3b). In length, the belt measures 740 km, and a maximum width of 76 km is attained in the vicinity of Walvis Bay. The continuity of the mud belt is broken at one place, Palgrave Point, where wave activity at the crest of a 50 m deep shoal has prevented settling of diatomaceous debris onto the sea floor. The concentration of opal is greatest along the landward flank of the diatomaceous mud belt, and maximum values (88.0%) are found in the vicinity of Walvis Bay. Besides this extensive well-defined belt of diatomaceous mud, a few small deposits occur elsewhere on the continental margin, but none of them possess more than 8% opal by weight.

The source of the opal is almost entirely due to diatom frustules which rain to the sea floor in great profusion. Nutrient-rich water necessary to maintain a dense standing-crop of phytoplankton is induced at the sea surface by upwelling. The intensity of this upwelling is greatest near Lüderitz. The upwelled water is then carried northward by the Benguela Current into a broad, coastal embayment at Walvis Bay where productivity is greatest.

Another source of opal in the surficial sediments is radiolarians. Silicious skeletal remains of these organisms are most prolific on the upper slope below -500 m (Fig. 3b). Although the opal content of the sediment is only about 1%, the volume percentage of radiolarians in the sand fraction is a maximum (21%) off Rocky Point. This radiolarian presence on the upper slope is tentatively attributed to vigorous shelf-break upwelling. Additional information on the distribution of opal on the continental margin may be found in Bremner (1978; 1980a).

The Diatomaceous Mud Belt

Age. Numerous divergent opinions have been expressed about the course of sea level rise during the Flandrian transgression, e.g., Godwin, Suggate and Willis (1958), Shepard (1960), Fairbridge (1961), Milliman and Emery (1968) and Guilcher (1969). However, there is consensus that sea level rose fairly rapidly at first, and that its present height was attained between 4,000 and 6,000 yrs B.P. Thus, despite the fact that with rising sea level the seaward flank of the diatomaceous mud belt would have been deposited prior to the landward flank, the age difference between them is likely to have been small due to the rapidity of the early transgression.

Several independent lines of evidence suggest that the depth of sea level lowering off South West Africa during the Würm II pleniglacial was of the order of -110 m (Bremner, 1978). Using wave parameters characteristic of 'local storms' that occur today (period T = 8 seconds; deep water wave length $L_o = 1.56T^2 = 99.8$ m; wave base $L_o/2=50$ m) puts the wave base 160 m below present sea level. This means that the seaward edge of the mud belt (average depth = 130 m; Bremner, 1980a) was located 30 m above wave base - a situation unlikely to have favored survival of the mud because of its location in the Würm II surf zone. Therefore, a realistic speculation for age is around 5,000 years, that is, when sea level was restored to its current position.

Supporting evidence for the recentness of the deposit has been obtained from radiocarbon dating of C-org and $CaCO_3$. Veeh, Calvert and Price (1974) measured an age of ±2880 years for mud at an average depth of 88 cm in 4 cores. Another core taken in the same area as one of the earlier ones by Morris and Calvert (1977), however, gave an age of only ±800 years for sediment at 57 cms depth. They attributed this difference to slumping, or reworking of the sediment. Although slumping is not considered to be a common phenomenon in the mud because of the low shelf gradient (Bremner, 1978), sediment reworking by wave-induced bottom currents or rising H_2S gas bubbles is suggested by contortions in some color-distinctive layers in cores, and by occasional relict mollusc shells at or near the surface of the deposit.

Composition. Terrigenous detritus (mainly quartz and mica) contaminates the landward flank of the diatomaceous mud belt, and $CaCO_3$ and organic matter are the most common diluents along the seaward flank. Fig. 6 illustrates the rate of change of these components along the length of the deposit between trace, 5% and 50% opal isopleths. From the diagram it is clear that terrigenous material is the most important of the opal dilutents - this being expressed by a strong negative correlation between the two variables on the trace and 5% graphs. This observation is also underscored by ratios between average opal and average terrigenous concentrations (as shown

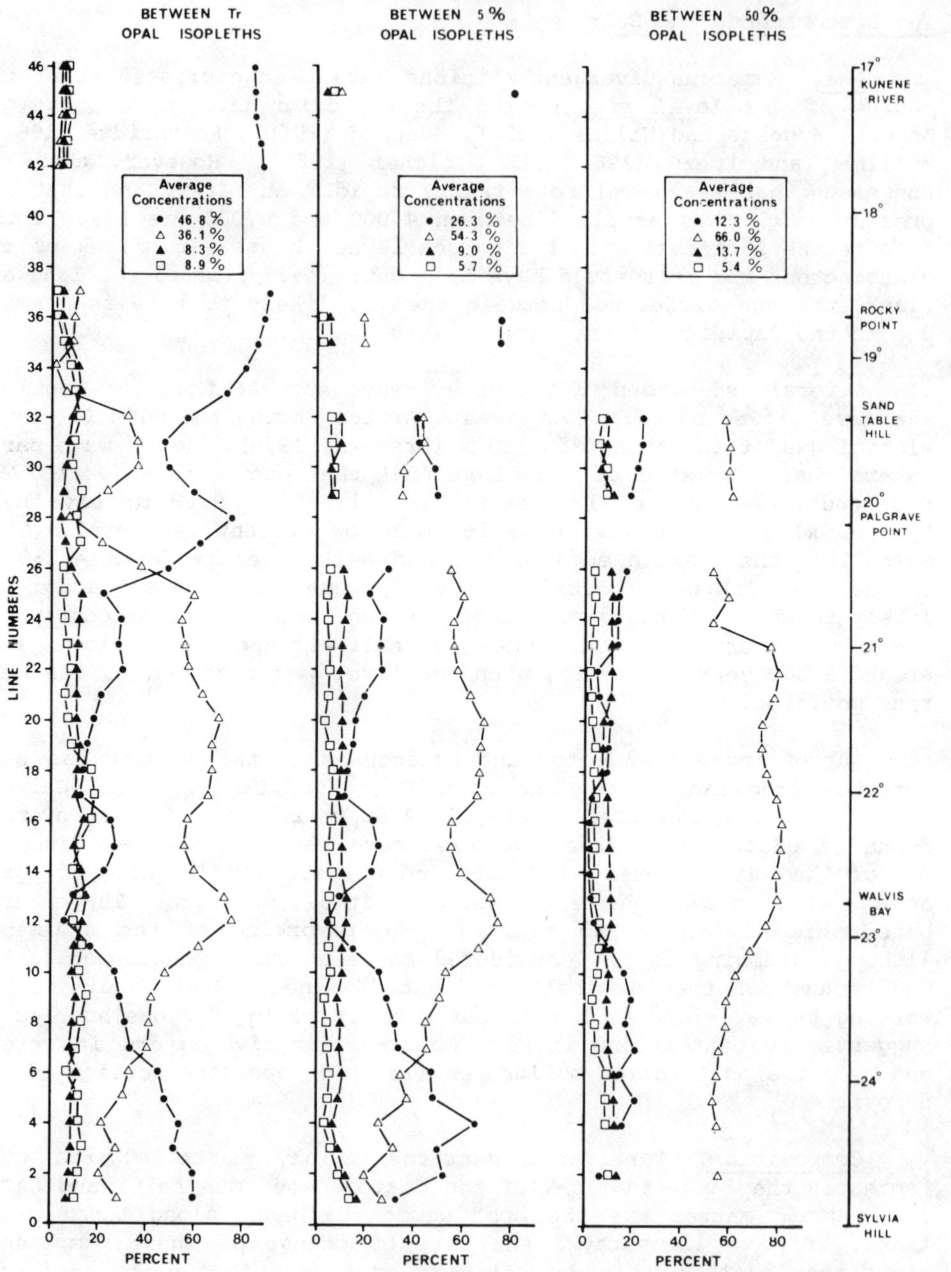

Fig. 6. Variations in the concentration of terrigenous detritus (●), opal (Δ), organic matter (▲), and $CaCO_3$ (□) along 46 sample lines oriented transverse to the length of the diatomaceous mud belt. The sample lines (not shown) were spaced approximately 16 km apart. They extended east-west between the landward and seaward boundaries of the three mud-belt zones depicted. The numbers of samples on each line varied between 1 and 8.

in Fig. 6) for the three zones considered, Tr : 5% : 50% = 0.8 : 2.1 : 5.4. In fact, for the area between trace opal-isopleths, opal dominates over terrigenous only between Palgrave Point and Conception Bay (24°S) which is roughly half of the length of the diatomaceous mud belt. Other striking relationships between the two components are the high and low opal concentrations at Walvis Bay and Swakopmund (22°40'), respectively. The former is due to intense diatom productivity around Pelican Point and the latter to periodic flash flooding of the Swakop River.

Terrigenous material is introduced to the diatomaceous mud in three ways: wave action in the nearshore, discharge from rivers during flash flooding, and aeolian transport. By virtue of its continuous process, the first of the three methods is probably most important and, because of the seasonal frequency of easterly 'bergwinds', the latter is most likely to be least significant. The terrigenous material consists of rounded to very well-rounded, fine to very coarse, sand and pebbles; and sub-angular to angular, very fine sand and silt. Its composition is predominantly quartz, mica (mostly biotite), minor feldspar and rock fragments, and traces of almandine garnet, magnetite and ferro-magnesian minerals. The boundary between quartzose and diatomaceous sediment is approximately 3 km in width, and quite often, this transition zone is marked by an abundance of biotite (up to 40%, e.g., between Cape Cross and Swakopmund). It is also noteworthy that besides opal and terrigenous material along the landward flank of the mud belt, phosphorite concretions, fish scales and fish bones are more common here than elsewhere in the deposit (Bremner, 1980b).

$CaCO_3$ and organic matter are roughly equiproportional between trace opal isopleths (Fig. 6). In the more central parts of the mud belt, as defined by the 5% and 50% opal isopleths, organic matter is invariably more concentrated than $CaCO_3$. This reflects an intrinsic relationship between the organic matter and opal, whereas the $CaCO_3$ is simply an independently fluctuating diluent. Whereas patches of sediment along the seaward flank of the mud belt contain very high levels of organic matter (max. 24.6%), $CaCO_3$ tends to increase progressively from about 5% near the middle to about 50% at the seaward edge. This is due to influx of planktonic foraminifers in the central region, and to the presence of benthic foraminifers and relict mollusc shells along the seaward edge.

Silica Budget - Supply. Heath (1974) has described four processes whereby dissolved silica may be supplied to the oceans. These are: the escape of silica from interstitial waters of pelagic sediments, river influx, volcanic influx, and submarine weathering of basalt. Of these, he has shown that the first two supply about 90% of the soluble silica to the ocean. Off southwest Africa, only the first of Heath's four processes are considered to be of importance since rivers draining the hinterland are intermittent and submarine basalt and active volcanism are non-existent. Calculation of a silica budget for the area is, however, complicated by the fact that no long-term studies have been conducted to monitor the upwelling of nutrients. A number of basic assumptions have therefore been made in an attempt to derive a reasonable estimate for the area.

A hypothetical, right-angled, triangular cross-section is positioned on the inner shelf off Sylvia Hill (latitude 25°S), so as to separate the region of maximum upwelling in the south (near Lüderitz) from the region of maximum productivity in the north (near Walvis Bay). The dimensions of the cross-section are chosen to correspond with the average dimensions of the mud belt (Bremner, 1980a) because standing crops of phytoplankton have been shown to reflect the opal content of sediments in the Gulf of California (Calvert, 1966) and off Chile/Peru (Zhuze, 1971). The height of the triangle measures 130 meters (mean depth of the trace opal-isopleth along the seaward edge of the mud belt), and its base is 46 kilometers (mean width of the mud belt between trace opal-isopleths). The area of the right-angled triangle is, therefore:

$$0.5 \times (46 \times 1000) \times 130$$

$$\text{Area} = 2.99 \times 10^6 \text{ m}^2.$$

The amount of water passing through the section every hour due to the Benguela Current is:

$$2.99 \times 10^6 \times 1852 \text{ m}^3 \quad (1 \text{ knot} = 1.852 \text{ km/hour})$$

$$\text{Volume} = 5.54 \times 10^9 \text{ m}^3.$$

In one year, the volume of water amounts to:

$$5.54 \times 10^9 \times 8760 \text{ m}^3 \quad (8760 \text{ hours} = 1 \text{ year})$$

$$\text{Annual flow} = 4.85 \times 10^{13} \text{ m}^3 \text{ or } 4.85 \times 10^{16} \text{ liters}$$

The silica content of surface and bottom waters passing through the triangular cross-section is based on a nutrient survey undertaken by Calvert and Price (1971) in spring 1968 (October). Mean values of the results obtained by them along five survey lines came to 1 µg atom Si/ℓ for surface water moving northwestward away from the coast, and 32 µg atoms Si/ℓ for South Atlantic Central water being upwelled northeastward into the euphotic zone. The amount of silica exported due to water passing through the top half of the hypothetical cross-section is therefore:

$$0.5 \; (4.5 \times 10^{16}) \times 1 \times 28 \times 10^{-6} \text{ g Si/year}$$

$$\text{Silica export} = 6.79 \times 10^{11} \text{ g Si/year.}$$

The amount of silica gained due to water advecting through the bottom half of the cross-section is:

$$0.5\ (4.85 \times 10^{16}) \times 32 \times 28 \times 10^{-6}\ \text{g Si/year}$$

$$\text{Silica import} = 2.17 \times 10^{13}\ \text{g Si/year}$$

Thus, net gain of silica to the environment by oceanic supply, is therefore:

$$(2.17 \times 10^{13}) - (6.79 \times 10^{11})$$

$$= \underline{2 \times 10^{13}\ \text{g Si/year}}$$

Silica Budget - Removal. The removal of silica from shelf waters due to biogenic sedimentation is also determined using the average dimensions of the diatomaceous mud belt (Bremner, 1980a). The length of the deposit is 740 km, its mean breadth is 33 km, and the mean thickness is 5.1 m. The volume of the deposit is therefore:

$$10^{6}(740 \times 33) \times 5.1$$

$$\text{Sediment volume} = 1.25 \times 10^{11}\ m^3$$

Since the age of the mud belt is estimated to be 5,000 years, the annual sediment accumulation by volume is:

$$2.49 \times 10^{7}\ m^3.$$

The mass accumulation rate of this sediment, based on the average dry density of samples containing more than 5% opal (0.8 g/cm^3; Bremner, 1980a), therefore, is:

$$2 \times 10^{13}\ \text{g/year}$$

Of this amount only 54.3% is due to opal such that the annual SiO_2 accumulation is:

$$1 \times 10^{13}\ \text{g } SiO_2\text{/year}$$

Disregarding a water content of ∿10% in opal (Calvert, 1966), the amount of silicon lost from the environment due to biogenic sedimentation is:

$$= 1.08 \times 10^{13} \times 28/60\ \text{g Si/year (at wts of Si = 28 and } SiO_2 = 60)$$

$$= \underline{5 \times 10^{12}\ \text{g Si/year}}$$

Although much uncertainty exists regarding assumptions made in determining the supply and removal rates of silica from the hydrographic environment, the difference between the rates is sufficiently large (by a factor of 4.2) that the concept of a purely oceanic source for the silica on the inner shelf is adequately accounted for.

Comparison with the Gulf of California. Calvert's (1966) detailed work on silicious deposits of the semi-enclosed Gulf of California affords some interesting comparisons with that of the open-shelf situation off South West Africa.

The combined rate of supply of silica from rivers (mainly the Colorado) and the ocean is 4.8 times greater in the Gulf than is the ocean supply off South West Africa. Moreover, the rate of removal due to biogenic sedimentation is 3-times greater in the Gulf than off South West Africa. These greater supply and removal rates have, not unexpectedly, given rise to a more extensive diatomaceous deposit in the Gulf. In area, it measures 3×10^4 km^2 for sediment containing >10% opal versus 2.4×10^4 km^2 for sediment containing >5% opal off South West Africa. Conversely, the purity of diatomaceous sediment in the Gulf is much less than that off South West Africa; samples with >10% opal in the Gulf contain an average of 25% opal whereas sediments containing >5% opal off South West Africa contain an average of about 54% opal. Finally, the maximum rate of sedimentation in the Gulf (Guaymas Basin) is somewhat less than off South West Africa (maximum thickness = 15 m), 2 mm/year versus 15,000 mm/5000 years = 3 mm/year, respectively.

Diatom and Radiolarian Assemblages

In a recently published taxonomic study of southwest African phytoplankton, Kruger (1980) has shown that diatoms outnumber all other groups and have the greatest species diversity. He identified 184 species in 61 genera for diatoms versus 158 species in 34 genera for dinoflagellates, 95 species in 34 genera for tintinnids, 40 species in 31 genera for radiolarians, 25 species in 19 genera for foraminifers, 15 species in 10 genera for coccolithophorids and 10 species for silicoflagellates. Over 46% of the recognized diatoms were found to be cosmopolitan, 42% were tropico-temperate, 4% were tropical, and the remaining 8% were endemic to specific areas or were not properly known. In an earlier study, Hart and Currie (1960) identified 95 diatom species from the euphotic zone and pointed out that *Chaetoceros* sp. and *Fragilaria karstenii* were exceptionally common in the coastal waters around Sylvia Hill.

The distribution of diatom taphocoenoses has been examined by Kruger (pers. comm., 1975) and Richert (pers. comm., 1976) on samples provided by the writer. The assemblage present in the diatomaceous mud was found to be dominated by *Chaetoceros* sp., *Actinocyclus ehrenbergii*, *Coscinodiscus* sp., *Fragilaria karstenii*, *Thalassionema*

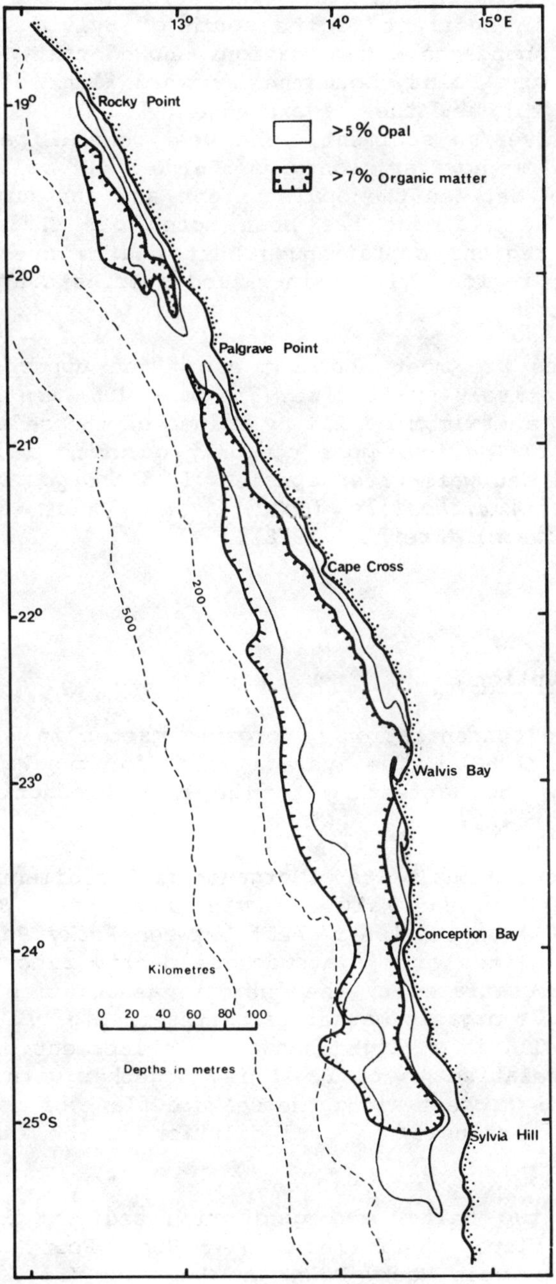

Fig. 7. Relationship between the distributions of opal and organic matter on the inner shelf.

nitzchioidea and *Raphoneis surirelliodes*. Although present in coastal waters as well as in the mud, *Chaetoceros* sp. and resting spores were not found in sediment to the south of Sylvia Hill or on the outer margin. Abundance determinations showed that, except for the area around Pelican Point, both the landward flank of the mud belt ($\sim$30 to 80 m depth), and the seaward edge ($\sim$130 to 140 m) contain <50 x 10^6 diatom valves/mg sediment. The west-central part of the mud belt ($\sim$80 to 130 m) however, contains 50-150 x 10^6 valves/mg. A similar discrepancy between the opal content and the number of diatom frustules in the sediment has been noted off Chile/Peru (Zhuze, 1971). There, regions containing robust diatom species have high abundances, but regions with thin-walled species contain substantially fewer diatoms.

Radiolarians are most abundant along the upper slope, and despite their relatively small size (75 µm - 400 µm; Siesser, 1975) they constitute a maximum of 21% by volume of the sediment off Rocky Point (Fig. 4b). The four most commonly occuring radiolaria are a mixture of cold and warm-water species viz., *Spongotrochus glacialis* (cold), *Hexacontinium hostile* (cold), *Actinoshpaera* sp. (warm) and *Actinomina* sp. (warm) (Bremner, 1978).

ORGANIC MATTER

Regional Distribution

The average concentration of organic matter in shelf sediments is 0.5% (Emery, 1960). The average value for the Walvis margin is 7.1% (Table 1), thus attesting to the high productivity of shelf waters in this area.

Three discontinuous belts of organic-rich sediment extend parallel to the coast on the Walvis margin (Fig. 3c). The richest of these (max. 24.6%) on the inner shelf between Rocky Point and Sylvia Hill, has accumulated from diatomaceous debris almost exclusively. This is well demonstrated by the intimate association shown between the 5% opal and 7% organic matter isopleths in Fig. 7. Noticeable in the figure as well, is a slight seaward displacement of the organic-rich sediment relative to opal. This, together with the fact that organic matter is richest along the seaward flank of the mud belt, is due more to the intensity of opal dilution on the landward than on the seaward flank.

The other two belts of organic-rich sediment lie on the mid shelf and upper slope, the former between Rocky Point and Walvis Bay, and the latter between Rocky Point and Lüderitz (Fig. 3c). The mid-shelf deposit (max. 12.5%) extends along the seaward side of the mid-shelf break at $\sim$250 m depth, and the upper-slope deposit (max. 8.8%) coincides roughly with the 1000 m isobath (Bremner, 1978). Both de-

Table 1. Average concentrations (%) of three biogenic components in the surficial sediments of the South West African continental margin[a]

	Kunene Margin[b]	Walvis Margin[b]	Orange Margin[b]	Entire Margin
Calcium Carbonate				
Mean	8.3	35.5	31.5	32.1
Standard Deviation[c]	7.6	30.8	28.2	29.7
Range	0.4-35.7	0-96.6	0-93.0	0-96.6
Opal				
Mean	0.6	17.7	--	10.6
Standard Deviation	1.6	29.5	--	24.3
Range	0-8.0	0-88.0	--	0-88.0
Organic Carbon				
Mean	3.2	7.1	4.1	5.9
Standard Deviation	1.3	4.5	3.2	4.5
Range	1.3-8.3	0-24.6	0.2-23.8	0-24.6

[a] The continental margin refers to the shelf and the upper slope to a depth of about 1500 m.

[b] Based on shelf morphology and/or sediment composition, the continental margin is divided into the following zones:
Kunene margin - 17°00'S to 18°40'S (Kunene River to Rocky Point)
Walvis margin - 18°40'S to 25°40'S (Rocky Point to Spencer Bay)
Orange margin - 25°40'S to 28°40'S (Spencer Bay to Orange River)

[c] The standard deviation occasionally exceeds the mean concentration indicating highly skewed distributions.

posits correlate closely with the distribution of fecal pellets indicating that benthic scavengers, such as polychaete worms and benthic foraminifers, tend to concentrate in areas of copious food supply. The main food sources are diatoms and planktonic foraminifers which most likely are prevented from settling in quantity onto the mid- and outer-shelf breaks, respectively, by the Ekman friction layer and currents related to upwelling (Bremner, 1978).

Associations

Opal. Disregarding the fact that the settling time of diatom frustules is greatly reduced when aggregated in fecal pellets

(Calvert, 1974), the time taken for settling along the landward edge of the mud belt (mean depth, 31 m; Bremner, 1980a) is theoretically 11 days in still water (10-20 µm frustules settle at ~1000 m/y; Calvert, 1966, p. 591). Along the seaward edge (mean depth, 130 m), the theoretical settling time would be 47 days. As a consequence of settling through the water column, a protective organic coating on the cell walls is removed and the frustules, particularly fragile ones, undergo rapid dissolution and recycling. Organic matter within the frustules, however, is to some extent protected from oxidation and bacterial decomposition. Therefore, a critical sedimentation rate is necessary to insure high concentrations of organic matter in surficial sediments. This is achieved along the seaward flank of the diatomaceous mud belt where only robust diatom frustules survive dissolution in transit to the bottom. On the landward flank of the mud belt, both robust and fragile tests accumulate thus diluting the associated organic matter to a greater extent than on the seaward flank, e.g., 15 km north of Walvis Bay, sample 4048 contains 82% opal and only 8.6% organic matter.

An estimate for the amount of organic matter associated with opal in the diatomaceous mud has been derived from the volume of the mud belt (calculated earlier in establishing the silica budget) as follows:

$$\text{Sediment volume} = 1.25 \times 10^{11}\ m^3$$

The average concentration of organic matter in the mud is 9.8% (= organic carbon x 1.8) and the dry density 0.8g/cm^3; therefore, the amount of organic matter is:

$$(1.25 \times 10^{11}) \times 9.8/100 \times (0.8 \times 10^6)$$

$$\text{Weight organic matter} = 10^{16} g$$

According, the annual supply rate of organic matter, based on an age of 5000 years of the mud belt, is:

$$\text{Annual organic matter deposition} = 2 \times 10^{12} g$$

Fecal Pellets. Whereas anoxic conditions at the surface of the diatomaceous mud have prevented its occupation by benthic metazoan life (other than anaerobic bacteria), oxygen-respirating invertebrates such as polychaete worms and benthic foraminifers play an active role in scavenging organic matter on the mid-shelf and upper slope.

The regional distribution of $CaCO_3$ (Fig. 3a), unlike that of fecal pellets (Fig. 8), shows little similarity to the linear trends of organic matter (Fig. 3c). The reason for this is that $CaCO_3$ source material comprises several different components, some of it

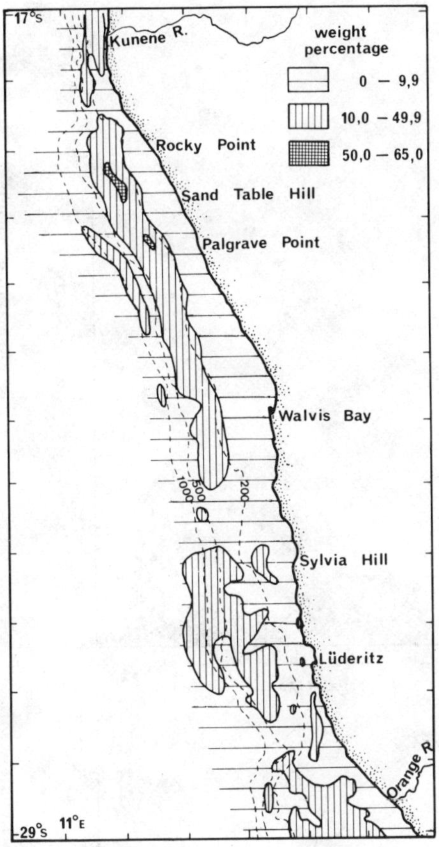

Fig. 8. Distribution of fecal pellets.

being relict, e.g., molluscs, which contribute nothing to the modern organic content of the sediment, and some of it, e.g., benthic foraminifers, being too low in abundance to significantly affect the distribution of organic matter.

A number of interesting aspects emerge from a bivariate plot of organic carbon vs. fecal pellets (Fig. 9a). It should be understood at the outset that because sand-size fecal pellets do not occur in the diatomaceous mud (no benthic metazoans), the plot only includes 'calcareous' samples (with >50% $CaCO_3$) and 'remaining' samples (terrigenous/authigenic with <5% opal and <50% $CaCO_3$). A solid regression line has been drawn through the calcareous points (x), and the two dashed lines through the remaining points (●).

1. The calcareous line and remaining line I exhibit a gradual increase in C-org concentration with increasing abundance of fecal

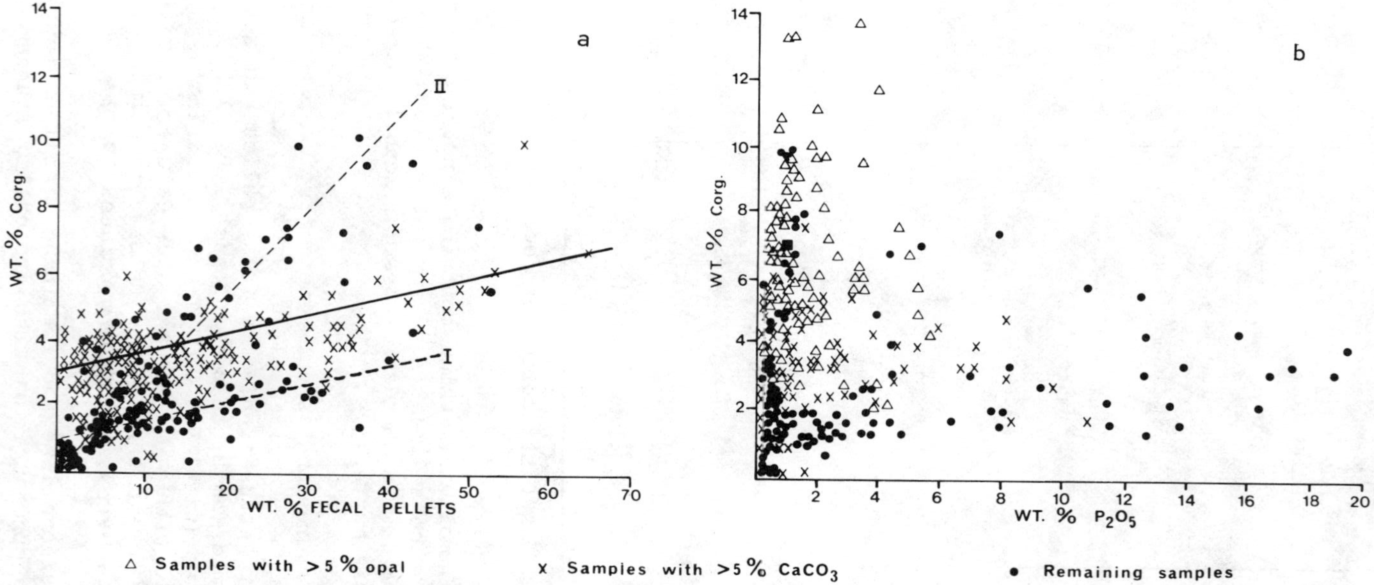

Fig. 9. Bivariate plots of:
a. Weight percentage of organic carbon vs. weight percentage fecal pellets.
b. Weight percentage of organic carbon vs. weight percentage P_2O_5.

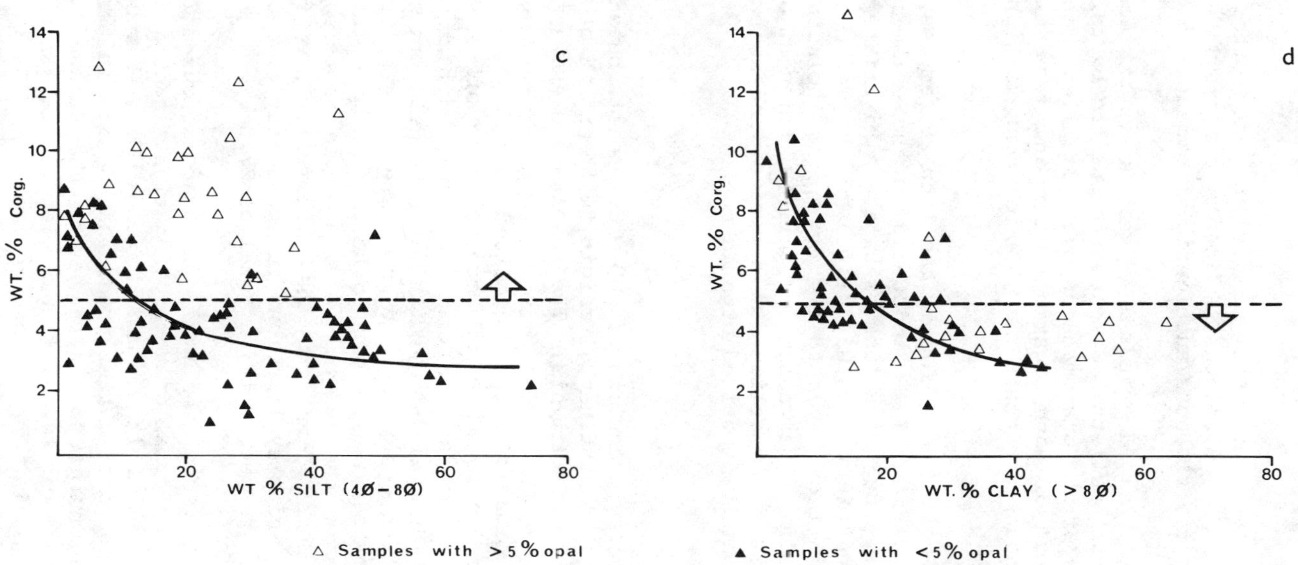

Fig. 9 (cont.)
c and d. Weight percentage of organic carbon in silt and clay vs. weight percentages of silt and clay in the sediment, respectively.

pellets. This means that polychaete worms tend to concentrate in areas of greatest organic matter supply (at $\sim$250 m and $\sim$1000 m depths).

2. The same two lines are parallel to each other indicating that, in general, the calcareous sediments contain $\sim$2% more C-org than the remaining (terrigenous/authigenic) sediments.

3. Remaining line II exhibits a very sharp increase in C-org with increasing fecal pellets. Samples represented by this line come from the two edges of the diatomaceous mud belt, i.e., nearshore terrigenous sand devoid of both organic matter and fecal pellets, and mid-shelf terrigenous/authigenic sands which contain high concentrations of both.

Phosphate. Senin (1970) found, based on 28 samples, that there was no correlation between P_2O_5 and C-org in the sediment, and suspected that this was due to a lack of data. A bivariate plot of the 555 samples examined in this study (Fig. 9b), however, confirms the absence of a correlation, even to the extent of including the Holocene diatomaceous mud. The reason for this is that most of the phosphorites on the shelf are relict lag deposits, and as such, have no bearing on the recently-deposited organic matter.

Size Distribution

The silt and clay fractions of 100 widely distributed subsamples from all depths on the continental margin to the north of Sylvia Hill were selected for C-org analysis. The results are shown in Fig. 10.

Most pelagic organic matter deposited on the shelf is silt and clay-sized material with values of both dropping off sharply on the upper slope. This decrease in organic matter with depth reflects oxidation of the material during its long transit onto the slope. The reverse situation occurs on the inner shelf where low concentrations of organic matter were shown to be due to opal dilution along the landward flank of the mud belt. An example of this is provided by sample 3426. Here, C-org for the bulk sample was found to be 13.7% whereas silt and clay fractions independently gave C-org values of 12.6% and 14.4%, respectively, a total of 27.0%.

Organic matter associated with the diatomaceous mud is mainly silt size with clay material only becoming significant along the seaward flank of the mud belt (Figs. 10a and 10b). Farther offshore however, clay-size organic matter predominates with values of >4% on the mid shelf continuing northward beyond the Kunene River mouth. Bivariate plots of C-org vs. silt and clay (Figs. 9c and 9d) show that diatomaceous silt always contains more than 5% C-org whereas diatomaceous clay material invariably contains less. Moreover, both the silt and clay fractions of non-diatomaceous sediment (<5% opal)

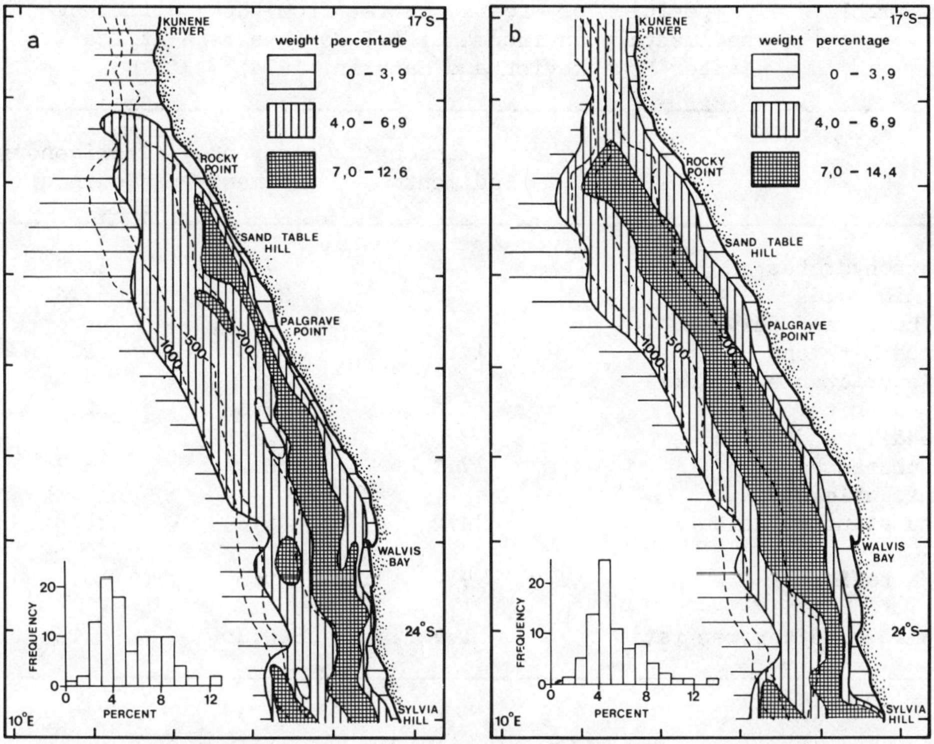

Fig. 10. Distribution of organic carbon in (a) the silt fraction of the sediment, (b) the clay fraction of the sediment.

exhibit an inverse relationship to the concentration of C-org, i.e., as the weight percentages of silt and clay increase, the abundance of organic matter in both size fractions decreases.

Composition

Bacteria change the state of organic matter during its passage through the water column, as well as when it comes to rest at the sediment-water interface. Once it is buried however, Boon, de Leeuw and Schenck (1975) have shown for diatomaceous mud that the number of bacteria decrease with depth, and slow diagenetic transformation takes place by reaction with the surrounding pore fluids. The principle diagenetic change affecting the organic matter with depth in the mud (Morris and Calvert, 1977) involves a decrease in the amount of amino acids (protein) and extractable lipids (free bitumens), and an increase in the amount of fulvics (humic acid).

Table 2 is a listing of the main components of organic matter in biogenic sediment from Walvis Bay, and from terrigenous sediment in

Table 2. Composition of organic matter from three different sedimentary environments off southwestern Africa (after Romankevich and Baturin, 1974)

	Diatomaceous Sediment	Calcareous Sediment	Terrigenous Sediment
Carbohydrates	7.9	7.6	11.8
Humic acids (humic and fulvic)	40.8	37.2	52.4
Free bitumens (extractable lipids)	14.8	8.0	2.9
Bound bitumens	3.0	2.3	2.7
Readily hydrolyzable substances	17.3	6.9	6.3
Insoluble organic substances (kerogen)	24.2	45.6	35.8
C/N ratio	9.4	9.6	10.9
Humic acid/kerogen ratio	1.7	0.8	1.5

the vicinity of the Kunene River. An important aspect of these results is the high humic acid/kerogen ratio, which apparently indicates a low state of diagenetic transformation.

Morris and Calvert (1977) have shown that diagenetic reactions promote the transfer of certain trace metals from pore-fluids to the organic matter. In addition, they showed that nickel, which commonly forms the nucleus of prophyrin pigments in petroleum, is somewhat enriched in the kerogen. Calvert and Price (1970) and Bremner (1978) have found that, besides nickel, copper and zinc and to some extent lead, are concentrated in the carbon-rich seaward flank of the mud belt.

DISCUSSION AND CONCLUSIONS

In comparison with continental margin sediments of the east coast of southern Africa, those of South West Africa are characterized by an extraordinary abundance of biogenic detritus. The factor primarily responsible for productivity in the region is the prevailing southerly wind. It reaches greatest intensity in the vicinity of Lüderitz where upwelling is most vigorous and persistent, and nutrients brought to the surface are then transported northwards by the Benguela Current into the broad coastal embayment at Walvis Bay.

Table 3. Correlation matrices between three biogenic sedimentary components and depth for each of the geographic regions of the continental margin

	$CaCO_3$	Opal	Organic Matter	Depth
		Kuenene Margin		
$CaCO_3$	1.00			
Opal	-.10	1.00		
Organic matter	.39	.38	1.00	
Depth	.52	-.17	.54	1.00
		Walvis Margin		
$CaCO_3$	1.00			
Opal	-.59	1.00		
Organic matter	-.08	.40	1.00	
Depth	.60	-.38	-.13	1.00
		Orange Margin		
$CaCO_3$	1.00			
Opal	--	--		
Organic matter	.21	--	1.00	
Depth	.61	--	.25	1.00
		Entire Margin		
$CaCO_3$	1.00			
Opal	-.39	1.00		
Organic matter	.06	.44	1.00	
Depth	.55	-.29	-.04	1.00

Here, productivity reaches its acme, and may equal or even exceed the productivity of Antarctic waters at certain times of the year (Hart and Currie, 1960).

Biogenic debris is not uniformly distributed on the margin, however, this being the basis for dividing it into three distinct latitudinal zones. In the south, the Orange margin supports high concentrations of predominantly foraminiferal $CaCO_3$ on its outer part; the centrally located Walvis margin carries a rich deposit of opaline mud on its inner part, and an abundance of mollusc/foraminiferal $CaCO_3$ on its outer part. The northerly Kunene margin, which underlies the largely sterile waters of the Angola Current, is practically deficient in all biogenic components.

Notwithstanding the low biogenic concentrations in Kunene margin sediment, Table 3 illustrates that calcium carbonate and organic matter, like that on the Orange margin, both correlate with depth. This means that the organic matter is derived mainly from planktonic foraminifers, which increase in abundance across the shelf towards the upper slope. Opal on the other hand, correlates negatively with depth on the Kunene and Walvis margins, thus reflecting its shallow water depositional environment. In the latter geographic region, organic matter correlates with opal rather than with $CaCO_3$, thereby reflecting the exceptionally high values recorded for the two components in the nearshore diatomaceous mud. Moreover, when the entire margin is considered, the opal-organic matter association completely overshadows the $CaCO_3$-organic matter association despite the fact that $CaCO_3$ is more widespread and attains higher concentrations than the opal.

In conclusion, then, biogenic-rich sediments on the southwest African continental margin are clearly a productivity response brought about by coastal upwelling in the Benguela Current. The distribution of the various facies closely reflects standing crops of phyto- and zooplankton. Benthic organisms are limited in their distribution by water depth, the oxygen content of the bottom water, and hydrodynamic processes. Descriptions given here of the individual distributions of various biogenic components will, it is hoped, be of value in planning future studies of smaller scale and greater detail.

ACKNOWLEDGEMENTS

The Director of the Geological Survey, Mr. L.N.J. Engelbrecht is thanked for obtaining permission for me to attend the Advanced Research Institute Conference on Upwelling which was held in Portugal in September, 1981, and also for allowing me to publish this paper. To Drs. E. Suess and J. Thiede and the Advanced Research Institute, I express my gratitude for inviting me to attend the Conference, and for making the trip financially possible. I am also indebted to Professor R.V. Dingle of the Geology Department, University of Cape Town, and Dr. U. von Stackelberg of the Federal Geological Survey, West Germany, both of whom read an early draft of the manuscript and made numerous helpful suggestions. "Biogenic sediments on the southwest African continental margin" is reproduced in accordance with the Government Printers copyright concession No. 778 of 10/3/1982.

REFERENCES

Bang, N.D., 1971, The southern Benguela Current region in February, 1966, Part II: Bathythermography and air-sea interactions, Deep-Sea Research, 18:209-224.

Bang, N.D., 1976, On estimating the ocean mass flux budget of lateral and cross circulations of the southern Benguela upwelling system, First Interdisciplinary Conference on Marine Freshwater Research in South Africa, Fiche 6G4-7:All.

Baturin, G.N., 1971, Stages of phosphorite formation on the South West African Shelf, Nature Physical Science, 232:61-62.

Boon, J.J., de Leeuw, J.W. and Schenck, P.A., 1975, Organic geochemistry of Walvis Bay diatomaceous ooze - I. Occurrence and significance of the fatty acids, Geochimica et Cosmochimica Acta, 39: 1559-1565.

Bremner, J.M., 1978, "Sediments on the Continental Margin off South West Africa between Latitudes 17° and 25°S," Ph.D. Dissertation, University of Cape Town, Republic of South Africa, 300 pp.

Bremner, J.M., 1980a, Physical parameters of the diatomaceous mud belt off South West Africa, Marine Geology, 34:M67-M76.

Bremner, J.M., 1980b, Concretionary phosphorite from South West Africa, Journal of the Geological Society of London, 137:773-786.

Brongersma-Sanders, M., Stephan, K.M., Kwee, T.G., and de Bruin, M., 1980, Distribution of minor elements in cores from the South West Africa shelf with notes on plankton and fish mortality, Marine Geology, 37:81-132.

Calvert, S.E., 1966, Accumulation of diatomaceous silica in the sediments of the Gulf of California, Bulletin of the Geological Society of America, 77:569-596.

Calvert, S.E., 1974, Deposition and diagenesis of silica in marine sediments, in: "Pelagic Sediments on Land and Under the Sea," K.J. Hsü and H.C. Jenkyns, eds., Blackwell Scientific Publications Ltd., Oxford, U.K., 273-329.

Calvert, S.E. and Price, N.B., 1970, Minor metal contents of Recent organic-rich sediments off South West Africa, Nature, 227:593-595.

Calvert, S.E. and Price, N.B., 1971, Upwelling and nutrient regeneration in the Benguela Current, October, 1968, Deep-Sea Research, 18:505-523.

Carmack, E.D. and Aargaard, K., 1977, A note on volumetric considerations of upwelling in the Benguela Current, Estuarine and Coastal Marine Science, 5:135-142.

Emery, K.O., 1960, "The Sea off Southern California," J. Wiley and Sons, New York, 366 pp.

Fairbridge, R.W., 1961, Eustatic changes in sea level, Physics and Chemistry of the Earth, 4:99-185.

Godwin, H., Suggate, R.P. and Willis, E.H., 1958, Radiocarbon dating of the eustatic rise in ocean-level, Nature, 181:1518-1519.

Guilcher, A., 1969, Pleistocene and Holocene sea level changes, Earth-Science Reviews, 5:69-97.

Hart, T.J., and Currie, R.I., 1960, The Benguela Current, Discovery Reports, XXXI:123-298.

Heath, G.R., 1974, Dissolved silica and deep-sea sediments, in: "Studies in Paleo-oceanography," Society of Economic Paleontologists and Mineralogists, Special Publication, 20:77-93.

Hülsemann, J., 1966, On the routine analysis of carbonates in unconsolidated sediments, Journal of Sedimentary Petrology, 36:622-625.

Kensley, B.F., 1973, "Sea Shells of Southern Africa," Maskew Miller, Cape Town.

Kruger, I., 1980, A checklist of South West African marine phytoplankton, with some phytogeographical relations, Fisheries Bulletin of South Africa, 13:31-53.

Lees, A., 1975, Possible influence of salinity and temperature on modern shelf carbonate sedimentation, Marine Geology, 19:159-198.

Milliman, J.D. and Emery, K.O., 1968, Sea levels during the past 35,000 years, Science, 162:1121-1123.

Morgans, J.F.S., 1956, Notes on the analysis of shallow soft substrata, Journal of Animal Ecology, 25:367-387.

Moroshkin, K.V., Bubnov, V.A. and Bulatov, R.P., 1970, Water circulation in the eastern South Atlantic Ocean, Oceanology, 10:27-37.

Morris, R.J. and Calvert, S.E., 1977, Geochemical studies of organic-rich sediments from the Namibian shelf. I: The organic fractions, in: "George Deacon 70th Anniversary Volume," M. Angel, ed., 647-655.

Rogers, J., 1977, "Sedimentation on the Continental Margin off the Orange River and the Namib Desert," Ph.D. Dissertation, University of Cape Town, Republic of South Africa, 212 pp.

Romankevich, Y.A. and Baturin, G.N., 1974, The biogeochemical composition of the sediments on the West African shelf, Oceanology, 14:529-533.

Schell, I.I., 1968, On the relation between winds off South West Africa and the Benguela Current, and Agulhas Current penetration in the South Atlantic, Deutsche Hydrographische Zeitschrift, 21:109-117.

Senin, Yu.M., 1968, Characteristics of sedimentation on the shelf of southwestern Africa, Lithology and Mineral Resources, 4:476-479.

Senin, Yu.M., 1970, Phosphorous in bottom sediments of the South West African shelf, Lithology and Mineral Resources (Consultants Bureau Translations), 25:8-20.

Shepard, F.P., 1960, Rise of sea level along Northwest Gulf of Mexico, in: "Recent Sediments, Northwest Gulf of Mexico," F.P. Shepard, P.B. Phleger and T.H. van Andel, eds., American Association of Petroleum Geologists, Tulsa, 338-344.

Siesser, W.G., 1975, Micropaleontological techniques used in Pleistocene climatology, South African Archaeological Bulletin, Godwin Series, 2:29-36.

Stander, G.H., 1964, The Benguela Current off South West Africa, Administration of South West Africa, Investigational Report No. 12, 122 pp.

Summerhayes, C.P., 1972, Studies of the mineralogy and geochemistry of unconsolidated sediments from around southern Africa. Part I: Aspects of the mineralogy and geochemistry of Agulhas Bank sediments, SANCOR/UCT Marine Geology Programme, Technical Report No. 4:64-82.

van Andel, T.H., 1964, Recent marine sediments of the Gulf of California, in: "Marine Geology of the Gulf of California," T.H. van Andel and G.G. Shor, eds., American Association of Petroleum Geologists, Memoir 3:216-310.

Veeh, H.H., Calvert, S.E. and Price, N.B., 1974, Accumulation of uranium in sediments and phosphorites on the South West African shelf, Marine Chemistry, 2:189-202.

Visser, G.A., 1969, The oxygen-minimum layer between the surface and 1000 m in the north-eastern South Atlantic, Bulletin of the Division of Sea Fisheries of South Africa, 6:10-22.

Visser, G.A., Kruger, I. and Coetzee, D.J., 1973, Environmental studies in South West African waters, Abstracts volume, South African National Oceanographic Symposium, Cape Town, 6-10 August 1973, 25-26.

Zhuze, A.P., 1971, Diatoms in the surface layer of sediments in the vicinity of Chile and Peru, Oceanology, 12:697-705.

THE MODERN UPWELLING RECORD OFF NORTHWEST AFRICA

Dieter K. Fütterer

Geologisch-Paläontologisches Institut
Olshausenstrasse 40-60, D-2300 Kiel
Present Address:
Alfred-Wegener-Institute for Polar Research
Columbus Center
D-2850 Bremerhaven
Federal Republic of Germany

ABSTRACT

Several biological and geochemical indicators have been suggested to characterize coastal upwelling processes and to document them in the underlying sediments. These criteria apparently work well in the intensive coastal upwelling regimes off Peru and Namibia. Surface sediments from shelf and slope of the northwest African coastal upwelling area between Cape Bojador, at about 26°N, and the Casamance River, at about 13°N, were examined to evaluate the fate of potential biological upwelling indicators in a moderate upwelling regime. Off northwest Africa no textural upwelling indicators were found in shallow shelf sediments directly beneath the upwelling centers. Due to oceanographic conditions they are rather winnowed out, transported beyond the shelf edge and dispersed on the continental slope by near-bottom currents or grain-by-grain downslope transport. During this transfer, potential upwelling indicators become altered or entirely destroyed by passing through various "filtering" mechanisms. Modern coastal upwelling conditions off northwest Africa are, therefore, not recorded in the underlying shelf sediments and are only badly documented in the adjacent slope sediments. This suggests that simple sediment component analyses may be misleading if used as the only tool to reconstruct paleoceanographic conditions.

INTRODUCTION

Coastal upwelling systems have for a long time attracted increasing attention as areas of high biological production not only from oceanographers and biologists but from geologists as well. This interest results from different causes, however; the main geological question is: "Do the sediments underlying an upwelling system reflect the processes occurring in the upwelled water mass by having characteristic facies patterns which can be recognized in the fossil record?" If the answer is "yes", then we would have an ideal tool to reconstruct special paleoceanographic conditions and environments from the fossil record.

The process of coastal upwelling develops when the wind-driven, alongshore surface water currents result in an offshore Ekman transport. This flow is compensated for by transport of nutrient-rich subsurface water into the euphotic layer where optimal light conditions for phytoplankton favor an increasing biological activity. The main characteristics that distinguish coastal upwelling areas from other regions of the ocean are: (1) High nutrient content which causes an enhanced primary production; (2) lowered temperatures at the sea surface; and (3) lowered oxygen content in the surface layer (Barber and Smith, 1981). These upwelling characteristics produce a unique ecosystem, the remains of which should be documented in some way or other in the underlying sediments. The direct relation, however, is strongly complicated by the fact that the upwelling processes are merely short-termed events in the order of days or weeks, including additional seasonal shifts, whereas the sediments show a long-termed memory in the order of hundreds to thousands of years. Depending on the sedimentation rate, only very mixed and averaged oceanographic signals are likely to be preserved in the sediments.

Various sedimentary parameters are discussed in detail as upwelling indicators by Diester-Haass (1978). She suggests that several biological and chemical indicators, like high biogenic opal content, abundant fish debris, low plankton/benthos ratios of foraminifers, presence of cool water foraminifers, high content of organic matter, etc., characterize the sedimentary record of upwelling processes. But, more precisely, such indicators do not stand for "upwelling" alone, but for water masses different from the nearby normal oceanic waters. In fact, these indicators tell us about the temperature, salinity, oxygen or nutrient content, but their individual presence does not necessarily prove that upwelling has occurred.

The main aim of the present paper is to describe the modern sedimentary record along the northwest African continental margin off the Saharan desert in order to detect the influence of the northwest African coastal upwelling system. Mainly, textural parameters and sediment constituents are considered; geochemical aspects, like phosphorite genesis, minor metals and isotope studies, are excluded, as are paleontological species analyses. The investigated sediments come from sample transects across the continental margin between Cape Bojador, at about 26°N, and the Casamance River, at about 13°N (Fig. 1).

MORPHOLOGICAL SETTING

The shelf in the studied area is generally narrow, about 45-55 km wide, with the exception of the Banc d'Arguin. There the shelf is about 140 km wide and can be divided into a broader, shallow inner shelf with less than 20 m water depth, and a deeper outer shelf area. In both areas the shelf break occurs at water depths of 110-150 m (Piessens and Chabot, 1977).

The continental slope is about 45 km wide with an average inclination of between 2° and 3°. Numerous small and several larger canyons are located off the shelf edge throughout the area, and par-

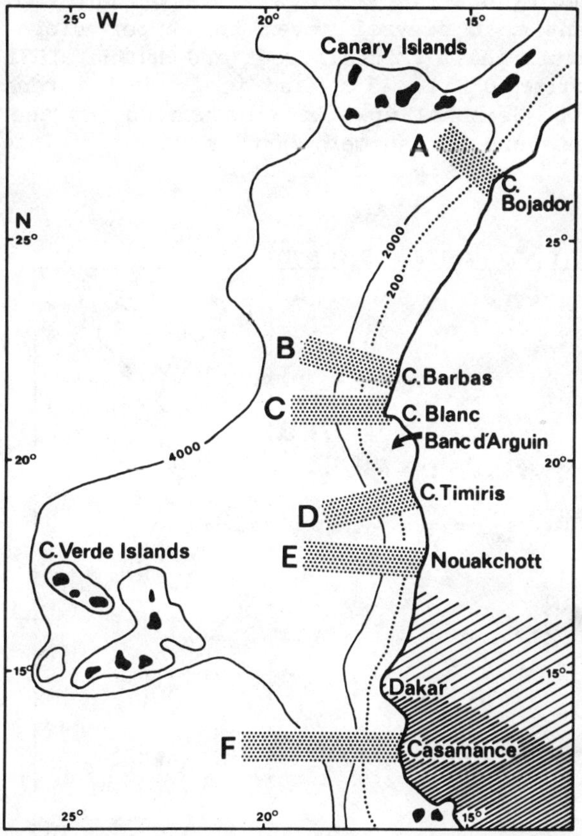

Fig. 1. Northwest African continental margin showing location of investigated sample lines (A,B = "Meteor" cruise 25-1971; C = "Meteor" cruise 39-1975; D,E,F = "Valdivia" cruise 10-3-1975). Densely hatched continental area = winter-dry tropical savannah; widely hatched = winter-dry hot steppe; blank = dry, hot region of Saharan desert.

ticularly south of Cape Bojador (Rust and Wieneke, 1973; Seibold and Hinz, 1974; Arthur et al., 1979), and off the Banc d'Arguin, where one canyon may even be traced on the shelf as well (Bein and Fütterer, 1977). Very fine-grained mud blankets found in some of these canyons appear to indicate that there are no strong currents at present.

THE UPWELLING SYSTEM AND OCEANOGRAPHIC SETTING

Off northwest Africa, coastal upwelling is caused by a complex interaction of the nearly longshore Trade Winds and the southward flowing cool Canary Current. The Ekman principle results in a seaward transport of the surface water layer. The nearshore surface layer is replaced by nutrient-rich water masses upwelled from a poleward undercurrent that prevails over the upper slope and at some places on the outer shelf (Mittelstaedt and Hamann, 1981). Upwelling is year round from 20°N to 25°N, and it is most intensive off Cape Blanc (Fig. 2A). Seasonal upwelling occurs during the winter south to Sierra Leone, and in summer farther north up to Morocco and Portugal.

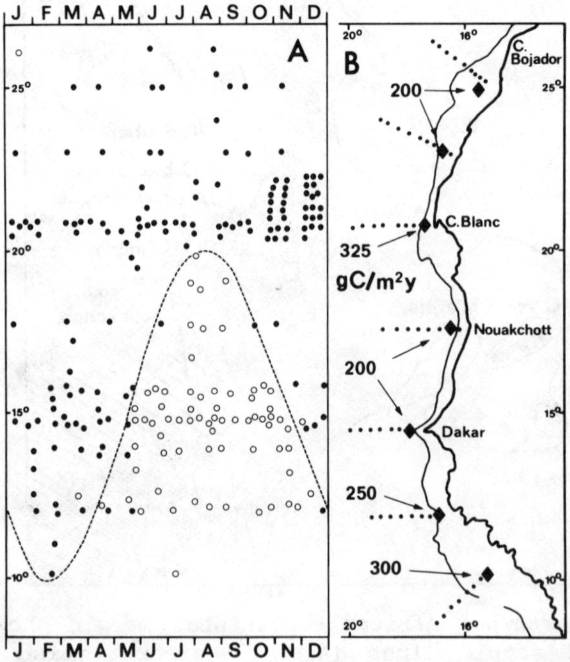

Fig 2. Variation of upwelling off northwest Africa in space and time (A) (dots = observed upwelling at given time) and maximum values for annual primary production (B); simplified from Schemainda, Nehring and Schulz, 1975.

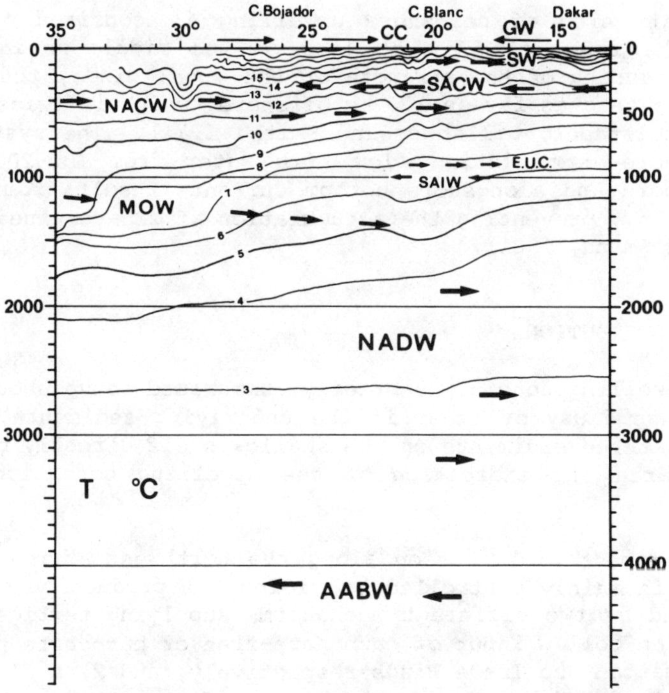

Fig. 3. Stratification of bottom water temperature and ocean circulation (arrows) near the bottom along the northwest African continental margin. AABW = Antarctic Bottom Water, NADW = North Atlantic Deep Water, MOW = Mediterranean Outflow Water, SAIW = South Atlantic Intermediate Water, E.U.C. = Equatorial Undercurrent, NACW = North Atlantic Central Water, SACW = South Atlantic Central Water, SW = Surface Water, CC = Canary Current, GW = Guinea Water Mass; simplified from Lutze, 1980 and Sarnthein et al., 1982; depth in (m).

The maximum values for the annual primary production in that region show a similar pattern (Fig. 2B). Highest values off Cape Blanc coincide with year round upwelling; decreasing values to the south are due to seasonal upwelling. The higher values off Sierra Leone are caused by an additional nutrient input by rivers. The relatively low values north of Cape Blanc can be explained by differences in stratification of the oceanographic circulation system (Fig. 3). Off Cape Blanc, at about 21°N to 23°N, the north flowing undercurrent of South Atlantic Central Water (SACW) provides a characteristically nutrient-rich source water for this upwelling region. North of about 23°N the SACW is replaced by the relatively nutrient poor North Atlantic Central Water (NACW) as the main source for the upwelled waters (Fraga, 1974; Codispoti and Friedrich, 1978). These principal differences in nutrient content of the source waters feeding the upwelling system are most likely the reason for the corresponding regional differences in primary production.

The main area of nearshore upwelling is confined to a water depth in the range of 50 m or less on the middle or inner shelf. However, depending on the strength of the local winds, the upwelling system may extend as far as 100 km offshore, which is twice the shelf width (Mittelstaedt, Pillsbury and Smith, 1975). The system is fed by an onshore compensation flow with water from 100-200 m depth. Strong onshore and alongshore bottom currents result from this compensation flow, preventing the accumulation of fine-grained sediments (Barber and Smith, 1981).

SEDIMENT DISTRIBUTION

The upwelling conditions briefly summarized above should be documented in some way or other in the underlying sediments. However, if we look at the sediments on the shallow shelf directly beneath the upwelling area, no indication of the upwelling conditions are observed.

The sediment distribution along the northwest African continental margin is mainly controlled by biological production of skeletal material and by two different mechanisms supplying terrigenous material: (1) an eolian input of carbonate-free or carbonate-poor detritus by Harmattan and Trade Winds respectively, and 2) a fluvial input of fine-grained muds in the southern part of the area by the Senegal River. The overall sediment distribution pattern reflects these predominant supply mechanisms (Fütterer, 1980). A general view of the diverse sedimentary lithofacies present in the surface sediments of that area is given by the carbonate distribution pattern (Fig. 4).

The shelf area off Cape Blanc which is the most productive upwelling area, and the shelf to the north is covered far beyond the shelf edge by coarse, in large part relict, molluscan and algal carbonate sands. Downslope, the carbonate content rapidly decreases to minimum values on the lower continental slope, and increases again on the continental rise. Farther to the south the carbonate content on the shelf decreases as the content of coarse-grained quartz increases. Fine-grained shelf muds are confined exclusively to the pro-delta area of the Senegal River. There is no indication on the northwest African margin of a shallow water mud lens like those reported from the shelf or upper slope of the upwelling areas off Namibia and Peru, respectively. This becomes further evident by a closer look at the grain-size distribution of the sediments. Off Cape Barbas, at about 22°N (Fig. 5), a very coarse-grained sand covers the shelf and extends far beyond the shelf edge onto the upper slope. The sand shows a distinct, continuous shift to smaller grain sizes with pronounced maxima in the silt-size range with increasing water depth. This striking silt-size maximum is even more obvious farther to the south, off Cape Timiris at 19°N (Fig. 6), and off Nouakchott at about 18°N. These areas are located beneath the center of the

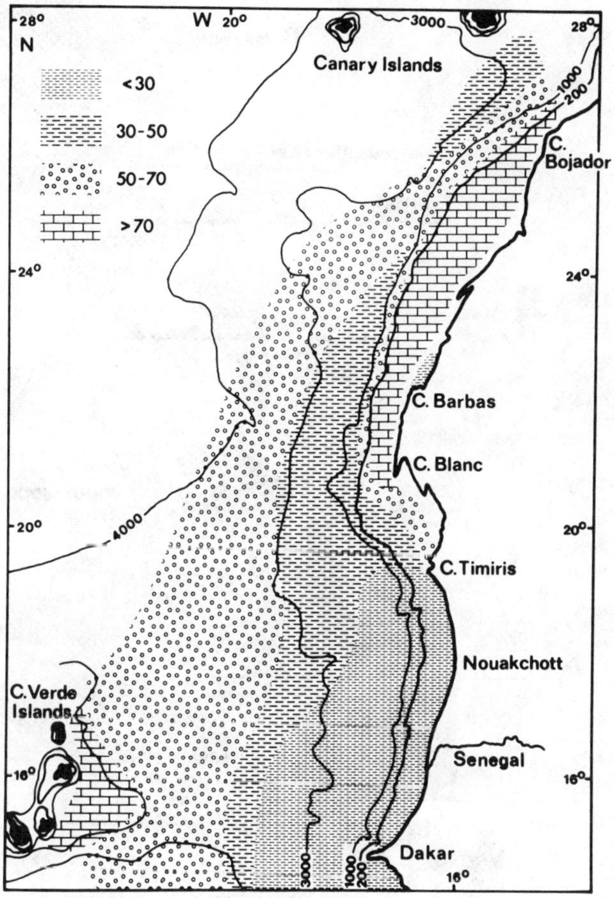

Fig. 4. Carbonate content in weight percent of surface sediments off northwest Africa.

eolian dust supply from the Saharan desert which is assumed to cause this characteristic grain size distribution. There is a sharp decrease in sand content, or a sharp increase in silt content, between 300-500 m water depth south of Cape Blanc, and at somewhat greater water depth of 500-800 m off and north of Cape Blanc. This pattern may be explained by the deflection to the west of the south flowing Canary Current system off Cape Blanc. This current may be deep enough and strong enough to prevent sedimentation of fine-grained material on the upper slope off and north of Cape Blanc, whereas the north flowing undercurrent originating from SACW and centered at 150-400 m water depth shapes the slope south of Cape Blanc.

We interpret the grain size distributions as reflecting oceanic currents and wave action which intensively retard sedimentation on

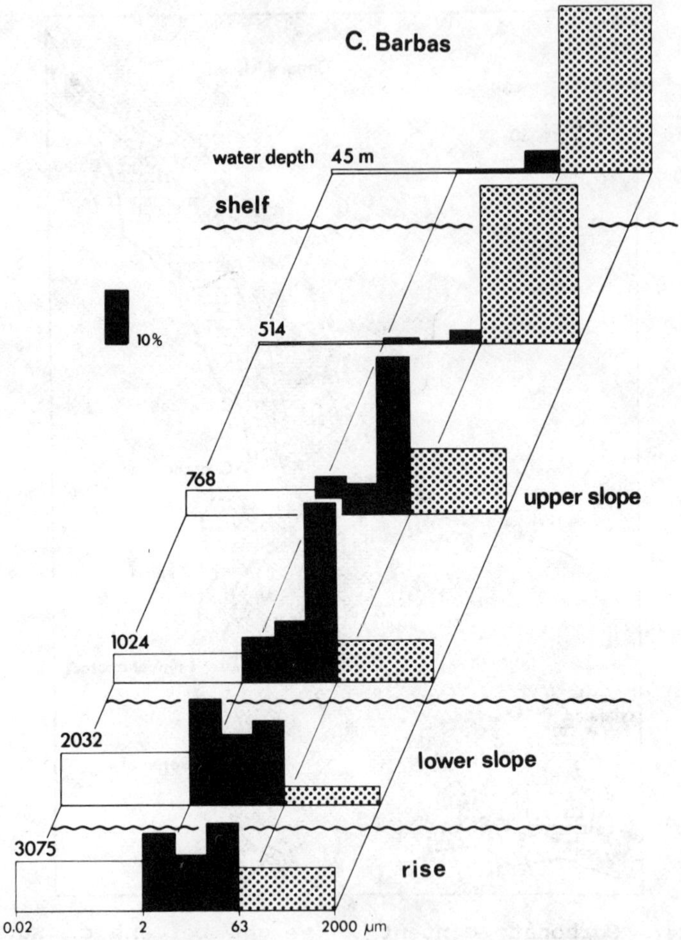

Fig. 5. Grain size distribution of surface sediments off Cape Barbas (sample line B in Fig. 1).

the shelf and upper continental slope at present. These agents may even be strong enough to rework and erode the shelf sediments in places. As a result, a substantial fraction of the sediment supplied to the shelf crosses the shelf edge and is transported downslope in suspension by a nepheloid layer system or grain-by-grain. The latter mechanism accounts for the abundant occurrence of relatively larger shallow water constituents in deeper waters downslope (Bein and Fütterer, 1977; Fütterer, 1980).

At this point we have to state that at present no record of upwelling can be recognized in the modern sediments directly underlying the upwelling area on the shelf and upper slope. All upwelling

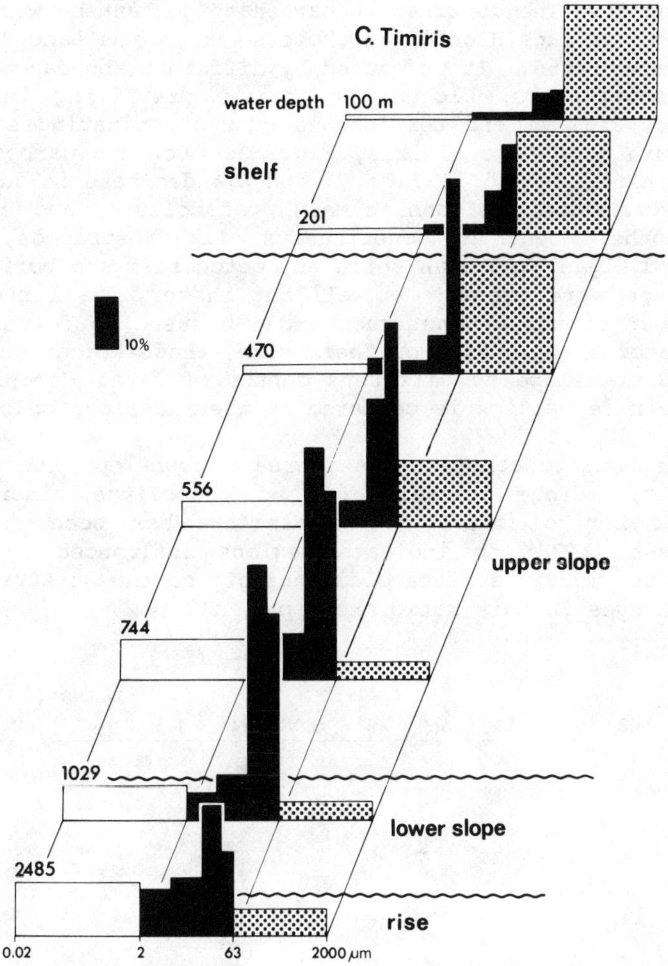

Fig. 6. Grain size distribution of surface sediments off Cape Timiris (sample line D in Fig. 1) showing prominent silt sized maximum caused by high eolian input.

indicators (Diester-Haass, 1978) have disappeared, and for the most part have been transported beyond the shelf edge and spread over the deeper continental slope and rise.

SEDIMENT CONSTITUENTS

To trace the shallow water material farther downslope, a quantitative analysis of the sediment components has been carried out by which the conditions described above become evident in more detail

(Figs. 7 and 8). The contrast in carbonate content between the areas off and north of Cape Blanc, and those south of the Banc d'Arguin, is clearly demonstrated. It is caused by different abundances of relict sediment constituents, biogenic carbonate prevailing in the north (Cape Blanc, Fig. 7) and terrigenous quartz dominating south of the Banc d'Arguin (off Cape Timiris, Fig. 8). Benthic carbonates other than foraminifers ("Oc" in Figs. 7 and 8), decrease in quantity from the shelf downslope and consist mainly of molluscs and echinoderms. However, other benthic constituents like ostracods, siliceous sponges, red algae, ascidian spicules, octocorals and boring chips of boring sponges are present as well but in very small numbers only. The last four of these constituents can be used as downslope transport indicators. Depending on their size, these tracer particles are transported downslope to different depths of final deposition. The depth of main deposition is centered at the mid-slope below 1000 m.

As the benthic constituents decrease downslope, the planktonic constituents, mainly foraminifers and coccoliths, increase. The plankton to benthos ratio of foraminifers has been suggested by Diester-Haass (1978) to indicate regions influenced by upwelling. But within the modern surface sediments off northwest Africa, no significant changes in this ratio are apparent.

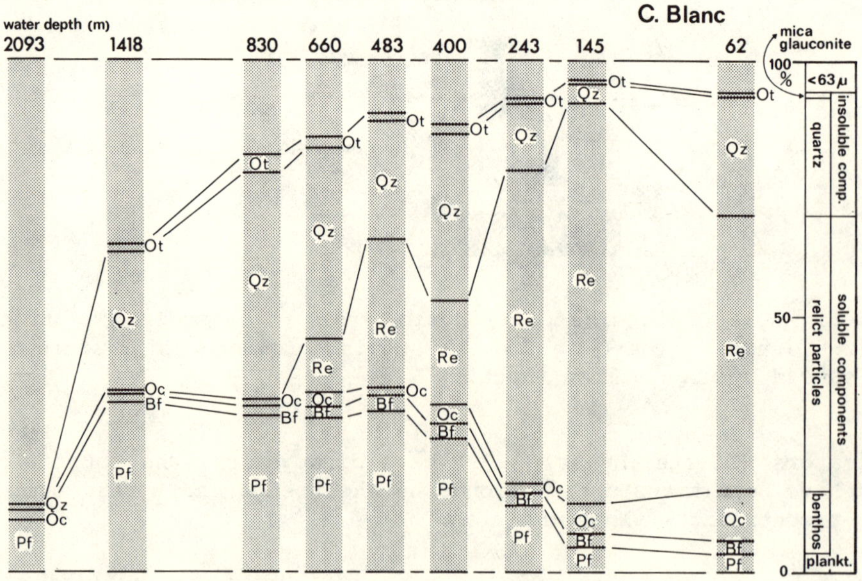

Fig. 7. Composition of coarse-grained (>63 μm) sediment components of surface sediments off Cape Blanc (sample line C in Fig. 1). Pf = planktonic foraminifers, Bf = benthic foraminifers, Oc = other biogeneous carbonate grains (mainly molluscs and echinoderms), Re = relictic biogenic material, Qz = quartz, Ot = other terrigeneous material (mainly mica and glauconite).

Radiolarians and diatoms occur very rarely in the slope sediments and are not abundant enough to be depicted as compositional units in Figs. 7 and 8. Both of these components are believed to indicate increased fertility. Diester-Haass (1977) found radiolarian to planktonic foraminiferal ratios to be highest in the center of the actual upwelling area off Cape Blanc, and used this ratio to propose increased upwelling during the Würm in this region. The abundance of radiolarian tests in the surface sediments in this area, however, seems to be linked more to the high nutrient content of subsurface waters, and reflects the distribution pattern of the SACW (in contrast to the NACW) rather than to an upwelling process (Labracherie, 1980). Diatom frustules or fragments likewise are extremely rare in the slope sediments and do not show any significant distribution pattern. The low abundance of diatoms in the sediments may be explained by the observation that most of the dominant species living in this upwelling system are only weakly silicified, a fact which favors their dissolution during settling and at the sea bottom (Richert, 1975). A detailed survey on the content of bulk biogenic opal in the slope sediments has been carried out by Koopmann (1979; 1981). As expected, the opaline silica content is very low (Fig. 9) showing relatively higher values nearshore. There is no significant latitudinal variation, although there may be slight maxima off Cape Blanc

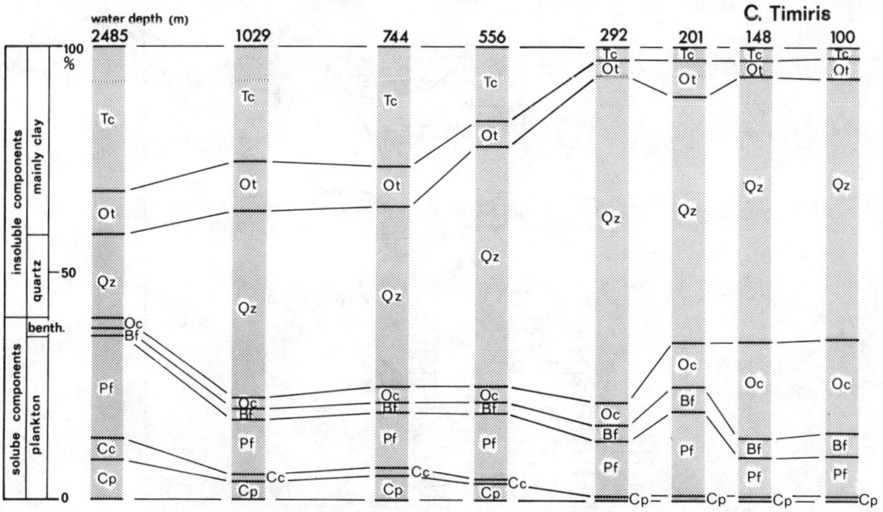

Fig. 8. Quantitative composition of sediment components off Cape Timiris (sample line D in Fig. 1, total sample). CP = <2 µm carbonate particles (derived mainly from coccoliths), Cc = silt-sized coccoliths, Pf = planktonic foraminifers, Bf = benthic foraminifers, Oc = other biogenic carbonate grains (mainly molluscs and echinoderms), Qz = quartz, Ot = other terrigeneous material (mainly mica and glauconite), Tc = <2 µm terrigeneous particles (mainly clay).

and Cape Barbas, and south of Dakar. With our background knowledge of the existing oceanographic conditions, we might interpret this pattern as upwelling induced off Cape Blanc, and river influenced south of Dakar. However, despite the low abundance of diatoms, fresh water diatoms, e.g., *Melosira granulata* which is a strongly silicified species and well known from eolian dust samples, are relatively abundant here (Fütterer, 1980). This implies that the small maxima of biogenic opal in the area off Cape Blanc reflects the input of eolian dust rather than increased fertility induced by upwelling. In contrast, biogenic opal is abundant in the very shallow and protected Baie du Levrier, on the innermost part of the Banc d'Arguin (Koopmann, Sarnthein and Schrader, 1978). Here the upwelling triggered plankton production from the outer shelf is moved inshore by currents and trapped in a small coastal basin.

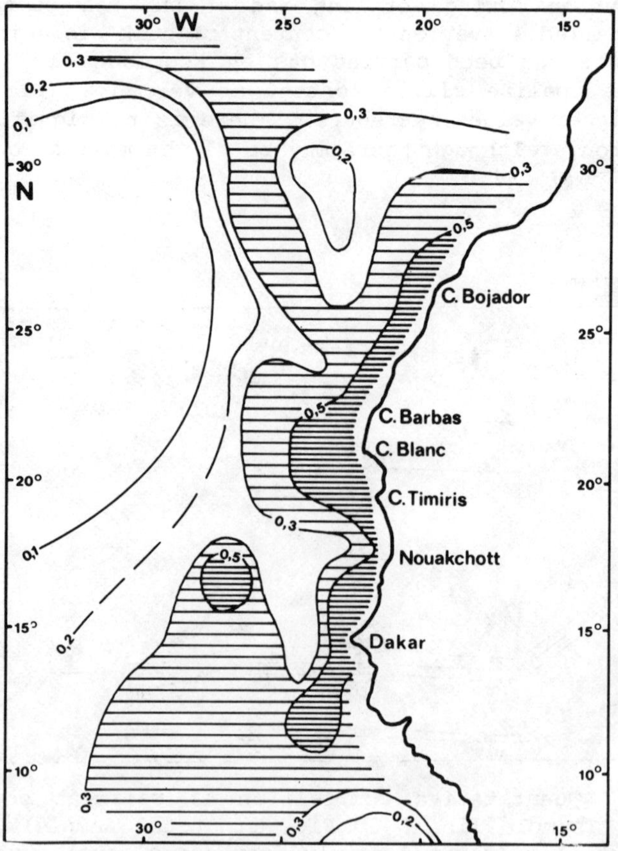

Fig. 9. Concentrations of biogenic opal (weight percent of >6 µm fraction) in the bulk carbonate-free surface sediments off northwest Africa; modified from Koopmann, 1979.

The pattern of distribution of fish debris, which is another generally accepted indicator for nutrient-rich waters, is like that of diatoms. Fish debris is only occasionally present in the slope sediments, but is enriched in the shallow water of the Baie du Levrier (Koopmann et al., 1978). The organic matter content of the sediments is only a very rough indicator of higher fertility or productivity and, hence, of upwelling. It depends in a complex way on grain-size, sedimentation rate, and organic matter production (Müller and Suess, 1979). On the continental slope off northwest Africa there are two areas of abundant organic matter (Fig. 10): 1) at about 16°N off the Senegal River, and 2) at about 21°N off Cape Blanc, which is the main area of modern upwelling activity. Both areas are at water depths of about 1000 m to 2000 m, which is where the maximum deposition of fine-grained material takes place and forms a mid-slope mud lens. The high organic matter content of this mid-slope mud lens at about 21°-22°N is most probably caused by the coastal upwelling system off Cape Blanc, whereas farther to the south, at about 15°-16°N, it is due to an increase in fertility caused by river input of nutrients. From the distribution of bulk organic matter alone it is impossible to tell which of these two areas is upwelling or river controlled. Better results come from organic geochemistry showing highest marine organic matter contribution (with high H/C ratios) off Cape Blanc, and highest terrestrial organic matter contribution (with low H/C ratios) off the Senegal River, in the mid-slope mud belt (Pelet, 1979).

Other arguments for differentiating between these environments come from textural parameters. The fluvial input of the Senegal River is characterized by a basically bimodal grain size distribution with an excess of clay sizes (Sarnthein et al., 1982), whereas the eolian input off Cape Blanc is dominated by silt sizes. Similar

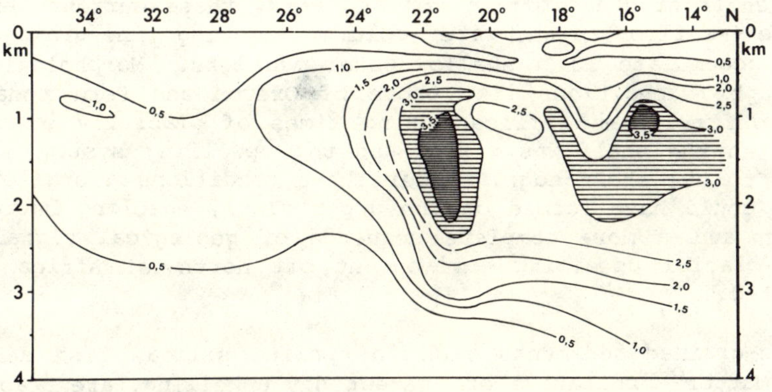

Fig. 10. Distribution of organic carbon (as weight percent) in surface sediments along the northwest African continental margin; modified from Sarnthein et al., 1982.

amounts of organic matter to those in the mid-slope mud lens are found in the fine-grained shallow water prodelta muds of the Senegal River (Domain, 1977), which are associated with considerable amounts of plant debris.

Intensive calcium carbonate dissolution along the continental margin is indicated by the disappearance of pteropod shells with increasing depth (Diester-Haass and Müller, 1979; Fütterer, 1980). The critical depth, assumed to be the aragonite compensation depth (ACD), shallows from 2500 m at the Cape Verde Rise, and from 2000 m on the continental slope off Cape Bojador, to about 500 m approaching the slope off Cape Blanc (Fütterer, 1980). This coincides with increased fragmentation of foraminiferal tests on the slope off Cape Blanc (Diester-Haass and Müller, 1979).

CONCLUSIONS

From textural and compositional analyses of the surface sediments underlying the coastal upwelling area off northwest Africa, we can conclude that modern coastal upwelling conditions are not recorded in the underlying sediments of the shelf region.

In comparison with other upwelling areas, such as off Oregon or Peru, where a shallow water upwelling facies is present in the sediments, the unique conditions off northwest Africa become evident. The shelf width and cross-shelf bathymetry (Fig. 11) are such that the poleward undercurrent is centered over the upper slope, only occasionally affecting the outer shelf (Barber and Smith, 1981; Mittelstaedt and Hamann, 1981). During upwelling, the onshore-offshore and alongshore compensation flows originating from undercurrent water masses create strong bottom currents which prevent sedimentation of fine-grained material on the shelf. Because of the steeper shelf gradient off Oregon and Peru, these currents are well above the sea floor in the water column, allowing fine-grained material to accumulate in a shallow water mud lens. Morphological and oceanographic conditions like those of Oregon and Peru today were present off northwest Africa during times of glacially lowered sea level, when the shelf was narrow and the upwelling system was more confined to the shelf edge. Under these conditions a shallow water mud lens could have formed on the upper slope, enabling better preservation and a more complete sequence of geological signals from which a glacial upwelling environment off northwest Africa can be interpreted.

Fine-grained sediments rich in opaline silica, fish debris or organic matter, indicative of present day upwelling, are trapped and preserved only locally in small nearshore basins on the shelf, such as in the Baie du Levrier of the inner Banc d'Arguin. Most of the upwelling-derived material is transported beyond the shelf edge

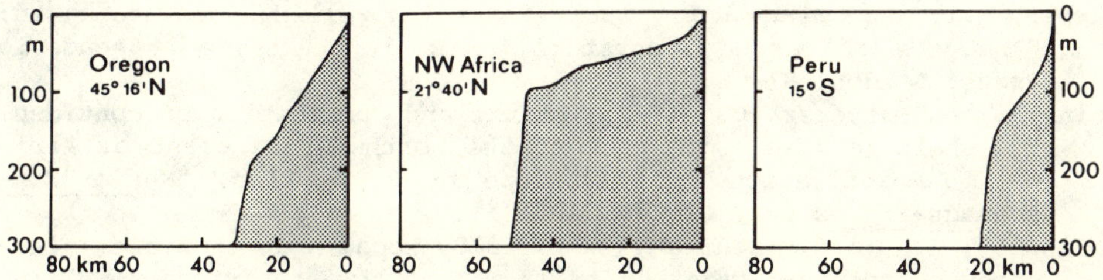

Fig. 11. Cross-shelf bathymetry of the upwelling regions off Oregon, Peru, and northwest Africa; modified from Barber and Smith, 1981. By permission from Analysis of Marine Ecosystems, Copyright (c) Academic Press, Inc.

and spread over the continental slope. During this transport the material has to pass diverse filter mechanisms which affect the various components in different ways. Much of the organic matter becomes oxidized, whereas most of the skeletal material is fragmented mechanically during bottom current transport on the shallow shelf or destroyed by dissolution during settling or at the sediment-water-interface. On the slope, the remaining material is transported farther downslope by mechanisms that we do not yet fully understand. Final accumulation takes place in a mid-slope mud lens at water depths below 1000 m. The upwelling "memory" of these slope sediments, however, is weak and in many cases ambiguous (i.e., may be attributed to factors other than upwelling).

From the point of view of the geologists who would like to reconstruct paleo-environments from fossil records, it has to be stated that the textural and compositional signals obtained from the modern slope sediments off northwest Africa are not clear enough to deduce an upwelling environment in the overlying water column without background information from modern oceanography.

ACKNOWLEDGEMENTS

I have to thank my colleagues G. Ganssen, P. Müller and M. Sarnthein for numerous valuable discussions and helpful advice. N. Exon, Canberra City, and C. Summerhayes, Houston, critically reviewed the manuscript and gave suggestions for its improvement. This work was partially supported by the Deutsche Forschungsgemeinschaft.

REFERENCES

Arthur, M.A., von Rad, U., Cornford, C., McCoy, F.W., and Sarnthein, M., 1979, Evolution and sedimentary history of the Cape Bojador continental margin, northwestern Africa, in: "Initial Reports DSDP," 47, Part I, U. von Rad, W.B.F. Ryan, et al., eds., U.S. Government Printing Office, Washington, 773-816.

Barber, R.T. and Smith, R.L., 1981, Coastal upwelling ecosystems, in: "Analysis of Marine Ecosystems," A.R. Longhurst, ed., Academic Press, London, 31-68.

Bein, A. and Fütterer, D., 1977, Texture and composition of continental shelf to rise sediments off the northwestern coast of Africa: An indication for downslope transportation, "Meteor"Forschungs-Ergebnisse, C27:46-74.

Codispoti, L.A. and Friedrich, G.E., 1978, Local and mesoscale influences on nutrient variability in the northwest African upwelling region near Cabo Corbeiro, Deep-Sea Research, 25:751-770.

Diester-Haass, L., 1977, Radiolarian/planktonic foraminiferal ratios in a coastal upwelling region, Journal of Foraminiferal Research, 7:26-33.

Diester-Haass, L., 1978, Sediments as indicators of upwelling, in: "Upwelling Ecosystems," R. Boje, and M. Tomczak, eds., Springer-Verlag, Berlin, 261-281.

Diester-Haass, L., and Müller, P.J., 1979, Processes influencing sand fraction composition and organic matter content in surface sediments off West Africa (12-19°N), "Meteor" Forschungs-Ergebnisse, C31:21-47.

Domain, F., 1977, Description de la sédimentation fine et des formations rocheuse du plateau continental ouest-africain de 17°N á 12°N, Association Sénégalese Étude Quaternaire Afrique Bulletin de Liaison, Sénégal, No. 50:11-22.

Fraga, F., 1974, Distribution des masses d'eau dans l'upwelling de Mauritanie, Téthys, 2:13-52.

Fütterer, D., 1980, Sedimentation am NW-afrikanischen Kontinentalrand: Quantitative Zusammensetzung und Verteilung der Siltfraktion in den Oberflächensedimenten, "Meteor"Forschungs-Ergebnisse, C33:15-60.

Koopman, B., 1979, "Saharastaub in den Sedimenten des subtropisch-tropischen Nordatlantik während der letzten 20,000 Jahre," Ph.D. Thesis, Fachbereich Mathematik-Naturwissenschaften, Universität Kiel, 107 pp.

Koopmann, B., 1981, Sedimentation von Saharastaub im subtropischen Nordatlantik während der letzten 25,000 Jahre, "Meteor"Forschungs-Ergebnisse, C35:23-59.

Koopman, B., Sarnthein, M., and Schrader, H.-J., 1978, Sedimentation influenced by upwelling in the subtropical Baie du Levrier (West Africa), in: "Upwelling Ecosystems," R. Boje and M. Tomczak, eds., Springer-Verlag, Berlin, 282-288.

Labracherie, M., 1980, Les Radiolaires temoins de l'évolution hydrologique depuis le dernier maximum glaciaire au large du Cap Blanc (Afrique du Nord-Ouest), Palaeogeography, Palaeoclimatology, Palaeoecology, 32:163-184.

Lutze, G.R., 1980, Depth distribution of benthic foraminifera on the continental margin off NW Africa, "Meteor"Forschungs-Ergebnisse, C32:31-80.

Mittelstaedt, E. and Hamann, I., 1981, The coastal circulation off Mauritania. Results of the upwelling experiment "Auftrieb 77" during January and February 1977, Deutsche Hydrographische Zeitschrift, 34:81-118.

Mittelstaedt, E., Pillsbury, D., and Smith, R.L., 1975, Flow patterns in the Northwest African upwelling area. Results of measurements along 21°40'N during February-April 1974, JOINT-I, Deutsche Hydrographische Zeitschrift, 28:145-167.

Müller, P.J. and Suess, E., 1979, Productivity, sedimentation rate, and sedimentary organic matter in the oceans - I. Organic carbon preservation, Deep-Sea Research, 26A:1347-1362.

Pelet, R., 1979, Géochimie organique des sédiments marins profonds au large de la Mauritanie et du Sénégal: Vue d'ensemble, "Géochimie Organique des Sédiments Marins Profonds - ORGON III, Mauritanie, Sénégal, Iles du Cap-Vert", M. Arnould and R. Pelet, eds., Editions du Centre National de la Recherche Scientifique, Paris, 425-441.

Piessens, P. and Chabot, A.G., 1977, Bathymetry and sediments of the Arguin Platform, Mauritania, West Africa, Mémoires de l'Institut géologique de l'Université de Louvain, 29:369-379.

Richert, P., 1975, "Die räumliche Verteilung und zeitliche Entwicklung des Phytoplanktons mit besonderer Berücksichtigung der Diatomeen im N.W.-afrikanischen Auftriebswassergebiet," Ph.D. Thesis, Fachbereich Mathematik-Naturwissenschaft, Universität Kiel, 140 pp.

Rust, U. and Wieneke, F., 1973, Bathymetrische und geomorphologische Bearbeitung von submarinen "Einschnitten" im Seegebiet vor Westafrika. Ein methodischer Versuch, Münchener Geographische Abhandlungen, 9:53-68.

Sarnthein, M., Pflaumann, U., Thiede, J., Erlenkeuser, H., Fütterer, D., Koopmann, B., Lange, H., and Seibold, E., 1982, Atmospheric and oceanic circulation patterns off Northwest Africa during the past 25 million years, in: "Geology of the West African Continental Margin," U. von Rad, K. Hinz, M. Sarnthein, and E. Seibold, eds., Springer-Verlag, Berlin, 545-604.

Schemainda, R., Nehring, D., and Schulz, S., 1975, Ozeanologische Untersuchungen zum Produktionspotential der nordwestafrikanischen Wasserauftriebsregion 1970-1973, Geodätische und Geophysikalische Veröffentlichungen, Reihe IV, Heft 16, Berlin, 88 pp.

Seibold, E. and Hinz, K., 1974, Continental slope construction and destruction, West Africa, in: "The Geology of Continental Margins", C.A. Burk and C. L. Drake, eds., Springer, Berlin, 176-196.

BIOGENIC SEDIMENTARY STRUCTURES IN A MODERN UPWELLING REGION: NORTHWEST AFRICAN CONTINENTAL MARGIN

Andreas Wetzel

Institut und Museum für Geologie und Paläontologie der Universität, Sigwartstrasse 10
D-7400 Tübingen, Federal Republic of Germany

ABSTRACT

Biogenic sedimentary structures reflect changes in sediment grain size, benthic food content, sedimentation rate, and oxygen content of the bottom water which are all related to climatic factors influencing upwelling intensity. In deposits accumulated during glacial times, when upwelling intensity was stronger, a higher trace fossil diversity and an increasing differentiation in bathymetric zonation of trace fossil associations is recognized. In these trace fossil associations biogenic structures have been grouped according to their mode of co-occurrence, depth of penetration, and behavioral patterns. The depth of penetration controls the preservational potential of the trace fossils. In order to evaluate the preservation process, a model has been formulated that divides the bioturbated zone into five levels. The preservation potential increases with increasing depth of penetration. The distribution of biogenic structures in sediments off northwest Africa can be considered "uniformitarian", because the ichnogenera are similar to those known from ancient bathyal and abyssal ichnocoenoses.

INTRODUCTION

Biogenic sedimentary structures are autochthonous indicators of environmental conditions. In many cases, a biotope can be better characterized by its biogenic sedimentary structures than by other organic constituents. Trace fossils primarily reflect adaptations to the biotope and are less influenced by the taxonomic position of the organisms producing them (Hertweck, 1972).

For a long time biogenic structures have been studied only in fossil examples. Based on a classification of organism behavior in relation to the biotope, Seilacher (1953; 1967) suggested a relative depth zonation for trace fossil communities (=ichnofacies). Starting with shallow water biotopes, however, additional observations on animal trace communities were made in modern marine environments by Richter (1924; 1926) and Schäfer (1956). Later, such studies were expanded to include deep-sea sediments (Chamberlain, 1975; Berger, Ekdale, and Bryant, 1979; Ekdale, 1980), but they were largely limited to a description of the trace fossil communities and preservation processes. Little is known as yet about ichnofaunal responses to environmental changes such as food supply, sediment grain size, or oxygen content of bottom water in the deep-sea biotope. For example, Kitchell et al. (1978) and Kitchell (1979) have studied surface trails in different deep-sea regions, and Wetzel (1979) has correlated short-term variations in C-org content, sediment grain size, and oxygen content of bottom water with changes in the ichnofauna.

In this study I will attempt to characterize an upwelling biotope, the northwest African continental margin, by its trace fossil content. In order to facilitate application of these characteristics from the recent environment to fossil sediments, I shall largely use the ichnological nomenclature (Werner and Wetzel, in press).

MATERIALS AND METHODS

Sedimentary structures have been investigated and described from 86 cores of 61 sampling stations (Fig. 1). This material was retrieved during several cruises by RV "Meteor" and RV "Valdivia" (Seibold, 1972; Seibold and Hinz, 1976).

The sediment cores, sampled mainly by box or gravity corer, measure between 10 and 50 cm in diameter and up to 11 m in length. Sedimentary structures were examined by means of x-ray radiographs of the entire core cross section using 8-mm thick sediment slices. For such investigations this method is far superior to visual assessment because it reveals significantly more biogenic sedimentary structures. In order to obtain a three-dimensional representation of certain distinct structures, additional x-ray radiographs were taken in serial sections or various orientations including stereo-pair sections. Methods are described in more detail in Wetzel (1979).

The different trace fossils were identified by means of typical cross sections since only small parts of a burrow can be observed in the x-ray radiographs. Because such classifications may be influenced by subjective evaluations, the observed trace fossils have been defined only at the level of ichnogenera as described by Häntzschel (1965; 1975). Sometimes a subdivsion of one ichnogenus into different types has been possible (Wetzel, 1981).

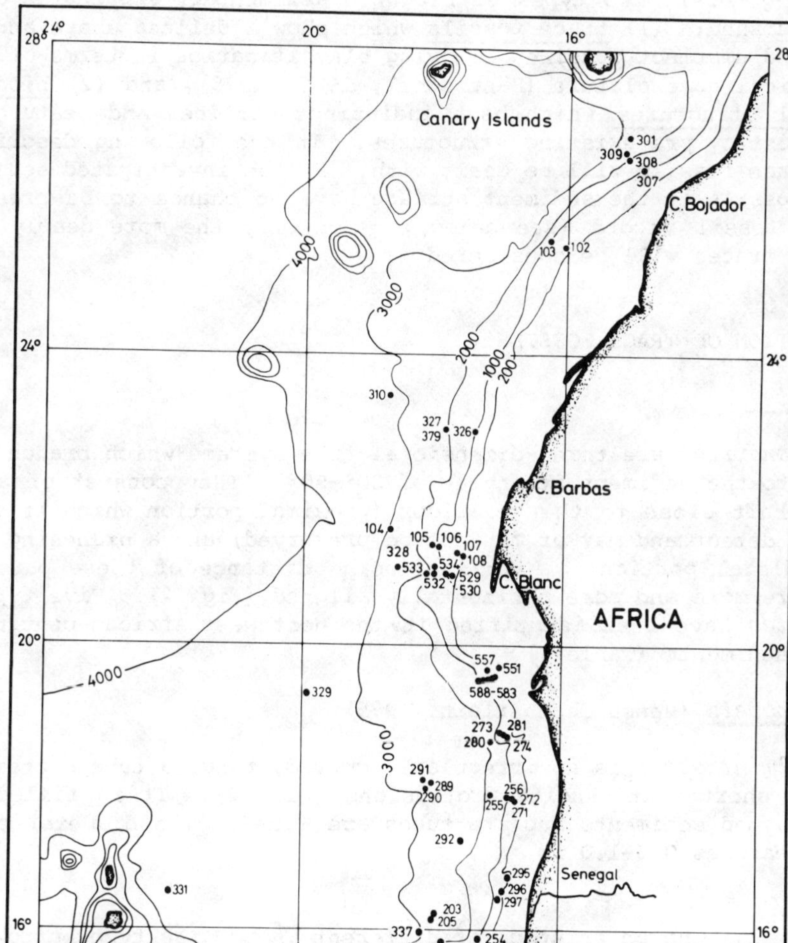

Fig. 1. Core locations (additional details see Wetzel, 1981).

TYPES OF DOMINANT BIOGENIC SEDIMENTARY STRUCTURES

In general, two types of biogenic sedimentary structures can be distinguished: (1) trace fossils which show a defined shape and have sharp and distinct outlines allowing classification in terms of paleontological nomenclature (Häntzschel, 1965; 1975), and (2) biodeformational structures which have indistinct outlines and features and which destroy pre-existing structures. In the following description only trace fossils will be dealt with. In the investigated sediments trace fossils at the sediment surface have no chance to be preserved in the fossil record (see below), hence only the more deeply penetrating traces will be considered.

DESCRIPTION OF TRACE FOSSILS

Chondrites

Chondrites are three-dimensional tube systems which branch downward into the sediment at angles of 30°-60°. They consist of a vertical shaft close to the sea floor (proximal portion which is difficult to detect and may or may not be preserved) and a branching lower part (distal portion). With increasing distance of these parts the tubes are more and more horizontally aligned (Fig. 2). Five types of *Chondrites* have been identified in the northwest African continental margin sediments (Table 1).

Helminthopsis (sensu Chamberlain, 1975)

Helminthopsis is an irregularly curved, tangled tube system with several short blind-ending protrusions (Fig. 2). It is filled with fine-grained sediments and its tubes are lined. The diameter of the tubes measures 0.5-1.0 mm.

Table 1. Characteristics of different *Chondrites* type burrows

Type	Diameter mm	Lining	Tube Filling	Shape
A	3	Absent	Coarse-grained	Plant-like dendritic
B	2	Absent	Fine-grained	Plant-like dendritic
C	1	Present	Fine-grained	Various
D	1	Present	(Fine-grained) empty	Various
E	1	Present	Fine-grained	Tassel-like

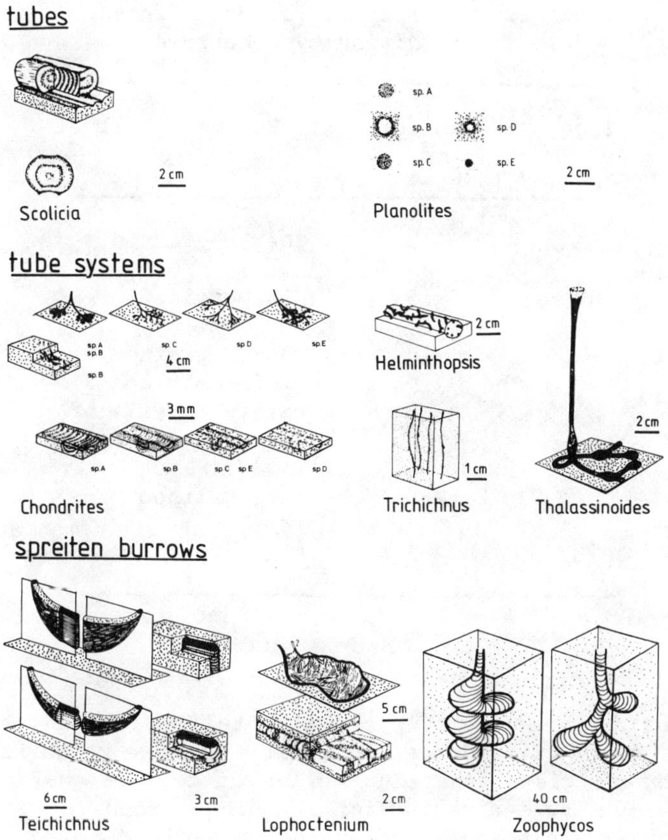

Fig. 2. Trace fossil types in cores off northwest Africa.

Lophoctenium

This trace fossil resembles bunches of closely spaced inwardly bent "twigs" with comb-like branches that join to form a main axis. In cross section it typically shows an irregular back-fill of sediment with crescentic laminae alternating with homogeneous parts. This type of back-fill is in contrast to the more regular structures of *Zoophycos* (see below). Sometimes a "halo" is observed surrounding the main axis (Fig. 2). The height of the burrow measures 1-2 cm.

Planolites

Planolites is normally unbranched, straight or gently curved, cylindrical and more or less horizontally oriented. The back-fill sediment is structureless (Fig. 2). Five types of *Planolites* have been distinguished by their sediment fills and their tube diameters (Table 2).

Table 2. Characteristics of different *Planolites* type burrows

Type	Diameter mm	Sediment Fill
A	8-15	Coarse-grained
B	6-15	Fine-grained
C	6-15	Coarse-grained, preferably containing shells of micro-organisms
D	2-5	Fine-grained
E	2-5	Like type C

Scolicia

Scolicia is a more or less horizontally oriented, bilaterally symmetrical burrow with typical internal structure which is formed by pinnated, "watch glass"-shaped lamellae of coarser grained material than the surrounding sediment (Fig. 2). This trace is very probably dug by irregular echinoids (Bromley and Asgaard, 1975).

Teichichnus

This trace fossil is a spreiten-burrow, formed by the vertical upward shift of a flat U-tube. Its configuration resembles stacked roof gutters with a tube on top (Fig. 2).

Thalassinoides

Thalassinoides is composed of cylindrical tubes that branch to form a three-dimensional horizontal network connected to the sediment surface by more or less vertical, often very extended, shafts. Typical swellings at points of branching or elsewhere are observed (Fig. 2). Fill structures result from active filling by the burrowing organism or collapse of the burrow walls. The tubes are 5-20 mm in diameter.

Trichichnus

Trichichnus is a thread-like and rarely branching cylindrical burrow, more or less vertically oriented, >20 cm in length, and 0.1-

1.0 mm in diameter. The tubes usually show a distinct lining and are empty or contain a structureless sediment fill (Fig. 2).

Zoophycos

Zoophycos consists of horizontally to obliquely oriented spreiten that surround a vertically oriented central axis in distinct levels or coilings (Fig. 2). This burrow shows a regular back-fill structure that appears as crescentic en-echelons in vertical sections. The individual spreiten structures are up to 1 cm in height. The organisms generating these structures are able to burrow as deep as 1 m. The total diameter of an entire burrow system can measure up to 1 m. *Zoophycos* trace fossils found off northwestern Africa have been described in detail by Wetzel and Werner (1981).

WATER DEPTH ZONATION OF TRACE FOSSILS

Trace fossils are useful to reconstruct the relative water depth of ancient sediments (Seilacher, 1967). Variations in water depth influence only the hydrostatic pressure which is ecologically of minor importance to organisms on the sea floor below shelf depth (Tait, 1971). Ecologically more important parameters such as sedimentation rate, food and texture are often related to distance from shore, and are thereby indirectly related to water depth (Thiel, 1975).

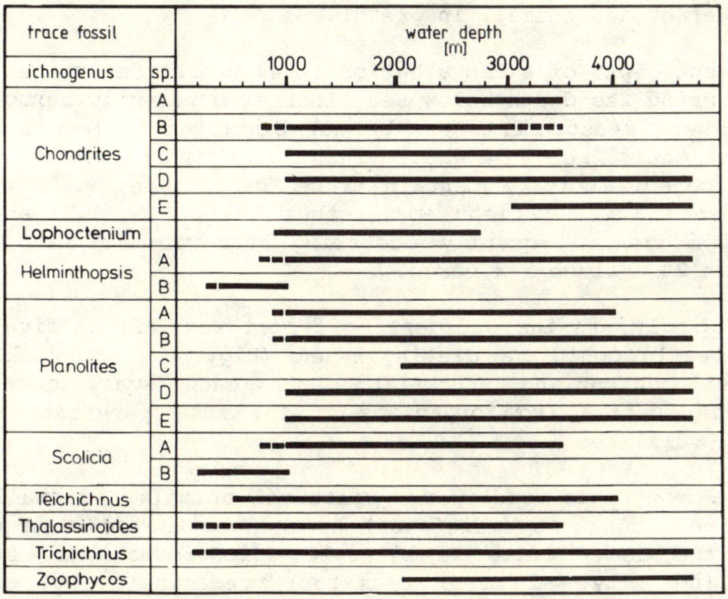

Fig. 3. Bathymetric distribution of trace fossils in sediments off northwest Africa.

Figure 3 shows the water depth distribution of the trace fossils found off northwest Africa. It illustrates well the changing composition of the ichnofauna with water depth. It should be kept in mind, however, that this distribution is based on observations from only one distinct area. Probably different depth zonations of trace fossils will be found in other areas. The conspicuous sharp boundaries of the trace fossil occurrences may result from biased sampling of cores.

RELATIONSHIPS BETWEEN DIFFERENT BIOGENIC STRUCTURES

In order to adequately interpret the trace fossil content in a sediment section, the relationship between the different burrow types has to be evaluated. The presence of a specific biogenic structure depends on a selection process which keeps it from being destroyed by other burrowing animals, rather than upon the dominance of a specific digging organism during its lifetime. Deeper dug burrows displace pre-existing structures and have therefore a higher probability of being preserved in the fossil record. This selection process can best be appreciated when subdividing the burrowed zone into bioturbational storeys or levels. Under steady-state conditions, a sequence of storeys moves gradually upwards in correspondence with the build-up of the sediment surface. In this model then, each bioturbational level contains trace fossils made by organisms able to penetrate to the same depth. Thus, the preservation rate of a certain trace fossil type is controlled by the total burrowing activity within deeper storeys (Werner and Wetzel, in press).

Sediment depth of a bioturbation level can be determined directly by measuring its depth below sea floor from recently burrowed sediments. These measurements can be substantiated in fossil deposits by means of their "relative age." This is an indication of the preservation potential for a certain trace fossil type, i.e., its penetration depth in comparison with other burrows. The penetration depths of modern, frequently occurring burrows were thus measured directly in 24 surface box cores.

Accordingly, in the sediments off northwest Africa five bioturbational levels could be distinguished (Fig. 4). They differ in trace fossil content and penetration depth; each level, however, contains trace fossils from organisms of the same penetration depth. The five levels are:

I. Homogeneous top layer: Epifaunal animals and small infaunal organisms (meiofauna) homogenize the sediment close to the sea floor. Small-scale, biodeformational structure (normally <<1 cm in diameter) predominate, but artifacts imposed during retrieving of the core or in preparation of x-ray radiograph slices cannot be excluded.

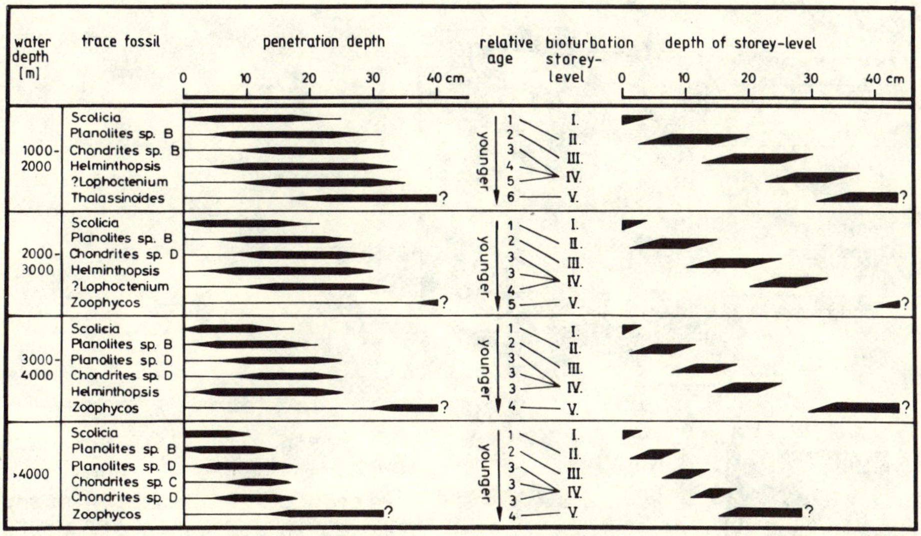

Fig. 4. Arrangement of typical trace fossil assemblages in bioturbational levels in sediments off northwest Africa. These levels are derived from the observed penetration depth of the burrows in surface cores, and the "relative ages" (controlled by burrowing depth) belong to the same bioturbational level. The following levels can be distinguished; I. homogeneous, containing surface trails and small-scale, biodeformational structures; II. *Scolicia* level; III. *Planolites* level; IV. *Chondrites-Helminthopsis-Lophoctenium* level; and V. *Zoophycos* level.

II. *Scolicia* level: Dominated by *Scolicia*, sometimes associated with vertically oriented tubes and/or U-shaped burrows.

III. *Planolites* level: Contains the trace fossils of *Planolites* types A, B, and C.

IV. *Chondrites-Helminthopsis-Lophoctenium* level: Defined by these trace fossils and additionally by *Thalassinoides* or *Planolites*, type D and E, or *Teichichnus* which may or may not be present.

V. *Zoophycos* level: Defined by the dominance of this trace fossil which sometimes may be associated with *Trichichnus*.

As a general feature, successive bioturbational levels become telescoped and condensed with increasing water depth as shown in Fig. 4. This is probably due to lower benthic food supply and sedimentation rates (Wetzel, 1980). The five bioturbation levels as characterized above are schematically illustrated in Fig. 5(A). However, five bioturbation levels may not always be present because the burrowing activity of the benthic animals depends on ever-changing environmental conditions as well.

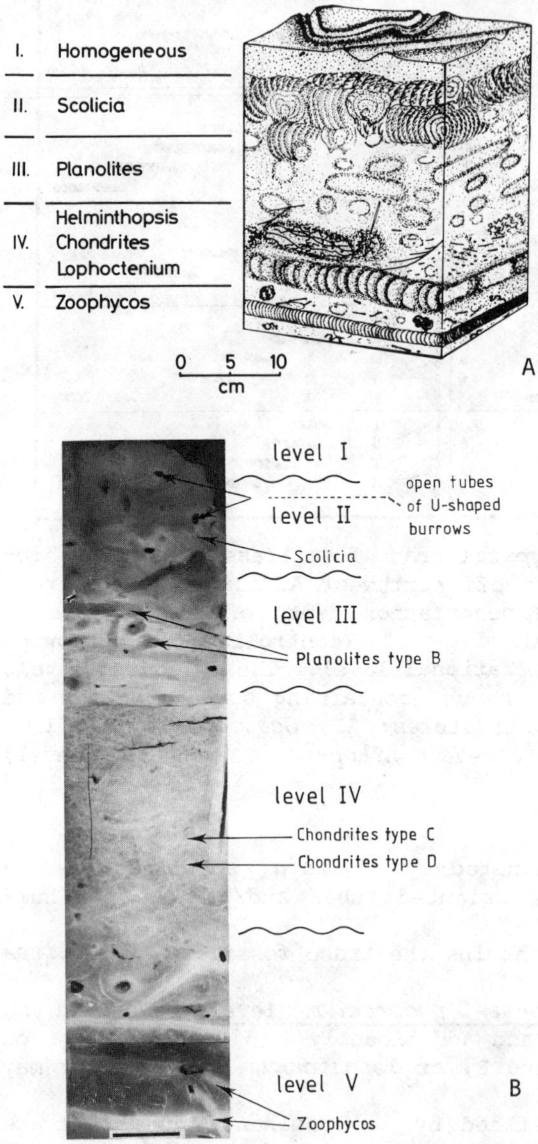

Fig. 5. (A) Schematic diagram of the arrangement of biogenic sedimentary structures in sediments off northwest Africa. The level zonation is shown on the left side, the ichnocoenose shown here is typical for water depths between 2000 and 3500 m. (B) X-ray radiography (negative) of core 211 from surface to 70 cm in the sediment (scale = 6 cm).

ENVIRONMENTAL CONDITIONS AND FORMATION OF BIOGENIC STRUCTURES

The composition of the trace fossil assemblages in a sediment volume reflects the burrowing activity of the infaunal organisms. Their behavior is influenced by grain-size distribution of the deposits, oxygen content within the bottom waters, benthic food supply and, hence, the sediment accumulation rates (Müller and Suess, 1979).

Variations in Grain-Size Composition

The grain sizes of soft surface sediments strongly influence their physical properties. Therefore, most of the infaunal organisms are adapted to a particular grain-size spectrum. Consequently, variations of grain size of a sediment are reflected by its ichnofauna (Fig. 6). Furthermore, the tolerance of the infauna to variations in grain size decreases with water depth. In the deep sea, coarser sediments are less intensely bioturbated than in upper slope regions. For instance, turbidite deposits are completely reworked by infauna down to water depths of 700-1000 m, while trace fossils are sparsely distributed in coarse-grained sequences of turbidites below 2000 m of depth (Wetzel, 1981).

Oxygen Content

The oxygen content in the respiration water for the burrowing organisms is often clearly reflected by the occurring trace fossil types (Table 3). Respiratory water is equivalent to overlying bottom

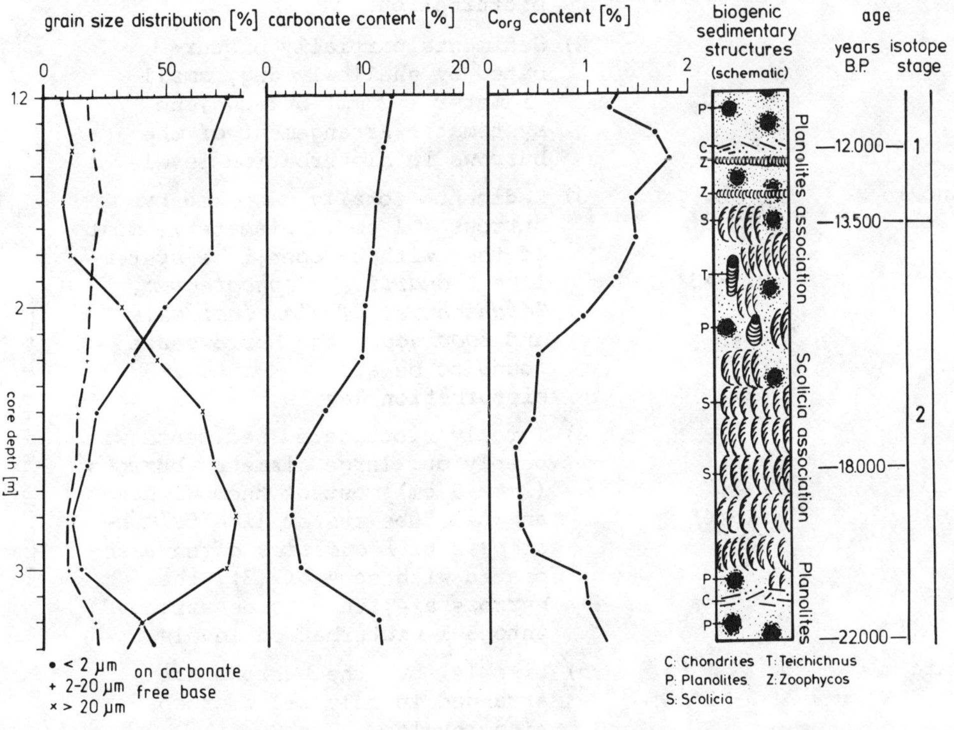

Fig. 6. Dependence of ichnofauna on grain size; *Scolicia* and *Planolites* association, core 347, 120-330 cm in sediment. The core was taken off the mouth of the Senegal River in 2576 m water depth. The *Scolicia* association is typical for high content of fraction >20 µm (on carbonate-free base) and low carbonate and C-org contents.

water for most animals. However, deeply digging organisms or such without an open tube system, e.g., *Planolites*, may also be affected by negative Eh-values within the sediment. Nevertheless, the following data refer to the oxygen content of the bottom water.

Table 3. Ichnofauna in relation to oxygen content of the bottom water and benthic food content off NW Africa

Oxygen Content		Biogenic Sedimentary Structures: Benthic Food Content	
Relative	Absolute ($m\ell$ O_2/ℓ)	C-org <2% → C-org >2%	
Euxinic	<0.1	1) Laminated sediments without bioturbation.	
	~0.2	2) Sediments partially bioturbated by shallowly dug, small diameter (∿2 mm) burrows, no systematic arrangement of the burrows in bioturbation levels.	
Anoxic	0.2->3*	3) Sediments totally reworked by burrows (∿1 cm in diameter), most of them with an open tube system like *Chondrites*, *Lophoctenium*, *Teichichnus*, *Thalassinoides*, and *Zoophycos*, the burrows are found to be arranged into 2-5 bioturbation levels.	Biodeformational structures
	1->4*	4) Totally bioturbated sediments with deeply dug large diameter burrows (2 mm-5 cm) most of them without an open tube system like *Helminthopsis* or *Planolites* often associated with some of (3); the burrows are found to be arranged into 5-1 bioturbation levels.	
Oxic	3->5*	5) Like (4) but the burrows are arranged in only 3-1 bioturbation levels.	

* Approximate values for oxygen content from literature (Weichart, 1974) in relation to the distribution of ichnofauna as found in typical trace fossil associations.

In euxinic environments (<0.1 mℓ O_2/ℓ) the sediments are usually not reworked by macro-organisms. In low oxygen environments (∿0.2 mℓ O_2/ℓ) small diameter burrows are normally found in partly undisturbed (laminated) sediments. Unfortunately, above this level oxygen concentration has not yet been investigated together with biogenic sedimentary structures. Thus, Table 3 is based on O_2 data from the literature (Weichart, 1974) and observations on the first occurrence of iron sulfides in the core sections. Using these data, it can be assumed that off northwest Africa bioturbation like type (4) in Table 3 occurs above the 1 - >4 mℓ level (O_2/ℓ) and type (5) above 3 - >5 mℓ (O_2/ℓ).

Nutrient Content

Because the benthic food content of sediments cannot be precisely determined, the organic matter content is used as a rough index for the available nutrients. The amount of C-org buried in the sediment is controlled by primary production rate at the sea surface and the sediment accumulation rate (Müller and Suess, 1979). In order to evaluate the influence of different nutrient contents, sediments with different C-org content and bioturbational features are compared (Table 4). Off northwest Africa two general types of anoxic sediments can be distinguished with respect to their biogenic structures.

(1) In deposits with >2% C-org large-scale, biodeformational structures --normally >1-2 cm in diameter-- are found which represent a low degree of behavioral specialization. They do not show an obvious search pattern and feeding does not result in an effective grain size sorting which is normally indicated by sharp outlines of the burrow or a fecal string. Furthermore, large-scale, biodeformational structures are not restricted to any one of the typical bioturbation levels. These observations are supported by the fact that large-scale, biodeformational structures are found only in sediments with a

Table 4. Relation between organic matter content and biogenic sedimentary structures

C-org Content	Biogenic Sedimentary Structures
>2%	Biodeformational structures
<2%	Trace fossils
> 0.3% (predominantly anoxic)	>3 bioturbational levels
< 0.2% (predominantly oxic)	1 bioturbational level

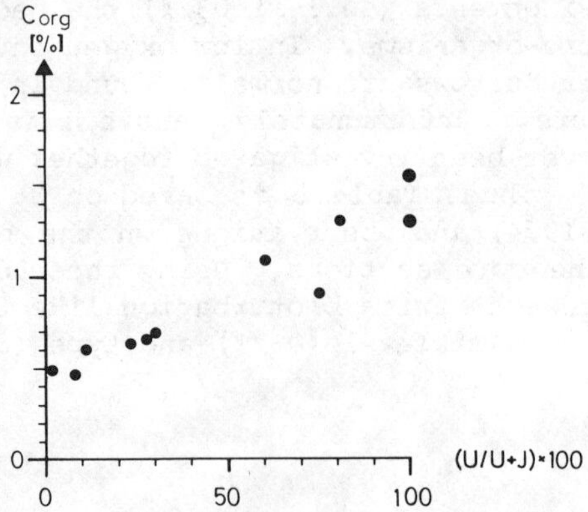

Fig. 7. Organic carbon content of sediments versus ratio between U- and J-shaped *Zoophycos* types. Ratio (U/U+J) x 100 determined in 12 core sections each containing about 10 feeding spreiten with marginal tubes; C-org are mean values measured on these core sections.

high organic matter content that makes behavioral specialization superfluous.

(2) In anoxic sediments off northwest Africa with <2% C-org, 3-5 bioturbational levels are typically present. The anoxic chemical environment of these deposits is indicated by the presence of iron sulfides at 10-100 cm below the sediment surface. In contrast, trace fossils in strongly oxic sediments are arranged in only one bioturbational layer. Sediments with such features have been cored in the central Pacific Ocean and are probably typical for most abyssal pelagic clays. They contain about 0.3% C-org in surficial sediments and less than 0.1% C-org at depth (Müller, 1975) but carry the same inventory of trace fossils as off northwest Africa. There is no subdivision into different bioturbational levels, however, and they seem to be condensed into one bioturbated zone. This can be interpreted as a reaction of the infauna nutrients being limited to a comparatively small surface sediment layer. In respect to the trace fossil diversity, the same trend has been found as reported by Kitchell et al. (1978) for surface trails; with increasing C-org accumulation rates the diversity of the trace fossils also increases (see below). However, for trace fossils dug within the sediment an upper nutrient level has been found above which only biodeformational structures have been formed.

CORRELATION OF SEDIMENTS BY ICHNOFAUNA

Off northwest Africa the same sediment volume is repeatedly reworked by organisms. The bioturbational patterns and the sensitivity of the infauna to environmental changes make it possible to establish an ecologic stratigraphy based on biogenic structures. For this purpose a total of six different associations of biogenic sedimentary structures have been defined. They refer to different behavior programs of the infaunal animals, e.g., digging of tubes, tube systems or spreiten, or physical adaptations as digging organs controlling penetration depth. Thus, the associations are related to the five bioturbational levels outlined above where level IV has been divided into two associations in respect to the different behavior programs found in this level. Using Seilacher's (1967) ichnofacies scheme the sediments off northwest Africa as a whole belong to the *Zoophycos* facies, but have been subdivided for a more detailed ecological analysis into six trace fossil associations. The three-dimensional distribution of the ichnofaunal associations is given in Fig. 8.

The biodeformational association is dominated by large-scale, biodeformational structures. Off northwest Africa in water depths greater than 500 m they characterize a biotope rich in benthic food (>2% C-org; Table 3). At two sites the biodeformational association has been observed: (1) in areas of coastal upwelling off Cape Blanc and Cape Barbas (Fig. 8); here they are restricted to coarse sediments accumulated during the last glacial period, when sea level was lower and upwelling stronger than today (Diester-Haass, Schrader and Thiede, 1973; Thiede, 1977), (2) at sites located off the mouths of the Senegal River, Gambia River, and Casamance River, where, biodeformational structures occur only in the finer-grained sediments accumulated during humid interglacial periods when the input of fluvial sediments was stronger. Large-scale, biodeformational structures in both cases of more intense upwelling and increased river input cannot be distinguished directly from one another. However, the trace fossils occurring below and above may be useful indicators to distinguish upwelling (*Scolicia* association) and non-upwelling situations (*Planolites* association) in areas similar to the northwest African continental margin.

The *Planolites* association is composed of *Planolites* type A, B and C between 1000 and 2000 m water depth; at greater depths *Planolites* type D and E occur with increasing frequency. *Helminthopsis* is found associated with these different *Planolites* types. *Planolites* is common in fine-grained sediments having 40-50% clay-sized material (<2 μm) on a carbonate-free basis. This facies was preferentially deposited close to the continent during the warmer climatic periods (interglacials) and in the deep sea throughout Pleistocene times. The *Planolites* association grades into the *Scolicia* or vertical spreiten association, when the sediments become coarser grained (Figs. 6 and 8).

The *Scolicia* association which is characterized by the dominance of the trace fossil *Scolicia* also contains a few *Planolites* and *Helminthopsis* burrows. It is preferably found in coarse-grained sedi-

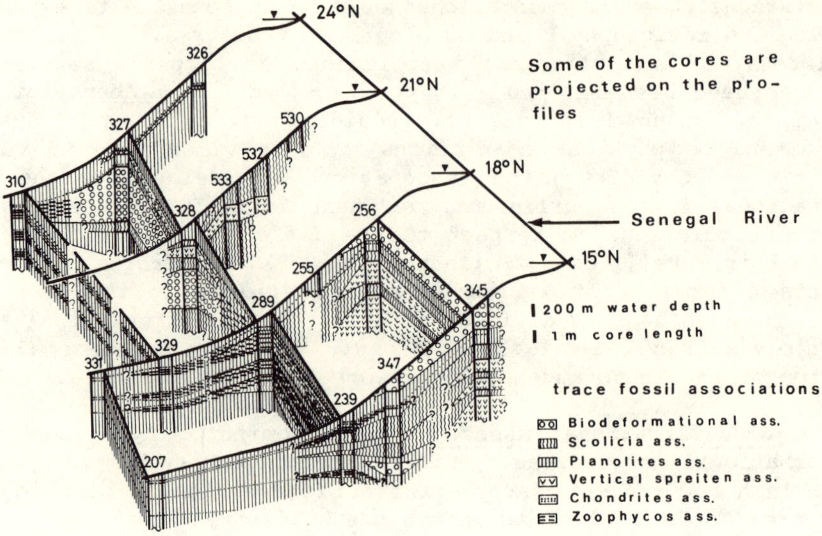

Fig. 8. Correlation of cores based on trace fossil associations.

ments having >50% material >20 μm (on a carbonate free basis) at depth >1000 m. The sediment texture results from the high input of terrigeneous material of up to 90% eolian silt (Sarnthein and Koopmann, 1980). This leads to a dilution and thus to relative minima in organic carbon and carbonate contents (Fig. 6). More deeply penetrating burrows are sparse in the *Scolicia* association because the mechanical properties of coarser sediments prevent the construction of feeding burrows by animals otherwise adapted to softer and fine-grained sediments. The *Scolicia* association occurs between 14° and 23°N at depths >1000 m and at distances of up to 250 km offshore (Fig. 8). From S to N the frequency of its occurrence seems to increase; between 14° and 19°N it is found only in sediments deposited during glacial periods, while north of 19°N it predominates over long sections because eolian dust was introduced during warm as well as cold periods.

The vertical spreiten association is characterized by the dominance of obliquely oriented *Lophoctenium* and vertical *Teichichnus* and related forms. It is found at water depths of between 500 and 3000 m with a maximum at about 1000 m (Fig. 9). Sediments accumulated during cold periods, mostly the coarser-grained material close to the continent, normally contain the vertical spreiten association. The ecological interpretation depends on burrow morphologies. The organisms show highly specialized behavior patterns including exploration patterns, grain size sorting, and open-tube systems, presumably for respiration, but penetrate the sediment only to comparatively shallow depths. In this case, oxygen content of bottom waters seems to be the limiting factor because these sediments otherwise contain sufficient food (>1% C-org). Possibly, during glacial periods the oxygen minimum layer, today located at ca. 200 to ca. 800 m of water depth (Weichart, 1974), was thicker or displaced to a greater depth as a result of lowering of sea level, changing water circulation or higher organic carbon input (Seibold et al., 1976).

The *Chondrites* association is preferably found in sediments accumulated during times of climatic change, i.e., at the end and the onset of cold phases. Farther offshore, however, it extends over the entire duration of the cold climatic periods. Therefore, depending on water depth, the *Chondrites* association occurs close to either the *Zoophycos*, the vertical spreiten, or the *Scolicia* associations. The organisms responsible for the *Chondrites* association fail to rework the sediment completely.

Off northwest Africa *Zoophycos* is normally found below 2000 m with maximum frequencies between 3000 and 4000 m of depth (Fig. 9).

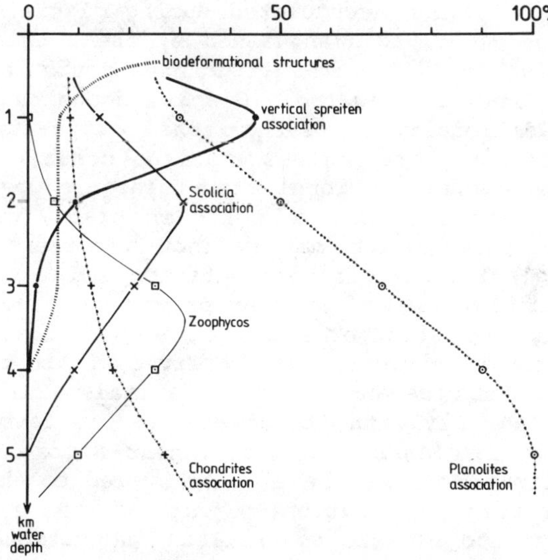

Fig. 9. Depth distribution of trace fossil associations off northwest Africa. Based on length of bioturbated interval by each trace fossil association in relation to the total core length; mean values for 1000 m water depth intervals below 500 m were calculated and connected to yield a frequency diagram. This represents a quantitative approach to map the optimal distribution of each trace fossil association. Sometimes the total abundance of structures is more than 100% within one interval because *Chondrites* and *Zoophycos* animals fail to rework a sediment section completely.

The organisms producing *Zoophycos* burrows prefer clayey-silty anoxic sediments with 0.3-1.8% C-org; they also fail to rework the sediment entirely. Additional information about the ecological conditions is provided by the shape of the dwelling tube. U-shaped tunnels, assumed to be typical for low oxygen environments, allow a more continuous and more effective supply of oxygenated water for respiration than do J-shaped tunnels with blind ends. These are assumed to be typical for more oxic environments. The frequency of U- and J-type burrows seems also to be related to the C-org content of the sediments (Fig. 7; for more details see Wetzel and Werner, 1981). In 80-90% of the examples, *Zoophycos* was found in sediments accumulated during glacial periods. In reality, however, taking into account penetration depth and sedimentation rate, the *Zoophycos* organism lived 4000-20000 years later. Such considerations show that the occurrence of *Zoophycos* is controlled more by sediment facies than by other factors (Wetzel, 1981).

RECORD OF ENVIRONMENTAL CHANGES OFF NORTHWEST AFRICA

The temporal and spatial distribution of biogenic structures is shown in Fig. 8; sediments accumulated during glacial and interglacial periods can be clearly distinguished by their trace fossil content. Farther offshore, *Planolites* structures predominate throughout the sediment sections investigated. Glacial deposits are marked by *Chondrites* and additionally, closer to the continent, by *Zoophycos* and *Scolicia*. Vertical spreiten associations occur off Cape Barbas and Cape Blanc and biodeformational structures are observed as well. In general, glacial sediments show a higher trace fossil diversity than interglacial ones, which may reflect important environmental changes. During glacial periods sedimentation and C-org-accumulation rates were higher because of higher primary productivity due to stronger upwelling and stronger input of eolian dust. Thus, more "nutritious" organic matter could be preserved within the sediment to be used later by *Chondrites* and *Zoophycos* animals. In addition, the deposition of eolian silt with its coarse texture favored the *Scolicia* association, while closer to the continent a more vertically extended oxygen minimum zone may have possibly led to the formation of the vertical spreiten association. Most of these environmental changes are related to climate controlled fluctuations in upwelling intensity (Diester-Haass, this volume; Thiede, 1977).

For a more objective description, it is necessary to quantify the ichnofaunal data. Based on percentages of core sections bioturbated by each trace fossil association, mean values for 1000-m water depth intervals below 500 m were calculated in relation to the total core length. To emphasize the tendencies of the distribution patterns, these mean values have been connected by lines (Fig. 9). A typical offshore succession of trace fossil associations can be defined based on their optimal range of occurrence reflecting the envi-

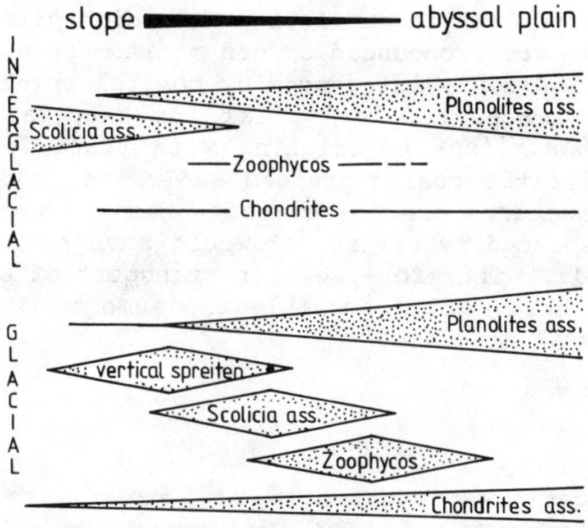

Fig. 10. Typical offshore successions of trace fossil associations in sediments of interglacial and glacial periods as found off northwest Africa. Biodeformational structures are not included.

ronmental factors as discussed above. According to the varying trace fossil diversity in sediments of interglacial and glacial periods, it is also possible ecologically to distinguish offshore profiles for these periods (Fig. 10). The successions shown in Figs. 9 and 10 (glacial) are typical for deposits influenced by upwelling. In upwelling areas which are not situated off sandy deserts, the *Scolicia* association should be absent.

CONCLUSIONS

(1) The distribution of trace fossils in recent sediments can be considered "uniformitarian" because the ichnogenera are similar to those known from ancient bathyal and abyssal ichnocoenoses. (2) Trace fossils penetrate to different depths within the sediment. A model based on different bioturbational levels within a community of burrowers helps to estimate the potential of ichnocoenoses for becoming preserved in the fossil record. Each level represents certain behavioral and ecological characteristics. (3) Trace fossil associations established during this study correspond to the different bioturbation levels. They are also sensitive to variations in grain size, sedimentation rate, nutrient and oxygen content at the sea floor. (4) Since each trace fossil association shows an optimal range of occurrence, successions perpendicular to the coast can be recognized. Sediments accumulated during glacial periods show higher trace fossil diversity and differentiation into 4-5 associations compared with only 2-3 associations in interglacial sediments. (5) The higher diversity during cold periods may be caused by increased rates

of sedimentation and C-org accumulation from high primary productivity, by a possibly more pronounced oxygen minimum zone, and by coarser sediments. (6) Sediments influenced by coastal upwelling off northwest Africa show trace fossils typical for deep-water environments (*Zoophycos* and *Chondrites*) co-existing with trace fossils typically found in non-turbiditic coarse-grained sediments. Occasionally biodeformational structures are associated. Had the coarse-grained material been transported by rivers, it would have been retained at or close to the shelf. Therefore, eolian transport of silt by offshore winds is assumed here, quite possibly the same wind system that is also forcing coastal upwelling.

ACKNOWLEDGEMENTS

The investigations presented in this paper were carried out in the Geological Institute of Kiel University. E. Seibold (now at Bonn) suggested this study; F. Werner (Kiel) advised and contributed valuable help during the investigations as well as stimulating discussions; A. Seilacher (Tübingen) reviewed the manuscript. Financial support was made available by the Deutsche Forschungsgemeinschaft. All these contributions are gratefully acknowledged.

REFERENCES

Berger, W.H., Ekdale, A.A. and Bryant, P.F., 1979, Selective preservation of burrows in deep-sea carbonates, Marine Geology, 32: 205-230.

Bromley, R.G. and Asgaard, U., 1975, Sediment structures produced by a spatangoid echinoid: a problem of preservation, Bulletin of the Geological Society of Denmark, 24:261-281.

Chamberlain, C.K., 1975, Trace fossils in DSDP cores from the Pacific, Journal of Paleontology, 49:1074-1096.

Diester-Haass, L., Schrader, H.J. and Thiede, J., 1973, Sedimentological and paleoclimatological investigations of two pelagic ooze cores of Cape Barbas, North-West Africa, "Meteor"Forschungs-Ergebnisse, C16:19-66.

Ekdale, A.A., 1980, Graphoglyptid burrows in modern deep-sea sediment, Science, 207:304-306.

Häntzschel, W., 1965, "Vestigia Invertebratorum et Problematica", in: "Fossilium Catalogus I: Animalia pars," 108, W. Junk, s'-Gravenhage, 142 pp.

Häntzschel, W., 1975, "Trace fossils and problematica", 2nd ed., in: "Treatise on Invertebrate Paleontology, Part W," Miscellanea, Supplement 1, C. Teichert, ed., The Geological Society of America, Boulder, Colorado and University of Kansas, Lawrence, Kansas, 269 pp.

Hertweck, G., 1972, Distribution and environmental significance of lebensspuren and *in situ* skeletal remains, Senckenbergiana maritima, 4:125-167.

Kitchell, J.A., 1979, Deep-sea foraging pathways: an analysis of randomness and resource exploitation, Paleobiology, 5:107-125.
Kitchell, J.A., Kitchell, J.F., Johnson, G.L., and Hunkins, K.L., 1978, Abyssal traces and megafauna: comparison of productivity, diversity and density in the Arctic and Antarctic, Paleobiology, 4:171-180.
Müller, P.J., 1975, Diagenese stickstoffhaltiger organischer Substanzen in oxischen und anoxischen marinen Sedimenten, "Meteor" Forschungs-Ergebnisse, C22:1-60.
Müller, P.J. and Suess, E., 1979, Productivity, sedimentation rate and sedimentary organic matter in the oceans, I. Organic carbon preservation, Deep-Sea Research, 26:1347-1362.
Richter, R., 1924, Flachseebeobachtungen zur Paläontologie und Geologie, VII-XI, Senckenbergiana, 6:119-165.
Richter, R., 1926, Flachseebeobachtungen zur Paläontologie und Geologie, XII-XIV, Senckenbergiana, 8:200-224.
Sarnthein, M. and Koopmann, B., 1980, Late Quaternary deep-sea record of Northwest African dust supply and wind circulation, in: "Paleoecology of Africa 12," M. Sarnthein, E. Seibold, and P. Rognon, eds., Balkema, Rotterdam, 239-253.
Schäfer, W., 1956, Wirkungen der Benthos-Organismen auf den jungen Schichtverband, Senckenbergiana, 37:183-263.
Seibold, E., 1972, Cruise 25/1971 of R.V. "Meteor": Continental margin of West Africa, General report and preliminary results, "Meteor"Forschungs-Ergebnisse, C10:17-38.
Seibold, E. and Hinz, K., 1976, German cruises to the continental margin of NW Africa in 1975: general reports and preliminary results from "Valdivia" 10 and "Meteor" 39, "Meteor"Forschungs-Ergebnisse, C25:47-80.
Seibold, E., Diester-Haass, L., Fütterer, D., Hartmann, M., Kögler, F.-C., Lange, H., Müller, P.J., Pflaumann, U., Schrader, H.-J., and Suess, E., 1976, Late Quaternary sedimentation off NW Africa, Annales de Academia brasileira de Ciencas, 48, suplemento: 287-296.
Seilacher, A., 1953, Studien zur Palichnologie, I. Über die Methoden der Palichnologie, Neues Jahrbuch für Geologie und Paläontologie, Abhandlungen, 96:421-451.
Seilacher, A., 1967, Bathymetry of trace fossils, Marine Geology, 5:413-428.
Tait, R.V., 1971, "Meeresökologie", Thieme, Stuttgart, 305 pp.
Thiede, J., 1977, Aspects of the variability of the Glacial and Interglacial North Atlantic eastern boundary current (last 150.000 years), "Meteor"Forschungs-Ergebnisse, C28:1-36.
Thiel, H., 1975, The size structure of the deep-sea benthos, Internationale Revue der gesamten Hydrobiologie, 60:575-606.
Weichart, G., 1974, Meereschemische Untersuchungen im nordwestafrikanischen Auftriebsgebiet 1968, "Meteor"Forschungs-Ergebnisse, A14:33-70.
Werner, F. and Wetzel, A., in press, Palecological interpretation of biogenic structures in North Atlantic sediments, Bulletin de l'Institut de Géologie du Bassin d'Aquitaine, 31

Wetzel, A., 1979, "Bioturbation in spät-quartären Tiefwasser-Sedimenten vor NW Afrika," Ph.D. Dissertation, University of Kiel, 111 pp.

Wetzel, A., 1980, Bioturbation in Late-Quaternary deep-sea sediments off NW Africa, International Association of Sedimentologists 1st European Meeting, Bochum, 1980, 56-58.

Wetzel, A., 1981, Ökologisch und stratigraphische Bedeutung biogener Gefüge in quartären Sedimenten am NW-afrikanischen Kontinentalrand, "Meteor"Forschungs-Ergebnisse, C34:1-47.

Wetzel, A. and Werner, F., 1981, Morphology and ecological significance of *Zoophycos* in deep-sea sediments off NW Africa, Palaeogeography, Palaeoclimatology, Palaeoecology, 32:185-212.

UPWELLING RECORDS IN RECENT SEDIMENTS FROM SOUTHERN PORTUGAL: A RECONNAISSANCE SURVEY

Jose H. Monteiro, Fatima G. Abrantes,
Joao M. Alveirinho-Dias and Luis C. Gaspar

Servicos Geologicos de Portugal
Rua Academia das Ciencias, 19-2°
1294 Lisboa Codex, Portugal

ABSTRACT

Data from laboratory analysis of 400 samples of bottom sediments were used to detect records of coastal upwelling off Portugal. Empirical groups of samples based on sediment properties were generated from 25 variables and 339 sample points using the K-Means algorithm (Hartigan, 1975). Two of these (3 and 4) were considered to represent "upwelling facies" because of high biogenic, glauconite and organic carbon contents.

Microscopic counts of diatoms, nannofossils, benthic and planktonic foraminifers and organic carbon were considered as possible "upwelling indicators". The spatial pattern derived from the diatom counts seems to be similar to the patterns identified from remote sensor images of sea surface temperatures. The association of glauconite and apatite, much reported in the literature, is perhaps a record of the Eh gradient history of upwelling processes. On the Portuguese shelf glauconite is associated with relatively high values of P_2O_5. The wide range in redox potentials within a narrow segment of the water column associated with the thermo-halocline and the opposing fluxes of the vertical advective gradient and downward diffusion gradient of oxygen is perhaps recorded in the authigenic sediment component.

INTRODUCTION

Several authors have reported the impact of the upwelling situations on the recent sediment record in the well-known areas of coastal upwelling, but no reports are known from the Portuguese shelf and slope. Coastal upwelling has been recognized off the Portuguese coast since the pioneer work of Ramalho and Dentinho (1928) and Boto (1945). More recently the Oceanography Group of the University of Lisbon has conducted several cruises and systematic airborne remote sensing surveys, reported in Fiuza (1980), Fiuza, Macedo and Guerreiro (1982), and Fiuza (this volume).

To investigate the impact of the upwelling phenomena on the recent sediment cover of the Portuguese shelf, we used sediment samples collected over the entire shelf and upper slope as part of a project of reconnaissance and inventory of the mineral resources carried out by the Portuguese Geological Survey (Monteiro, Gaspar and Dias, 1976). The area south of Cape Espichel was selected based on reconnaissance of the intensity maxima of upwelling in the region off Sines (Fiuza, 1980). This southern half of the Portuguese western shelf is protected from the northwest Atlantic by the Roca and Espichel capes and receives a small contribution of land-derived sediment carried by rivers. For these reasons the upwelling-derived patterns in the sediments on this portion of the shelf probably will be more clearly evident than elsewhere.

FIELD AND LABORATORY METHODS

During the cruises AC 75/1, AC 76/1 and AC 77/2 (Fig. 1), samples of the superficial bottom sediments were collected with a large van Veen grab sampler and, in a few cases, with a Shipek sediment sampler. They were analyzed by methods currently in use at the Marine Geology Laboratory of the Geological Survey. After treatment with hydrogen peroxide to remove the organic matter and careful washing with distilled water, the samples were wet sieved to separate sand (>63 μm) from silt and clay (<63 μm). A standard pipette analysis was made of the fine fraction to determine the clay and silt content. The sand and gravel fractions were then sieved at 1 phi intervals and examined at a binocular microscope. The composition was determined by point-counting of 100 grains of each size class. Grain-size parameters were determined for all samples. The carbonate content of the total sample and of the sand and mud fraction (<2 mm) was determined using a gasometric method (Hülsemann, 1966). Total carbon determinations were made using a LECO Carbon Determinator; the organic carbon content was determined by calculating the difference between total carbon and carbonate carbon. Nitrogen was determined by the micro-Kjeldahl technique.

The fine fraction (<20 μm) was analyzed to search for upwelling indicators. The fine material (<63 μm) was dispersed with sodium hexametaphosphate (0.1 N). Twenty mℓ of the <20 μm fraction were then withdrawn by the pipette method, centrifuged, and decanted to concentrate the samples. Microscope slides were made of a subsample taken immediately after it had been shaken. The subsamples were

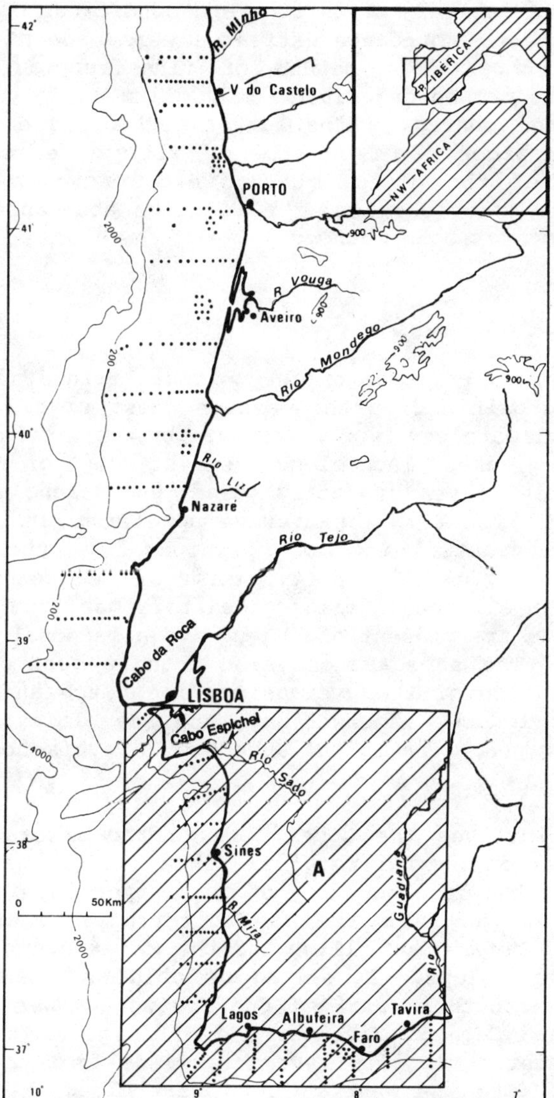

Fig. 1. Sample locations with identification of the study area.

spread on the glass slides in small circles and evaporated at 40°C. Two or three drops of oil (nc = 1.515) were added before observation with a polarizing microscope.

The percentages of all constituents were determined by visual evaluation of their relative abundances in each field of view along lines 2 mm apart. Each slide was scanned at 400-x magnification in order to note the general composition and conditions of the skeletons and other particles. Abundance estimates were made at 1000-x magnification. To compare our method of fine-fraction treatment and counting with the method described by McCollum (1975), a few samples were treated and counted by the latter method. The results showed that the small diatoms, which in some of our samples constituted the great majority of the diatom phase, were removed by the McCollum method. For the other samples, the relative abundances of the diatoms are the same for both methods.

OCEANOGRAPHY

The location of Portugal on the northern fringe of the subtropical anticyclonic belt and on the eastern coast of a large ocean determines the climatology and oceanography of its coastal ocean. Coastal upwelling takes place along the west coast of Portugal during the summer and is driven by fairly steady and strong northerly winds of the Portuguese trades. The average upwelling conditions are restricted to the coast, being more pronounced to the south of Cape Carvoeiro (39°20'N), with intensity maxima in the coastal region off Sines (38°N). In all the climatological averages from several years, upwelling indices are present and they have a seasonal pattern (Fiuza et al., 1982). The same authors also found a good correlation between the evaluation of the average sardine catch and the variation of the physical indices of upwelling, and they infer the importance of the upwelling phenomena to the abundance of this component of the food chain.

Upwelling involves the upper few hundred meters of the water column over the continental margin. Fiuza (1980) states that, generally speaking, the deep boundary of this upper layer may be set at the beginning of the influence of Mediterranean outflow (MW), but some interaction between upwelling and the MW exists over the Portuguese continental slope. Upwelling of cold and less saline water occurs along the southern part of the Portuguese west coast between Cape Espichel and Cape S. Vicente, and off Algarve. It is directly induced by persistent locally favorable winds from the north on the west coast, and from the west and northwest along the Algarve coast. Temperature and salinity analyses made by Fiuza (1980) show that upwelling brings to the surface slope water from depths between 60 and 120 m. South of Cape Espichel the influence of land runoff is negligible, but the shelf water is strongly affected by mixing with warmer and saltier surface waters of offshore origin. Below 100 m (or even reaching the surface in winter) and down to 300-400 m, lies a water mass which has thermohaline characteristics corresponding to the Eastern North Atlantic Water (ENAW) defined by Fiuza and Halpern (1982) for the Canary Current in the area between Cape Bojador and Cape Blanc. ENAW is apparently formed by the mixing of Subtropical Water (STW--24.5°C, 37.3°/₀₀) and remnants of Antarctic Intermediate Water (AAIW--2.2°C, 33.8°/₀₀) in the southwest of the Azores.

There are very few published data about the density structure and chemical characteristics of the shelf water. Boto (1945) states that there is a permanent pycnoline occurring at about 300 m in the winter. During the year secondary pycnolines form at shallower depths. The phosphate concentration in the upper layer between zero and 300 m varies during the year from total depletion at the surface to 0.6 µg-at P/ℓ. There is an intermediate maximum at 400 to 700 m reaching 0.8 µg-at P/ℓ followed by a minimum between 500 and 1000 m, probably related to the MW. Recent surveys conducted by the Oceanography Group of the University of Lisbon showed that sea surface temperatures indicate that the coastal ocean responds quickly with upwelling events to strengthening cycles of the local north wind stress, that upwelling is stronger right at the coast, and that the extension of the upwelled waters is well correlated with bottom topography.

The Fiuza paper presented at this meeting showed differences of upwelling patterns between Cape Espichel and Sines and south of Sines. The topographic irregularities related to the Tejo-Setubal canyons determine a three-dimensionality of the upwelling circulation. South of Sines the pattern of upwelled water is more regular, and the steep morphology of the shelf compresses the thermal gradients towards the shore. Near Cape S. Vicente on the Algarve coast there is an easterly extension of the equatorward coastal upwelling from the west coast. Westerly winds also generate upwelling on the Algarve coast which can even extend to the west coast. There is also considerable dynamic interaction between cold upwelled water and warm-core mesoscale eddies.

SEDIMENT FACIES OF THE PORTUGUESE SHELF

Empirical groups of samples were formed based on sedimentary properties using the K-means algorithm (Hartigan, 1975). Six clusters were generated from the 25 variables and 339 sample points for the entire shelf from the Minho River to the Guadiana River (Monteiro et al., 1980). Mean, standard deviation, maximum and minimum values for each variable are reported in Table 1, and the spatial distribution of each cluster is represented in Fig. 2. The clusters have a certain degree of spatial continuity, and they are in a sense true sediment facies of the shelf sediment cover. The characteristics of the different clusters are given below.

a. Cluster 1 corresponds to a very coarse facies composed mainly of terrigenous material. Samples of this cluster have the highest percent of gravel and very coarse sand (-1 to 0 phi). The carbonate content is very low and biogenic material in the sand fraction is practically non-existent. Quartz grains and other terrigenous grains dominate the sand fraction.

Table 1. Clusters generated by the K-means algorithm

VARIÁVEIS		CLUSTER 1				CLUSTER 2				CLUSTER 3				CLUSTER 4				CLUSTER 5				CLUSTER 6			
		$\overline{X}$	σ	Xm	XM	$\overline{X}$	σ	Xm	XM	$\overline{X}$	σ	Xm	XM	$\overline{X}$	σ	Xm	XM	$\overline{X}$	σ	Xm	XM	$\overline{X}$	σ	Xm	XM
% Gravel	GRVE	25.6	1.4	1.0	87.0	0.1	0.2	0	1.0	3.2	1.3	0	27.0	0.1	0.3	0	4.0	9.3	1.5	1.0	77.0	16.1	2.2	0	59
% Sand	SAND	60.4	0.5	11.0	99.0	94.5	0.1	66.0	99.0	63.7	0.7	3.0	96.0	60.7	0.9	6.0	99.4	78.5	0.3	22.0	99.0	69.1	0.3	4.0	99
% Silt	SILT	0.3	0.5	0	4.0	2.1	1.5	0	34.0	14.8	1.1	2.0	70.0	15.5	1.5	0	86.0	2.6	1.4	0	22	1.0	0.8	0	8
% Clay	CLAY	0.1	0.2	0	1.0	0.6	0.7	0	8.0	1.4	2.3	0	63.0	2.2	2.5	0	74.0	0.3	0.5	0	3	0.8	0.5	0	3
% Sand Fr -1-0ϕ	SND 1	37.9	0.6	2.0	65.0	0.1	0.4	0	3.0	4.0	1.4	0	43.0	0.6	0.8	0	10.0	21.9	1.3	1	66	32.2	1.7	1	94
% " " 0 a 1ϕ	SND 2	38.2	0.4	11.0	60.0	0.8	0.8	0	8.0	8.6	1.5	0	46.0	2.5	1.5	0	28.0	39.2	0.5	6	70	31.0	0.6	5	60
% " " 1 a 2ϕ	SND 3	12.7	0.9	1.0	62.0	6.3	2.0	0	76.0	13.4	4.1	1.0	50.0	9.5	1.4	1.0	58.0	17.2	0.8	3	65	8.7	2.0	0	58
% " " 2 a 3ϕ	SND 4	2.0	0.8	0	11.0	52.2	0.7	10.0	90.0	27.3	0.9	5.0	80.0	30.9	0.7	6.0	76.0	4.2	1.4	1	28	2.6	1.3	0	17
% " " 3 a 4ϕ	SND 5	0.3	0.7	0	16.0	18.2	1.9	0	86.0	25.0	0.9	3.0	85.0	35.2	1.2	2.0	91.0	1.1	1.3	0	11	0.7	0.9	0	8
%Biog.gravel	BGRV	5.9	1.6	0	61.0	0.3	1.9	0	99.0	66.1	1.4	2.0	99.0	0	0	0	0	32.7	2.1	1	99	71.8	1.4	0	99
%Quartz in sand	SQTZ	72.5	0.1	44.0	91.0	51.1	0.5	8.0	87.0	20.7	1.4	0	59.0	17.6	1.5	0	58.0	39.0	0.6	11	73	1.6	1.5	0	22
%Terrig. " "	STER	16.0	0.5	2.0	29.0	21.8	0.6	1.0	42.0	10.2	4.1	0	49.0	9.5	1.3	0	28.0	12.8	0.5	4	27	1.2	1.0	0	8
%Aggregates "	SAGR	2.0	0.8	0	7.0	0.1	0.4	0	4.0	0.1	0.3	0	2.0	0.0	0.2	0	2.0	1.2	0.8	0	8	0.4	0.9	0	11
%Mica "	SMCA	0.4	0.6	0	5.0	2.0	1.9	0	29.0	0.6	1.0	0	27.0	0.8	1.1	0	41.0	0.0	0.3	0	4	0.1	0.2	0	1
%Glauconite "	SGLA	0.0	0.2	0	2.0	0.4	1.3	0	47.0	1.9	1.5	0	34.0	2.7	2.1	0	51.0	0.8	1.5	0	40	0.2	0.6	0	4
%Planct.Foram "	SFRP	0.0	0.2	0	3.0	0.3	0.7	0	8.0	4.4	1.4	0	37.0	7.2	1.2	0	35.0	0.5	1.0	0	12	0.8	1.1	0	10
%Bent.Foram. "	SFRB	0.1	0.4	0	6.0	1.3	1.2	0	21.0	11.1	0.9	1.0	35.0	16.2	0.8	1.0	60.0	2.5	1.5	0	20	5.3	1.6	0	24
%Moluscs "	SMOL	2.9	1.2	0	22.0	3.3	1.1	0	21.0	12.8	0.9	1.0	44.0	7.2	1.3	0	30.0	14.1	0.9	3	46	30.1	0.8	7	96
%Equinoderm "	SEQU	0.2	0.4	0	3.0	0.8	0.6	0	4.0	2.8	0.9	0	12.0	1.9	0.8	0	8.0	1.4	0.8	0	7	3.6	0.9	0	11
%Other biog. "	SBIO	0.5	0.7	0	13.0	2.3	1.2	0	23.0	6.9	8.8	1.0	36.0	7.2	1.3	0	45.0	4.3	1.0	0	25	24.6	1.6	1	89
%n/ident.biog "	SNID	0.4	0.8	0	10.0	1.4	1.7	0	13.0	5.1	4.3	0	43.0	2.4	1.4	0	24.0	4.1	2.3	0	73	3.0	3.0	0	46
%C_aCo_3 in total sample	CAB 1	4.8	0.8	0	19.0	7.8	1.8	0	45.0	34.7	0.6	4.0	80.0	30.6	0.7	6.0	80.0	27.6	0.7	8	74	67.8	1.3	0	94
%Nitrogen	NTRO	0	0	0	0	0.6	1.2	0	11.0	4.0	1.7	0	17.0	3.4	2.0	0	16.0	0.3	1.0	0	7	0	0	0	0
%Organic carbon	CORG	0	0	0	0	0.2	0.6	0	6.0	2.1	1.5	0	13.0	2.3	1.5	0	13.0	0.2	0.7	0	6	0	0	0	0
%$CaCo_3$ in the sand fraction	CAB 2	4.7	0.9	0	29.0	8.1	1.7	0	45.0	32.7	0.7	4.0	81.0	30.6	0.7	6.0	80.0	25.6	0.8	7	70	67.2	1.3	0	90
Number of samples		62				52				90				63				44				28			

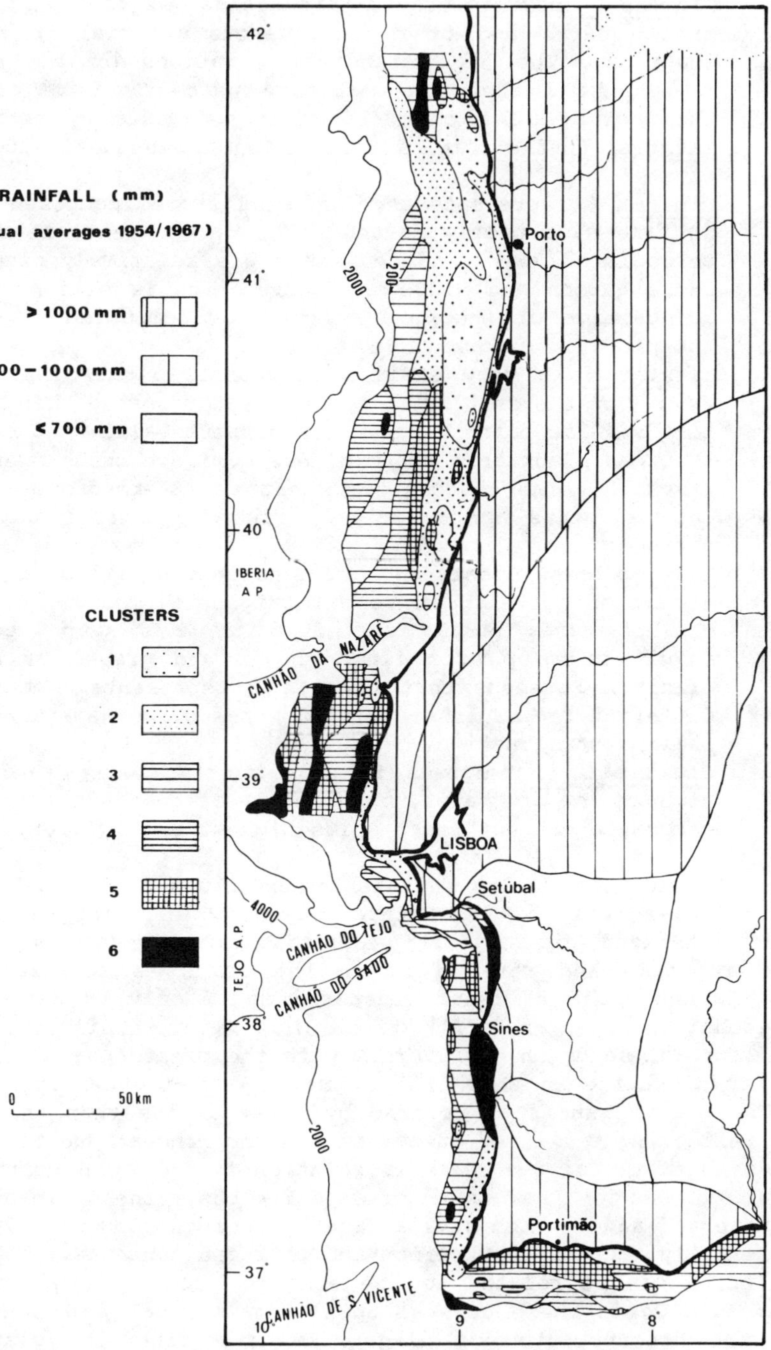

Fig. 2. Map of the clusters generated by the computer using the K-means algorithm.

b. Cluster 2 corresponds to fine sands (2 to 3 phi) rich in mica, quartz and other terrigenous material in the sand fraction. The high standard deviations of the biogenic components and of the carbonate percentage are caused by the lack of dilution of biogenic sediments by terrigenous material in the samples situated in the seaward side of the cluster.

c. Cluster 3 represents a silty sand facies composed of sand or fine sand poorly sorted. The carbonate content is intermediate (30%), and sediments are relatively rich in organic carbon and nitrogen. Glauconite is present but the composition of the sand fraction is dominated by molluscs and planktonic foraminifers.

d. Cluster 4 is very similar to cluster 3, also being silty sand, but the sand is finer and better sorted than in cluster 3. Glauconite reaches its highest value, and both benthic and planktonic foraminifers dominate the biogenic sand fraction. The carbonate content is intermediate (30%) and the sediments are relatively rich in organic carbon and nitrogen.

e. Cluster 5 represents another sand facies with a significant amount of gravel. It is a terrigenous, medium- to coarse-grained sand, poorly sorted, with a biogenic component mainly made up of mollusc shells and gravel of biogenic origin. It seems that cluster 5 represents a mixture of different facies in transition zones or perhaps a relict or palimpsest situation.

f. Cluster 6 corresponds to well sorted coarse sand with a high amount of gravel, both mainly of biogenic origin. The carbonate content reaches its highest value (90%).

The terrigenous facies represented by clusters 1 and 2 are the dominant facies north of the Nazare Cañyon. The relief and humidity of the hinterland differ north and south of the canyon, causing greater rainfall and river flow in the north as compared to the south. In the south we found cluster 1 and 2 only in narrow belts from Setubal to Sines and south of the Mira River. This reflects the coarse littoral sands in equilibrium with the present wave conditions and river discharge in this area of relatively low terrigenous supply. The Algarve shelf is covered by finer facies than the western shelf, reflecting the lower energy of the environment due to the different orientation of the coast in relation to the open North Atlantic. In the study area, south of Cape Espichel, the dominant facies are clusters 3 and 4 which have a high biogenic component, glauconite and relatively high organic carbon and nitrogen contents. We believe these facies can be related to the eastern boundary environment of the Atlantic Ocean where coastal upwelling is usually present. This causes the precipitation of diagenetic iron minerals (glauconite) near the sediment-water interface and the development of plankton blooms which result in the accumulation of organic-rich sediments.

DIATOM TESTS IN THE <20 μm SEDIMENT FRACTION

Moderate to high diatom concentrations are present in belts running parallel to the coast from Cape Espichel to north of Cape S. Vicente, and a tongue of lower values seems to penetrate from the south (Fig. 3). On the southern coast, immediately east of Cape S. Vicente and near Faro, zones of moderate to high values are also present. Koopmann, Sarnthein and Schrader (1978) state that marine

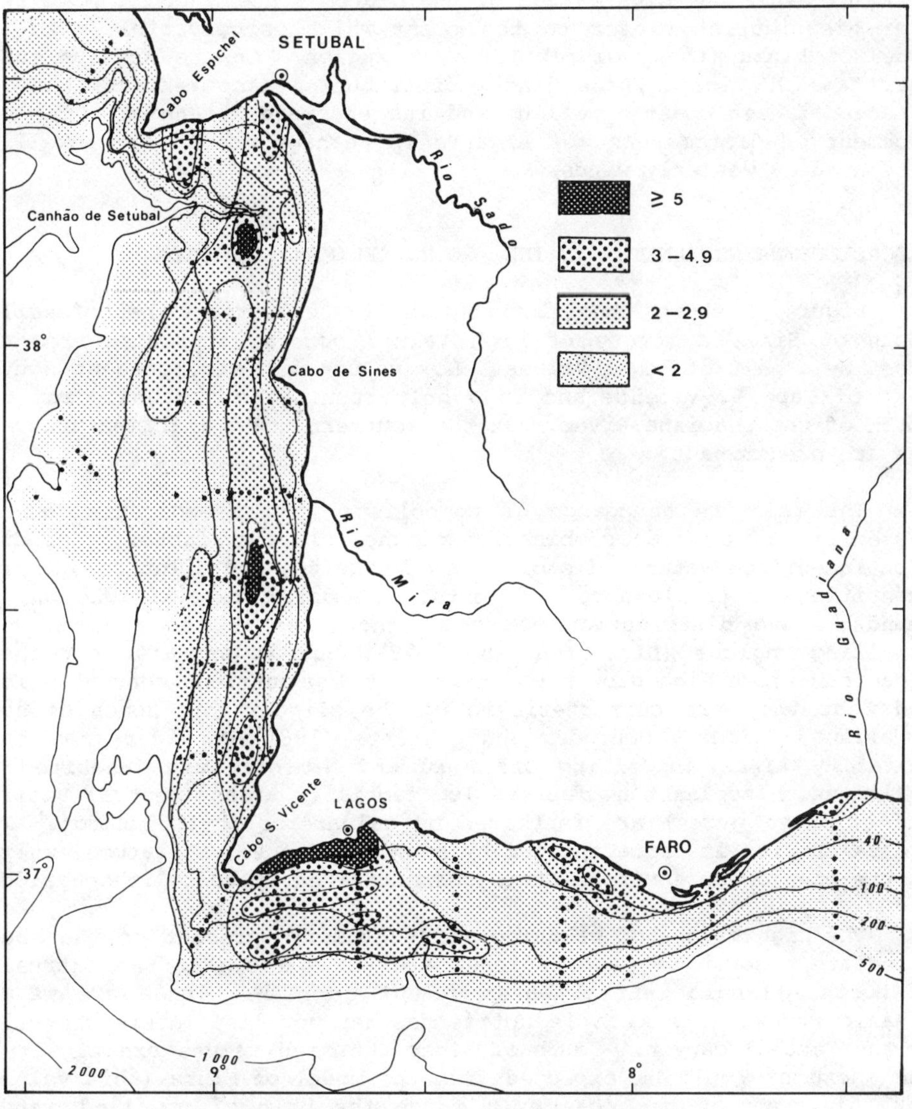

Fig. 3. Percent counts of diatoms in the <20 μm sediment fraction.

sediments rich in diatom tests reflect the productivity of the photic zone. The content of diatoms in the sediments of the continental shelf is generally related to the productivity of the water column, and their relative abundance is perhaps related to the upwelling patterns. It appears that immediately south of the prominent capes Espichel and S. Vicente, there are enrichments in the diatom content. This bifurcated distribution between Cape Espichel and Sines may indicate that the tridimensional effects of upwelling mentioned by Fiuza (this volume) for this region are recorded in the sediments. South of Sines the high values of the diatom content have a more regular distribution closer to the coast which is compatible with the model of Fiuza (this volume) for this region. On the Algarve coast near Cape S. Vicente, the diatom distribution also reflects the influence of west coast upwelling and its eastward extension. The enrichment of diatoms east of Algarve is perhaps due to the upwelling generated by westerly winds.

CALCAREOUS NANNOPLANKTON IN THE <20 μm SEDIMENT FRACTION

Figure 4 shows the distribution of calcareous nannofossils. South of Sines a tongue of high values occurs, separated from the coast by a belt of lower values. Low values are also present southeast of Cape S. Vicente and in a belt running to the west from the mouth of the Guadiana River. On the southern coast intermediate values are predominant.

The relative abundance of coccoliths in the sediments can be caused by all or some combination of the following factors: production in surface waters, dissolution of calcite, and dilution by large quantities of terrigenous sediments. No data are available on the abundance and distribution of coccolithophores in the sediments of upwelling regions (Diester-Haass, 1978), but it appears that these algae can reach high growth potentials at low nutrient concentrations and that they are characteristic of the oligotrophic zones of some upwelling regions. Schrader and Schuette (1981) consider that high fertility areas (upwelling off Peru and Namibia) are inhabited by siliceous phytoplankton, whereas low fertility areas (Central Pacific and Atlantic gyres) are inhabited by calcareous phytoplankton. The coastal regime is characterized by high abundances of diatoms, whereas the open ocean environment is characterized by coccolithophorids.

The high values of calcareous nannofossils present on the shelf immediately south of Cape Espichel are perhaps due to the intrusion of North Atlantic Central Water (NACW) or ENAW. These are warmer oceanic waters less rich in nutrients than upwelled waters channeled by the Setubal canyon. South of Sines the high values are closer to the shore as would be expected from the model of Fiuza (this volume) for this part of the coast, in which the belt of upwelled waters would be narrower than to the north. On the southern coast the low values of the calcareous nannoplankton extend along the shelf break and close to the shore east of Cape S. Vicente. This may occur as a result of upwelling due to the propagation of upwelled waters from the west coast and upwelling off the southern coast due to the westerly winds as described by the same author. The high values west of Faro are perhaps due to oceanic waters trapped in the inner shelf when winds calm down.

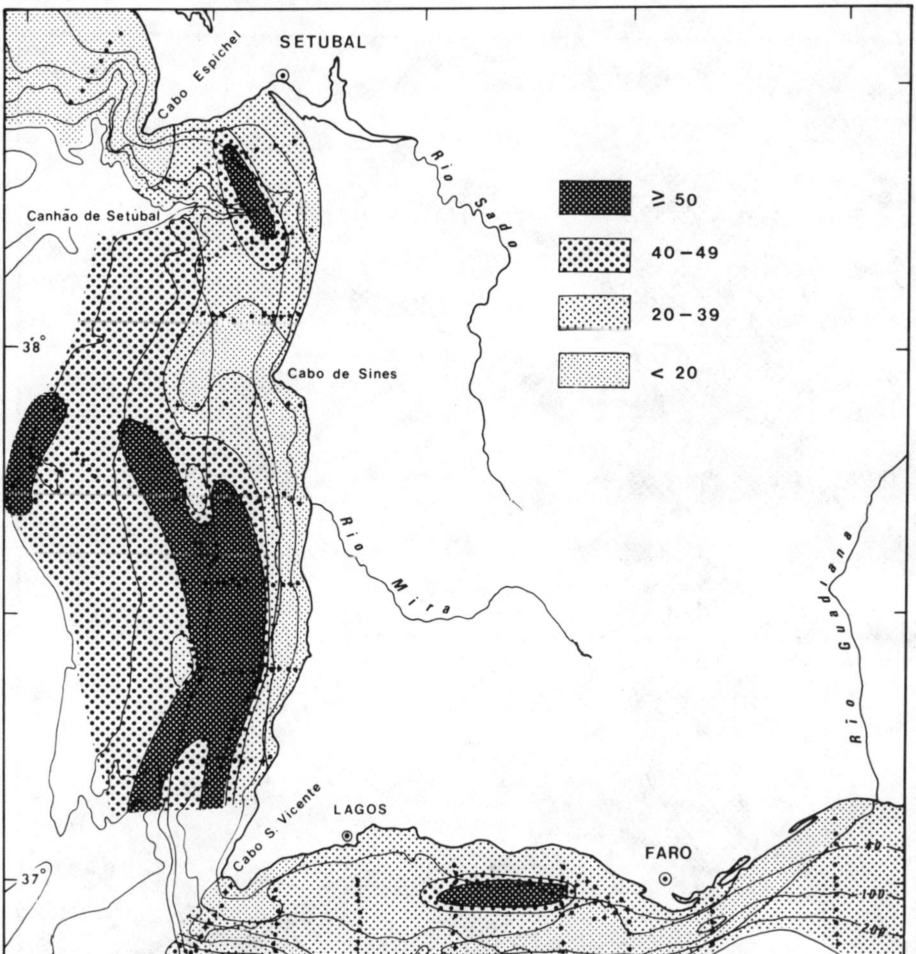

Fig. 4. Percent counts of calcareous nannoplankton in the <20 μm sediment fraction.

BENTHOS-PLANKTON RATIO OF FORAMINIFERS IN THE SAND FRACTION

Figure 5 shows the distribution pattern of the benthos/plankton ratio of foraminifers. Values greater than 80% are present near the coast from Cape Sines to Cape S. Vicente and in a belt parallel to the coast that occurs offshore of the same area. On the southern

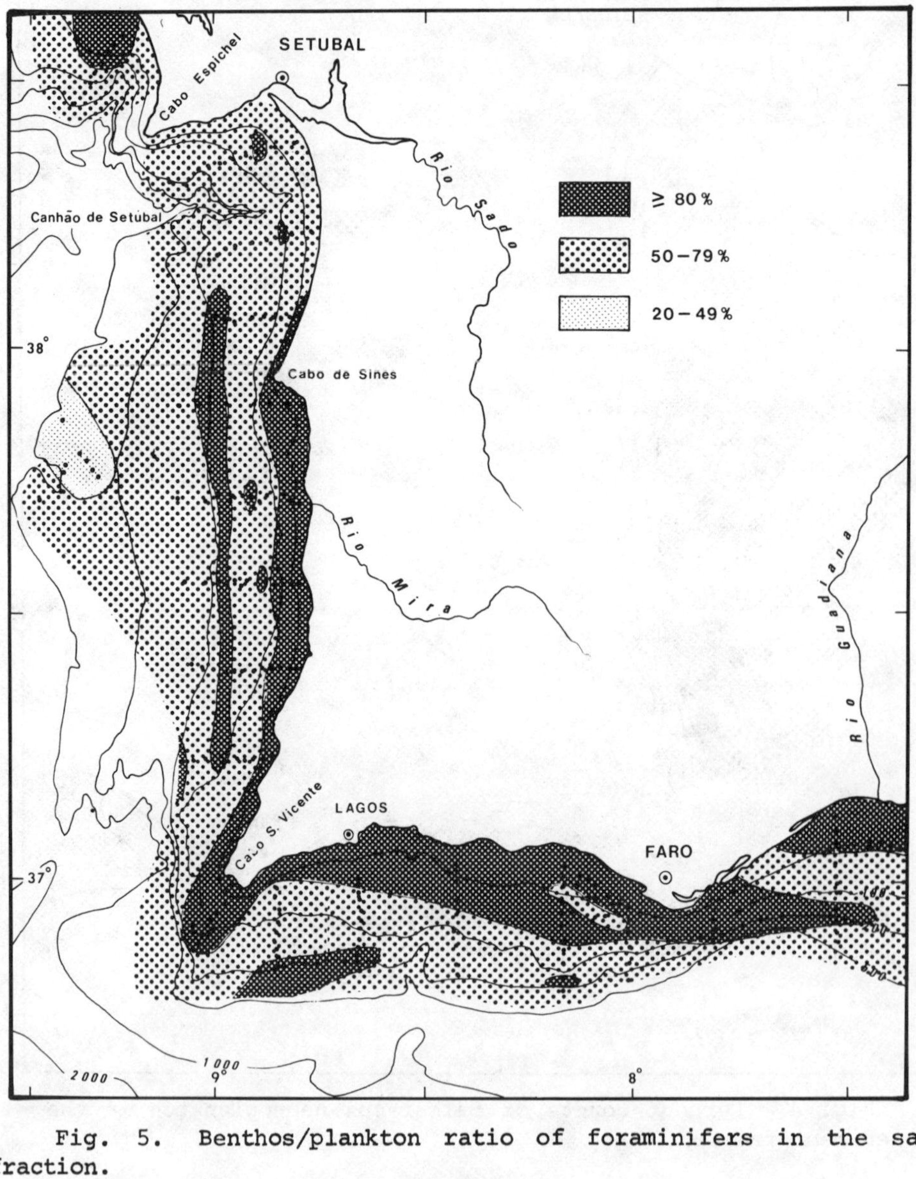

Fig. 5. Benthos/plankton ratio of foraminifers in the sand fraction.

coast the littoral band is wider but the offshore belt is more discontinuous. Most of the continental shelf and upper slope is covered by surface sediments in which the benthos/plankton ratio of foraminifers reaches 50 to 79%. The minimum values were found in the samples collected seaward of the 500-m depth contour, west of Cape Sines. High benthic/planktonic ratios of foraminifers have been related to fertile regions (Diester-Haass, 1978; Diester-Haass and Schrader,

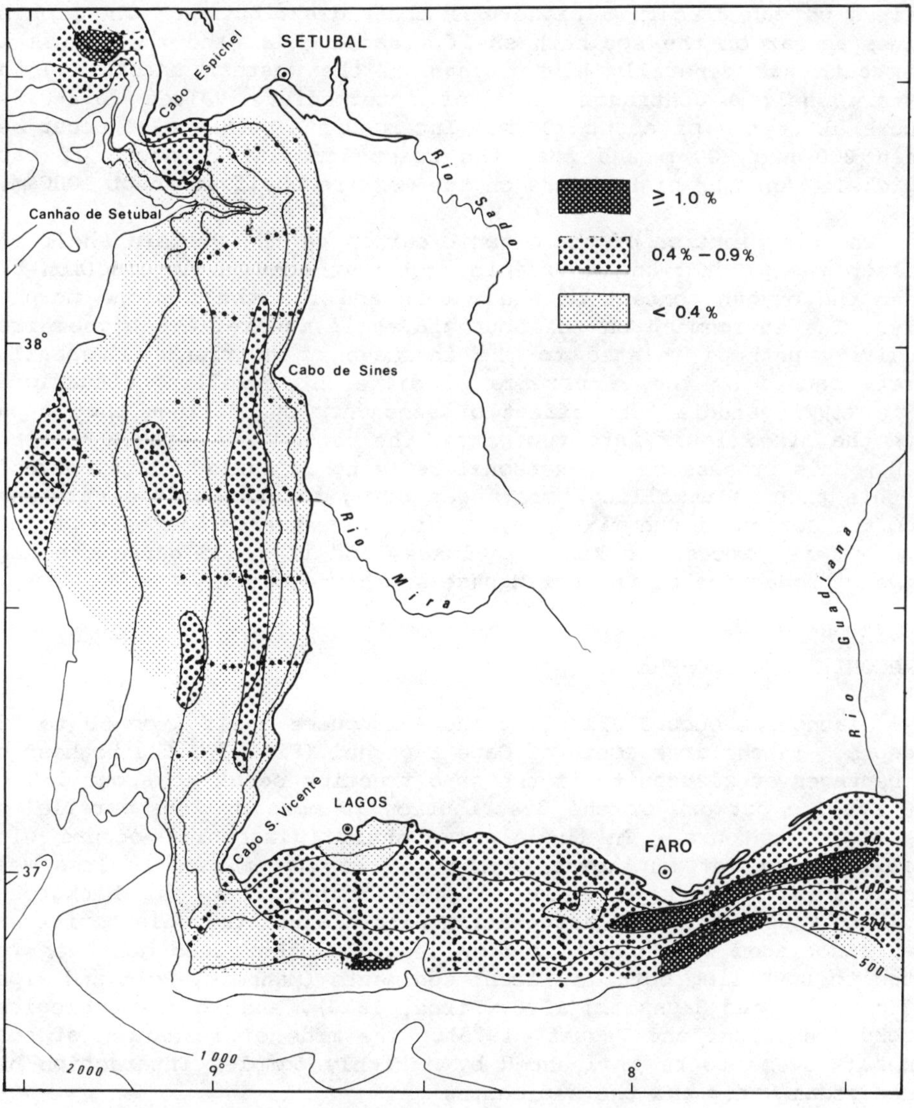

Fig. 6. Percent of organic carbon in the total sample.

1979). The littoral belt may be related to the absence of planktonic species near shore, but the high proportion of benthic foraminifers in the outer belt may be related to the high productivity of upwelling fronts.

ORGANIC CARBON

The percent values of organic carbon in the total sample and their distribution appear in Fig. 6. The values are low but nevertheless we can discern a pattern in their distribution. The highest values appear on the southern shelf, reaching 1%, and the values in the south are generally higher than on the western shelf. On the western shelf a continuous belt of intermediate values (0.4-0.9%) occurs at depths of about 100 m. Intermediate values also occur between 200 and 500 m and near the coast immediately south of Cape Espichel. Another high occurs on the western shelf at about 1000 m.

The distribution of the organic carbon on the western shelf may reflect the production of organic matter related to interaction between the oxygen content of the waters and the shelf slope morphology. The enrichment on the southern shelf may reflect higher productivity perhaps related to the increase of nutrients in shallow waters caused by the recurrence of different upwelling situations. Lewis (1981) studied the effect of secondary pycnoclines that separate the mixed layer into two parts, the bottom one being a nutrient trap. This process of non-seasonal remixing is called "Atelomixes". By this process upwelling events can bring to the surface nutrients from the bottom of the mixed layer which may enhance biologic production. This process is known in lakes, and it could cause the high organic production of the Portuguese southern coast.

GLAUCONITE AND PHOSPHATE

Glauconite occurs all along the Portuguese shelf from 50 to 500 m depth. In the area south of Cape Espichel (Fig. 7), the highest of percentages of glauconite in the sand fraction occur at about 150 to 250 m. The pattern of the distribution is more or less parallel to the shelf contours. On the Algarve shelf, glauconite occurs with smaller values at shallower depths (50 to 200 m) and is less well distributed. Values of P_2O_5 slightly higher than the background (0.15 to 0.50%) occur in association with glauconite. Glauconite has been associated with phosphates by several authors and has been related to upwelling both in recent sediments (Manheim, Rowe and Jipa, 1975; Kohler and Haussler, 1978; Birch, 1979), and in the geological record (Debrabant and Paquet, 1975). The mode of formation of both minerals seems to be influenced by a highly complex interaction between ocean water and the sediments.

Arthur and Jenkyns (1981) suggested that phosphate precipitation is favored in regions of fluctuating redox conditions like those associated with oxygen minimum layers (or alternating anoxic-suboxic zones). Also, the formation of glauconite has been reported from around Eh=0, with some authors favoring a slightly oxidizing environment and others a slightly reducing environment (McRae, 1972). The glauconite of the Portuguese shelf has been interpreted mainly as detrital, but its continuity along the shelf implies a mechanism of concentration and sorting difficult to understand. The different depth ranges to the west and south are also difficult to explain.

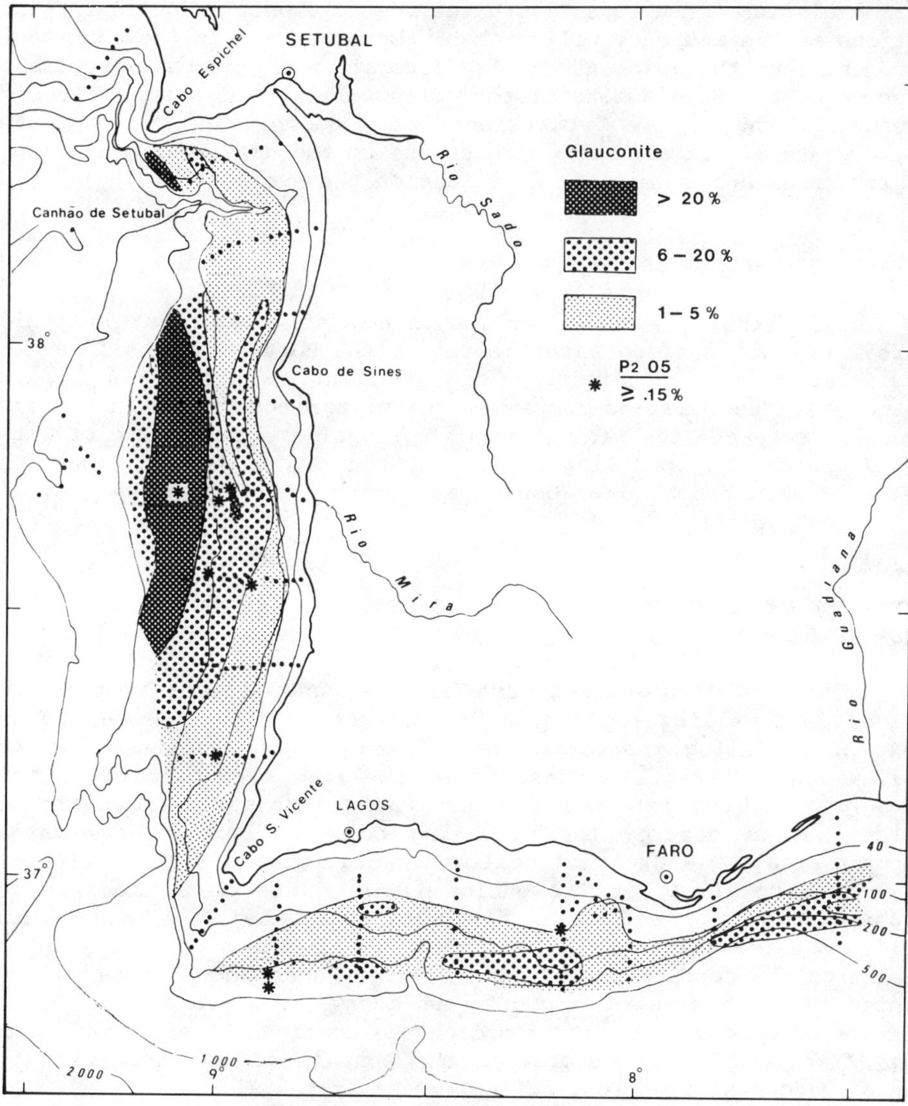

Fig. 7. Percent counts of glauconite and $P_2O_5 \geq 15\%$.

One possible explanation is that the glauconite, perhaps associated with phosphate, is the result of the hydrographic structures related to the upwelling processes.

Degens and Stoffers (1976) pointed out that the physicochemical phenomena established at the boundary between two different states characterizes these two states. The thermo-haloclines and the opposing fluxes of the vertical advective gradient and the downward diffusion gradient of oxygen will cause a wide range in redox potentials within a narrow segment of the water column. Furthermore, this thermo-halocline structure fluctuates due to upwelling pulses. Fluctuations of the boundary will probably be preserved in the authigenic component, and the association of glauconite and phosphate is perhaps a record of the Eh gradient history of upwelling processes. The differences in the depth of occurrence of glauconite on the west and south coasts of our area are perhaps due to the different hydrographic structures and upwelling situations on these coasts.

TEMPERATURE INDICATORS

Thiede (1980) reported the occurrence of foraminiferal species typical of cold surface water masses on the Portuguese shelf, and he considered this an indication of upwelling. Thiede (1977) also stated that the polar-to-temperate assemblages occurring in the surface sediments off the west coast of Portugal are indicative of water masses inside the upwelling regions. Occurrences of cold water species were also reported by Ubaldo and Otero (1978) from samples off the Portuguese coast.

COMPARISON OF UPWELLING SEDIMENT RECORDS OF THE PORTUGUESE SHELF WITH OTHER REGIONS

Indicators of upwelling have been recognized over most of the shelves in upwelling regions. Off northwest Africa, between 20 and 25°N, no upwelling indicators were found in the sediments of the shallow shelf directly beneath the upwelling area (Fütterer, this volume). Fütterer felt this absence was due to winnowing by bottom currents which destroy the upwelling record. For the same area, Koopmann et al. (1978) found opal sedimentation and other indicators of coastal upwelling in the enclosed Baie du Levrier which are not documented in the sediments directly underlying the continental shelf. Diester-Haass (1978) stated that indicators of upwelling are present on the continental slope. Diester-Haass and Schrader (1979) related the existence of a continuous downwelling front with the occurrence of a good sediment record on the continental shelf of southwest Africa, and they associated the absence of a record with the lack of a downwelling front off northwest Africa.

On the Portuguese shelf a series of features in the different distributions of indicators appears to be related to the upwelling aspects reported by Fiuza (this volume). Based on several sediment variables, an eastern-boundary type of facies appears to be identifiable. Faunal temperature indicators, organic carbon, diatom content, calcareous nannoplankton and the benthos/plankton ratio of foraminifers were studied and appeared to give patterns related to the coastal upwelling. Glauconite and phosphate may indicate the fluctuations of the Eh gradient history of upwelling phenomena. The models for the western shelf reported by Fiuza are different for areas north and south of Sines. This can also be observed in the sediments from the upwelling indicators considered here. The effects of capes on the upwelling process are also apparent in the sediment record. On the Algarve the sediments show an enrichment of organic matter relative to the western shelf. This enrichment could be the result of a more complex upwelling situation, but further studies will be necessary to understand it.

The reconnaissance of upwelling patterns in this Portuguese offshore area and the lack of such patterns on the northwest African shelf are perhaps a reflection of the lower energy of the Portuguese region due to the protection by capes and the different orientation of the coast.

REFERENCES

Arthur, M.A. and Jenkyns, H.C., 1981, Phosphorites and paleoceanography, Oceanologica Acta, no. S.P. 83-96.

Birch, G.F., 1979, The association of glauconite and apatite minerals in phosphatic rocks from the South African margin, Transactions Geological Society of South Africa, 82:43-53.

Boto, R.G., 1945, Contribuicão para os estudos de oceanografia ao longo da costa de Portugal-fosfatos e nitratos, Travaux de la Station de Biologie Maritime de Lisbonne, no. 49, 102 pp.

Debrabant, P. and Paquet, J., 1975, L'association glauconites- phosphates-carbonates (Albien de la Sierra de Espuña, Espagne meridionale), Chemical Geology, 15:61-75.

Degens, E.T. and Stoffers, P., 1976, Stratified waters as key to the past, Nature, 263:22-27.

Diester-Haass, L., 1978, Sediments as indicators of upwelling, in: "Upwelling Ecosystems," R. Boje and M. Tomczak, eds., Springer-Verlag, 261-281.

Diester-Haass, L. and Schrader, H., 1979, Neogene coastal upwelling history off Northwest and Southwest Africa, Marine Geology, 29: 39-55.

Fiuza, A.F., 1980, The Portuguese coastal upwelling system, preprints of the Seminar on Present Problems of Oceanography in Portugal, Lisboa, 19-20 November, 1980, 31 pp.

Fiuza, A. and Halpern, D., 1982, Hydrographic observations of the Canary Current between 21°N and 25.5°N during March April 1974,

Conseil Permanente International pour l'Exploration de la Mer, Rapport, 180:58-64.

Fiuza, A., Macedo, M. and Guerreiro, R., 1982, Climatological space and time variation of the Portuguese coastal upwelling, Oceanologica Acta, 5(1):31-40.

Hartigan, J.A., 1975, "Clustering Algorithms," John Wiley and Sons, New York, 355 pp.

Hülsemann, J., 1966, On the routine analysis of carbonates in unconsolidated sediments, Journal of Sedimentary Petrology, 36: 622-625.

Kohler, E.E. and Haussler, H., 1978, Zur Entstehung von Phosphorit- und Glaukonitvorkommen der Mittel- und Oberkreide im Helvetikum des Allgäus, Geologisches Jahrbuch, A.46:69-91.

Koopmann, B., Sarnthein, M. and Schrader, H.J., 1978, Sedimentation influenced by upwelling in the subtropical Baie de Levrier (West Africa), in: "Upwelling Ecosystems," R. Boje and M. Tomczak, eds., Springer-Verlag, 282-288.

Lewis, W.M., Jr., 1981, "Zooplankton Community Analysis Studies on a Tropical System,". Springer-Verlag, New York, 163 pp.

Manheim, F., Rowe, G.T. and Jipa, D., 1975, Marine phosphorite formation off Peru, Journal of Sedimentary Petrology, 45(1):235-251.

McCollum, D.W., 1975, Diatom stratigraphy of the southern ocean, in: "Initial Reports DSDP," 28, D.E. Hayes, L.A. Frankes et al., U.S. Government Printing Office, Washington, 515-571.

McRae, S.G., 1972, Glauconite, Earth Science Reviews, 8:397-440.

Monteiro, J.H., Gaspar, L. and Dias, J.A., 1976, Avaliacao dos recursos minerais da margem continental metropolitana, Boletim de Minas, 13(4):187-197.

Monteiro, J.H., Dias, J.A., Gaspar, L, and Possolo, A.M., 1980, Recent marine sediments of the Portuguese continental shelf, Seminar on Present Problems of Oceanography in Portugal, Lisboa, 19-20 November, 1980, preprint.

Ramalho, A. and Dentinho, L., 1928, Nota sobre as condicoes oceanograficas ao largo da costa de Portugal, Travaux de la Station de Biologie Maritime de Lisbonne, no. 15, 16 pp.

Schrader, H.J. and Schuette, G., 1981, Marine diatoms in: "The Oceanic Lithosphere," C. Emiliani, ed., THE SEA, Vol. 7, Wiley and Sons, New York, 1179-1232.

Thiede, J., 1977, Aspects of the variability of the glacial and interglacial North Atlantic eastern boundary current (last 150,000 years), "Meteor"Forschungs-Ergebnisse, C:28:1-36.

Thiede, J., 1980, The late Quaternary marine paleo-environments between Europe and Africa, Paleoecology of Africa and the Surrounding Islands, 12:213-225.

Ubaldo, M.L. and Otero, M.R., 1978, Foraminiferos da costa su-sudoeste de Portugal, Garcia da Horta, 2(2):77-130.

ENVIRONMENTAL CONTROLS ON SEDIMENT TEXTURE AND COMPOSITION IN LOW OXYGEN ZONES OFF PERU AND OREGON

Lawrence A. Krissek* and Kenneth F. Scheidegger

School of Oceanography
Oregon State University
Corvallis, Oregon 97331, U.S.A.
*Present address:
Department of Geology
Ohio State University
Columbus, Ohio 43210, U.S.A.

ABSTRACT

To assess the potential for the preservation of a strong coastal upwelling signal in sediments deposited beneath upwelling zones off Peru and Oregon, an index of major environmental parameters has been calculated for five areas on the Peru margin and three on the Oregon margin. Parameters included in the index are surface primary productivity, intensity of the O_2-minimum and its position relative to the sedimentary deposit, and regional terrigenous sediment input. The calculations indicate that the most favorable environment for a biogenic signal exists on the Peru margin at 13°S, where a strong O_2-minimum impinges directly on the shelf and upper slope, primary productivity is high, and dilution by terrigenous components is minimal. Areas at 3°S and 43°, 44°, and 46°N are ranked as least favorable mainly because biogenic components are significantly diluted by fluvially transported terrigenous material.

Clay content of bulk sediments, skewness values calculated for the silt fractions, and the organic carbon/carbonate carbon composition of the underlying sediments appear to be sensitive indicators of the different environmental conditions. Samples from 13°S consist of >60% clay and the minor silt fraction exhibits strong negative skewness; such terrigenous components are bound in organic-carbon-rich fecal pellets produced by nekton in the overlying water column. Samples from 8°, 10°, and 11°S are coarser grained and dominated by calcareous biogenic debris. Silt fractions of these samples exhibit positive skewness, suggesting that much of the fine-grained terrigenous fraction has been winnowed by bottom currents in these areas. Samples from 3°S, and 43°, 44°, and 46°N have low organic-carbon and carbonate content, intermediate clay content, and the silt fraction exhibits little pronounced skewness. Rapid accumulation of fluvially derived terrigenous sediments appears to be responsible for the observed characteristics in these four areas.

Four components may be considered to contribute to sediments beneath upwelling centers in the study areas: biogenic siliceous, biogenic calcareous, organic-carbon/fine silt and clay terrigenous aggregates, and coarser terrigenous material. Deposition of the first three is influenced by chemical (carbonate compensation depth, O_2-minimum zone) and physical (current activity) processes, and the feeding activities of the zooplankton and nekton, but each still carries information about the environment of formation within the upwelling zone. The fourth acts as a dilutant. Therefore, in order to identify present and past upwelling-influenced sediments, tracers from all these groups must be used together; no single marker is diagnostic in all cases.

INTRODUCTION

We have recently reported the presence of an anomalous fine-grained, organic-carbon-rich mud lens on the upper continental slope off central Peru (Krissek, Scheidegger and Kulm, 1980). The chemical signature of these sediments indicates a strong coastal upwelling influence and the deposit occurs where an intense O_2-minimum layer impinges on the shelf and upper slope. To the north of the mud lens, the strength of the coastal upwelling influence diminishes as the intensity of the O_2-minimum zone weakens, the primary productivity in the water column is somewhat less and the influx of terrigenous sediment is greater. In order to further investigate these environmental interrelationships and their possible influences on the underlying sediments, we have re-examined sediment inputs for the Peru margin between 3°and 13°S. In addition, we have examined comparable data from the Oregon continental margin because it, too, has upwelling-influenced hydrographic regimes and associated enhanced productivity. In this paper we examine the following questions: What factors or combination of factors are most influential in creating a depositional environment where the biogenous products of coastal upwelling can be best preserved and can dominate the resulting sediments? What compositional parameters reflect this influence?

The strength of the upwelling signature in sediments may depend on:

1. surface water productivity,
2. strength and persistence of the O_2-minimum zone and its location relative to the sediment accumulation,

3. magnitude of terrigenous sediment input to the region, and
4. bottom current activity and location of the carbonate compensation depth (CCD).

Biogenous products should become more dominant in sediments as productivity in the overlying water column increases, and as the accumulation rate of terrigenous sediment decreases. Similarly, if the O_2-minimum zone becomes stronger and impinges directly on the sediments, organic matter produced in the overlying water column should be better preserved. In this paper we first describe eight depositional environments, five on the Peru margin and three on the Oregon margin. For each area a parameter is calculated which summarizes the respective sedimentary environment's favorability for the production, preservation, and predominance of upwelling-related components in the underlying sediments. The composition and texture of these sediments are then examined in light of the overall environment to decide which sediment characteristics are sensitive to environmental changes and how those characteristics vary in response to such enviromental changes. Finally, the important environmental controls and their

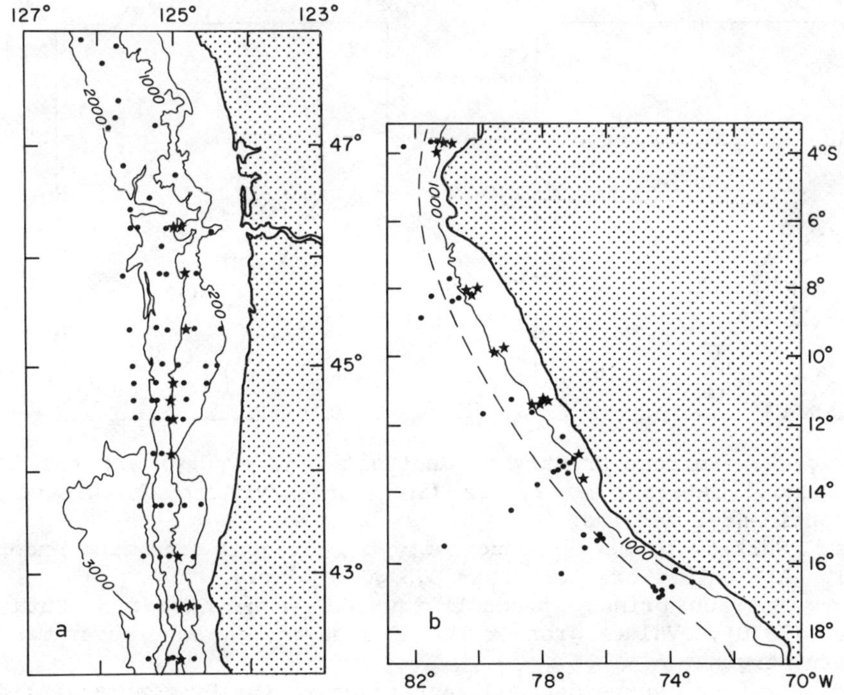

Fig. 1. Generalized bathymetry of the Oregon and Peru continental margins, showing locations of all cores available (dots) and of those cores used for this study (stars). Contours are in meters. Sample groups are from 46°, 44°, and 43°N, and 3°, 8°, 10°, 11°, and 11-14°S.

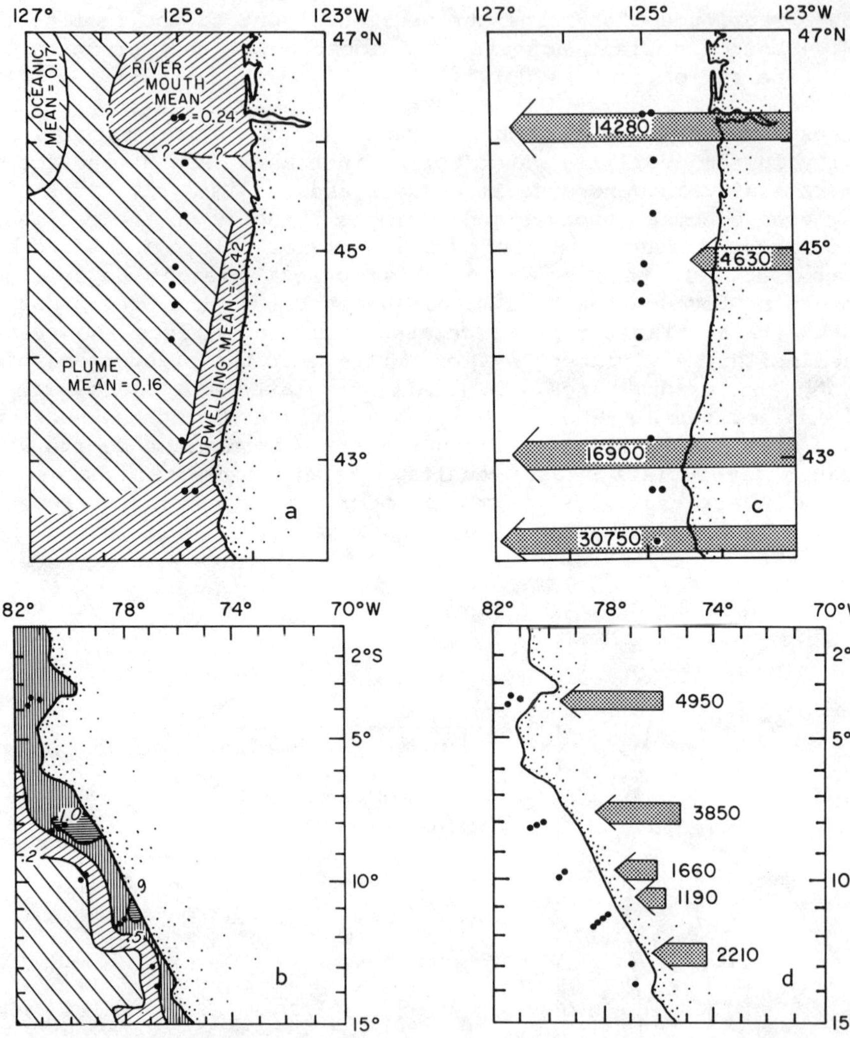

Fig. 2. Surface primary productivity (gC/m^2/day) and the input of suspended material by rivers (10^3 tonnes/yr) on the Oregon and Peru continental margins.

a. Surface primary productivity off Oregon; data from Anderson (1964). All values are less than 0.5 gC/m^2/day.

b. Surface primary productivity off Peru; data from Zuta and Guillen (1970). Values greater than 0.5 gC/m^2/day occur over most of the Peru margin.

c. Regional suspended sediment input to the Oregon margin; data from Karlin (1980). Extremely large amounts of material are derived from coastal rivers of the Pacific Northwest.

d. Suspended sediment input to the Peru margin; estimated from discharge values of Zuta and Guillen (1970). Values generally decrease from north to south, and are much less than those from the Oregon margin.

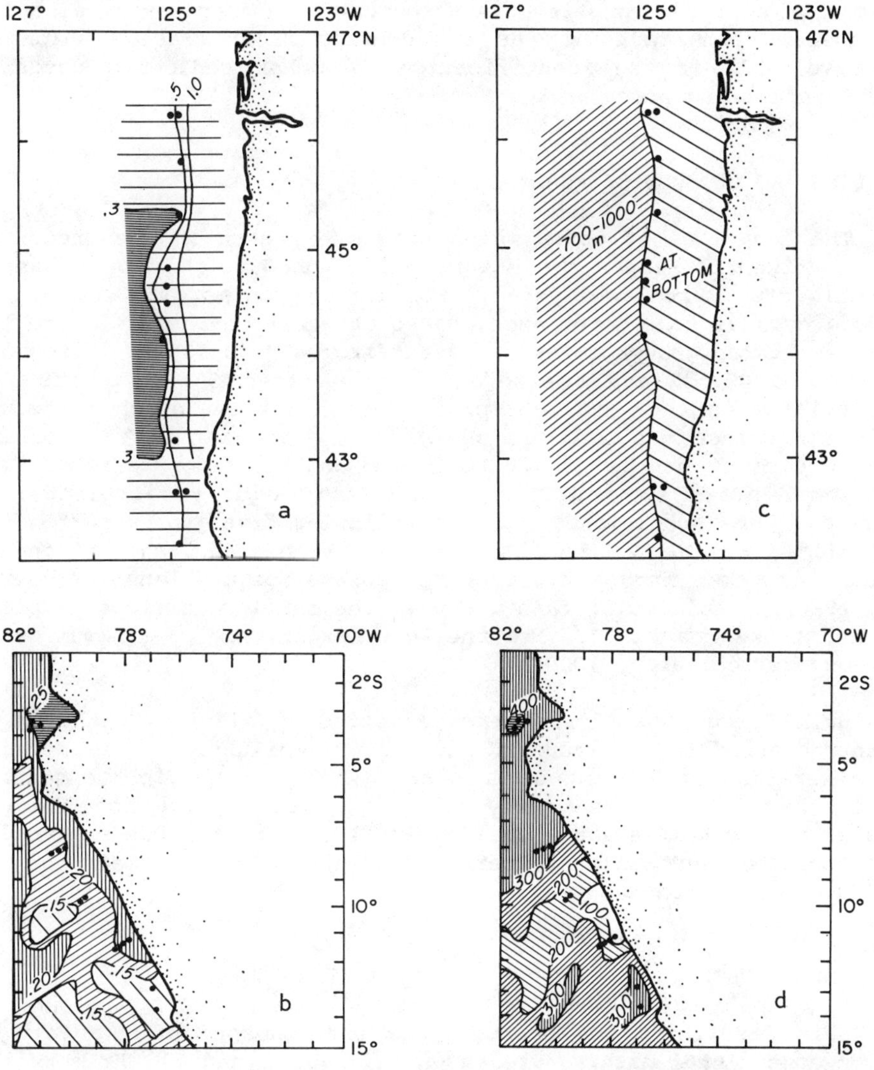

Fig. 3. Intensity and topography of the O_2-minimum zones on the Peru and Oregon continental margins.

a. Minimum O_2 concentrations on the Oregon margin; data from assorted Oregon State University reports (see text for references). Lowest minimum values approximately 0.3 ml O_2/ℓ.

b. Minimum O_2 concentrations on the Peru margin; data from Zuta and Guillen (1970). The O_2-minimum becomes more intense from north to south, and is everywhere less than 0.3 ml O_2/ℓ.

c. Topography of the O_2-minimum off Oregon; see text for references. Above 700-1000 m, lowest O_2 concentrations occur at the bottom; the O_2 minimum occurs at 700-1000 m in deeper water.

d. Topography of the O_2 minimum off Peru; data from Zuta and Guillen (1970). The O_2 minimum shoals from north to south.

effects on sediment constituents through time and space are discussed in general terms, allowing us to identify major tracers which may prove valuable in the identification of other sediments deposited beneath upwelling zones.

SAMPLING AND METHODS OF ANALYSIS

The locations of cores which have been taken on the Oregon and Peru continental slopes are shown in Fig. 1, and the cores sampled for this study are highlighted. For the Oregon margin, we selected three groups of four cores each, taken at water depths of 750 to 1050 m. These three groups cover the areas from 42° to 43°N, 44° to 45°N, and 45° to 46.5°N and are referred to as the 43°, 44°, and 46°N groups later in the text. Samples from this water depth range were selected because, as will be shown later, the regional O_2-minimum layer occurs at depths of 700 to 1000 m on the Oregon margin. Fourteen samples from above 1000 m on the Peru margin were selected for study and were subdivided into the following five groups: 3°S, 8°S, 10°S, 11°S, and 13°S. The 13°S is a composite group, and it consists of samples taken from a distinctive upper slope mud lens (Krissek et al., 1980). Included in this group is the shallowest slope sample at 11° and those near 14°S. For the Peru margin, the O_2-minimum generally occurs well above 1000 m.

Bulk textural analyses were performed by sieving, settling and decanting according to standard techniques, while silt size distributions were measured using sedimentation balance and pipette methods. Results have been presented by Krissek et al. (1980) for the Peru samples and by Krissek and Scheidegger (1980) for the Oregon samples. Organic carbon amd carbonate carbon concentrations were measured with a LECO WR-12 carbon analyzer.

DATA AND RESULTS

Data describing the hydrographic environments of the Peru and Oregon continental margins are shown in Figs. 2 and 3. Primary productivity values off Oregon were obtained from Anderson (1964) and represent measurements made over a period of two years. The primary productivity off Oregon (Fig. 2a) is approximately .2 $gC/m^2/day$, with higher values along the continental shelf. Primary productivity data for the Peru margin are from Zuta and Guillen (1970) and are a summary of over 1,000 measurements made over a 10-year period (1960-1970); many of these stations were reoccupied several times during this period. The Peru margin is marked by higher productivity (Fig. 2b) than the Oregon margin, with values commonly greater than 0.4 $gC/m^2/day$, and with localized zones of very high productivity at 8°S and 11°S. Because of the nature of the data used for this map, we can infer that this pattern of localized regions of high productivity is persistent on a time scale of at least decades during interglacial periods (i.e., when upwelling is occurring off the Peru coast, highest primary productivity is consistently found in these areas).

Estimates of fluvial suspended sediment to the Northwest Pacific are from Karlin (1980) and were calculated using water-discharge records from 2 to 19 years in length. The margin off Oregon is subject to extremely large terrigenous inputs from several major source areas (Fig. 2c; Karlin, 1980; Krissek and Scheidegger, 1980) reflecting the abundant precipitation and the physiography of this region. Input of suspended material by Peruvian coastal rivers was estimated from water discharge records covering 13 to 50-year periods (Zuta and Guillen, 1970) by assuming that one half of the yearly discharge occurred as flood events, and by assuming flood concentrations to be 1.0 g/ℓ (Renard and Lane, 1975). Due to the extremely arid nature of the Peruvian coastal plain (Johnson, 1976), fluvial inputs to the Peru margin are estimated to be smaller by a factor of 3 to 10 (Fig. 2d). A decrease in the annual rainfall from north to south (Krissek et al., 1980) causes a similar drop in the regional suspended sediment input pattern (Fig. 2d), though the presence of two major streams produces a slight increase south of 11°S. Thus, in terms of the input of sediment components, the Oregon margin appears to have a consistently low input of biogenous material and a consistently high input of terrigenous material, while the Peru margin has high but variable biogenous input and lower but variable terrigenous input.

Data on the intensity and topography of the O_2-minimum layer off Oregon (Figs. 3a and 3c) were compiled from 12 years of repeated measurements published in assorted technical reports at Oregon State University (Wyatt and Kujala, 1962; 1963; Wyatt and Gilbert, 1967; Wyatt et al., 1967; 1970; 1971; 1972; Barstow et al., 1968; Barstow, Gilbert and Wyatt, 1969a; 1969b). As was the case for the productivity analyses off Peru, the O_2-minimum data (Figs. 3b and 3d) were obtained from Zuta and Guillen (1970) and represent measurements made over a ten-year period (1960-1970). Minimum O_2 values in excess of 0.3 mℓ/ℓ are found off Oregon (Fig. 3a); a more intense O_2-minimum exists off Peru (Fig. 3b) where concentrations decrease from .25 mℓ/ℓ at 3°S to less than 0.15 mℓ/ℓ at 11°-15°S. Off Oregon (Fig. 3c), this O_2-minimum lies at a depth of 700 to 1000 m. At shallower sites along the continental slope, somewhat higher O_2 concentrations (>0.5 mℓ/ℓ) occur at the sediment-water interface. In contrast, the O_2-minimum off Peru shoals from approximately 400 m at 4°S to 100 m or less at 10°-12°S (Fig. 3d). As a result, poorly oxygenated waters lie at the sediment-water interface on the upper continental slope along most of the Peru margin, especially in the region from 10°S to 13°S.

Because bottom currents may also influence the sedimentary deposits which form under zones of high productivity, we have compiled the available physical oceanographic observations from the Peru and

Oregon continental slopes. Brockmann et al. (1980) have reported results of current meter measurements on the Peru margin. They found that maximum poleward flow was at a depth of about 100 m at both 5° and 15°S, although such flow was stronger and more persistent at 5°S than at 15°S. They also concluded, from a comparison of the means and standard deviations of velocity data taken between 200 m and 800 m, that the undercurrent is stronger (0.2 to 10.8 cm/sec) off northern Peru (5° to 12°S) than off central and southern Peru (0.2 to 5.0 cm/sec at 12°S to 15°S). Unpublished current meter data obtained from the Oregon slope by Oregon State University personnel (A. Huyer, pers. comm., 1982) show near-bottom speeds of 5-10 cm/sec above 500 m,

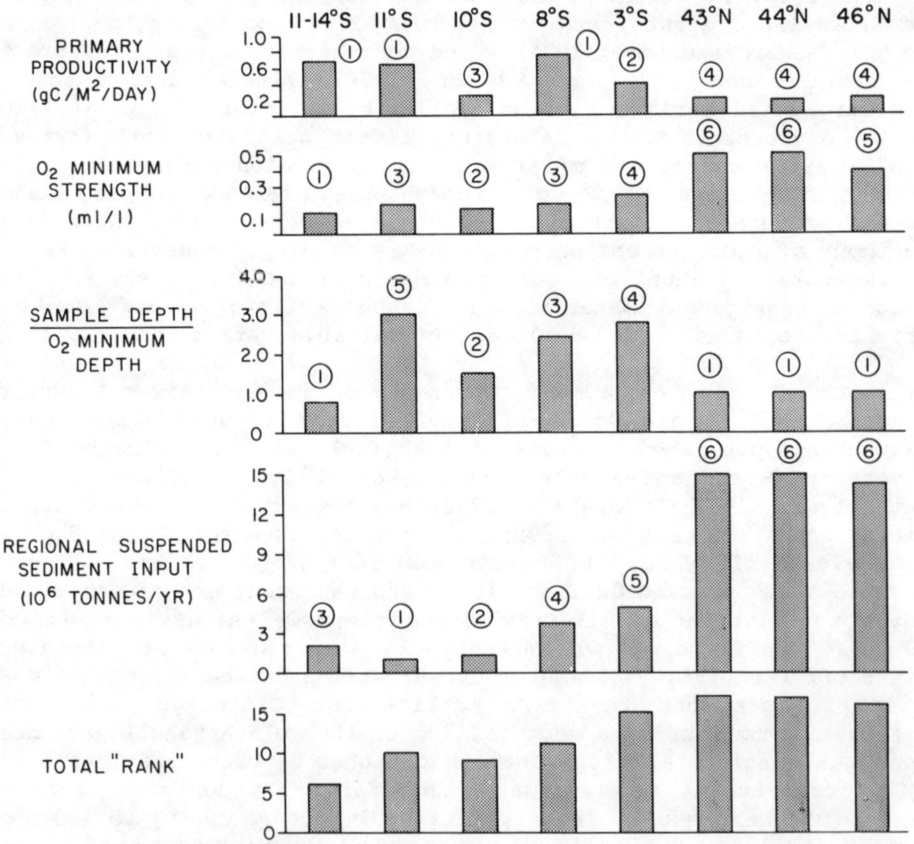

Fig. 4. Ranking (numbers shown in circles) of the eight sets of samples for four environmental parameters and their total favorability for the production, preservation, and predominance of an upwelling signal in the underlying sediments. Because of uncertainties associated with the initial data, values which differ by less than 15% are assigned equal ranks. Low values imply most favorable conditions, high values imply least favorable conditions.

2-5 cm/sec at 500 m to 1000 m water depth, and <5 cm/sec below 1000 m year-round. Hickey (1979) shows similar near-bottom speeds for the Washington continental slope. Thus, the northern Peru margin appears to be influenced by stronger near-bottom currents than the central and southern Peru and the Oregon continental slopes at bottom depths below 500 m.

To examine the influence of the total environment on the production, preservation and predominance of soft organic and hard biogenic components in the underlying sediments, it is necessary to assess the combined effect of these previously discussed environmental factors. To perform such an assessment, we began by ranking the eight regions relative to one another for each environmental parameter considered, with a value of one given for most favorable conditions and higher values assigned in order of decreasing favorability. Thus, values of one were assigned to the regions with highest primary productivity, most intense O_2-minimum, closest correspondence of sample depth and O_2-minimum location, and minimum regional suspended sediment input. The primary productivity is an indicator of the initial input of organic and biogenic material to a region, while the O_2-minimum concentration and O_2-minimum location relative to sample depth play an important role in their preservation in the water column and at the sediment-water interface by limiting bacterial and benthic degradation. Regional input of suspended material is an indicator of potential dilution of the upwelling signal by terrigenous sediment components.

The values which resulted from this ranking are shown in Fig. 4 for the four environmental characteristics of each region. Rankings for each region were then summed to give the total "rank" shown. Such a combination scheme is purely empirical (i.e., unsupported on any theoretical grounds), but it provides a simple and direct characterization of the relative favorability of these eight regions for the production, preservation, and predominance of soft organic and hard biogenic sediment components. As a result of this characterization, the region at 13°S appears to be most favorable for an organic/biogenic signal, the regions at 8°, 10°, and 11°S are slightly less favorable but similar among themselves, and the regions at 43°, 44° and 46°N and 3°S are comparable and least favorable for an organic/biogenic signal.

With these environmental summaries available, we then examined the textural and compositional signature of the underlying sediments in an attempt to discern systematic variations related to the relative difference in the nature of the depositional environments. The results of this investigation are presented and discussed here, dealing first with sediment composition and bulk textural characteristics and then considering environmentally sensitive changes in the silt-size fraction. If certain compositional and textural signatures can be shown to characterize modern high productivity/low oxygen environ-

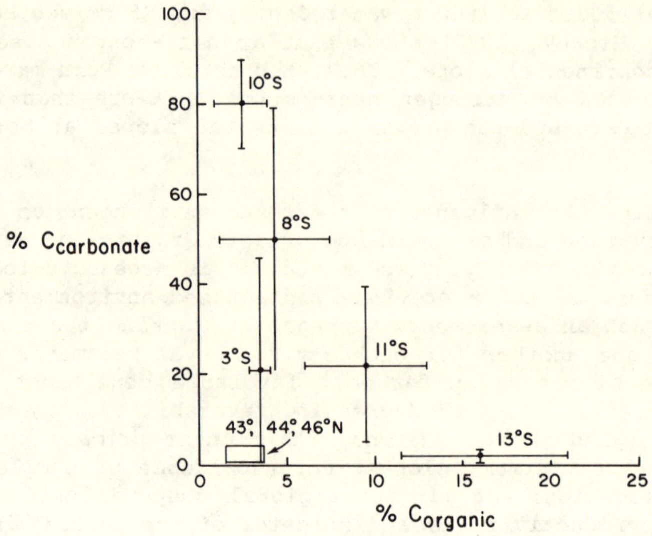

Fig. 5. Plot of organic carbon and carbonate carbon contents for the eight sample sets, showing an inverse correlation. Terrigenous-dominated samples (3°S, 43°, 44°, and 46°N) have both low carbonate carbon and low organic carbon contents.

ments, these distinctive signatures may provide a tool for identifying similar environments in older sediments and the rock record independent of micropaleontologic and geochemical methods. In each of the following plots, the data are shown as the mean ±1 standard deviation for each area.

Figure 5 illustrates the strong inverse relationship which exists between organic-carbon and carbonate-carbon contents in these samples. Samples from 13°S have extremely high organic-carbon contents, reflecting the favorable depositional environment in this area (Fig. 4). Samples from 8°, 10° and 11°S have low to intermediate organic-carbon contents and intermediate to high carbonate-carbon contents, the result of increased O_2 concentrations and related organic matter decomposition at the sediment-water interface. The least favorable environments (3°S, 46°, 44°, and 43°N) have relatively low contents of both organic-carbon and carbonate-carbon, a result believed to be associated with reduced productivity, increased degradation at the sediment-water interface, and dilution by terrigenous components.

The sand-sized material in these samples is predominantly calcareous, reflecting its biogenic origin. The coarsest, most carbonate-rich samples occur at 10°, 8° and 11°S as a result of the relatively high productivity in these areas and the location of this material below the core of the O_2-minimum zone. This may be enhanced by the winnowing action of contour-type currents, especially along the northern Peru margin (Brockmann et al., 1980). Finer-grained, carbonate-poor samples at 3°S and 43°, 44° and 46°N reflect dilution by fine terrigenous material, while clay-rich carbon-poor sediments at 13°S are a result of the predominance of aggregates of organic and fine-grained terrigenous material within the upper slope mud lens (Krissek et al., 1980).

Relationships of increasing clay and increasing organic-carbon contents similar to that shown in Fig. 6 have been described previously in other environments (Gross et al., 1972; Lisitzin, 1972). In this case, the most favorable region (13°S, as described in Fig. 4) has highest organic-carbon and clay abundances with decreasing values for the intermediate regions (8°, 10° and 11°S). The least favorable regions (3°S and 43°, 44° and 46°N) have moderately low organic-carbon contents but are enriched in clay relative to other samples with similar organic-carbon values. The samples from 13°S reflect the favorable environmental conditions (high productivity samples taken from the core of an intense O_2-minimum) and are strongly influenced by the mechanism of sedimentation for this material. As discussed in Krissek et al. (1980) and Scheidegger and Krissek (1982), these sediments consist of abundant organic-carbon-rich fecal pellets that contain fine-grained terrigenous material. In addition, the work of Staresinic et al. (this volume) suggests that the bulk of these fecal

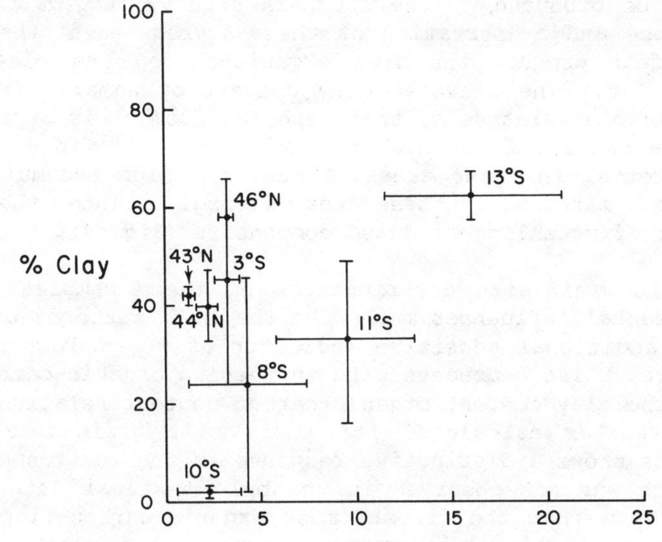

Fig. 6. Plot of clay vs. organic carbon content, showing positive correlation. Terrigenous-dominated samples contain excess clay relative to organic carbon due to importance of fluvial material.

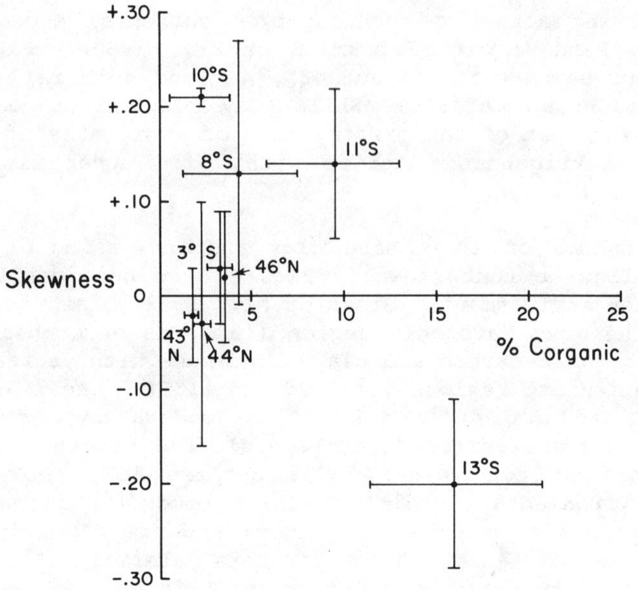

Fig. 7. Plot of silt frequency distribution skewness vs. organic carbon content, showing the distinctive textural character of organic-rich and carbonate-rich sample sets.

pellets may be produced by the southern anchovy, *Engraulis ringens*. The prevalence and preservation of these aggregates at the sediment-water interface produce the high organic-carbon and clay contents shown in Fig. 6. The elevated clay content of samples from 3°S and the Oregon margin relative to that expected from their organic-carbon content is a result of the higher fluvial input of fine-grained terrigenous material in these areas. Thus, the high accumulation rate of terrigenous material in these areas strongly dilutes the relatively low input of upwelling-produced components (Fig. 4).

The silt grain-size distributions of these samples respond to the environmental influences much like the bulk textures do and thus provide an additional sensitive indicator of the sediment response. Mean silt grain size decreases with increasing organic-carbon content similar to the clay content-organic-carbon content relationship. The skewness parameter calculated for these silt grain-size frequency distributions shows a distinctive response to the environment of deposition which was not observed in the bulk textural data (Fig. 7). These data show that the finest, most organic-carbon-rich material, found in the most favorable environment (13°S), is characterized by strongly negative skewness values, indicating the presence of a tail of coarse silt. Such a distribution may reflect the efficiency of the biological filter-feeders and associated nekton in removing materials from suspension, especially the finest silts, and packaging it into fecal pellets. Samples from the intermediate environments at 8°, 10° and 11°S are the coarsest in texture, lowest in organic-carbon contents, and have low to strongly positive skewness values indicating a fine silt tail on their frequency distributions. These are carbonate-rich deposits, and their silt size fraction is dominated by the presence of coarse biogenic carbonate debris. In this case, much of the fine silt and clay may have been winnowed by the relatively strong bottom currents reported by Brockmann et al. (1980). Samples from the least favorable environments (3°S, 43°, 44°, and 46°N) have intermediate mean silt values, intermediate organic-carbon contents, and mean skewness values in the range +0.03 to -0.03. Such characteristics reflect the strong influence of dilution by fluvially-transported terrigenous material with little opportunity for selective removal of coarse or fine silt components by either biological or physical oceanographic processes.

DISCUSSION

The preceding data have shown that interrelationships between sources of sediment components (biogenic/organic vs. terrigenous) and dissolved oxygen characteristics of the water column exert a strong control on the texture and composition of the material which is deposited. In addition, we briefly discussed the effect of bottom current activity on the sediments, especially in the case of the coarse-grained carbonate-rich material at 8°, 10° and 11°S. So far, however, we have known the environmental characteristics of the areas of interest; therefore, we have been able to explain sediment textural and compositional variations in terms of our understanding of their environments of formation and deposition. Such environmental data are not always available, especially in studies of older deposits. In addition, certain environmental controls (i.e., the CCD, strong bottom currents, or a weak O_2-minimum zone) may act to completely remove one or more of the characteristic tracers of deposition beneath an area of upwelling. As a result, each of the four major classes of material which can originate in the surface water layer of an upwelling zone (biogenic silica, biogenic carbonate, aggregates of organic and fine-grained terrigenous silts and clays [including fecal pellets], and coarser terrigenous material) must be investigated independently to ensure that an upwelling signature, if present, is recognized.

Figure 8 presents a listing of the four major sediment components and their characteristics which may be used to detect an upwelling influence within a deposit, and describes their relative usefulness under various environmental conditions. This matrix was constructed for when productivity influence is greater than the input of detrital material. As such, it is probably more relevant to the 13°, 11°, 10° and 8°S areas considered in this study than to the regions more dominated by higher detrital input at 3°S and 43°, 44° and 46°N.

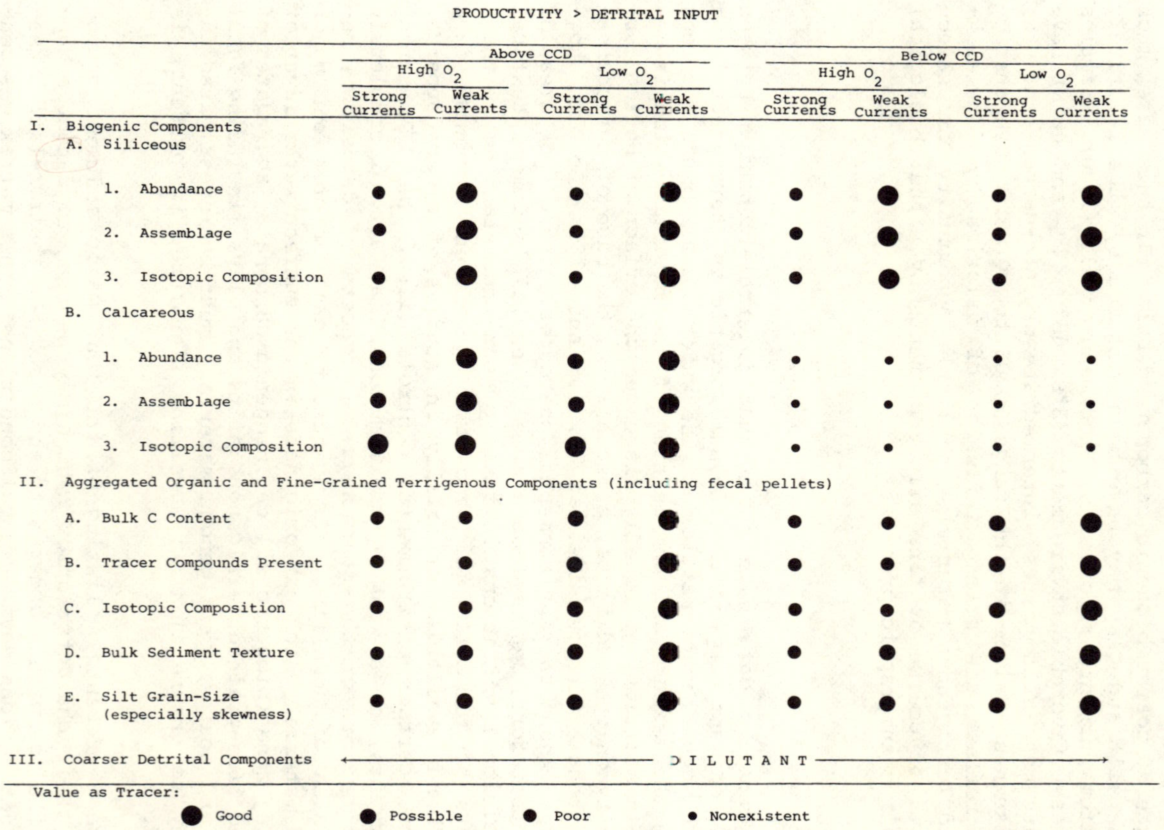

Fig. 8. Summary of the importance of upwelling-sensitive constituents in sediments and their capability to act as a marker of an upwelling influence under various environmental conditions. The strongest and most complete upwelling signal is preserved above the calcium carbonate compensation depth (CCD), under low O_2 conditions, and under the influence of weak bottom currents.

With a higher detrital input, dilution of the organic and biogenic constituents would occur, reducing their abundances and perhaps also posing analytical problems for the extraction and measurement of minor upwelling-associated sediment components (siliceous microfossils, biogenic and organic phase isotopic compositions, and biogeochemical tracer compounds).

The abundance of biogenic siliceous components is controlled by both the relative magnitudes of biogenic and detrital inputs and by the winnowing effect of strong bottom currents. Their total assemblage is best preserved and interpreted in low energy environments (Schuette, 1980); in addition, the isotopic signal of oxygen in diatom silica (Juillet, Labeyrie, and Schrader, this volume) should be most representative in a total upwelling assemblage, such as is found under low energy conditions.

The upwelling signal carried by the calcareous sediment components is controlled strongly by both the relative importance of biogenic and detrital inputs and the location of the CCD relative to the sediment-water interface. If the site of deposition is below the CCD, the entire carbonate signal will be removed, thereby providing no evidence of the zones of high productivity. However, if deposition occurs above the CCD, then the carbonate constituents provide distinctive abundance, assemblage, and isotopic tracers (Labracherie et al.; Ganssen and Sarnthein, both this volume). The abundance signal of the carbonate material is intensified by the winnowing effects of strong bottom currents, as we have seen for the study areas at 8°, 10°, and 11°S on the Peru continental slope. Under such high-energy conditions, in fact, the carbonate signal may provide the only evidence of an upwelling influence due to the lack of siliceous and organic phases.

Aggregates of organic material and fine-grained terrigenous detritus, especially in the form of fecal pellets, provide a distinctive indicator of the coexistence of an upwelling influence and favorable environmental conditions (Krissek et al., 1980; Scheidegger and Krissek, 1982). Under these favorable conditions of low O_2 concentration at the sediment-water interface and calm bottom waters, bulk organic-carbon contents are high (Reimers, 1981), allowing the identification of characteristic organic compounds (Brassell and Eglinton; Gagosian, Volkman, and Nigrelli; Simoneit, all this volume) and their isotopic composition (Reimers, 1981). As the O_2-minimum becomes less intense or ceases to impinge on the sediment-water interface, or as current activity increases, bulk organic-carbon abundances decrease (Krissek et al., 1980; Summerhayes, this volume), providing a less representative organic signal for trace-compound and isotopic analyses. In addition, lower organic-carbon abundances may cause analytical difficulties in the extraction and measurement of these minor constituents. Bulk sediment texture provides a distinctive upwelling signal only under favorably low-O_2 and calm bottom-water conditions, when the fine silt-clay-rich material bound in fecal pellets remains in large aggregate forms. When this material becomes disaggregated, it can easily be winnowed by bottom currents, removing its signal from the sediments. However, the coarse texture of these winnowed deposits (as in the samples from 8°, 10°, and 11°S) provides evidence of such modification by winnowing, thereby revealing valuable environmental information. Silt grain-size distributions, and especially the skewness parameter, are also sensitive to near-bottom conditions.

Finally, coarser detrital components act as a dilutant within the system and provide little or no information on surface-water processes. However, they may reveal the dominant processes acting along the bottom (turbidity flows, mass movements, or strong onshore/offshore vs. alongshore currents).

In summary, we see that the most complete upwelling signal occurs in sediments from calm, low-O_2 sites above the CCD when the productivity influence exceeds the input of detrital material. The calcareous signal disappears below the CCD, the organic signal disappears in high-O_2 waters, and the fine-grained signal (both detrital and biogenic) disappears in current-influenced high-O_2 environments.

Therefore, when attempting to classify recent or older sediments as upwelling-influenced or not, it is necessary to examine all of these constituent groups, since unknown environmental conditions at or after the time of deposition may have destroyed the upwelling signal carried by an individual component. As a result, such an examination must then involve a major commitment and cooperation of researchers with various analytical techniques.

ACKNOWLEDGEMENTS

The Peru margin samples and analyses used in this study were obtained with support by the Office of Naval Research through contract N00014-76-C-0067, while the Oregon portion of the study was supported by the National Science Foundation through contract OCE-7819825. Analyses of the Peru samples were performed by K.K. Klaffke (sediment texture) and P.A. Price (C-organic and $CaCO_3$). Organic-carbon analyses of the Oregon samples were performed by K. Weliky. M. Ledbetter provided a thoughtful review of the manuscript.

REFERENCES

Anderson, G.D., 1964, The seasonal and geographic distribution of primary productivity off the Washington and Oregon coasts, Limnology and Oceanography, 9:284-302.

Barstow, D., Gilbert, W., Park, K., Still, R., and Wyatt, B., 1968, Hydrographic data from Oregon waters, 1966, Data Report No. 33, Ref. 68-34, Oregon State University, Corvallis, 109 pp.

Barstow, D., Gilbert, W. and Wyatt, B., 1969a, Hydrographic data from Oregon waters, 1967, Data Report No. 35, Ref. 69-3, Oregon State University, Corvallis, 77 pp.

Barstow, D., Gilbert, W. and Wyatt, B., 1969b, Hydrographic data from Oregon waters, 1968, Data Report No. 36, Ref. 69-6, Oregon State University, Corvallis, 84 pp.

Brockmann, C., Fahrbach, E., Huyer, A., and Smith, R.L., 1980, The poleward undercurrent along the Peru coast: 5 to 15°S, Deep-Sea Research, 27A:847-856.

Gross, M.G., Carey, A.G., Jr., Fowler, G.A., and Kulm, L.D., 1972, Distribution of organic carbon in surface sediment, northeast Pacific Ocean, in: "The Columbia River Estuary and Adjacent Ocean Waters," A.T. Pruter and D.L. Alverson, eds., University of Washington Press, Seattle, 254-264.

Hickey, B.M., 1979, The California current system--hypotheses and facts, Progress in Oceanography, 8:191-279.

Johnson, A.M., 1976, The climate of Peru, Bolivia, and Ecuador, in: "World Survey of Climatology: Climates of Central and South America," Vol. 12, W. Schwerdtfeger, ed., Elsevier Scientific, New York, 147-218.

Karlin, R., 1980, Sediment sources and clay mineral distributions off the Oregon coast, Journal of Sedimentary Petrology, 50:543-559.

Krissek, L.A. and Scheidegger, K.F., 1980, Composition, sources, and dispersal of hemipelagic sediments on the Oregon continental margin, Geological Society of America, Annual Meeting, 12:466 (abstract).

Krissek, L.A., Scheidegger, K.F. and Kulm, L.D., 1980, Surface sediments of the Peru-Chile continental margin and the Nazca Plate, Geological Society of America, Bulletin, Part 1, 91:321-331.

Lisitzin, A.P., 1972, Sedimentation in the world ocean, Society of Economic Paleontologists and Mineralogists, Special Publication No. 17, 218 pp.

Reimers, C.E., 1981, "Sedimentary Organic Matter: Distribution and Alteration Processes in the Coastal Upwelling Region off Peru," Ph.D. Dissertation, Oregon State University, Corvallis, 219 pp.

Renard, K.G., and Lane, L.T., 1975, Sediment yield as related to stochastic model of ephemeral runoff, in: "Present and Prospective Technology for Predicting Sediment Yields and Sources," Ref. ARS-1-40, U.S. Department of Agriculture, Agricultural Research Station, Southern Region, New Orleans, 253-263.

Scheidegger, K.F. and Krissek, L.A., 1982, Dispersal and deposition of eolian and fluvial sediments off Peru and northern Chile, Geological Society of America, Bulletin, 93:150-162.

Schuette, G., 1980, "Recent Marine Diatom Taphocoenoses off Peru and off Southwest Africa: Reflection of Coastal Upwelling," Ph.D. Dissertation, Oregon State University, Corvallis, 115 pp.

Wyatt, B. and Kujala, N.F., 1962, Hydrographic data from Oregon coastal waters, June 1960 through May 1961, Data Report No. 7, Ref. 62-6, Oregon State University, 77 pp.

Wyatt, B. and Kujala, N.F., 1963, Hydrographic data from Oregon waters, June through December 1961, Data Report No. 12, Ref. 63-33, Oregon State University, 36 pp.

Wyatt, B. and Gilbert, W.E., 1967, Hydrographic data from Oregon waters, 1962 through 1964, Data Report No. 24, Ref. 67-1, Oregon State University, 175 pp.

Wyatt, B., Still, R., Barstow, D., and Gilbert, W., 1967, Hydrographic data from Oregon waters, 1965, Data Report No. 27, Ref. 67-28, Oregon State University, Corvallis, 56 pp.

Wyatt, B., Gilbert, W., Gordon, L., and Barstow, D., 1970, Hydrographic data from Oregon waters, 1969, Data Report No. 42, Ref. 70-12, Oregon State University, Corvallis, 155 pp.

Wyatt, B., Tomlinson, R., Gilbert W., Gordon, L., and Barstow, D., 1971, Hydrographic data from Oregon waters, 1970, Data Report No. 49, Ref. 71-23, Oregon State University, Corvallis, 134 pp.

Wyatt, B., Tomlinson, R., Gilbert W., Gordon, L., and Barstow, D., 1972, Hydrographic data from Oregon waters, 1971, Data Report No. 53, Ref. 72-14, Oregon State University, Corvallis, 77 pp.

Zuta, S. and Guillén, O., 1970, Oceanografia de las aguas costeras del Peru [Oceanography of the coastal waters of Peru], Instituto del Mar del Peru Boletin, 2:161-323.

COASTAL UPWELLING, ITS INFLUENCE ON THE GEOTECHNICAL PROPERTIES AND STABILITY CHARACTERISTICS OF SUBMARINE DEPOSITS

George H. Keller

School of Oceanography
Oregon State University
Corvallis, Oregon 97331, U.S.A.

ABSTRACT

Studies along the continental margins of Peru indicate that the coastal upwelling process, by concentrating organic matter in the underlying and nearby bottom sediments, indirectly contributes significantly to the alteration of the geotechnical properties and stability characteristics of these deposits. Those sediments in close proximity to areas of intense upwelling possess distinctly different mass physical properties than do those of comparable sediment type some distance away. The ability of organic matter to adsorb water and to cause clay-size particles to aggregate to form an open fabric appears to result in unusually high water contents (853% by dry weight), and plasticity as well as exceptionally low bulk densities (1.09 Mg/m^3). Undrained shear strength of these sediments is also unexpectedly high, resulting apparently from some form of bonding of the sediment by organic agents. Sensitivity (ratio of the natural to the remolded or disturbed shear strength) is uniquely high (21) indicating high susceptability to failure should the sediments become severely disturbed. All the organic-rich sediments studied behave as if they are overconsolidated. The higher the organic content the greater the degree of overconsolidation. In some areas the degree of overconsolidation is on the order of 6 to 7 times that of other slope deposits with similar characteristics but differing only in that they have relatively low organic contents. This suggests that organic-matter related interparticle bonding may be responsible for this unusual degree of apparent overconsolidation. Owing to the unique geotechnical properties of these continental margin deposits, their stability characteristics are considered to be poor under circumstances in which excess pore pressures are generated. Disturbance or generation of excess pore pressure may be by a number of means such as rapid sedimentation, cyclic loading by waves or seismic shock. Failure of these deposits appears to take the form of a fluidized flow somewhat similar to the landslides in the quick clays of Norway and Canada. Such submarine failures are likely generators of turbidity flows.

INTRODUCTION

The occurrence of upwelling and high primary productivity along with correspondingly high concentrations of organic carbon in the bottom sediments is a common association found at many of the ocean margins, especially along the western margins of bordering land masses. This relationship of relatively high organic carbon content in bottom sediments and the upwelling process on a regional scale is especially pronounced along the western margin of North and South America as well as off southwest Africa (Premuzic, 1980).

The concentration of organic matter in bottom deposits underlying or in close proximity to areas of intense upwelling is often 10 to 20 times greater than is commonly found in submarine sediments. It might be expected that such abnormal concentrations of organic carbon would influence or alter not only the chemistry of the sediment into which it is incorporated, but also the physical or geotechnical properties of these sediments (Söderblom, 1966; Pusch, 1973; Rashid and Brown, 1975). Indirectly, therefore, there is an expectation that sediments underlying or adjacent to areas of intense upwelling will display geotechnical properties quite different from deposits having similar textural properties, but which are far removed from the influence of the upwelling process. As with most generalities, exceptions are found when other factors are involved such as dilution by large local supplies of clastic sediments entering an area, e.g., off northwest Africa, or when there is no well-developed oxygen minimum zone intersecting the bottom to prevent oxidization of the organic carbon.

ENVIRONMENTAL SETTING

The intensity of upwelling and the associated levels of primary production along the Peru margin are among the highest reported (Zuta and Guillen, 1970). Rowe (1971) found that productivity in this area of intense upwelling is considerably higher (10 to 50 times) than in areas not displaying upwelling. This high productivity in turn increases the biomass significantly such that organic carbon concentrations within the bottom sediments are abnormally high, up to 20% (Busch and Keller, 1981). Owing to the unusually high concentrations of organic carbon found along the Peru margin in association with areas of intense upwelling, the outer shelf and continental slope deposits off Peru were selected for study to examine the influence of organic matter on the geotechnical properties and stability characteristics of the submarine deposits.

In addition to the oceanographic and biological factors which lead to the occurrence of high concentrations of organic matter in the bottom sediments, the physiographic configuration of the Peruvian continental margin, primarily due to its width, is also an important factor in making the area ideal for this study. Although the continental shelf varies in width from 6 to 125 km, more commonly it is less than 35 km wide and is essentially flat. The continental slope is relatively narrow (50 to 90 km) and steep, with an average gradient of 2° (Masias-Echegaray, 1976). Bordering the margin is the prominent Peru-Chile Trench which forms a pronounced topographic barrier to the seaward migration of sediment by turbidity currents (Schweller and Kulm, 1978) and also possibly inhibits a similar transport of suspended material in the bottom nepheloid layer.

Sediment cover on the Peru margin is relatively thin, usually on the order of 50 m or less (Krissek, Scheidegger and Kulm, 1980; Busch and Keller, 1981). These surface and near-surface deposits do not form a continuous blanket over the shelf and slope, but comprise various mud layers and lens interspersed among regions barren of sediment. Krissek et al. (1980) have presented a more detailed discussion on the geometry of sediment accumulation along the Peru margin. On the steeper lowermost flanks of the continental margin, sub-bottom profiles indicate that sediments are confined to local basins (Prince and Kulm, 1975). Sedimentation rates on the outer shelf and uppermost slope, where upwelling and high productivity prevail, commonly

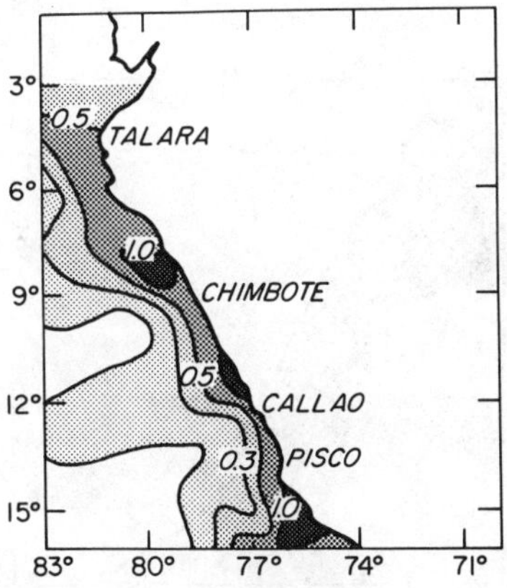

Fig. 1. Distribution of total primary production in $g \cdot m^{-2} \cdot day^{-1}$ over the Peruvian margin based on surface water measurement during 1960 to 1970 from Zuta and Guillén (1970) (modified from Reimers, 1981).

exceeding 50 cm 1000 y^{-1}, appear to coincide with isolated centers of intense upwelling. Continental slope areas less directly influenced by upwelling and high productivity have sedimentation rates ranging from 6 to 47 cm 1000 y^{-1} (Reimers and Suess, this volume).

Sediment distribution patterns and compositional characteristics along the Peru margin are strongly influenced by the general surface and subsurface currents. The impact of coastal upwelling is, however, undoubtedly the most pronounced process of the various elements making up the dynamic regime along the upper Peruvian margin. Coastal upwelling is very distinct north of 20°S and occurs most intensely in a number of centers spaced along the coast which extend as far as 50 km offshore (Zuta, Rivera and Bustamante, 1978; Jones et al., this volume; Smith, this volume). A pronounced oxygen minimum zone occurs along the Peru margin with its greatest intensity commonly found at the depth interval of 200 to 300 m. This condition is attributed to *in situ* oxygen consumption by bacteria and animals as organic matter settles, and to the poleward two-layered flow which prevails along this margin (Barber and Smith, 1981).

SEDIMENT PROPERTIES

Organic Carbon

The occurrence of the upwelling cells or centers and the transport of nutrient-rich waters to the surface accounts for the high local biological productivity (Fig. 1) which in turn leads to the observed local accumulations of organic-rich sediments. The combination of the large quantity of available organic matter, its rapid deposition, most commonly in the form of fecal pellets, and the lack

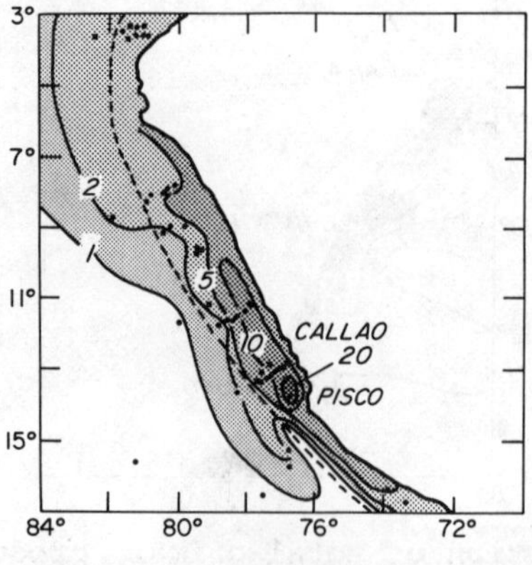

Fig. 2. Total organic carbon content in continental margin surface sediments. Dashed line denotes the axis of the Peru-Chile Trench (modified from Reimers, 1981).

or near lack of oxygen in the bottom waters results in the uniquely high concentration of organic matter in the submarine deposits of the Peru margin (Fig. 2). Although there is no doubt that the above mentioned relationship accounts for the concentration of organic matter in the bottom sediments, a close examination of Figs. 1 and 2 reveals that the areas of highest organic carbon content do not always coincide with the areas of greatest productivity. Reimers (1981) attributes this discrepancy to the effects of dilution from local sources of sediment. By using actual bulk sedimentation rates along the margin, Reimers was able to determine that the highest accumulations of organic carbon do in fact coincide with the areas of greatest productivity.

Texture and Clay Mineralogy

For the most part, the surface sediments of the Peruvian continental shelf and slope can be classed as silty clays with a few localized areas of clayey silt and clay. Between 8°S and 10°S there is little if any contemporary sedimentation and cemented biogenic sands with phosphorites, and fish debris comprise the upper slope deposits in this area (Fig. 3). In a closer examination of the silt and clay-size fractions, Krissek and others (1980) found a general fining in grain size from the south to the north and in an offshore direction. In comparing the muddy sediments which comprise most of the outer shelf and upper slope deposits there is texturally relatively little difference between those that are organic rich and those considerably lower in organic content. Typically, both types of deposits are composed of 5 to 10% sand, 25 to 40% silt, and 50 to 70% clay.

Considerable variance exists between the clay mineralogy of the organic-rich deposits and that of sediments from the Peruvian margin

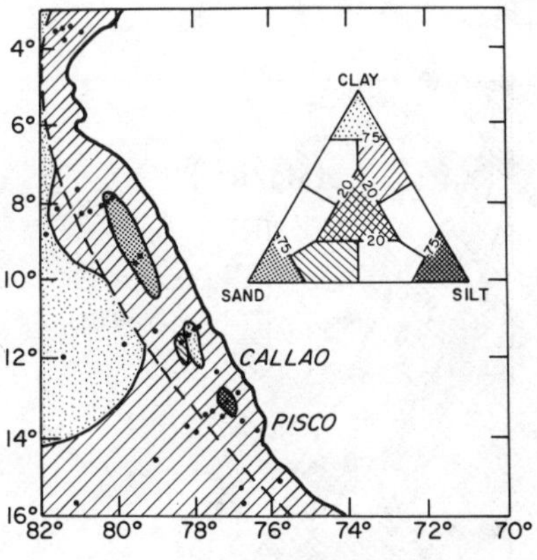

Fig. 3. Texture of surface sediments along the Peruvian continental margin. Dots represent core locations. Axis of Peru-Chile Trench shown as dashed line, modified from Busch and Keller (1981).

where the organic contents are considerably smaller. For the most part, the organic-rich sediments are high in illite (45 to 60%) and chlorite (30 to 40%), but low in smectite (5 to 20%) whereas the other Peruvian margin deposits show a reversal in this relationship (Rosato and Kulm, 1981; Busch and Keller, 1982). Smectite concentrations increase away from the organic-rich muds to the north, south and west as illite decreases. This distribution of clay minerals is attributed to variations in the continental source rocks (Rosato and Kulm, 1981).

Geotechnical Properties

As a whole, the slope and outer shelf deposits of the Peruvian continental margin possess unusually high percentages of organic carbon (2 to 20%) (Fig. 2). Unique among these deposits are those of the outer shelf and upper slope between 11°S and 14°S where organic contents of 10 to 20% are considerably higher than elsewhere along the margin. As will be shown in the following discussion, these organic-rich deposits display equally unique geotechnical properties and stability characteristics in contrast to the other Peruvian margin deposits. Geotechnical properties, such as wet bulk density, water content, and Atterberg limits, discussed in the following sections, were measured using the procedures outlined by Richards (1962). Shear strengths were obtained with a laboratory vane shear apparatus using a vane rotation rate of 60 degrees per minute.

Wet bulk density. Wet bulk density characterizes the wet weight per unit volume of a sediment and, as is shown in Fig. 4., considerable variation is found within the continental margin deposits. Mean wet bulk densities for the 0-2 m subseafloor interval commonly range from 1.20 to 1.69 Mg/m^3 with specific horizons within this interval being as low as 1.09 Mg/m^3. The organic-rich sediments possess dis-

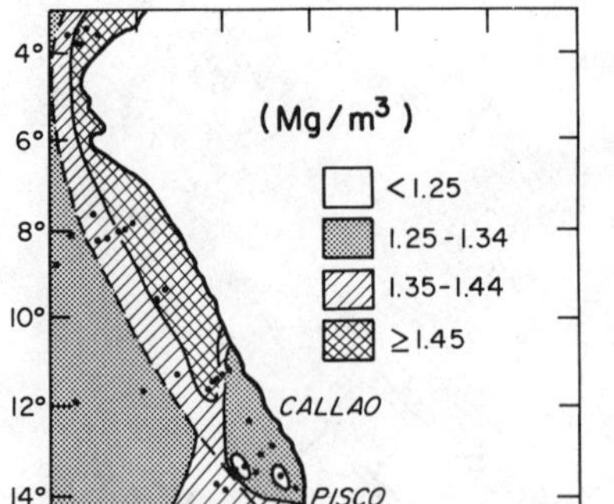

Fig. 4. Wet bulk density, mean values for the upper 2 meters of sediment.

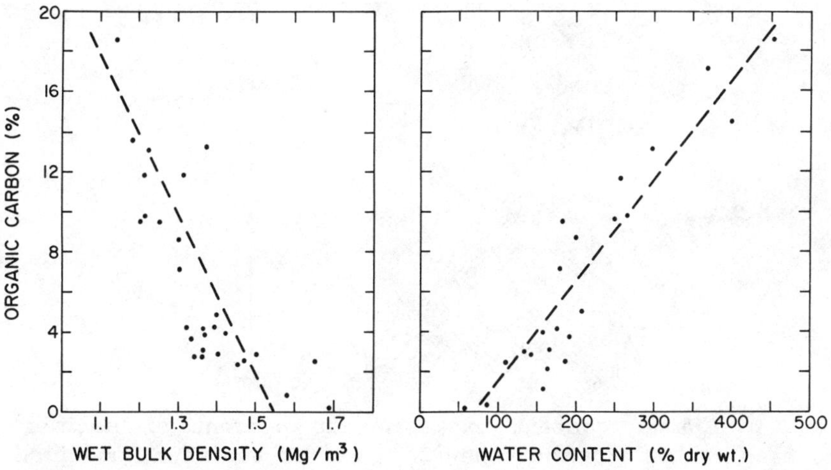

Fig. 5. Relationship of wet bulk density (a) and water content (b) to total organic content; Peruvian margin sediments, 0 to 2 meter interval.

tinctly lower mean-bulk densities (1.20 to 1.30 Mg/m^3) than do the other Peruvian deposits (1.42 to 1.69 Mg/m^3). In considering the margin deposits overall, it is clear that as the organic carbon content increases, the wet bulk density decreases (Fig. 5a). This is not surprising when it is recalled that an almost pure organic deposit such as peat may have a wet bulk density on the order of 0.5 to 1.20 Mg/m^3.

Water content. Water content, expressed as the percent of the sediment dry weight of the Peruvian deposits is also found to vary significantly in proportion to the organic content. The organic-rich sediments between 11°S and 14°S have unusually high mean water contents (>200%) in contrast to the majority of the Peruvian deposits (Fig. 6). Within the organic-rich sediments, water contents as high as 853% were measured in specific intervals. These are some of the highest values yet reported for submarine deposits. The anomalously low mean water contents (150-199%) in the vicinity of 13.6°S are attributed to the exposure of slightly more consolidated sediment as a result of the erosion of some of the more recent overburden (Busch and Keller, 1982). Relatively low water contents along the shelf and upper slope to the north are indicative of coarser-grained material and the more consolidated silty clays which comprise the bottom deposits in this area. This rather clear relationship between organic carbon content and water content is evident when a comparison of these two parameters is made for the Peruvian margin deposits as a whole (Fig. 5b).

Plasticity characteristics. Atterberg limits refer to water contents that correspond to various states of consistency of a re-

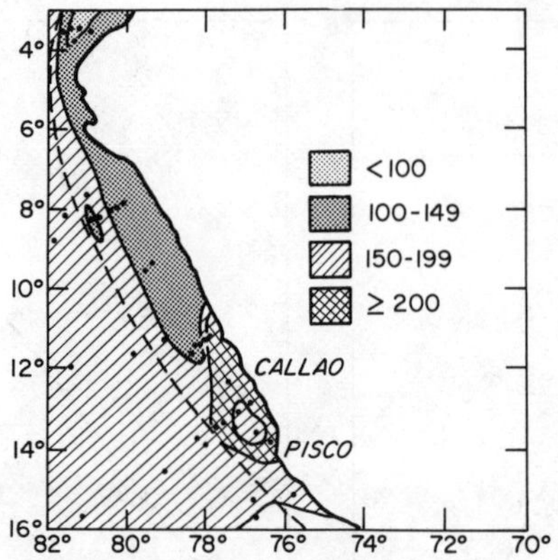

Fig. 6. Water content expressed as percent of sediment dry weight, mean values for the upper 2 meters of sediment, modified from Busch and Keller (1981).

molded sediment. Liquid limit and plastic limit are the lower-most bounds of water contents at which a sediment behaves as a liquid and plastic material, respectively. The difference betweeen these two values, the plasticity index, represents the range of water contents over which a sediment behaves as a plastic material. Using a system devised by Casagrande (1948) which classifies a sediment by its plasticity index and its liquid limit, the Peruvian margin deposits can be readily differentiated from one another. The organic-rich sediments possess distinctly higher plasticity and liquid limits than do any of the other sediments along the Peru margin or on the adjacent abyssal plain (Fig. 7). The overlap of values shown on the plastici-

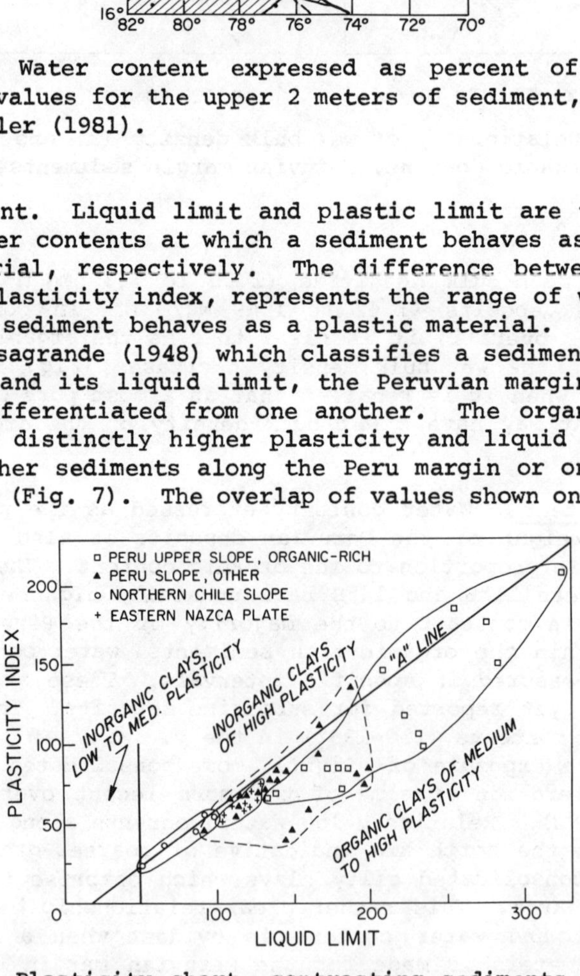

Fig. 7. Plasticity chart, contrasting sediments from the Peruvian organic-rich deposits to sediment away from centers of intense upwelling with much less organic carbon. Values are from subsamples taken from the entire cored intervals (0.5 to 3 meters). By permission from J. Sed. Petrol., Copyright (c) 1981, SEPM, Tulsa.

ty chart between the organic-rich and other Peruvian slope deposits reflects the effects of decreasing plasticity with depth of burial for the organic-rich muds and the high plasticity of the adjacent slope deposits.

Liquid and plastic limits provide a measure of the ability of a sediment particle to attract and hold water to its surface. Odell, Thornburn and McKenzie (1960) have shown that these two limits and the plasticity index increase with both increasing clay and organic content. Since textural variations of the blanketing muds are relatively small between those that are organic rich and those relatively low in organic content, the differences in the plastic indices appear to be primarily related to differences in the organic content. In considering only the clay mineralogy, higher plasticity indices and water contents would be expected in the smectite-rich clays occurring outside of the area of intense upwelling relative to the illite chlorite clays of the organic-rich deposits between 11°S and 14°S. This relationship is not found because the overwhelming effect of organic matter on these properties appears to mask the influence of the clay mineralogy. With increasing organic content, the liquid and plastic limits are found to increase (Fig. 8).

Nearly all of the Peru margin silty clay deposits are found to have a natural water content that exceeds the liquid limit. This implies that the particle-to-particle fabric is of sufficient strength to allow the natural sediment to exist as a coherent, firm mass at a water content which would cause the sediment to flow if it was in a remolded state. This property is particularly important when considering the stability characteristics of these deposits as will be discussed later.

<u>Shear strength</u>. Considerable variability in the mean natural shear strength (2 to 26 kPa) of the 0 to 2 m interval exists in the

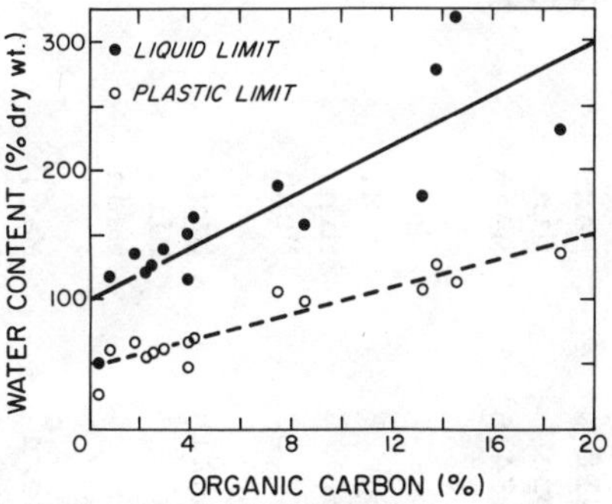

Fig. 8. The relationship between Atterberg limits and organic carbon content of Peruvian margin surface sediments. Lines are least square fits through the data. Scatter in the data reflects in part the variation in the clay content (from Busch and Keller, 1981). By permission from <u>J. Sed. Petrol.</u>, Copyright (c) 1981, SEPM, Tulsa.

Table 1. Comparison of geotechnical properties from the Peru margin to the Pacific basin and the slope and rise deposits off the northeastern United States

Region		Organic Carbon Content (%)	Water Content (% Dry Wt.)	Wet Bulk Density (Mg/M^3)[f]	Shear Strength (kPa)[g]	Sensitivity	Over-Consolidation Ratio
Peru[a]							
Outer Shelf &	Max.	21	853	1.73	34	21	17
Upper Slope	Min.	10	57	1.09	0	1	7
(11°S to 14°S)	Mean	15	215	1.30	13	9	13
Peru[a]							
Upper Slope	Max.	10	214	1.74	32	21	3
Outside of	Min.	2	48	1.44	3	2	1
11°S to 14°S	Mean	6	126	1.42	14	6	2
Pacific Basin[b]	Max.		423	2.03	56	57	[d] 11
	Min.		24	1.15	0	1	1
	Mean	[e] .3	175	1.45	5	4	1
Northeastern[c]	Max.	1.5	165	1.97	24	12	
United States	Min.	.30	33	1.31	1	1	
	Mean	.75	88	1.52	8	3	

[a] Busch and Keller (1982)
[b] Keller and Bennett (1970)
[c] Keller, Lambert and Bennett (1979)
[d] Richards and Hamilton (1967)
[e] Premiuciz (1980)

[f] in units of 10^6 gram per cubic meter
[g] in units of kilo Pascal

Peruvian continental margin sediments (Busch and Keller, 1981). Although the mean values for the organic-rich sediment do not differ appreciatively from those of the adjacent margin deposits, the margin sediments as a whole display relatively high mean shear strengths which possibly may reflect their higher than normal organic content (Table 1).

In the organic-rich sediments of the upper slope and outer shelf, the remolded shear strength, or the strength of thoroughly disturbed material, is found to decrease markedly as the organic content increases (Fig. 9). From earlier studies, there is some indication that humic acid or organic molecules such as cellulose in the presence of polyvalent cations form gel complexes which appear to have a bonding effect on the sediment particles (Pusch, 1973). Such a strengthening would impart a higher than normal natural shear strength to the sediment in a manner similar, but to a much lesser degree, to that found in calcium carbonate-rich deposits. Pusch (1973) found that when a sediment containing humic acid was disturbed, the remolded shear strength was markedly lower than that of a sediment of similar grain size and mineralogy, but with no, or considerably less, humic acid. These findings may in part explain why the Peruvian slope deposits outside of the area of high organic enrichment do not show a discernable relationship between organic-carbon content and the remolded shear strength. Although not yet clearly documented, there are indications that sediments containing less than 4 or 5% organic carbon do not show a distinct interrelationship between organic carbon and remolded shear strength (Pusch, 1973; Rashid and Brown, 1975).

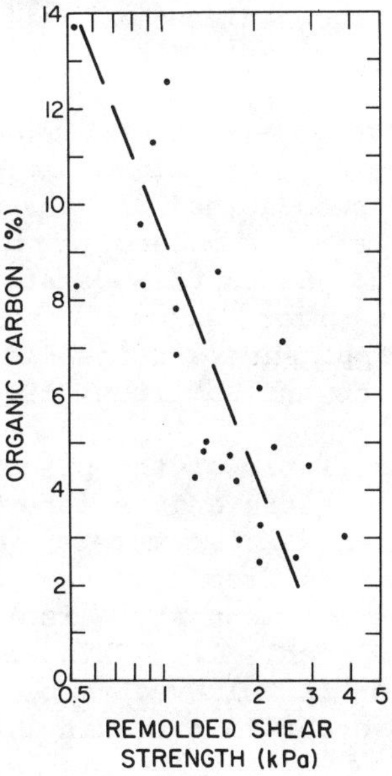

Fig. 9. Relationship of remolded shear strength to organic carbon content. Values are taken from the entire cored intervals (0.5 to 3 meters).

The combination of relatively high natural shear strength and low remolded strength of the organic-rich deposits between 11°S and 14°S results in the sediment having a high sensitivity, the ratio of the natural or undisturbed shear strength to the remolded shear strength (Skempton, 1952). The higher the sensitivity, the more strength the sediment loses when disturbed. Most submarine sediments have a mean sensitivity of 3 to 4 indicating a loss of 50 to 75% of the sediment's natural strength when disturbed. The organic-rich deposits off Peru have a mean sensitivity of 9 which indicates a strength loss of about 88% when disturbed (Richards, 1962) (Table 1). This property of highly organic deposits is particularly significant when considering the stability characteristics of such material.

Consolidation characteristics. Consolidation as used here refers to the compression or reduction in sediment volume due to loading, which in the case of saturated submarine deposits also requires the expulsion of interstitial or pore water. Loading in the natural state is normally attributed to the effects of overburden, but Skempton (1970) suggested that the process of consolidation is a complex one which can include desiccation, cementation, and interparticle bonding. Borrowing from soil mechanics, sediments are classified according to their degree of consolidation as being either normally, under-, or overconsolidated. Normally consolidated sediments are those that display a loading history in which the preconsolidation pressure or the maximum past load is equal to that generated by the present effective overburden pressure. If the sediment possesses a preconsolidation pressure, determined by means of a conventional consolidation test (Taylor, 1948), which is lower than the load imposed by the existing overburden, the material is classed as being underconsolidated. Should the preconsolidation pressure be greater than the present effective overburden pressure, the sediment is said to be overconsolidated.

An overconsolidated state normally occurs when erosion has removed previously deposited material or desiccation has taken place. Exclusive of areas of rapid sedimentation such as deltas, the majority of submarine sediments within the first few meters of the sea floor display a state of overconsolidation. Since this "state" does not appear to be the result of erosion or a prior load in most cases or attributed to desiccation, the term "apparent overconsolidation" is commonly given to such sediments (Richards and Hamilton, 1967).

The degree of overconsolidation is the ratio of the preconsolidation pressure to the present overburden pressure and is referred to as the overconsolidation ratio. For the upper few meters of most submarine deposits, this ratio commonly ranges from 1 to 4. In the case of the organic-rich Peruvian deposits, the mean ratio is 13, one of the highest noted in submarine sediments (Table 1). This unusually high degree of apparent overconsolidation is believed to be due to a strengthening of the sediment by a means other than loading, and in

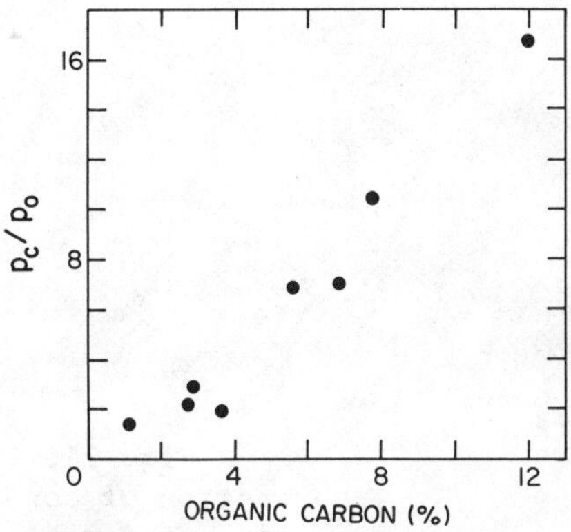

Fig. 10. Comparison of the overconsolidation ratio to total organic carbon content. Values are from cored intervals ranging from 1 to 3 meters.

this case it is proposed that the unusually high organic-carbon content of the sediment has played a role in this strengthening process. Such strengthening may be the result of bonding or cementing of particles by the organic matter. Pusch (1973) has found that humus in sediment, in the presence of polyvalent cations such as Mg^{2+} and Ca^{2+}, causes gel complexes to form which then serve to hold the natural sediment particles together. The presence of organic sheaths covering very fine-grained material has been found in these Peruvian deposits (Reimers, 1981). Some support for this hypothesis is found when the overconsolidation ratio is contrasted with the organic content for a number of samples from the Peru margin (Fig. 10).

SEDIMENT STABILITY

Very little evidence has been found to indicate that slumping, in the form of massive coherent blocks, prevails to any great extent along the Peru margin. In light of the frequency of seismic activity in this area, this finding was somewhat surprising at first. However, upon examination of the geotechnical properties, particularly the strength characteristics, of the organic-rich outer shelf and upper slope deposits, it was found that they possess unusually high sensitivities. As noted above, sediments with a high sensitivity will lose most of their natural strength and fail when disturbed. Such failures commonly occur when, for some reason, the pore water pressures become excessive and literally force the sediment particles apart and transfer the overburden load to the water from the particles themselves. Such excess pore pressure may result from a number of circumstances such as rapid sediment loading, cyclic loading by waves, both surface and internal, and the seismic shock of earthquakes. Although sedimentation rates of 100 to 140 cm 1000 y^{-1} are

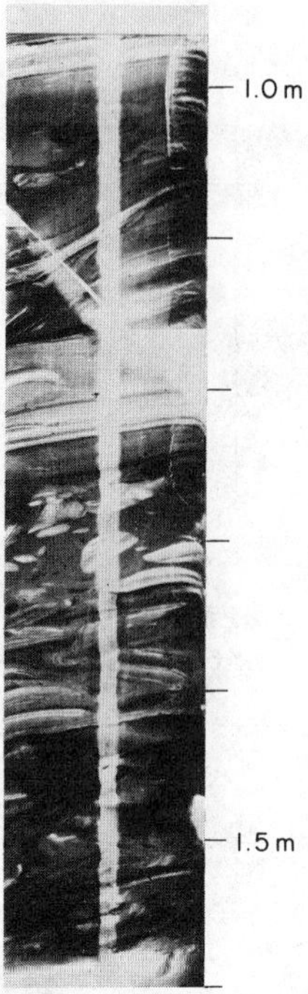

Fig. 11. X-radiograph of a section of core from the organic-rich sediments showing laminated mud interclasts and contorted laminations. The vertical light band is the result of a slight indentation in the core due to the design of the core cutting head (from Busch and Keller, 1982).

common on the Peru margin, they are not believed to be high enough to generate excess pore pressures that might reduce the resistance to slumping (Busch, 1980). Cyclic loading due to surface waves is generally only important in water depths of less than 150-200 m and is probably not a significant factor in sediment stability off Peru. The influence of internal waves, however, has not really been assessed owing to a lack of information on these waves off Peru. Seismic activity, on the other hand, is a common occurrence along the Peru margin (Gutenberg, 1939; Lee and Monge, 1968) with magnitudes as high as 7.75 on the Richter scale occurring in the offshore areas (Cluff, 1971). The intensity of such earthquakes is capable of increasing the pore pressure gradient sufficiently so as to cause failure (Morgenstern, 1967). The combination of having sediments of very high water content, low bulk density and unusually high sensitivity in an environment where relatively frequent earthquakes can generate excess pore pressures is most conducive to a fluidize failure rather than the failure of a coherent slump block. This type of fluidize failure may be somewhat similar to the massive landslides reported from Scandinavia where highly sensitive, "quick clays" sediments literally fluidize in a matter of minutes causing many thousands of cubic meters of sediment to flow downslope (Gregersen, 1981).

Evidence indicates that the landslide form of failure may be a prevalent mode of failure in the organic-rich deposits off Peru. The presence of flow features such as contorted bedding and the occurrence of well rounded clay balls within an otherwise well laminated undisturbed sediment core from off Pisco suggests that a flow-like failure took place in the highly organic sediments of that area (Fig. 11). Based on isotopic age dates from a number of cored intervals in the Peruvian organic-rich deposits, Busch and Keller (1982) report the presence of a hiatus ranging from 3.5 to 7 m. Reimers (1981) attributes this loss of section to erosion by bottom currents. That some of the observed hiatuses along the continental slope are the result of fluidize failures cannot be ruled out, and perhaps a combination of both transport mechanisms may in fact have led to the removal of this material.

Additional evidence for the massive movement of sediment downslope is seen in high resolution sub-bottom profile records. A 3.5 kHz profile down the upper slope across the area of organic-rich sediments reveals a series of what are believed to be submarine slides, indicating that the uppermost slope material has moved downslope in a series of progressive events (Fig. 12).

Although not yet documented, it is suspected that sediment failures in the form of subaqueous flows may be quite prevalent along the

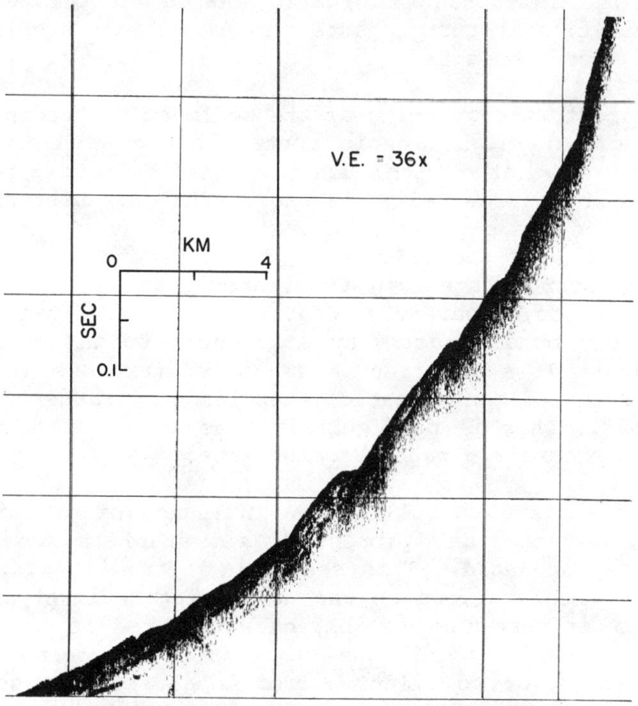

Fig. 12. Sub-bottom profile (3.5 kHz) down the Peruvian continental slope showing a series of slide deposits.

Peru margin. The leveling out or smoothing of the sea floor by such flows or slides may not, in many cases, be readily apparent from echo-sounding records which in turn might explain why this type of sea floor process is so poorly documented.

SUMMARY

It may be said that the process of coastal upwelling indirectly influences the geotechnical properties and stability characteristics of the underlying or nearby bottom sediments to some degree.

In a case such as off Peru where the combination of high productivity, very low rates of oxidation, and relatively low dilution by continental sediments leads to unusually high concentratons of organic carbon (up to 20%) in the bottom sediments, a strong relationship appears to exist between the organic content and the geotechnical properties of the bottom sediments. In considering the mean values determined for various geotechnical properties for the 0 to 2 m subseafloor interval, it is clear that the organic-rich deposits between 11°S and 14°S differ distinctly from sediments which, for the most part, are very much the same but lack the very high organic contents (Table 1). The influence of organic matter on the sediment mass physical properties in turn imparts somewhat unique stability characteristics to these deposits.

The apparent strengthening of the sediment by bonding or cementing because of the high organic content along with the very high water content associated with such deposits leads to a rather unstable condition when these sediments are disturbed. Disturbance causes the breakdown of the organic bonds as well as the natural structural or fabric strength of the sediment. The difference between the natural and the disturbed strength is greater than is normally found in submarine sediments, apparently because of the added natural strength gained from the bonding activity attributed to the organic matter. This significant loss of strength due to disturbance (up to 95% in some of the Peru sediments tested) then leads to rather unstable bottom conditions. This is particularly true in an area such as Peru where seismic shocks are rather commonplace.

In other areas of upwelling the influence of organic matter on the sediment geotechnical properties and stability characteristics may not be so pronounced. This seems to be true in areas where organic carbon concentrations in the underlying sediment are less than 4 to 5%, which in turn may possibly be attributed to the presence of oxygen and the oxidizing of the organics before being incorporated into the sediment, or to dilution from local sediment sources, or a combination of the two. The margin deposits of Oregon and northwest Africa, where upwelling is pronounced, seem to fit this situation and show somewhat less distinct relationships between the geotechnical properties and the organic content than have been observed in the Peruvian deposits.

In general, the following characteristics appear to typify cohesive submarine sediment in which the organic content is greater than 5%:

- very high water content,
- very high liquid and plastic limits,
- high natural shear strength,
- very high sensitivity,
- very high overconsolidation ratio,
- very low wet bulk density,
- high potential for failure in situations of excess pore pressure.

ACKNOWLEDGEMENT

The author acknowledges with appreciation funding support for this study from the Office of Naval Research, under contract number N00014-79-C-0004.

REFERENCES

Barber, R.T. and Smith, R.L., 1981, Coastal upwelling ecosystems, in: "Analysis of Marine Ecosystems," A.R. Longhurst, ed., Academic Press, New York, 31-68.

Busch, W.H., 1980, "The Physical Properties, Consolidation Behavior, and Stability of the Sediments of the Peru-Chile Continental Margin," Ph.D. Dissertation, Oregon State University, Corvallis, 149 pp.

Busch, W.H. and Keller, G.H., 1981, The physical properties of Peru-Chile continental margin sediments--the influence of coastal upwelling on sediment properties, Journal of Sedimentary Petrology, 51:705-719.

Busch, W.H. and Keller, G.H., 1982, Consolidation characteristics of sediments from the Peru-Chile continental margin and implications for past sediment instability, Marine Geology, 45:17-39.

Casagrande, A., 1948, Classification and identification of soils, American Society of Civil Engineers, Transactions, 113:901-931.

Cluff, L.S., 1971, Peru earthquake of May 13, 1970, Engineering Geology. Observations, Bulletin of the Seismological Society of America, 61:511-533.

Gregersen, O., 1981, The quick clay landslide in Rissa, Norway. The sliding processes and discussion of failure modes, Norwegian Geotechnical Institute, Publication 125:1-6.

Gutenberg, B., 1939, Tsunamis and earthquakes, Bulletin of the Seismological Society of America, 29:517-526.

Keller, G.H. and Bennett, R.H., 1970, Variations in the mass physical properties of selected submarine sediments, Marine Geology, 9:215-223.

Keller, G.H., Lambert, D.N. and Bennett, R.H., 1979, Geotechnical properties of continental slope deposits--Cape Hatteras to Hydrographer Canyon, in: "Geology of Continental Slopes," L.J. Doyle and O.H. Pilkey, eds., Society of Economic Paleontologists and Mineralogists, Special Publication No. 27, 131-151.

Krissek, L.A., Scheidegger, K.F. and Kulm, L.D., 1980, Surface sediments of the Peru-Chile continental margin and the Nazca Plate, Geological Society of America, Bulletin, 91:321-331.

Lee, K.L. and Monge, J., 1968, Effects of soil conditions on damage in the Peru earthquake of October 17, 1966, Bulletin of the Seismological Society of America, 58:937-963.

Masias-Echegaray, J.A., 1976, "Morphology, Shallow Structure, and Evolution of the Peruvian Continental Margin, 6°to 18°S, M.S. Thesis, Oregon State University, Corvallis, 92 pp.

Morgenstern, N.R., 1967, Submarine slumping and the initiation of turbidity currents, in: "Marine Geotechnique," A.F. Richards, ed., University of Illinois Press, Urbana, 189-219.

Odell, R.T., Thornburn, T.H. and McKenzie, L.T., 1960, Relationship of Atterberg limits to some other properties of Illinois soils, Proceedings of Soil Science of America, 24:297-300.

Premuciz, E.T., 1980, Organic carbon and nitrogen in the surface sediments of world oceans and seas: distribution and relationship to bottom topography, Brookhaven National Laboratory, Publication 51084, Upton, New York, 118 pp.

Prince, R.A. and Kulm, L.D., 1975, Crustal rupture and the initiation of imbricate thrusting in the Peru Trench, Geological Society of America, Bulletin, 86:1639-1653.

Pusch, R., 1973, Influence of organic matter on the geotechnical properties of clays, Statens institut för byggnads forskning, Document D11, Stockholm, 64 pp.

Rashid, M.A. and Brown, J.D., 1975, Influence of marine organic compounds on the engineering properties of a remoulded sediment, Engineering Geology, 9:141-154.

Reimers, C., 1981, "Sedimentary Organic Matter: Distribution and Alteration Processes in the Coastal Upwelling Region off Peru," Ph.D. Dissertation, Oregon State University, Corvallis, 219 pp.

Richards, A.F., 1962, Investigations of deep-sea sediment cores. II. Mass physical properties, U.S. Navy Hydrographic Office, Washington, Technical Report 106, 146 pp.

Richards, A.F. and Hamilton, E.L., 1967, Investigations of deep-sea sediment cores, III. Consolidation, in: "Marine Geotechnique," A.F. Richards, ed., University of Illinois Press, Urbana, 93-117.

Rosato, V.J. and Kulm, L.D., 1981, Clay mineralogy of the Peru continental margin and adjacent Nazca Plate: Implications for provenance, sea level changes and continental accretion, in: "Nazca Plate: Crustal Formation and Andean Convergence," L.D. Kulm, J.

Dymond, E.J. Dasch, D.M. Hussong, and R. Roderick, eds., Geological Society of America, Memoir, 154:545-568.

Rowe, G.T., 1971, Benthic biomass in the Pisco, Peru upwelling, Investigacion Pesquera, 35:127-135.

Schweller, W.J. and Kulm, L.D., 1978, Depositional patterns and channelized sedimentation in active Eastern Pacific trenches, in: "Sedimentation in Submarine Canyons, Fans, and Trenches," D.J. Stanley and G. Kelling, eds., Dowden, Hutchinson and Ross, Stroudsburg, Pennsylvania, 311-324.

Skempton, A.W., 1952, The sensitivity of clays, Gëotechnique, 1: 30-53.

Skempton, A.W., 1970, The consolidation of clays by gravitational compaction, Journal of the Geological Society of London, 125: 373-411.

Söderblom, R., 1966, Chemical aspects of quick-clay formation, Engineering Geology, 1:415-431.

Taylor, D.W., 1948, "Fundamentals of Soil Mechanics," John Wiley and Sons, Inc., New York, 700 pp.

Zuta, S. and Guillen, O., 1970, Oceanografia de las aquas Costeras del Peru, Institute de Mar Peru Boletin, 2:159-323.

Zuta, S., Rivera, T. and Bustamante, A., 1978, Hydrologic aspects of the main upwelling areas off Peru, in: "Upwelling Ecosystems," R. Boje and M. Tomczak, eds., Springer-Verlag, New York, 235-260.

UPWELLING ALONG THE WESTERN INDIAN CONTINENTAL MARGIN AND ITS GEOLOGICAL RECORD: A SUMMARY

M.G. Anantha Padmanabha Setty

National Institute of Oceanography
Dona Paula, 403 004, Goa
India

ABSTRACT

Upwelling along the western Indian continental margin is recorded each year from Cape Comorin to Cochin and further north up to Goa from June to September during the southwest monsoon. During this upwelling heavy phytoplankton blooms develop which contribute to the sediment compositions along the western continental margin. Here, under the influence of coastal upwelling and the resulting intense midwater oxygen minimum, a belt of laminated dark organic-carbon-rich muds is deposited along the upper slope.

These laminations are believed to represent the response of the depositional environment to the annually recurring upwelling events. Distributions of organic carbon, opaline silica, benthic and planktonic faunas, and the preservation of biogenic sediment components can be correlated to the characteristics of the water masses and, thereby, to the intensity and magnitude of upwelling. Benthic and planktonic foraminifers are particularly useful indicators; e.g., the invasion of immigrant populations of *Globigerina bulloides*, and of dextrally coiled *Neogloboquadrina pachyderma typica* through upwelling leave a record of the presence of such water masses in the sediments of that region. The very occurrence of *Globigerina bulloides* and the absence or irregular occurrence of *Globorotalia menardii* also imply upwelling in that region, as does the presence of benthic foraminifers *Hyalinea balthica* and *Ehrenbergina sp.*

UPWELLING ALONG THE WEST COAST OF INDIA

During the southwest monsoon (June-September) in the Arabian Sea and the northwest monsoon (December-March) in the Bay of Bengal,

Table 1. Oceanographic and meteorological data during pre-monsoon and southwest monsoon period (April-September) and upwelling along the Indian west coast (modified after Subrahmanyam and Sarma, 1960)

	April	May	June	July	August	September
Rainfall	Pre-monsoon showers	SW monsoon begins	Heavy	Heavy	Decreases	Diminishes
Humidity	74-76%	76-81%	88-93%	92-96%	88-93%	83-91%
Condition of the sea	Calm and clear	Rough, turbid	Rough, turbid	Rough, turbid	Rough, turbid	Calmer, clear
Wind & velocity (V)	From North V=8.2-9.5 mph	North or NW V=7.7-9.9 mph	West V=5.7-7.5 mph	From seaward V=5.7-7 mph	From sea V=5.3-6.5 mph	From seaward V=4.9-6.2 mph
Current	North to South	North to South	Varies	Varies	NE-SW	NE-SW
Temperature (seawater)	29.2-30.5°C	28.4-29.7°C	26.5-28.4°C	24.9-26.3°C	24.3-26.5°C	24.5-26.2°C
Salinity						
(a) Surface	33.87-35.89°/oo	34.74-35.65°/oo	22.67-34.62°/oo	20.74-31.90°/oo	28.35-32.77°/oo	31.64-34.96°/oo
(b) Bottom	33.97-36.08°/oo	34.69-35.72°/oo	34.03-35.45°/oo	33.02-34.53°/oo	34.08-34.88°/oo	34.31-34.87°/oo
Oxygen content	3.62-6.07 ml/l	3.49-4.8 ml/l	2.48-5.43 ml/l	3.51-6.03 ml/l	3.60-5.03 ml/l	2.97-4.97 ml/l
Phosphate/Phosphorus						
(a) Surface	0.322-0.886 μg at/l	0.334-0.805 μg/l	0.23-0.897 μg/l	0.564-1.369 μg/l	0.495-1.932 μg/l	0.690-1.205 μg/l
(b) Bottom	0.322-0.420 μg/l	0.460-1.334 μg/l	0.288-2.036 μg/l	0.513-2.415 μg/l	0.598-2.427 μg/l	0.851-1.564 μg/l
Nitrate/Nitrogen						
(a) Surface	1.36-6.70 μg at/l	0.66-34.00 μg/l	3.20-20.70 μg/l	1.28-26.20 μg/l	0.47-9.80 μg/l	2.37-3.90 μg/l
(b) Bottom	0.77-1.75 μg/l	2.49-8.20 μ/l	Not available	Not available	2.21-5.95 μg/l	0.36-1.20 μg/l
Silicate/Silicon						
(a) Surface	7.21-12.50 μg at/l	7.73-13.86 μg/l	12.50-29.50 μg/l	25.00-34.82 μg/l	22.16-47.22 μg/l	15.11-40.58 μg/l
(b) Bottom	9.17-14.71 μg/l	11.11-19.42 μg/l	10.42-40.47 μg/l	16.67-31.89 μg/l	20.77-33.54 μg/l	12.50-42.34 μg/l

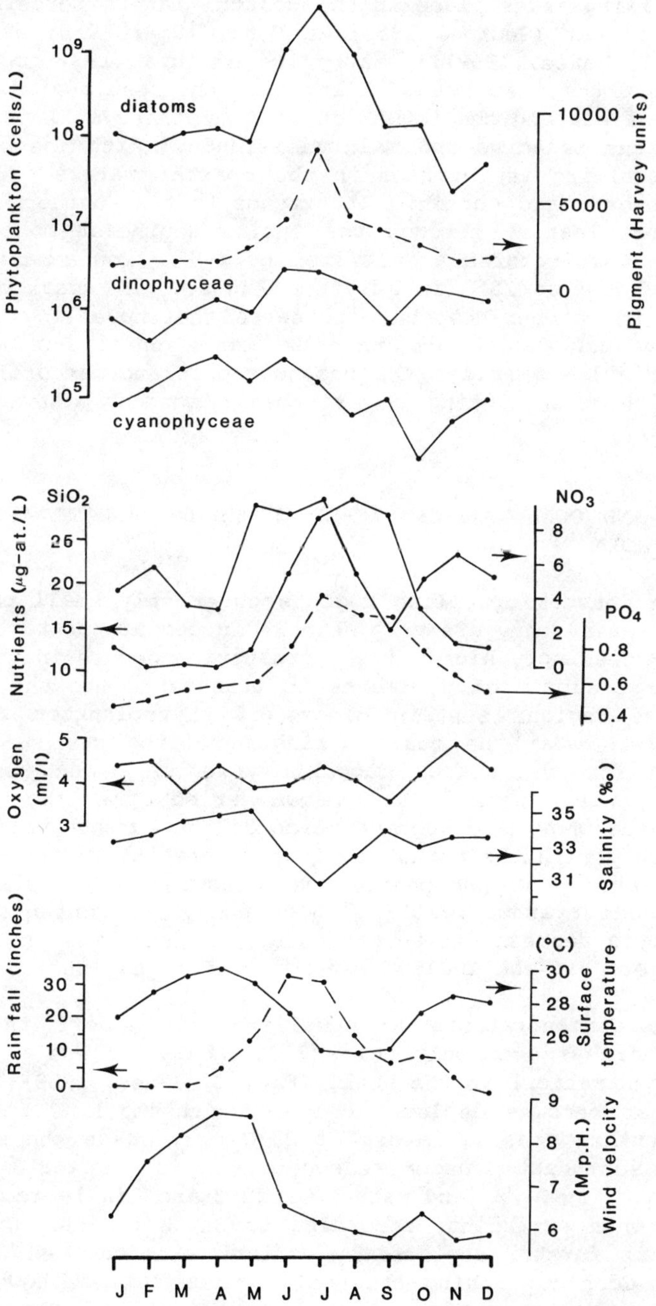

Fig. 1. Seasonal fluctuations in total phytoplankton (*Diatomaceae*, *Dinophyceae*, *Cyanophyceae*), nutrients and some meteorological data (after Subrahmanyam, 1958).

strong upwelling takes place in the coastal waters bordering the Indian subcontinent (George, 1953; La Fond, 1954; 1958; Jayaraman and Gogate, 1957; Banse, 1959). Sastry and d'Souza (1972) consider that upwelling off the west coast of India is the result of divergence in the current field and the formation of a cyclonic cell in the neighborhood of the Lacadive and Maldive islands. With the onset of the monsoon, upwelling is reported in the coastal waters of Sri Lanka, off Cape Comorin and Cochin. It extends to the southern part of Goa in the north, lasting through the entire southeast monsoon season. Sea surface temperatures are lowered by 4-5°C from a maximum at 29°C in April and May to 25°C in July and August. A remarkable increase to 29-30°C in October/November indicates that upwelling is strongest during July and August and that it lasts until October (George, 1953). Table 1 summarizes the oceanographic and meteorological conditions which exist during the southeast monsoon (Subrahmanyam and Sarma, 1960).

BIOLOGICAL AND OCEANOGRAPHIC IMPLICATIONS OF COASTAL UPWELLING OFF SOUTHWEST INDIA

Though coastal upwelling regimes cover only small parts of the total ocean area, they are very fertile in comparison to the adjacent regions. Exceedingly high primary production occurs in those regions where a large supply of nutrients is transferred to the surface waters. These regions sustain blooms of phytoplankton in the very shallow waters near the coast during upwelling periods, as can be illustrated for the coastal upwelling regime along western India (Fig. 1). There, in the post-monsoon period after the cessation of upwelling, the phosphate concentration of the surface waters seems to vary inversely with the amount of diatoms present. Concentrations of 2.4-3.5 μg-at./ℓ of phosphate were measured above the sea floor (Seshappa and Jayaraman, 1956). High phosphate contents, often accompanied by a decrease in temperature, were observed in the upwelled surface waters off the Indian west coast (George, 1953).

Oxygen concentrations in upwelling waters off the southwest coast of India are commonly below 0.25 mℓ O_2/ℓ (less than 5% of the normal concentration) on the shelf (Fig. 1; Banse, 1959). The oxygen in the water becomes depleted below the thermocline in wide areas along the entire coast. Waters of 23°C present on the shelf during the entire southeast monsoon season usually do not have oxygen concentrations >2 mℓ O_2/ℓ, and waters of 22°C are likely to be close to 1 mℓ O_2/ℓ when surfacing, according to Banse (1968). On the shelf, the oxygen is further depleted by contact with the sea floor as observed off Cochin. During the post-monsoon period, however, oxygen concentrations increase to 3.16-4.52 mℓ O_2/ℓ (Setty and Guptha, 1972).

The low oxygen but high nutrient concentrations of the cool upwelled water in the shelf region further augment phytoplankton development in the euphotic zone. Plant pigment concentrations measured for seven years off Calicut (Subrahmanyam and Sarma, 1967) and off Cochin (Banse, 1968) indicated an increase that was roughly ten times higher during the upwelling season than during the off-season of any year. This suggests that short-term massive productivity in the surface waters accounts for high organic carbon contents in the underlying sediments. Thus, the accumulating organic matter may contribute to the overall oxygen depletion which results in strong anoxic conditions near the sediment surface. This can, for example, promote the generation of hydrogen sulfide and the deposition of pyrite as infillings in tests of foraminifers. Such infillings are observed in samples collected off Mangalore (Setty, 1979).

UPWELLING AND SEDIMENTARY RECORDS

The general sediment zonation along the western Indian continental margin seems to reflect a response to the monsoonal upwelling regime (Fig. 2). According to von Stackelberg (1972), the sediments in the shallowest water are highly reworked and layered as a result of currents running parallel to the coast. The sediment is progressively finer from the coast to approximately 50 m of water depth. On the outer shelf it becomes coarse again in a zone containing relict deposits where fine-grained material is stirred up, whirled and transported away thus leaving large areas without any recent sediment cover. Relict sands with ooids were formed in shallow turbulent waters during Pleistocene times. The fine material carried away from the zone of relict sands of the outer shelf is believed to be found in deposits along the uppermost part of the slope where laminated fine-grained sediments accumulate, sedimentation rates are high, and low oxygen concentrations persist in the bottom waters. The laminations observed in these sediments are believed to reflect annual variations in deposition which may be attributed to the seasonality of the southeast monsoon. The lower boundary of the laminated muds on the continental slope coincides with the lower boundary of the intensive oxygen minimum layer.

This peculiar microenvironment results in intensive differential preservation of opaline and calcareous skeletal material (von Stackelberg, 1972). Total phosphorus reaches its highest concentrations (2.14-1.12 mg/g or 1.04-0.84 wt-% P_2O_5 on a carbonate-free basis) in the upper 7.5 cm of the slope sediments and attains its minimum concentrations (0.105-0.80 mg/g or 0.025-0.020 wt-% P_2O_5) in similar deposits on the shelf. The organic carbon fluctuates closely with the increase and decrease in phosphorus and ranges from 4.7 wt-% C-org with 1.04 wt-% P_2O_5 in the top part of cores from the slope to 0.4 wt-% C-org with 0.16 wt-% P_2O_5 in the bottom parts of those from the shelf. The close relationship between phosphorus and organic carbon contents is indicative of "fresh" organic matter believed to be due to upwelling, luxurient organic productivity, a non-depositional environment and the presence of abundant carbonates (Setty and Rao, 1972).

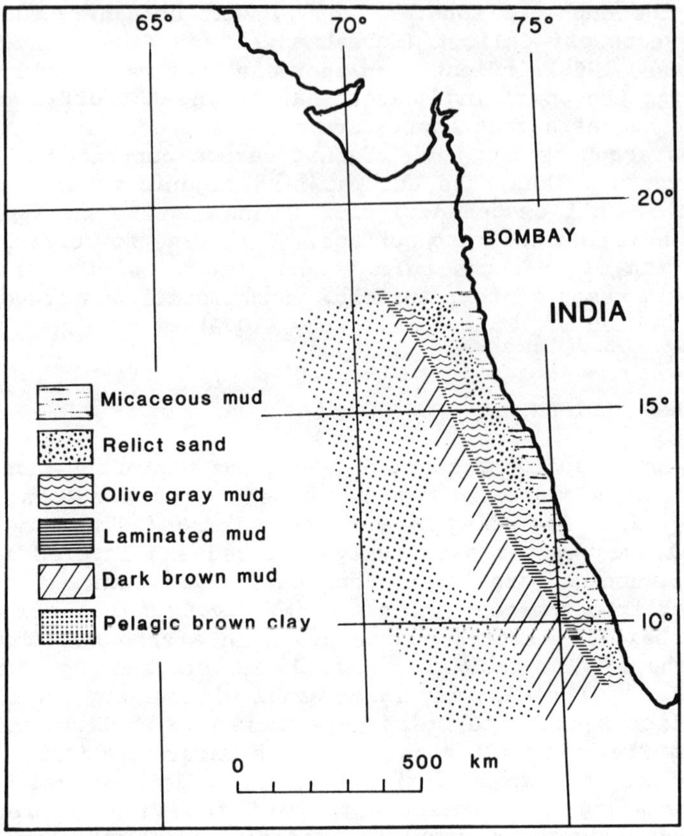

Fig. 2. Sediment distribution pattern on the shelf-slope region of Western India (modified after von Stackelberg, 1972).

The intensity and extent of the upwelling regime are preserved in the sediments as sedimentological, biogenic and isotopic signals (Duplessy, 1982), and biogenic particles especially provide some of the very best signals. The ratios of planktonic to benthic foraminifers, the organic carbon and opal contents, distributions of foraminifer, diatom, and radiolarian assemblages have all been used to identify and estimate the magnitude of past upwelling (Prell and Curry, 1981).

UPWELLING AND PLANKTONIC FORAMINIFERAL RECORDS

Zobel (1971; 1973) and Bé and Hutson (1977) recorded the occurrence of *Globigerina bulloides* in the upwelled waters of southwest India (Fig. 3), and Prell and Curry (1981) considered its occurrence

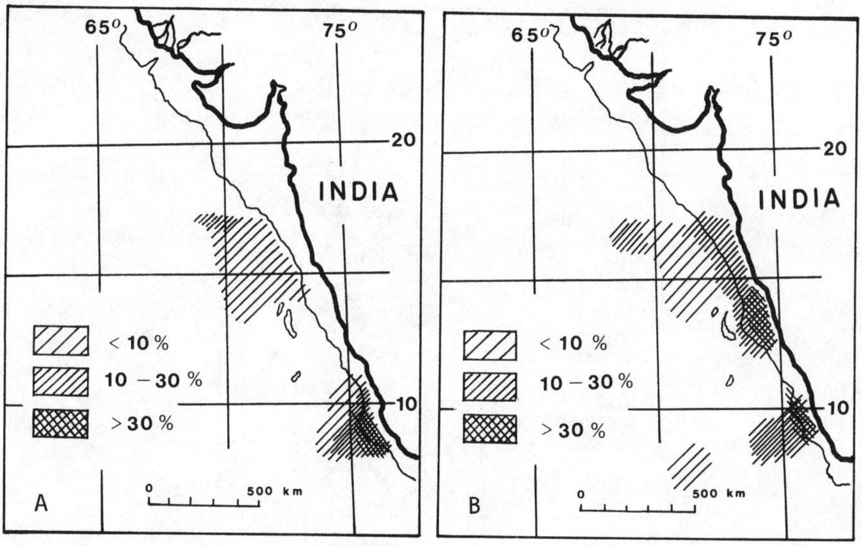

Fig. 3. Frequency distribution pattern of *Globigerina bulloides* (in total planktonic foraminifers): (A) in plankton tows from 200-0 m water depth; (B) in surface sediments along shelf-slope part of Western India (modified after Zobel, 1973).

as highly correlated to upwelling. It is noted from post-southwest monsoonal plankton tows that *Globigerina bulloides* and *Globigerinoides ruber* are most abundant, while *Globigerinoides quadrilobatus sacculifer*, *Globigerinella aequilateralis* and *Globorotalia menardii* are common (Setty, 1972). However, the complete absence of *G. menardii* from the surface sediments near the upwelling zone off Cochin attests to the result of the complexity of this region since it usually inhabits tropical water masses (Setty, 1972). Though *G. menardii* and its subspecies are definitely tropical planktonic foraminifers, they do live in deep or intermediate water masses (Bé and Tolderlund, 1971) and therefore show very irregular occurrences in upwelling regions (Thiede, 1975; Cifelli and Stern, 1976). The presence of *G. bulloides* in coastal sediment off western India (Fig. 3) seems in itself indicative of coastal upwelling in that region (Zobel, 1973; Bé and Hutson, 1977). Zobel also stated that this view coincides with the observations made earlier by Beljaeva (1964). Distribution of such planktonic foraminifers as *G. bulloides*, dextrally coiled *Neogloboquadrina pachyderma typica* s.s., and *Globoquadrina dutertrei* with *G. menardii* relates to tropical warm waters and to the areas of cool upwelled waters. Dextrally coiled *Neogloboquadrina pachyderma typica* s.s., a cold to cool water species which appears in very small numbers in sediments up to 23°N along the western Indian coast, is considered to originate in upwelled waters and be transported to the region and deposited in the sediments (Setty, 1977). Phleger (1960)

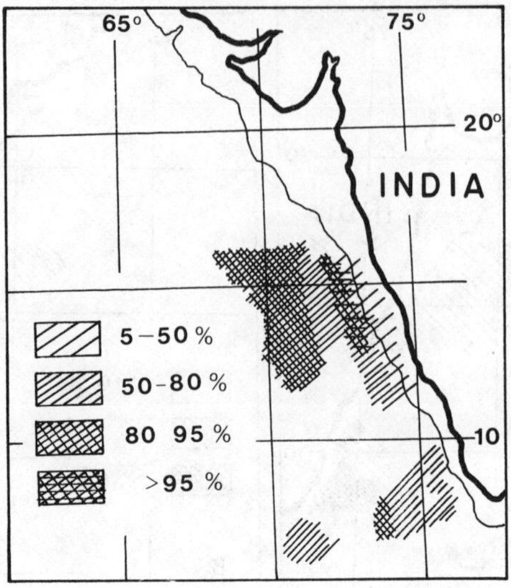

Fig. 4. Distribution of planktonic foraminifers (in percent of total foraminiferal faunas) in the surface sediments. Benthic foraminifers are extremely rare in a zone on the upper continental slope (modified after Zobel, 1973).

refers to such a condition as invasion of immigrant populations with the bottom water. This invasion leaves a record in the sediment of the presence of that water mass and may be an aid to understanding the mechanisms of water exchange and water movement.

UPWELLING AND BENTHIC FORAMINIFERAL RECORDS

Whereas planktonic foraminifers provide information about surface and subsurface water masses of upwelling regions, benthic foraminifers offer insight into the conditions at and close to the sea floor. The ratio of planktonic to benthic foraminifers changes drastically in response to the increasing water depth and changing oceanographic conditions under the western Indian upwelling regime. Figure 4 illustrates the frequency distribution of planktonic foraminifers and reveals a zone on the upper continental slope which is very poor in benthic foraminifers, probably due to the influence of the oxygen minimum layer. The benthic foraminiferal faunas along the western Indian continental margin is quite well studied, but often it is not quite clear how it is related to the upwelling regime in detail.

According to Zobel (1973) *Ehrenbergina* sp., which had a concentration of 17% in cores taken at 230 m depth off Karwar and 150 m depth off Cochin, shows a sharp decline when the upwelled waters with oxygen minima appear at this depth. *Ammonia* and its variants which are characteristic of shallow waters were observed to be reworked with *Operculina complanata*, an offshore form of the relict sediment present in the fine-grained laminated slope sediments. White, abraded and iron-stained larger foraminiferal species like *Amphistegina papillosa, Cellanthus craticulatus, Operculina ammonoides, Borelis schlumbergeri, Heterostegina suborbicularis* and *Gypsina globula* occur in the offshore relict sediments off Karwar. *Hyalinea balthica* a cold to temperate, normal marine form having a depth range from 0-1000 m was found with a concentration of 17% in the Cochin region (Zobel, 1973; Setty, 1974). Its occurrence is considered indicative of upwelling. *H. balthica* is living today in the Atlantic in a temperature gradient of 12°-5°C and also in a 10°-22°C gradient of the Persian Gulf.

DISCUSSION: SEDIMENT-WATER MASS-BIOGENIC COMPONENT RELATIONSHIPS

Parameters such as those discussed above which can be measured in the sediments can be correlated to the characteristics of the deep water or bottom water masses. In the oxygen minima laminated muds, complete or partial dissolution of siliceous tests is caused by the development of an alkaline environment, while in the oxygen high areas partial dissolution of calcareous tests result (von Stackelberg, 1972; Zobel, 1973; Setty, 1974). If frequency of benthic fossils and abundance of the organic carbon component is controlled by surface water dynamics, they may not correlate with the sediment (Diester-Haass, 1976), but their characteristics may reflect changes of the bottom water. Von Stackelberg (1972) recognized that organic carbon contents are comparable for the laminated muds of cores from the upper continental slope off Cochin (3.46%) and off Karwar (3.91%) as they represent the rhythm of the southwest monsoon.

The distributions of benthic faunas, opaline silica, and organic carbon may be influenced by bottom water rather than surface water (Pisias, Heath and Moore, 1975). The high percentages of quartz (>10%) along the west Indian margin appear to be confined to shallower depths and areas of relict Pleistocene sands at the outer shelf (von Stackelberg, 1972; Kolla et al., 1981). This high quartz abundance derived from Pre-Cambrian granitic gneisses and other metamorphics reflects terrigenous influxes during Pleistocene and Holocene times whereas kaolinite which extends north-south on the shelf is in part controlled by changing monsoonal winds (Kolla et al., 1981). In wide areas, ooide-rich relict sands are exposed which were formed in shallow waters at the turn of the Pleistocene to Holocene (von Stackelberg, 1972). This is evidenced by the occurrence of some white to translucent, abraded and iron-stained relict foraminifers off Karwar. These relict sands were exposed to violent wave action and areal erosion and were later gradually covered by recent material (von Stackelberg, 1972).

High organic carbon contents coincide with high concentrations of benthic foraminifers (*Uvigerina bellula, Cancris sagra, Cibicides refulgens, Bulimina denudata, Hyalinea balthica*, etc.) (Setty, 1974), and planktonic foraminifers (*Globigerina bulloides*), dextrally coiled *Neogloboquadrina pachyderma typica* s.s., *Orbulina universa*, off the Karwar region (Setty and Guptha, 1972). *Globigerina hexagona* and *Globoquadrina conglomerata* occur in offshore Cochin and Karwar (Setty, 1972; Setty and Guptha, 1972; Zobel, 1973). This flux of biogenic components into the sediment is controlled by productivity near the surface and certain hydrographic characteristics of the bottom water mass. The close correlation of foraminiferal abundances and organic carbon concentrations indicates that the bottom water characteristics differ considerably under glacial vs. interglacial conditions, as proposed by Weyl (1968) and Duplessy (1982).

The sea surface temperatures of the pre-monsoon period (April) for the Cochin-Karwar region, ranging from 29.2 to 30.5°C, vary considerably from those of the southwest monsoon and upwelling periods (June-September), ranging from 24.3° to 30.5°C, and will be preserved as an isotopic signal in the tests of foraminifers (Table 1). Table 1 further illustrates the other oceanographic and meteorological parameters prevailing during the southwest monsoon period.

Figure 2 illustrates the seasonal fluctuations of total phytoplankton, classes of algae correlating with certain oceanographic and meteorological factors. It indicates that a fall in temperature (optimum level 23°C-25°C), of salinity (not more than 10% of highest recorded values) and increase in nutrients (phosphate, nitrate and silicate) triggers a phytoplankton bloom, the maximum production taking place during the southwest monsoon and later declining. Diatoms (including *Dinophyceae, Cyanophyceae* and others) make the bulk of the crop.

For a precise interpretation of the effects of upwelling all along the coast of India, further collections of living specimens through planktonic tows, benthic foraminifers from the topmost layers of sediment, and cores from shelf through slope to the deep sea, collected at regular intervals during upwelling and pre- and post-monsoonal seasons, will help present a clearer picture.

CONCLUSIONS

1. Upwelling during June to September is a phenomena triggered by the southwest monsoon, along Cape Comorin to Cochin and south of Goa in the shelf-slope region of western India. It results in blooms of phytoplankton in the surface waters which contribute to the composition of the sediment.

2. Laminated muds formed where sedimentation rates are high and upwelling water is low in oxygen, suggesting depositional variations and the rhythm of the southwest monsoon. The biogenic components include very high proportions of planktonic foraminifers (and extremely poor benthic foraminifers) within this zone. Some larger foraminifers occur as reworked fauna from the relict sediments of the outer shelf, which were formed in shallow waters and exposed to erosion at the turn of the Pleistocene to Holocene.

3. The very occurrence of *Globigerina bulloides*, dextrally coiled *Neogloboquadrina pachyderma typica* s.s., and the absence or irregular occurrence of *Globorotalia menardii* implies upwelling in the region. Similarly, the presence of *Ehrenbergina* sp. and *Hyalinea balthica* in the sediments suggest an area of upwelling.

4. Periods of high organic carbon content and ratios of planktonic to benthic foraminifers in sediments off Cochin correlate with those off Karwar.

5. A more thorough understanding of the effects of upwelling would require: living foraminifers from plankton tows (200-0 m), benthic foraminifers from the topmost layers of sediment, and cores from the shelf-slope region from Cape Comorin north to Bombay during the southwest monsoon and pre- and post-monsoonal seasons.

ACKNOWLEDGEMENTS

I am thankful to K.F. Bowden, University of Liverpool and to J.S. Sastry for review and helpful comments on upwelling, and to the Director of the National Institute of Oceanography for permission to present this paper at the Symposium. I am also thankful to Shri Kamalakar G. Chitari in the preparation of illustrations and to Shri D.K. Narvenkar for preparation of the final manuscript.

REFERENCES

Banse, K., 1959, On upwelling and bottom trawling off the southwest coast of India, Journal of the Marine Biological Association of India, 1:33-49.

Banse, K., 1968, Hydrography of the Arabian Sea Shelf of India and Pakistan and effects on demersal fishes, Deep-Sea Research, 15:45-79.

Bé, A.W.H. and Tolderlund, D.S., 1971, Distribution and ecology of living planktonic foraminifera in surface waters of the Atlantic and Indian Oceans, in: "The Micropaleontology of the Oceans," B.M. Funnel and W.R. Riedel, eds., Cambridge, UK, 105-149.

Bé, A.W.H. and Hutson, W.H., 1977, Ecology of planktonic foraminifera and biogeographic patterns of life and fossil assemblages in the Indian Ocean, Micropaleontology, 23:369-414.

Beljaeva, N.V., 1964, Distribution of plantonic foraminifera in the water and on the floor in Indian Ocean (Russian, English summary), Trudy Institut Okeanology, Academii Nauk S.S.S.R., 68: 12-83.

Cifelli, R. and Stern, B.C., 1976, Planktonic foraminifera from near the west African Coast and a consideration of faunal parcelling in the North Atlantic, Journal of Foraminiferal Research, 6: 258-273.

Diester-Haass, L., 1976, Sediments as indicators of upwelling, in: "Upwelling Ecosystems," R. Boje and M. Tomczak, eds., Springer-Verlag, Berlin, 261-281.

Duplessy, J.C., 1982, Glacial to interglacial contrasts in the northern Indian Ocean, Nature, 295:494-498.

George, P.C., 1953, The marine plankton of the coastal waters of Calicut with observations on the hydrological conditions, Journal of the Zoological Society of India, 5:76-107.

Jayaraman, R. and Gogate, S.S., 1957, Salinity and temperature variations in the surface waters of the Arabian Sea off the Bombay and Saurashtra coasts, Proceedings of the Indian Academy of Sciences, B45:151-164.

Kolla, V., Kostecki, J.A., Robinson, F., Biscaye, P.E., and Ray, P.K., 1981, Distribution and origins of clay minerals and quartz in surface sediments of the Arabian Sea, Journal of Sedimentary Petrology, 51:563-569.

LaFond, E.C., 1954, On upwelling and sinking off the east coast of India, Andhra University Series No. 49:117-121.

LaFond, E.C., 1958, On the circulation of the surface layers of the east coast of India, Andhra University Series No. 62:1-21.

Phleger, F., 1960, "Ecology and Distribution of Recent Foraminifera," The John Hopkins University Press, Baltimore, 297 pp.

Pisias, N.G., Heath, G.R. and Moore, J.C., 1975, Lag times for oceanic responses to climatic change, Nature, 256:716-717.

Prell, W.L. and Curry, W.B., 1981, Faunal and isotopic indices of monsoonal upwelling: Western Arabian Sea, Oceanologica Acta, 4:91-98.

Sastry, J.S. and d'Souza, R.S., 1972, Upwelling and upward mixing in the Arabian Sea, Indian Journal of Marine Sciences, 1:17-27.

Seshappa, G. and Jayaraman, R., 1956, Observations on the composition of bottom muds in relation to the phosphate cycle in the onshore waters of the Malbar Coast, Proceedings of the Indian Academy of Sciences, B43:288-301.

Setty, M.G. Anantha P., 1972, Holocene planktonic foraminifera from the shelf sediments of Kerala Coast, Journal of the Geological Society of India, 13:131-138.

Setty, M.G. Anantha P. and Guptha, M.V.S.N., 1972, Recent planktonic foraminifera from the sediment off Karwar and Mangalore, Bulletin of the National Institute of Sciences of India, 38A(5-6): 148-160.

Setty, M.G. Anantha P. and Rao, C.M., 1972, Phosphate, carbonate and organic matter distribution in sediment cores off Bombay-Saurashtra coast, India 24 Session, International Geological Congress, 182-191.

Setty, M.G. Anantha P., 1974, Holocene benthonic foraminifera from the shelf sediments of Kerala Coast, Bulletin of Earth Sciences, 3:21-28.

Setty, M.G. Anantha P., 1977, Occurrence of *Neogloboquadrina pachyderma* new subspecies in the shelf slope sediments of Northern Indian Ocean, Indian Journal of Marine Sciences, 6:72-75.

Setty, M.G. Anantha P., 1979, Role of foraminifera in oceanographic events, Journal of Scientific and Industrial Research, 38:380-399.

Stackelberg, U. von, 1972, Fazieserteilung in Sedimenten des Indisch-pakistanischen Kontinentalrandes (Arabisches Meer), "Meteor" Forschungs-Ergebnisse, C9:1-73.

Subrahmanyam, R., 1958, Ecological studies on the marine phytoplankton on the west coast of India, Memoirs of the Indian Botanical Society, 1:145-151.

Subrahmanyam, R. and Sarma, A.H.V., 1960, Studies on the phytoplankton of the west coast of India, Part III. Seasonal variation of the phytoplankters and environmental factors, Indian Journal of Fisheries, 7:307-336.

Subrahmanyam, R. and Sarma, A.H.V., 1967, Studies on the phytoplankton of the west coast of India, IV. Magnitude of the standing crop for 1955-1962 with observations on nannoplankton and its significance to fisheries, Journal of the Marine Biological Association of India, 7:406-419.

Thiede, J., 1975, Distribution of foraminifera in surface waters of a coastal upwelling area, Nature, 253(5494):712-714.

Weyl, P.K., 1968, The role of the oceans in climatic change. A theory of the ice ages, Meteorological Monographs, 8:37-62.

Zobel, B., 1971, Foraminifera from plankton tows, Arabian Sea: Areal distribution as influenced by ocean water masses, in: "Proceedings II Planktonic Conference", A. Farinacci, ed., Rome, 1970, 1323-1335.

Zobel, B., 1973, Biostratigraphische Untersuchungen an Sedimenten des Indisch-pakistanischen Kontinentalrandes (Arabisches Meer), "Meteor"Forschungs-Ergebnisse, C12:9-73.

HIGH RESOLUTION HOLOCENE TIME SCALES

STABLE ISOTOPE RECORD OF UPWELLING AND CLIMATE FROM SANTA BARBARA BASIN, CALIFORNIA

Robert B. Dunbar

Scripps Institution of Oceanography, University of California, San Diego, La Jolla, California 92093, U.S.A.
and
Department of Geology, Rice University
Houston, Texas 77251, U.S.A.

ABSTRACT

The Santa Barbara Basin (California) is a unique natural laboratory for examination of the sediment record of climate and upwelling. The sediments deposited in the central anoxic part of the basin are layered, or varved, due to seasonal fluctuations in the supply of terrigenous material and the absence of bioturbation. Varving allows high resolution sampling and accurate dating of the sedimentary record. In the sediments of this coastal upwelling area, small specimens of planktonic foraminifers are enriched in ^{18}O relative to large individuals, opposite the "normal" trend in the deep sea. $^{18}O/^{16}O$ ratios in the planktonic foraminifer *Globigerina bulloides* sampled at three to five year intervals in the varved sediments show good correlation with the historical record of upwelling (sea level, water temperature, and air temperature) from Port Hueneme, Los Angeles, and San Diego. The $\delta^{18}O$ range between "Davidson Current events" and periods of strong upwelling over the past 230 years is nearly 1.5°/₀₀. There is a slight positive correlation between $\delta^{18}O$ and $\delta^{13}C$ which suggests that in a highly productive environment, $^{13}C/^{12}C$ ratios are controlled by changes in foraminiferal metabolism rather than by upwelling induced fluctuations in $\delta^{13}C$ of seawater bicarbonate. Tests of *G. bulloides* in a slightly dissolved sequence (about 10% carbonate loss) are enriched in ^{18}O by as much as 0.6°/₀₀ relative to specimens from undissolved sections. Carbonate content of the varved sediments reflects variations in benthic and planktonic productivity, redeposition, and dilution by terrigenous detritus. Carbonate accumulation rates track varve thickness indicating that carbonate productivity is linked to regional rainfall during stormy periods and/or detrital input from run-off. Both isotope and carbonate stratigraphies suggest that upwelling rates and surface and benthic productivity were all greater during the "Little Ice Age" (about 50 to 400 years ago) with cold peaks at ∿1900, 1870, 1810, and 1770 A.D.

INTRODUCTION

Marine carbonates have been used in a variety of ways as indicators of ocean "climate" and fertility. Carbonate concentrations (Arrhenius, 1952; and many others), stable isotopic composition (Emiliani, 1955; and many others), and trace element content (Cronblad and Malmgren, 1981) in marine sediments have all been used to study climatic variations. Nearly all Quaternary "climatic stratigraphies" based on marine carbonates reflect a complex combination of processes. The original signal of climate and productivity has been modified by dissolution and/or benthic mixing before becoming part of the sedimentary record. Major climatic events such as the glacial-interglacial cycles may be discerned although there is controversy over what each parameter ($\delta^{18}O$, $\delta^{13}C$, CO_3, Sr, Mg, faunal ratios, etc.) is actually measuring (temperature, productivity, salinity, ice volume, dissolution, mixing artifact, etc.). The record of short period (10's or 100's of years) climatic events is especially affected by sediment mixing. As a result, the marine carbonate record of the late Holocene has not yet been rigorously examined.

Ocean circulation and climate during the Little Ice Age, a period of cool temperatures and glacial advance occurring from 50 to 400 years ago, is of special interest for two reasons. 1) Baseline studies of recent natural climatic variation are needed in order to assess anthropogenic alteration of climate. 2) The Little Ice Age may serve as a useful analog for the study of the major climatic perturbations of the Pleistocene, the glacial-interglacial cycles.

I have examined box cores from the Santa Barbara Basin, an anoxic silled basin at the northern end of the Southern California Bight. The Santa Barbara Basin provides us with a unique natural laboratory with which to study recent climate and fertility in the northeast Pacific. The central part of the basin contains unmixed, varved sediments in which carbonate microfossils are well-preserved. The terrigenous and biogenous components have been studied in some detail by several investigators (Emery, 1960; Emery and Hülsemann, 1962; Hülsemann and Emery, 1961; Soutar and Isaacs, 1969; Berger and Soutar, 1970; Fleischer, 1972; Koide, Soutar, and Goldberg, 1972; Chow et al., 1973; Krishnaswami et al., 1973; Soutar and Crill, 1977; Heusser, 1978; Pisias, 1979).

The Santa Barbara Basin has a maximum depth of 580 meters. The sill depth (about 480 meters) coincides with the level of maximum development of the regional oxygen minimum zone (450 to 650 meters; Rittenberg, Emery, and Orr, 1955). As a result, replenishment of dissolved oxygen in basin bottom water by advection over the sill occurs very slowly. There is a large flux of organic matter to the seafloor from both surface production and terrestrial input. Degradation of organic material in the deep central part of the basin consumes dissolved oxygen to a point where dysaerobic conditions prevail in bottom water and anaerobic conditions develop at very shallow depth in the sediments. Low levels of oxygen prevent the migration of large benthic burrowing organisms into the central part of the basin. In the unmixed sediments, seasonal fluctuation in the supply of terrigenous and biogenous components produces annual layering.

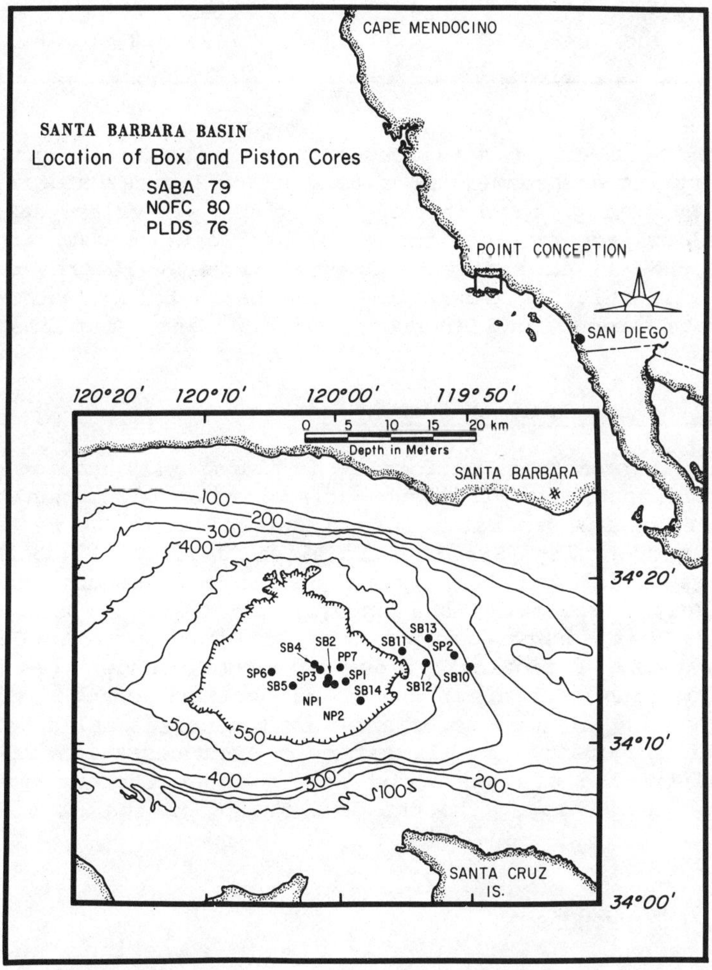

Fig. 1. Location map of box (xBx) and piston cores (xPx) collected in Santa Barbara Basin on the Pleiades (PLDS) April, 1976, cruise of the RV "Melville" (Pxx), the SABA August, 1979, cruise of the RV "New Horizon" (Sxx), and the NOFC July, 1980, cruise of the RV "Melville" (Nxx).

Table 1. Location of box cores collected during the SABA 79 Cruise of the RV "New Horizon"

Box Core	Length (cm)	Latitude (°N)	Longitude (°W)	Water Depth (m)
4	52	34 14.8	120 02.0	581
5	51	34 13.7	120 03.5	581
10	45	34 14.6	119 50.0	475
11	43	34 15.6	119 55.0	512
14	54	34 12.6	119 58.2	563

Each varve consists of a siliceous diatom-rich band produced during the late spring and summer and a clay mineral-rich band representing dilution by terrigenous material during the winter and early spring. Varving allows accurate dating and high resolution sampling. Carbonates in the varved sediments are relatively well preserved due to the high alkalinity of interstitial and basin bottom waters produced by sulfate reduction (Berner, Scott, and Thomlinson, 1970; Sholkovitz, 1972).

Upwelling and the Davidson Current

Along the coast of California the timing and duration of upwelling of cold, nutrient-rich, sub-surface water determines to a large extent the annual variation in sea surface temperature and phytoplankton productivity (Sverdrup and Allen, 1939). Off southern California, upwelling usually occurs during the months of February through July. Following the upwelling season, a warm, northerly-flowing coastal counter-current (the Davidson Current) develops in surface waters. Periodically, due to a poorly understood meteorological triggering mechanism, the normal pattern changes and winter-spring upwelling ceases, accompanied by the unseasonal development of the Davidson Current. I will refer to such cessation of upwelling and the intrusion of warm water as "Davidson Current events." As such, they are analogous to the development of the El Niño Current off Peru.

MATERIALS AND METHODS

Box cores were collected during July and August, 1979, on the SABA 79 cruise of the RV "New Horizon". The central plain and eastern flank of the basin (Fig. 1) were cored using a full size Unsell spade-corer modified with wooden "landing pads" to prevent overpenetration in soft sediment. Water depth at the core sites ranges from

475 to 581 meters (Table 1). Up to 8 sub-cores were taken from each box and maintained at 4°C in seawater during transport and storage. Because of high water content, cores 4 and 5 were frozen prior to lengthwise cutting.

The cores were sampled at one or two centimeter intervals for carbonate content measurement and stable isotope analysis. A portion of each sample was dried, weighed, and reacted in 2M HCℓ. The volume of CO_2 liberated was measured at ambient temperature and pressure. Carbonate contents were calculated by comparison with 100% $CaCO_3$ standards after correction for salt content.

The remaining sample was washed through a 62 μm screen following mild ultrasonification in a buffered calgon solution, dried, and sieved into different size fractions. Specimens of the foraminifers *Globigerina bulloides, Orbulina universa, Neogloboquadrina dutertrei, Bolivina argentea, Chilostomella ovoidea*, and *Cassidulinoides cornuta* from different size fractions from the surface sediment of Bx-14 were analyzed as part of a calibration study. Specimens of *G. bulloides* and *C. cornuta* from the 250 to 350 μm size fraction were picked for the downcore studies. The average number of specimens of *G. bulloides* and *C. cornuta* picked for isotopic analysis was 25 and 35, respectively. The 250 to 350 μm size fraction was used for faunal counting. Downcore faunal and benthic foraminifer isotope data are presented elsewhere (Dunbar, 1981).

Water samples from Santa Barbara Basin were collected in August, 1979, and again in July, 1980, in order to measure the variation in $^{18}O/^{16}O$ ratio of seawater with depth. Seawater was prepared for isotopic analysis by equilibration with CO_2 gas by standard analytical procedure (Epstein and Mayeda, 1953). Carbonates were reacted in 100% phosphoric acid at 50°C. Liberated CO_2 gas was analyzed with an "on line" V. G. Micromass 602 mass spectrometer. Results are reported in the standard δ notation where:

$$\delta^{18}O(°/_{oo}) = [(^{18}O/^{16}O)\text{sample}/(^{18}O/^{16}O)\text{standard} - 1] \times 1000.$$

Results are given relative to the Chicago PDB standard calculated by the usual procedure (Craig, 1957). The analytical precision for the NBS-20 carbonate standard was 0.07°/₀₀ and 0.05°/₀₀ for $\delta^{18}O$ and $\delta^{13}C$, respectively.

STRATIGRAPHY

Box cores 4, 5, and 14 consist of alternating olive-green and black layers. The surface of Bx-4 contains a white mucous-like material, probably the filamentous bacterial mat described by Soutar and Crill (1977) implying that the water-sediment interface was retrieved intact. Varves were well developed and clearly evident on observa-

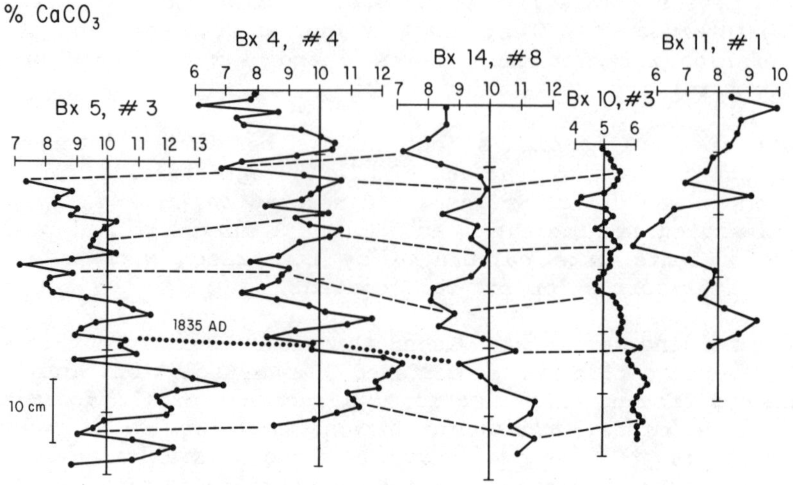

Fig. 2. Calcium carbonate % variations in box cores Bx-4, Bx-5, Bx-10, Bx-11, and Bx-14. The reproducibility for the % $CaCO_3$ determinations is better than 0.5%. Note the 1835 shell layer (marked by dots) used for correlation and time frame construction. The vertical scale interval is 10 cm. The dashed lines represent the best curve-to-curve correlation based on matching both the *G. bulloides* $\delta^{18}O$ and $\%CaCO_3$ stratigraphies.

tion through the core liner of Bx-4. The near-surface sediment of Bx-5 had a much higher water content than in Bx-4 and appeared disturbed, probably as a result of coring. Varves were least well developed in Bx-14. The olive-green layers are less dark than in Bx-4 or 5, and the sediment is firmer.

Bx-11 consists of medium brown, olive-green, and dark gray layers. The sediment is very stiff and contains layers enriched in pelecypod shell fragments at 12, 16 and 30 cm depth. Bx-10 consists of homogenous medium olive-green mud. Many large benthic organisms were present at the top of the core, including polychaete worms and deposit-feeding snails.

$CaCO_3$ content, $^{18}O/^{16}O$ ratios in the planktonic foraminifers *G. bulloides*, and a pelecypod shell layer are used for correlation of the cores. Shells of the pelecypod *Macoma leptonoidea* form a thin, consistent layer in the sediments of central Santa Barbara Basin (Soutar and Crill, 1977). This layer is found at 27 to 29 cm in Bx-5, 40 to 41 cm in Bx-4, and 43 to 45 cm in Bx-14 (a smaller concentration of shells is found at 37 to 39 cm in this core). Consistent downcore variation in $\%CaCO_3$ (Fig. 2) is evident in cores Bx-5, Bx-4, Bx-14, and Bx-10. Bx-11 exhibits fluctuations unlike any other core and is not used further for this study in view of the additional evidence for erosion and reworking (i.e., shell layers, unusual stiffness).

$^{18}O/^{16}O$ ratios in *G. bulloides* record conditions in the surface water off southern California. This provides an additional tool for correlation of cores Bx-5, Bx-4, Bx-14. *G. bulloides* is not present in Bx-10 in sufficient quantities for isotopic analysis.

I have used a combination of both $\%CaCO_3$ and *G. bulloides* $\delta^{18}O$ data (Figs. 2 and 3) to visually match horizons from core to core. Bx-4 contains the most complete record of recent sedimentation, in agreement with the observation of the surficial bacterial mat, an indicator of the sediment-water interface. Bx-14 and Bx-10 appear to be missing 2 to 4 and about 7 cm, respectively, from their tops. Bx-5 may be missing up to 10 cm at the top. Additionally, the upper 3 cm of this core appears disturbed and was not sampled. The sediments of Santa Barbara Basin have very high water contents, ranging from 90% at the surface to 70% at 20 cm (Soutar and Crill, 1977) and it is likely that sediment loss occurred by winnowing of the surface layers during coring. As a result, data for the upper few centimeters of each core should be interpreted with caution.

The consistency of $CaCO_3$ content variation (and total deposition rate) in the upper 50 cm of the sediment between cores Bx-5, Bx-4, Bx-14, and Bx-10 implies that both carbonate and terrigenous accumulation rates are fairly uniform over a large area of the basin floor. Either pelagic sedimentation dominates or there are bottom transport

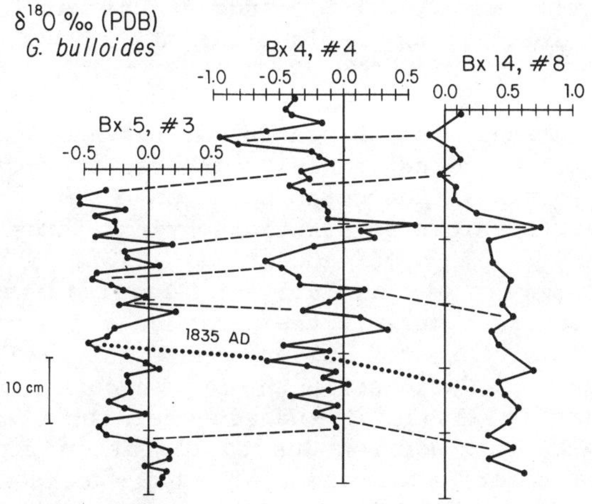

Fig. 3. $\delta^{18}O$ variations in the planktonic foraminifers *G. bulloides* in cores Bx-5, Bx-4, and Bx-14. The shell layer is marked by dots. The vertical scale interval is 10 cm. The dashed lines represent the best curve-to-curve correlation based on matching both the *G. bulloides* $\delta^{18}O$ and $\%CaCO_3$ stratigraphies.

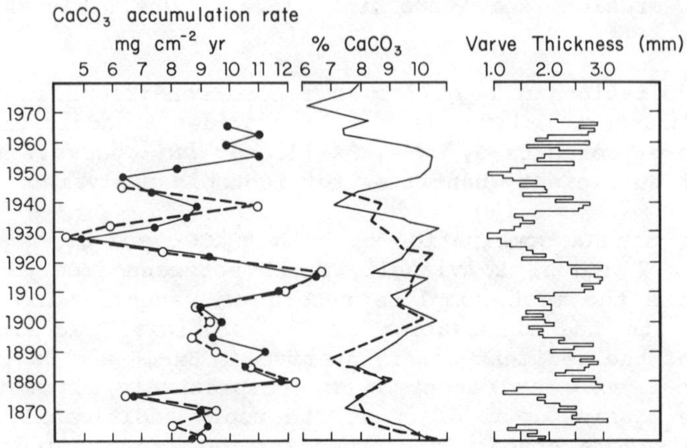

Fig. 4. $CaCO_3$ accumulation rate in mg cm^{-2} yr^{-1} and $CaCO_3$ from Bx-4 (solid) and Bx-5 (open and dashed), and average varve thickness stratigraphy from Soutar and Crill (1977, based on a 3-core "stacked" varve thickness stratigraphy given in units of standard deviation and converted to actual thickness using the given mean varve thickness of 2.03 m and standard deviation of 0.56 mm). Varve thicknesses are normalized to 60% water by weight.

processes at work which evenly distribute the terrigenous and biogenic materials. The results of several short-term sediment trap deployments in the Santa Barbara Basin (Soutar et al., 1977; Dunbar and Berger, 1981; Dymond et al., 1981) suggest that at least half of the total flux of material to the basin sediments arrives by the pelagic path.

Because of laminae disturbance in cores 4 and 5 caused by freezing, ages could not be directly assigned based on varve counts. Ages are assigned to levels in the cores based on the *M. leptonoidea* layer and an average varve thickness stratigraphy for three cores collected by Soutar (Soutar, 1975; Soutar and Crill, 1977). The pelecypod layer occurs approximately at the 1835 A.D. level (determined by varve counting in the Soutar cores) in the sediment and serves to mark the transition between bioturbated (older) and finely laminated (younger) deposits, indicating a decrease in the oxygen content of basin bottom water. Soutar and Crill (1977) suggest that the clams were concentrated at the 1835 A.D. horizon due to migration from a sub-bottom habitat during a period of decreasing oxygen content. One of the Soutar cores, C-239, is a well-dated box core collected from within 1.5 km of the Bx-4 core site. Soutar and Crill (1977) have shown that the average varve thickness for the last 140 years is fairly uniform over distances of several kilometers on the basin bottom. Varve thickness in core C-239 (corrected to 60 wt-% water) ranges from 1.12 to 3.81 mm with a mean of 2.03 mm. Ages are assigned to

Table 2. Calculated time frame for Bx-4 based on the varved thickness data of Soutar and Crill (1977) corrected for water content volume changes using the 1835 A.D. shell layer at 40 to 41 cm as the base point

Year A.D.	Actual Depth (cm)	Year A.D.	Actual Depth (cm)
1970	3.1	1850	36.4
1960	7.1	1840	38.5
1950	10.4	1830	40.9
1940	13.0	1820	43.4
1930	15.6	1810	45.6
1920	17.5	1800	47.9
1910	21.0	1790	51.1
1900	23.4	1780	52.4
1890	26.0	1770	54.6
1880	29.3	1760	56.9
1870	31.8	1750	59.1
1860	34.2		

different levels in Bx-4 (Table 2) by the following steps. 1) The varve dimension data of Soutar (average values for 3 cores given in units of standard deviation) is converted to wt. % water normalized thickness (Fig. 4) using a mean varve thickness of 2.03 mm and a standard deviation of 0.56 mm (as given for core C-239). 2) This wt. % water corrected thickness is converted to actual thickness using water contents measured in Bx-4, Bx-14, and data given by Soutar and Crill (1977). 3) Ages are assigned using the calculated varve thickness beginning at the *M. leptonoidea* layer (at 40 to 41 cm in Bx-4, assumed to occur at the 1835 A.D. level). A correction is made for the 7% expansion of Bx-4 caused by freezing. By this method, there is uncertainty associated with the age assignation at the top of the core, in the post-1967 deposits due to the high water content and unknown thickness of the varves.

As an independent check on the 1835 to 1967 time frame, the $\delta^{18}O$ anomaly at 6 to 8 cm in core Bx-4 (Fig. 3) which represents foraminiferal growth during the warm "Davidson Current event" of 1958 and 1959, occurs within 0.5 cm of the calculated depth. An estimation of the error in calculated age due to local sedimentation rate variations and bias associated with the 1 cm sampling interval ranges from 4 years in near-surface sediments to ±5 to 10 years at depth. Ages for Bx-5, Bx-14, and Bx-10 are estimated by comparison of the isotope and carbonate stratigraphies with Bx-4, again using the 1835 shell layer as a baseline. Below the 1835 shell horizon there are no varves. The age assignments for the bottom of core Bx-4 and the extension of Bx-5 to the 1750 level are calculated based on a constant accumulation rate of 110 mg cm^{-2} yr^{-1}. The approximate number of years averaged by 1 cm sampling ranges from 2 in near-surface sediments to ∿5 below 20 cm.

Oxygen Isotopes

The results of the calibration study using the surface sediment from Bx-14 are presented in Fig. 5. Larger specimens of *O. universa*,

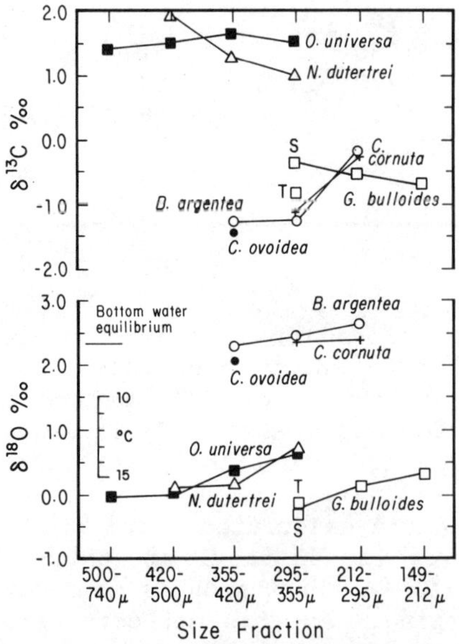

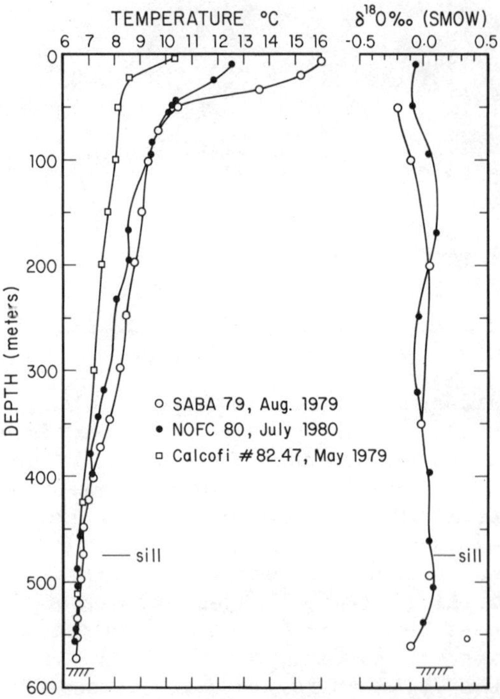

Fig. 5. Size vs. $\delta^{18}O$ and $\delta^{13}C$ variations in the planktonic foraminifers *G. bulloides*, *N. dutertrei*, *O. universa* and the benthic foraminifers *C. cornuta*, *B. argentea* and *C. ovoidea* from the surface sediment of Bx-14. For *G. bulloides*, "T" indicates a compact variety with the final chamber connected to several chambers of the previous whorl. "S" indicates an open variety with the terminal chamber connected only to the next chamber. Equilibrium $\delta^{18}O$ values for calcite precipitated at various temperatures (using the seawater $\delta^{18}O$ values from Fig. 6) are calculated using the temperature- $\delta^{18}O$ equation of Epstein et al. (1953).

Fig. 6. Temperature and $\delta^{18}O$ profiles of seawater in the central part of Santa Barbara Basin from the SABA 1979 and NOFC 1980 cruises and a May, 1960, CalCOFI cruise (1960b, temperature only).

N. dutertrei, and *G. bulloides* are depleted in ^{18}O relative to smaller specimens. Wefer, Dunbar and Suess (this volume) have observed a similar trend for *N. dutertrei* from the surface sediments of 6 box cores from beneath the Peruvian upwelling system. The opposite trend, i.e., depletion of ^{18}O in the smaller size classes, has been observed in specimens of *N. dutertrei* from the surface sediments of the Ontong-Java Plateau (Berger, Killingley and Vincent, 1978). This is the "normal" trend for most species from deep-sea sediments. In a coastal upwelling environment, it is likely that the size distribution of a foraminiferal population changes in response to fertility (Berger, 1969; Bé, 1980). During upwelling, when low surface temperatures prevail, large numbers of juveniles may be produced (as reproduction occurs earlier) thus biasing the isotopic temperature aspect of the small size classes in the sediment. Additionally, the larger size classes may include expatriate specimens from outside of Santa Barbara Basin where upwelling-related temperature changes are not as great.

Although very low oxygen concentrations prevail in the bottom water of the basin, there is a large standing stock of benthic foraminifers especially adapted to these conditions (Harman, 1964). The benthic foraminifers *Bolivina argentea* is slightly depleted in ^{18}O in the larger size classes while *C. cornuta* exhibits little variation (although only two sizes were examined).

Equilibrium calculations may be made based on Epstein et al. (1953) temperature - $\delta^{18}O$ equation for calcite. Fig. 6 shows the vertical profiles of temperature and seawater $\delta^{18}O$ (measured in the

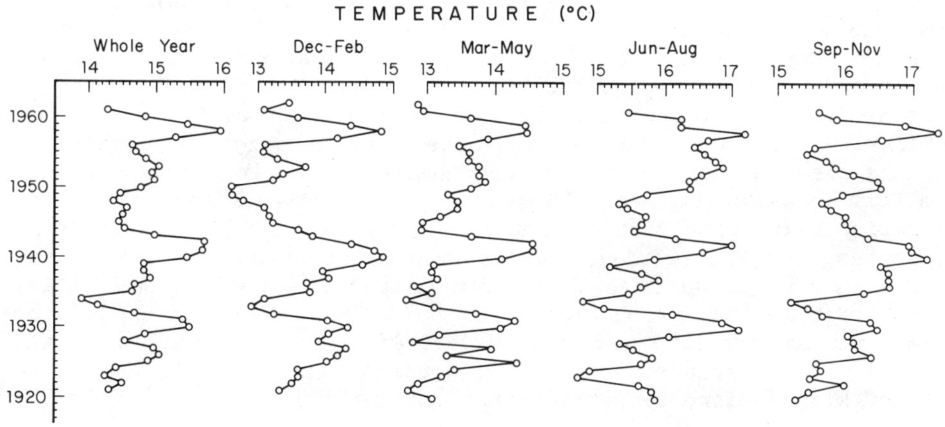

Fig. 7. Three-point moving averages of seasonal temperatures based on daily temperature measurements at the Port Hueneme tide station. Note the coherence of sea temperature anomalies (for example: the Davidson Current events of 1930, 1941, and 1958-1959) between all seasons of a given year.

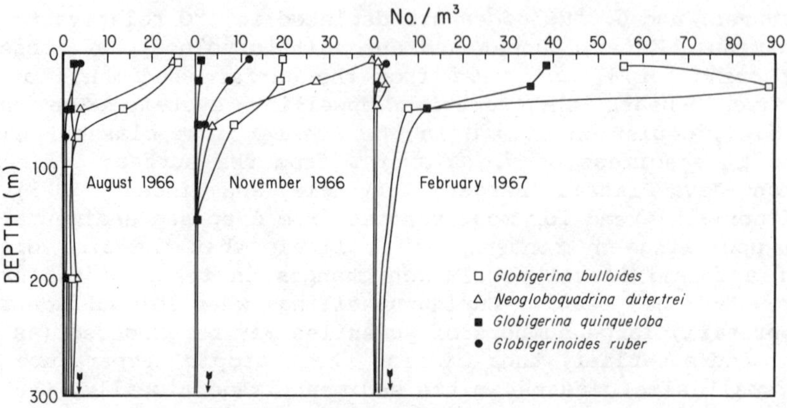

Fig. 8. Distribution of foraminifers in the size range 125-400 μm in surface water of the Santa Barbara Basin during August and November, 1966, and February, 1967 (foraminiferal counts supplied by Berger, pers. comm.).

lab) during cruises to Santa Barbara Basin in August, 1979 and July, 1980. A CalCOFI (1960b) temperature profile from the central part of the basin during a period of strong upwelling in May, 1960, indicates that the average temperature of the upper 100 m of the water column can drop as low as 9°C. The anomalous enrichment in $\delta^{18}O$ of seawater at 555 m probably represents an experimental error. The benthic foraminifers apparently precipitate calcite close to the equilibrium value of 2.35°/₀₀ at 6.5°C.

Using an average surface water $\delta^{18}O$ of -0.1°/₀₀ (SMOW), equilibrium values for calcite are plotted in Fig. 5 for surface temperatures ranging from 10 to 16°C. All three planktonic foraminifers exhibit $\delta^{18}O$ values ranging from 0.5 to 0.3°/₀₀ (13 to 14°C at equilibrium) in the smaller size classes, to less than 0.0°/₀₀ (>16°C at equilibrium) in the larger size classes. This range is consistent with the observed annual and interannual variation in sea temperature from Port Hueneme (Fig. 7), although several observations bear consideration. In Santa Barbara Basin, temperatures >16°C occur only in the surface layer (<20 m) during the late summer and fall. Net tow results (although we have data from only three short-time slices) suggest that maximum production of *G. bulloides* occurs in the winter while maximum production of *N. dutertrei* occurs in the late fall (Fig. 8). Low concentrations of all species were found in August, a month of high surface temperature but low fertility.

It is likely that *N. dutertrei* precipitates $CaCO_3$ close to isotopic equilibrium with seawater. *G. bulloides* may be slightly depleted in ^{18}O if maximum production occurs during the winter. The exact relationship between the rates of production and supply to the sediment and the annual change in size distribution for this species awaits clarification. The average degree of disequilibrium may be greater in a coastal upwelling environment relative to the open ocean because of higher nutrient concentrations and more rapid growth. Young, rapidly growing individuals tend towards the greatest disequilibrium (Berger et al., 1978) but are more abundant during the cold upwelling months of the winter and spring. Larger specimens precipitate $CaCO_3$ close to equilibrium but their shells may record growth during warmer months, thus resulting in relatively "light" $\delta^{18}O$ values.

For downcore studies, specimens of *G. bulloides* in the size range 250-350 µm were analyzed. *G. bulloides*, commonly referred to as an "upwelling species", records surface water conditions (although there may be some "vital effect" overprint) during the winter and spring, a key period in terms of climate and fertility.

The downcore profiles of *G. bulloides* (Fig. 3) in Bx-5 and Bx-4 show variations of up to 1.5°/₀₀ in $\delta^{18}O$. There is a general trend towards greater $^{18}O/^{16}O$ ratios towards the bottom of the cores. Using the calculated time frame, data from the upper parts of Bx-4 and Bx-5 is plotted with historical parameters measured between 1919 and 1975 (Fig. 9). Tont (1981) has shown that there is excellent correlation between sea level, sea temperature, and air temperature records along the coast of southern California. The isotopic curve

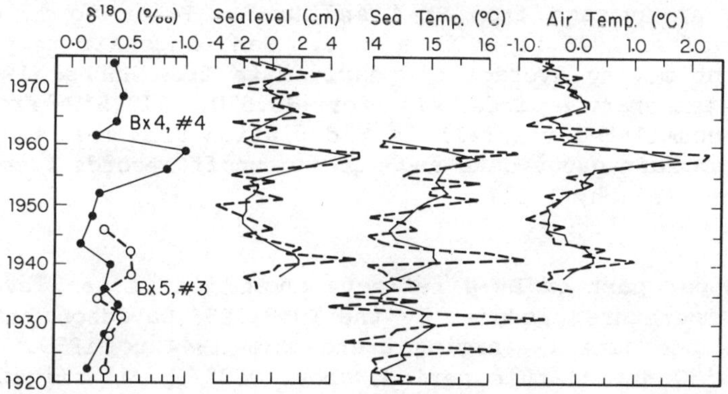

Fig. 9. $\delta^{18}O$ variation in °/₀₀ (PDB) of *G. bulloides* from the upper parts of Bx-4 and Bx-5 using calculated time frame with 1835 shell layers as the base, Los Angeles sea level anomaly from Tont (1978), sea temperature record from Port Hueneme tide station (eastern end of Santa Barbara Basin), and Los Angeles air temperature anomaly from Tont (1978). Dashed lines indicate yearly average values. Solid lines show a three-point moving average of yearly data after 1950, and a five-point moving average before 1950 to simulate the effect of sampling the sediment at 1 cm intervals.

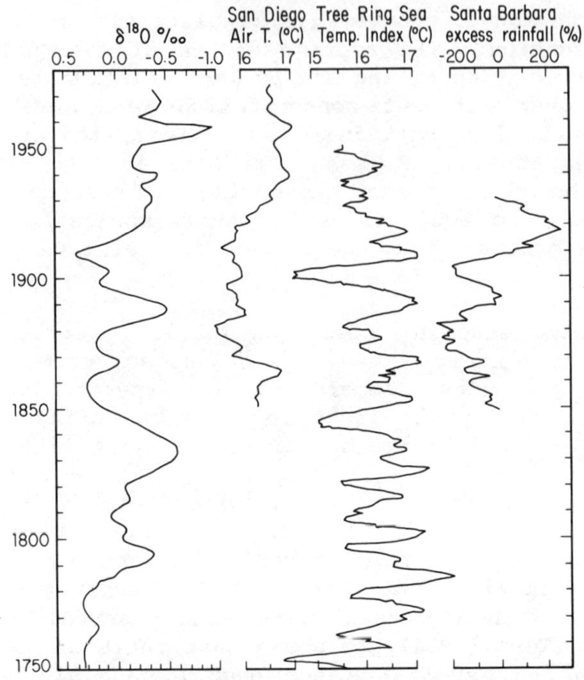

Fig. 10. Comparison of $\delta^{18}O$ variation in *G. bulloides* with historical air temperature and rainfall, and sea temperature index calculated from tree rings. Between 1790 and 1945, the isotope curve represents an average from Bx-4 and Bx-5. Prior to 1790, data is from Bx-5; after 1945 it is from Bx-4. San Diego air temperature is a five-point moving average of yearly data from Hubbs (1948). Tree ring sea temperature index is for 33.5°N, 118.5°W from Douglas (1976). Accumulated rainfall is % deviation from mean rainfall calculated from rain gauge data and stream runoff records from the Santa Barbara area (Lynch, 1931).

from the upper part of Bx-4 reflects anomalies in sea level and sea and air temperature produced by the 1958-1959 Davidson Current event as well as the general warming trend from 1944 to 1959. The total range in $\delta^{18}O$ during this period, about 0.9‰, is slightly greater than expected if the foraminiferal isotopes reflect changes in average temperature (1‰ = ~4°C).

In sediments deposited prior to ~1940, each 1 cm sample represents an average of about 5 years and greater attenuation of the isotopic record of short-term events is expected. In order to remove some of the effects of sample interval biasing, I have "stacked" or averaged the isotope records from the two well-preserved cores, Bx-4 and Bx-5 (Fig. 10). Also plotted is San Diego air temperature (Hubbs, 1948; Tont, 1978), calculated sea surface temperature from tree ring thicknesses for 33.5°N, 118.5°W (about 100 km southeast of Santa Barbara Basin from Douglas, 1976), and Santa Barbara area accumulated rainfall (Lynch, 1931). The $\delta^{18}O$ curve shows large variations through the last 230 years. *G. bulloides* is relatively enriched in ^{18}O during the period 1850 through 1910, and for a period of at least 30 years prior to 1780.

There is good correlation between the isotope curve and the historic record of air temperature since 1870. Both show the well-documented warming trend from 1910 to 1940 and discrete cool events at 1880, 1900 and 1910. Hubbs (1948) discusses the generally "cool" aspect of the fish fauna off southern California during the period 1860 to 1930. Hubbs also cites data collected from San Diego harbor during the 1850's which indicates that the yearly average temperature of the sea was slightly cooler during this period than the period 1916 through 1947. Specifically, winter months were approximately 1.2°C cooler while spring months were about 0.7°C warmer. Although sub-tropical fish were collected off San Diego in the 1850's, this may be the result of warmer sea temperatures during the spring. The *G. bulloides* $\delta^{18}O$ peak at 1855 is consistent with cooler water temperatures during the winter months. The lack of correlation between the San Diego air temperature record and the isotope record between 1850 and 1870 may be the result of erroneous air temperature measurements (Hubbs mentions that the "quality" of the records increased in October, 1871, when the Weather Service took over from the Army Medical Corps).

The large shift towards depleted $\delta^{18}O$ values centered around 1890 most probably represents a major Davidson Current event which is only slightly reflected in the averaged San Diego air temperature data. (The San Francisco air temperature record shows a more marked warming during this interval.) This was also a period of heavy rainfall in southern California, consistent with previous observations that major El Niños (analogous to Davidson Current events) of the past several centuries have been accompanied in Peru by torrential rainfall and flooding (M. Mosely, pers. comm.). Except for the 1890 event, the isotope curve shows little correlation with the Santa Barbara area rainfall data. The temperature index calculated from southern California tree ring thickness (Douglas, 1976), however, shows much better correlation with the rainfall data than with the air temperature record. This is consistent with the observation of Soutar and Crill (1977) that coastal temperature and rainfall in the Santa Barbara area exhibit only slight correlation while local tree ring thickness and rainfall are highly correlated (correlation coefficient = 0.79). It may be that during "normal" periods coastal rainfall and sea surface temperature are decoupled, but that major temperature anomalies may induce a significant response in terms of precipitation.

Carbon Isotopes

Different size classes of foraminifers from the surface of Bx-14 exhibit quite different $^{13}C/^{12}C$ ratios (Fig. 5). *N. dutertrei* is depleted in ^{13}C in the smaller size classes, the "normal" trend in deep-sea sediment (Berger et al., 1978) probably due to greater disequilibrium fractionation by rapidly growing young individuals. Temperature has little effect on $\delta^{13}C$ in marine carbonates (temperature dependency = 0.035°/₀₀/°C; Emrich, Ehhalt and Vogel, 1970).

O. universa exhibits no significant variation in $\delta^{13}C$ with size. Two varieties of *G. bulloides* were analyzed, compact forms with terminal chambers connected with the last whorl (T), and open forms in which the final chamber is only touching the next chamber (S). While there is no significant difference in $\delta^{18}O$, the specimens with separate final chambers are relatively enriched in ^{13}C by about 0.6°/₀₀, possibly due to varietal specific "vital effect". Only compact specimens of *G. bulloides* were picked for this study. *O. universa* and *N. dutertrei* exhibit carbon isotopic compositions which are relatively close to equilibrium values (within 1°/₀₀) while *G. bulloides* is relatively depleted in ^{13}C by about 2°/₀₀. This is consistent with the results of isotopic analyses of *G. bulloides* from sediment trap samples (Deuser et al., 1981) and deep-sea sediments (Deuser, 1978) which suggest that this species incorporates a significant amount of metabolic carbon during calcification.

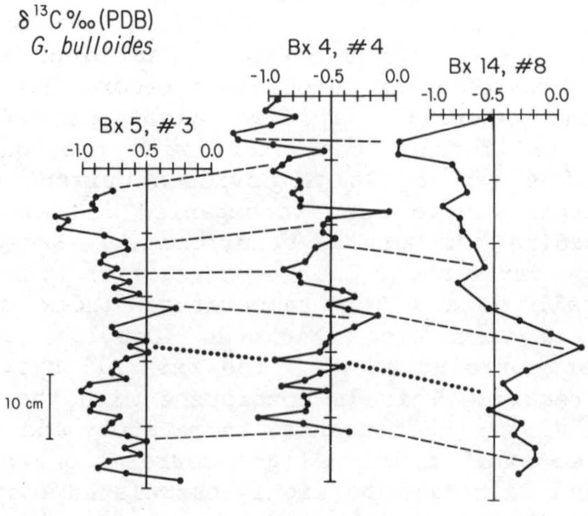

Fig. 11. $\delta^{13}C$ variations in °/₀₀ (PDB) of the planktonic foraminifers *G. bulloides* in cores Bx-5, Bx-4 and Bx-14. Dots indicate correlation levels based on $CaCO_3$ and $\delta^{18}O$ variations in *G. bulloides*. Vertical scale interval is 10 cm. The dashed lines represent the best curve-to-curve correlation based on matching both the *bulloides* $\delta^{18}O$ and %$CaCO_3$ stratigraphies.

Both *C. cornuta* and *B. argentea* are about 1°/₀₀ enriched in ^{13}C in the 212-295 μm size class relative to the 295-355 μm size class. A similar trend (although slightly less pronounced) has been observed in most species of benthic foraminifers in surface sediments recovered from within the oxygen minimum zone off northern Peru (Wefer et al., this volume). Foraminiferal reproduction (and growth rate) is linked to oxygen concentration in such environments. When oxygenation of the basin bottom water (and coincident increase in the $^{13}C/^{12}C$ ratio of seawater ΣCO_2) occurs, reproduction of some species of benthic foraminifers may be stimulated by the higher metabolic rates. Thus, the shells of small specimens record the conditions of the basin during periods of flushing while the larger adult specimens record slow growth during periods of relative stagnation (perhaps incorporating ^{13}C depleted CO_2 from the sediment).

The downcore profiles of $\delta^{13}C$ in *G. bulloides* are given in Fig. 11. All three cores exhibit a well-defined $\delta^{13}C$ maxima at about 6 cm above the 1835 shell layer (~1860) and a general trend downcore toward greater $^{13}C/^{12}C$ ratios. The oxygen and carbon isotope profiles for *G. bulloides* in each of the tree cores is nearly parallel. An inverse correlation is expected if foraminiferal $^{18}O/^{16}O$ and $^{13}C/^{12}C$ ratios respond directly to temperature and ΣCO_2-$\delta^{13}C$ variations related to upwelling events. The upper water column should be depleted in ^{13}C during periods of strong upwelling due to the upward advection of subsurface water enriched in ^{12}C from the breakdown of sinking organic matter. Foraminifers growing during an upwelling event should be enriched in ^{18}O (from cool temperatures) and depleted in ^{13}C.

The observed positive correlation may imply that: 1) temperature-related "vital effects" dominate the $\delta^{13}C$ signal in *G. bulloides*, 2) large (>250 μm) specimens may not record growth during

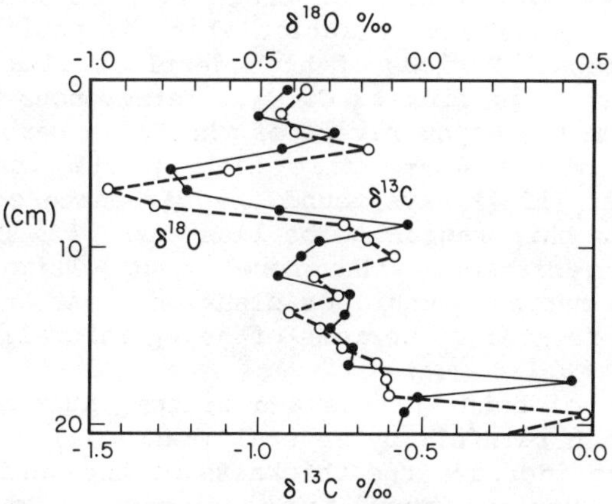

Fig. 12. $\delta^{18}O$ and $\delta^{13}C$ of *G. bulloides* from the upper 20 cm of Bx-4.

upwelling events, and/or 3) in such a highly productive environment such as the California current, nutrient supply may not be the ultimate growth limiting factor for certain foraminifers. Although nutrient supply is high during upwelling events, foraminiferal population growth may be related directly or indirectly to another factor such as temperature or light intensity. It is interesting to note that the $\delta^{13}C$ shifts appear to lag $\delta^{18}O$ shifts by about 1 to 2 cm (2 to 7 years) in the upper 20 cm of Bx-4 (Fig. 12). Although more surface sediments (where high resolution sampling is possible) from other Santa Barbara Basin box cores must be examined to verify this observation, the data implies a long-term (on the order of several years) lag between discrete temperature events and changes in foraminiferal productivity.

Carbonate Supply

The calcium carbonate content of marine sediments is controlled by benthic and planktonic productivity, redeposition, dissolution, and dilution by terrigenous detritus. The detailed time control for varved sediment allows the direct calculation of accumulation rates, which removes the effects of dilution by non-carbonate material (Fig. 4). However, the actual composition and size distribution of the carbonate fraction is not well characterized.

Soutar and Crill (1977) have shown that varve thickness variations primarily reflect fluctuations in the supply of terrigenous detrital material which are positively correlated with the historical record of rainfall and river runoff in the Santa Barbara area. Other components (mainly organic C, $CaCO_3$, and biogenic silica) make up less than 20% of the average central basin sediment. The % carbonate curves for Bx-4 and Bx-5 (Fig. 4) reflect dilution by terrigenous matter during periods when the varves are relatively thick. Because varve thickness varies by more than a factor of 3 and the $CaCO_3$/non-$CaCO_3$ ratio varies by a factor of about 1.5, carbonate accumulation rate exhibits a significant variation which is positively correlated with varve thickness. Periods of high detrital flux are accompanied by a corresponding high flux of $CaCO_3$. Terrigenous particles entering the ocean from the major rivers of the Santa Barbara area consist mainly of clay minerals and silt size quartz grains. Fleischer (1972) and Shiller (1982) have found the carbonate content of riverborne sediment in this region to be less than 2%. The carbonate accumulation rate variations, therefore, must ultimately be due to local changes in marine benthic or planktonic carbonate productivity and redeposition related to periods of heavy rainfall.

Productivity of both surface and bottom water may increase during periods of high rainfall by several means. 1) Major storms along the coast tend to increase the thickness of the surface mixed layer, thus increasing nutrient recycling and surface productivity. 2) River and stream run-off may supply nutrients to surface water. 3)

Clay minerals brought in by the rivers carry organic materials and may contribute significantly to the supply of food in both surface and bottom water. 4) The presence of clay minerals in surface waters also contributes to the efficient transfer of organic matter (as well as $CaCO_3$) to the sea floor by relatively robust, rapidly settling zooplankton fecal pellets. 5) Weather patterns related to "normal" rainfall anomalies (for example, periods characterized by numerous storms originating in the northeast Pacific) may produce an increase in the average rates of coastal upwelling. Increased productivity in surface and bottom water results in greater production of planktonic and benthic foraminiferal tests as well as other carbonate remains such as pteropods, pelecypods, and worm tubes (along the flanks of the basin).

During a sediment trap deployment in Santa Barbara Basin in March-April, 1978, a period of heavy rainfall and high detrital input, the trapped material consisted of >13% $CaCO_3$ (Dunbar and Berger, 1981), significantly greater than the average from the varved sediments of <10%. The pelagic carbonate flux consisted mainly of the free-settling tests of planktonic foraminifers with a small contribution of coccolith carbonate transported by fecal pellets. The trapped $CaCO_3$ flux is estimated as 6 mg cm^{-2} yr^{-1} compared with the average flux to the sediments (Bx-4) for the last 100 years of about 9.4 mg cm^{-2} yr^{-1}. It is likely that production of $CaCO_3$ by benthic foraminifers in the central basin sediments makes up a significant fraction of the total carbonate accumulation. Benthic to planktonic foraminiferal ratios (>250 µm fraction) in near-surface sediments of the oxygen depleted region range from 1/4 to 1/1.

An additional carbonate component in the central basin sediment is material derived from shallower areas by resuspension and downslope bottom transport. Very few upper slope benthic foraminifers were found in any of the varved sediment samples, however, suggesting that this transport path does not supply a significant amount of coarse fraction carbonate. Additionally, sediments from the flanks of the basin are subject to dissolution and exhibit relatively low carbonate contents (about 5% in Box-10). It is likely that material arriving at the central basin floor by bottom transport is relatively depleted in $CaCO_3$, although a high fine fraction carbonate flux from erosion of unusual submarine outcrops cannot be ruled out entirely. Barring such input, an increase in planktonic or *in situ* benthic foraminiferal production, or both, must therefore be responsible for the carbonate accumulation pulses during periods of high detrital flux.

I suggest that the production of biogenic carbonate in the surface water and the sediments of Santa Barbara Basin is influenced by as yet undetermined factors related to rainfall and/or sediment supply. This productivity signal may alter the responses of the $CaCO_3$ flux to variations in surface water fertility produced by changes in upwelling rate. Evaluation of the mechanisms involved as well as the possible influence of "relict" carbonate awaits detailed grain size analysis and description of the carbonate fossil remains in individual pairs of annual laminae.

Dissolution and Mixing

The effects of dissolution and mixing are clearly evident in the carbonate and stable isotope profiles from Bx-14 and Bx-10. Bx-10, from a depth of 475 meters, exhibits attenuation of carbonate content variation (Fig. 2) relative to the other cores, with a range of less than 2%. Based on a mean carbonate content of 5.5% and a detrital sedimentation rate equivalent to that on the central basin floor, the recent average carbonate accumulation rate for this core is 5.6 mg cm^{-2} yr^{-1}, compared with the flux to the varved sediments of about 9.4 mg cm^{-2} yr^{-1}. As benthic carbonate production is almost certainly greater at the shallower core site (this core contained numerous pelecypods), a minimum carbonate dissolution rate of 3.8 mg cm^{-2} yr^{-1} is indicated.

Dissolution produces a shift to greater $^{18}O/^{16}O$ ratios in specimens of *G. bulloides* from Bx-14, presumably due to preferential removal of ^{18}O depleted tests or parts of tests (Fig. 3). This shift averages 0.3°/₀₀ in post-1910 deposits and 0.6°/₀₀ in pre-1910 sediments. There is no difference in the $\delta^{13}C$ (Fig. 11) of *G. bulloides* between the cores in the post-1850 deposits and a slight enrichment in δ^{13}, about 0.3°/₀₀, before this period. The data suggest that, in general, the degree of dissolution increases downcore. The greater shift in $\delta^{18}O$ relative to $\delta^{13}C$ suggests that the $\delta^{18}O$ range among individual specimens of *G. bulloides* is greater than the $\delta^{13}C$ range.

DISCUSSION AND SUMMARY

Stable isotope and carbonate content variations in the sediments of Santa Barbara Basin provide both a unique record of recent climate and new insight into sedimentary processes. While the planktonic foraminifer *G. bulloides* appears to precipitate calcite depleted in ^{18}O with respect to equilibrium, $\delta^{18}O$ variations in box cores from the varved sediments reflect fluctuations in sea temperature during the past 230 years. Sea temperature anomalies generally reflect variations in the average rate of coastal upwelling and are therefore highly correlated with sea level and air temperature (Tont, 1978; 1981). The amplitude of the isotopic signal is greater than expected from sampling of 3 to 5 year time slices, which suggests that seasonal or annual foraminiferal production differences amplify the isotopic response to temperature of this species.

The isotope record indicates relatively cool conditions between 1850 and 1880, 1895 and 1915, and for a period of at least 30 years prior to 1780. It is interesting that this prolonged period of apparently very cool surface water from 200 to >230 years ago coincides with previous estimates for the timing of maximum glacial re-advance during the Little Ice Age (Ahlman, 1953, as quoted by Worthington, 1968; Bray, 1974; Bryson, 1974). It is likely that this period was characterized by an intensification and equatorward shift of both the trade and westerly wind belts. As a result, both low and mid-latitude coastal upwelling increased. From the isotope record in the Santa Barbara Basin, the oceanic expression of the Little Ice Age does not appear to be comprised of one single event, but rather a series of relatively cool periods separated by relatively warm periods, perhaps the result of major Davidson Current events. The warming trend from 1900-1910 on, which is reflected in the $\delta^{18}O$ curve of *G. bulloides*, represents the termination of the Little Ice Age.

At present, flushing of the basin bottom water occurs during periods of upwelling in the winter and spring. During one intense upwelling event, Sholkovitz and Gieskes (1971) measured an increase in basin bottom water oxygen content of up to 0.4 ml/liter. This relationship between surface upwelling and bottom water oxygenation plays an important role in production of the sedimentary record in Santa Barbara Basin. The $^{18}O/^{16}O$ ratios in *G. bulloides* indicate strong upwelling during the period before 1835, when benthic macrofauna was present at the basin bottom. Both Bx-4 and Bx-5 exhibit a marked decrease in $\delta^{18}O$ of *G. bulloides* at the 1835 level, indicating a temporary cessation of intense upwelling. Presumably, flushing of the basin with dense oxygen-rich shelf water decreased as a result, and the basin became stagnant, killing the macrofauna, and allowing laminated deposits to form. Subsequent upwelling which occurred after 1850 apparently lacked the intensity or duration necessary to raise the oxygen concentration to a level favorable to larger organisms.

The generally depleted $\delta^{18}O$ values of *G. bulloides* in post-1930 deposits indicate warmer surface water, greater stratification of the water column, and less intense upwelling. As a result, the basin may have become "more anaerobic." One expression of this may be an increase in the extent of the zone of laminated sediment formation on the basin bottom. The isotopic and faunal evidence from Bx-14 suggests that depth differences of 10 to 20 meters are critical in terms of oxygen concentration. The greater carbonate preservation at the surface of Bx-14 reflects a decrease in oxygen content (although a sparse macrofauna assemblage was found at the top of this core).

In summary:

1) $^{18}O/^{16}O$ ratios in the planktonic foraminifers *G. bulloides* from the varved sediments of Santa Barbara Basin apparently track temperature variations during the past 230 years. Relatively cool conditions, associated with the Little Ice Age, prevailed from 1895 to 1910, 1850 to 1880, and for a period of at least 30 years prior to 1780.

2) The $\delta^{18}O$ range in *G. bulloides* between "cool" events and "warm" events is nearly 1.5°/₀₀, greater than expected from historical temperature records. It is likely that seasonal and/or annual differential production (and possibly dissolution on the sea floor) amplifies the isotopic response to temperature of this species.

3) $\delta^{18}O$ and $\delta^{13}C$ in *G. bulloides* exhibit slight positive correlation, indicating that foraminiferal metabolic processes are more important than upwelling related changes in the $\delta^{13}C$ of seawater ΣCO_2 in controlling $^{13}C/^{12}C$ ratios in this species.

4) Dissolution may cause a significant shift in $\delta^{18}O$ of *G. bulloides* by preferential removal of those individuals relatively depleted in ^{18}O and thereby bias the record towards preservation of "cold" water events. The observation of a significant $\delta^{18}O$ shift (up to 0.6°/₀₀) in a core which is only slightly dissolved (<10% carbonate loss) illustrates the importance of accounting for this factor when interpreting isotopic profiles from deep-sea sediments (Berger and Killingley, 1977).

5) The size - $\delta^{18}O$ relationship for planktonic foraminifers in a coastal upwelling environment is opposite from the trend observed for most species from recent deep-sea sediments. While smaller specimens may precipitate calcite depleted in ^{18}O relative to equilibrium, they apparently reflect growth in cooler, more productive upwelled water and therefore exhibit greater $\delta^{18}O$ values relative to larger specimens.

6) Carbonate accumulation rates are high during periods when the varves are relatively thick, implying a direct (and not well understood) relationship between carbonate productivity and periods of high detrital input (assuming homogeneous supply over the entire basin).

7) The minimum carbonate dissolution rate averaged over the past 100 years is estimated as 3.8 mg cm^{-2} yr^{-1} based on comparison of sediments from a box core at 475 meters with those from the central part of the Santa Barbara Basin.

The varved sediments of the Santa Barbara Basin contain a unique record of climate. Calibration with the historical record of upwelling and climate reveals that $\delta^{18}O$ profiles of the planktonic foraminifers, *G. bulloides* may be used to estimate sea surface temperature variations which occur over periods as short as several years. The potential resolution of the climatic record in the varved sediments is on the order of seasonal or annual variation. Application of the isotopic technique reveals that during the Little Ice Age, coastal upwelling was stronger in the Santa Barbara area.

ACKNOWLEDGEMENTS

This work was part of a Ph.D. dissertation completed at the University of California, San Diego and was supported by grants from the National Science Foundation (Oceanography Section: Grant No. OCE78-25587) and the Exxon Production Research Company. I thank H. Thierstein, W.H. Berger and J. Killingley for manuscript review. K. McMurrer assisted in many phases of the laboratory work. The efforts of T. Walsh and S. Witherow produced a highly successful coring program during the SABA 79 cruise of the RV "New Horizon".

REFERENCES

Ahlmann, H.W., 1953, Bowman Memorial Lecture, American Geographical Society.

Arrhenius, G., 1952, Sediment cores from the East Pacific, Reports of the Swedish Deep-Sea Expedition, 1947-1948, 5:1-228.

Bé, A.W.H., 1980, Gametogenic calcification in a spinose planktonic foraminifera. *Globigerinoides sacculifer* (Brady), Marine Micropaleontology, 5:283-310.

Berger, W.H., 1969, Ecological patterns of living planktonic foraminifera, Deep-Sea Research, 16:1-24.

Berger, W.H. and Souter, A., 1970, Preservation of planktonic shells in an anaerobic basin off California, Geological Society of America, Bulletin, 81:275-282.

Berger, W.H. and Killingley, J.S., 1977, Glacial-Holocene transition in deep-sea carbonates: selective dissolution and the stable isotope signal, Science, 197:563-566.

Berger, W.H., Killingley, J.S. and Vincent, E., 1978, Stable isotopes in deep-sea carbonates: Box Core ERDC-92, West Equatorial Pacific, Oceanologica Acta, 1:203-216.

Berner, R.A., Scott, M.R. and Thomlinson, C., 1970, Carbonate alkalinity in the pore waters of anoxic marine sediments, Limnology and Oceanography, 15:544-555.

Bray, J.R., 1974, Glacial advance relative to volcanic activity since 1500 A.D., Nature, 248:42-43.

Bryson, R.A., 1974, A perspective on climatic change, Science, 184: 753-760.

CalCOFI Cruise Data Report, 1960, Physical and chemical data, Cruise 6005, station 82.47, SIO Ref. 62-7.

Chow, T.J., Bruland, K.W., Bertine, K.K., Soutar, A., Kodie, M., and Goldberg, E.D., 1973, Lead pollution records in southern California coastal sediments, Science, 181:551-52.

Craig, H., 1957, Isotopic standards for carbon and oxygen and correction factors for mass spectrometric analysis of carbon dioxide, Geochimica et Cosmochimica Acta, 12:133-149.

Cronblad, H.G. and Malmgren, B.A., 1981, Climatically controlled variation of Sr and Mg in Quaternary planktonic foraminifera, Nature, 291:61-64.

Deuser, W.G., 1978, Stable isotope paleoclimatology: a possible measure of past seasonal contrast from foraminiferal tests, New Zealand Department of Science and Industrial Research Bulletin, 220:55-60.

Deuser, W.G., Ross, E.H., Hemleben, C., and Spindler, M., 1981, Seasonal changes in species composition, numbers, mass, size, and isotopic composition of planktonic foraminifera settling into the deep Sargasso Sea, Palaeogeography, Palaeoclimatology, Palaeoecology, 33:103-127.

Douglas, A.V, 1976, "Past Air-Sea Interactions Over the Eastern North Pacific Ocean as Revealed by Tree Ring Data," Ph.D. Dissertation, University of Arizona.

Dunbar, R.B., 1981, "Sedimentation and the History of Upwelling and Climate in High Fertility Areas of the Northeastern Pacific Ocean," Ph.D. Dissertation, University of California, San Diego, 226 pp.

Dunbar, R.B. and Berger, W.H., 1981, Fecal pellet flux to modern bottom sediment of Santa Barbara Basin (California) based on sediment trapping, Geological Society of America, Bulletin, 92:212-218.

Dymond, J., Fischer, K., Clauson, M., Cobbler, R., Gardner, W., Sullivan, L., Soutar, A., Berger, W.H., and Dunbar, R., 1981, A sediment trap intercomparison study in Santa Barbara Basin, Earth and Planetary Science Letters, 53:409-418.

Emery, K.O., 1960, "The Sea Off Southern California: A Modern Habitat of Petroleum," John Wiley and Sons, New York, 366 pp.

Emery, K.O. and Hülsemann, J., 1962, The relationships of sediments, life and water in a marine basin, Deep-Sea Research, 8:165-180.

Emiliani, C., 1955, Pleistocene temperatures, Journal of Geology, 63:538-578.

Emrich, K., Ehhalt, D.H. and Vogel, J.C., 1970, Carbon isotope fractionation during the precipitation of calcium carbonate, Earth and Planetary Science Letters, 8:363-371.

Epstein, S. and Mayeda, T., 1953, Variation of ^{18}O contents of waters from natural sources, Geochimica et Cosmochimica Acta, 4:213-224.

Epstein, S., Buchsbaum, R., Lowenstam, H., and Urey, H.C., 1953, Revised carbonate-water isotopic temperature scale, Geological Society of America, Bulletin, 64:1315-1326.

Fleischer, P., 1972, Mineralogy and sedimentation history, Santa Barbara Basin, California, Journal of Sedimentary Petrology, 42:49-58.

Harman, R.A, 1964, Distribution of foraminifera in the Santa Barbara Basin, California, Micropaleontology, 10:81-96.

Heusser, L., 1978, Pollen in Santa Barbara Basin, California: a 12,000 year record, Geological Society of America, Bulletin, 89:673-678.

Hubbs, C.L., 1948, Changes in the fish fauna of western North America correlated with changes in ocean temperature, Journal of Marine Research, 7:459-482.

Hülsemann, J. and Emery, K.O., 1961, Stratification in recent sediments of the Santa Barbara Basin as controlled by organisms and water characteristics, Journal of Geology, 69:279-290.

Koide, M., Soutar, A. and Goldberg, E.D., 1972, Marine geochronology with Pb-210, Earth and Planetary Science Letters, 14:422-446.

Krishnaswami, S., Lal, D., Amin, S., and Soutar, A., 1973, Geochronogical studies in Santa Barbara Basin: Fe-55 as a unique tracer for particulate settling, Limnology and Oceanography, 18:763-770.

Lynch, H.B., 1931, Rainfall and stream run-off in Southern California since 1769, The Metropolitan District of Southern California Report, 30 pp.

Pisias, N., 1979, Model for paleoceanographic reconstructions of the California Current during the last 8,000 years, Quaternary Research, 11:373-386.

Rittenberg, S.C, Emery, K.O. and Orr, W.L., 1955, Regeneration of nutrients in sediments of marine basins, Deep-Sea Research, 3: 23-45.

Shiller, A.M., 1982, "The Geochemistry of Particulate Major Elements in the Santa Barbara Basin and Observations on the Calcium Carbonate-Carbon Dioxide System in the Ocean," Ph.D. Dissertation, University of California, San Diego, 197 pp.

Sholkovitz, E.R., 1972, "The Chemical and Physical Oceanography and the Interstitial Water Chemistry of the Santa Barbara Basin," Ph.D. Dissertation, University of California, San Diego.

Sholkovitz, E.R. and Gieskes, J.M., 1971, A physical-chemical study of the flushing of the Santa Barbara Basin, Limnology and Oceanography, 16:479-489.

Smith, B.N. and Epstein, S., 1971, Two categories of $^{13}C/^{12}C$ ratios for higher plants, Plant Physiology, 47:380-384.

Soutar, A., 1975, Historical fluctuations of climatic and bioclimatic factors as recorded in a varved sediment deposit in a coastal sea, Proceedings World Meteorlogical Organization, IAMAP Symposium on Long-term Climatic Fluctuations, Norwich, 18-23 August 1975, #421, 147-158.

Soutar, A. and Crill, P.A., 1977, Sedimentation and climatic patterns in the Santa Barbara Basin during the 19th and 20th centuries, Geological Society of America, Bulletin, 88:1161-1172.

Soutar, A. and Isaacs, J.D., 1969, History of fish populations inferred from fish scales in anaerobic sediments off California, California Marine Research Commission, CalCOFI, 13:63-70.

Soutar, A., Kling, S.A., Crill, P.A., Duffrin, E., and Bruland, K., 1977, Monitoring the marine environment through sedimentation, Nature, 26:136-139.

Sverdrup, H.U. and Allen, W.E., 1939, Distribution of diatoms in relation to the character of water masses off southern California in 1938, Journal of Marine Research, 2:131-144.

Tont, S.A., 1978, Sea level-air temperature correlations near a coastal zone, Nature, 276:171-172.

Tont, S.A., 1981, Temporal variations in diatom abundance off Southern California in relation to surface temperature, air temperature and sea level, Journal of Marine Research, 39:191-201.

Worthington, L.V., 1968, Genesis and evolution of water masses, Meteorological Monographs, 8:63-67.

Appendix 1

Bx-4, #4

Depth(cm)	%$CaCO_3$	*Globigerina bulloides* $\delta^{18}O$ ‰(PDB)	*Globigerina bulloides* $\delta^{13}C$ ‰(PDB)
0-1	7.9	-0.39	-0.92
1-2	7.8	-0.46	-1.02
2-3	6.1		
3-4	8.7	-0.40	-0.77
4-5	7.3	-0.17	-0.95
5-6	7.6	-0.61	-1.28
6-7	9.4	-0.98	-1.23
7-8	10.0	-0.82	-0.94
8-9	10.5	-0.25	-0.55
9-10	10.4	-0.21	-0.83
10-11	9.2	-0.10	-0.88
11-12	7.5	-0.35	-0.96
12-13	6.8	-0.29	-0.75
13-14	9.5		
14-15	10.7	-0.43	-0.73
15-16	10.0	-0.31	-0.81
16-17	9.7	-0.28	-0.73
17-18	9.4	-0.14	-0.75
18-19	8.2	0.11	-0.03
19-20	10.3	-0.13	-0.51
20-21	9.2	0.54	-0.57
21-22	9.7	0.11	-0.57
22-23	10.7	0.24	-0.47
23-24	10.3	-0.25	-0.63
24-25	9.3		
25-26	8.7	-0.62	-0.70
26-27	7.8	-0.47	-0.87
27-28	9.0	-0.35	-0.74
28-29	8.8	-0.35	-0.72
29-30	8.7	0.16	-0.41
30-31	8.2	-0.06	-0.32
31-32	7.5		
32-33	8.6	-0.11	-0.51
33-34	9.7	-0.44	-0.35
34-35	10.2	0.14	-0.12
35-36	11.7	0.33	-0.33
36-37	10.9		
37-38	9.2	-0.01	-0.52
38-39	8.3	-0.48	-0.54
39-40	9.9	-0.13	-0.57
40-41	9.8	-0.61	-0.95
41-42	12.1	-0.30	-0.39

Appendix 1 (cont.)

Bx-4, #4

Depth(cm)	%$CaCO_3$	*Globigerina bulloides* $\delta^{18}O$ °/₀₀ (PDB)	*Globigerina bulloides* $\delta^{13}C$ °/₀₀ (PDB)
42-43	12.7	-0.07	-0.57
43-44	12.4	-0.18	-0.67
44-45	12.2	0.03	-0.90
45-46	11.8	-0.11	-0.51
46-47	11.9	-0.52	-0.70
47-48	10.9	-0.07	-0.67
48-49	11.1	-0.24	-0.68
49-50	11.3	-0.10	-1.10
50-51	10.6	-0.08	-0.69
51-52	9.9	0.26	-0.35
52-53	8.5		
Bx-5, #3			
3-4	7.3	-0.35	-0.78
4-5	8.9	-0.58	-0.92
5-6	8.4	-0.58	-0.91
6-7	8.3	-0.17	-0.91
7-8	9.0	-0.10	-1.24
8-9	8.8	-0.54	-0.91
9-10	10.3	-0.84	-0.94
10-11	9.9	-0.45	-0.82
11-12	9.6	0.13	-0.66
12-13	9.5	-0.20	-0.62
13-14	9.4	-0.19	-0.84
14-15	10.3	0.07	-0.86
15-16	8.8	-0.43	-0.73
16-17	7.1	-0.45	-0.84
17-18	9.0	-0.30	-0.64
18-19	8.1	-0.21	-0.77
19-20	8.0	-0.05	-0.54
20-21	8.2	-0.23	-0.76
21-22	9.3	0.17	-0.28
22-23	10.4		
23-24	10.8		
24-25	11.4	-0.28	-0.77
25-26	9.6	-0.32	-0.74
26-27	9.1	-0.47	-0.49
27-28	8.9	-0.39	-0.65
28-29	10.6	-0.17	-0.47
29-30	10.4	-0.03	-0.67
30-31	11.0	0.11	-0.73

Appendix 1 (cont.)

Bx-5, #3

Depth(cm)	%$CaCO_3$	*Globigerina bulloides* $\delta^{18}O$ ‰(PDB)	$\delta^{13}C$
31-32	8.9	-0.18	-0.76
32-33	11.3	-0.16	-0.66
33-34	12.2		
34-35	12.8	-0.15	-1.01
35-36	13.9	-0.34	-0.82
36-37	12.3	-0.04	-0.89
37-38	11.6	-0.21	-0.94
38-39	11.6	-0.05	-0.70
39-40	11.8	-0.37	-0.77
40-41	12.2	-0.40	-0.83
41-42	11.9	-0.19	-0.65
42-43	9.9	0.02	-0.50
43-44	9.5	0.16	-0.68
44-45	9.0	0.12	-0.54
45-46	10.8	-0.05	-0.79
46-47	12.2	0.11	-0.88
47-48	11.7	0.09	-0.65
48-49	10.8	0.07	-0.23
49-50	8.8		

Bx-14, #8

Depth(cm)	%$CaCO_3$	Depth(cm)	*Globigerina bulloides* $\delta^{18}O$ ‰(PDB)	$\delta^{13}C$
0-1	8.6	0-1	0.13	-0.51
3-4	8.6	3-5	-0.11	-1.28
5-6	8.0	5-7	0.06	-1.31
7-8	7.2	7-9	0.12	-0.81
9-10	8.4	9-11	-0.03	-0.75
11-12	9.7	9-13		
13-14	9.9	11-13	0.10	-0.72
15-16	9.7	13-15	0.09	-0.92
17-18	8.5	15-17	0.25	-0.78
19-20	9.6	17-19	0.76	-0.76
21-22	9.4	19-21	0.35	-0.69
23-24	10.0	21-23	0.39	-0.59
25-26	9.8	23-25		
27-28	9.4	25-27	0.53	-0.82
29-30	8.2	27-29		
31-32	8.1	29-31	0.46	-0.56
33-34	8.9	31-33	0.54	-0.32

Appendix 1 (cont.)

Bx-14, #8

Depth (cm)	%$CaCO_3$	Depth (cm)	*Globigerina bulloides* $\delta^{18}O$ ‰ (PDB)	$\delta^{13}C$
35-36	8.3	33-35	0.39	-0.09
37-38	9.8	35-37	0.43	0.18
39-40	10.8	37-39		
41-42	9.0	39-41	0.67	-0.29
43-44	9.7	41-43	0.41	-0.47
45-46	10.2	43-45	0.48	-0.36
47-48	11.5	45-47	0.58	-0.56
49-50	11.3	47-49	0.51	-0.30
51-52	10.7	49-51	0.34	-0.34
53-54	11.5	51-53	0.55	-0.21
55-56	10.9	53-55	0.33	-0.20
		55-57	0.61	-0.33

DECADAL VARIATION OF UPWELLING IN THE CENTRAL GULF OF CALIFORNIA

Hans Schrader

School of Oceanography, Oregon State University
Corvallis, Oregon 97331, U.S.A.

Tim Baumgartner

Centro de Investigacíon Cientifica y Educacíon Superior
de Ensenada, Ensenada, Baja California, México

ABSTRACT

Decadal variations of productivity levels during the last 500 years in the central Gulf of California have been studied in four Kasten cores retrieved from the varved facies along the slopes of the Guaymas Basin. Silicoflagellate population changes were used to interpret productivity changes caused by varying degrees of upwelling phenomena. The chronology is based on successive varve counts. A comparison between the sediment record and a nearby tree-ring record shows a good correlation. During dry continental periods, with prevailing northerly wind, upwelling was enhanced, but during humid continental conditions and associated southerly winds, less upwelling and more oceanic conditions prevailed. At present, a decline in primary productivity levels is recognizable; the bottom waters are more oxygenated allowing a benthic foraminiferal fauna to thrive on the Baja side within the oxygen minimum zone between 400-800 m water depth. Between 1450 A.D. and the present, six major stages with increased productivity levels were recognized: 1490-1520, 1570-1580, 1660-1700, 1770-1810, 1840-1850 and 1920-1930. Periods of lower productivity levels occurred between 1540-1560, 1600-1650, 1720-1750, 1820-1840, and 1890-1910.

INTRODUCTION

Varved marine sediments are natural records of climatic and physical oceanographic changes affecting the local depositional environment. Such marine records have only been studied sporadically and their potential for documenting climatic changes over periods equivalent to those of historical records and continental dendrochronology has been largely untouched. Occurrences of varved marine sediments are rare but they are found along the slopes of the lower Gulf of California between 400 and 800 meters water depth (Calvert, 1964; 1966a; 1966b; Donegan and Schrader, 1981).

This paper is the result of initial studies to investigate and evaluate the temporal and spatial coherence of major and minor oceanographic changes over the last 500 years as recorded in sediments of the central Gulf. The primary focus is the variability of upwelling phenomena over the last 500 years and how it relates to proxy climatic records obtained from nearby areas. The purpose of this paper is to document changes of decadal or longer period duration by averaging the highly variable seasonal records over blocks of ten to fifteen years. Silicoflagellates were the primary microfossils used throughout this study to determine stratigraphic horizons, interpret fluctuating levels of primary productivity, and detect increased importance of Equatorial Pacific and California Current waters in the central Gulf.

GULF OF CALIFORNIA: CLIMATOLOGY, OCEANOGRAPHY AND VARVE DEPOSITION

The Gulf of California can be divided into two distinct topographic and oceanographic regions separated by the island chain of Angel de la Guarda, San Lorenzo, San Estaban, and Tiburón which are grouped along 29°N latitude. South of these islands, where the Guaymas Basin and the study areas are located, the Gulf is in open communication with the Pacific (Fig. 1). The composition and structure of sediments in the lower Gulf is related to the climatic and oceanographic parameters discussed below.

Oceanography

Summaries of the meteorology and oceanography of the Gulf are given by Roden (1958; 1964), Roden and Groves (1959), Alvarez-Borrego (in press), and Ayala Castañares, Phleger and Schwartzlose (in press). Exchange between water formed inside the Gulf and the open Pacific complicates the water mass structure (Roden, 1972b). Superimposed on these factors are seasonal and interannual variations which are still poorly understood. Notwithstanding this complexity, it is possible to distinguish three water masses above 200 meters (Griffiths, 1965; 1968; Stevenson, 1970; Warsh and Warsh, 1971; Roden, 1972b; Warsh, Warsh and Stanley, 1973; Alvarez-Sanchez, 1974). California Current Water is cool and has low salinities (<22°C, <34.6°/$_{oo}$). It flows southward along the west coast of Baja California and has been observed turning east around the tip of the peninsula and penetrating into the Gulf. The extent and dimension of this incursion depend upon the season and year of observation (Stevenson, 1970; Alvarez-Sanchez, 1974).

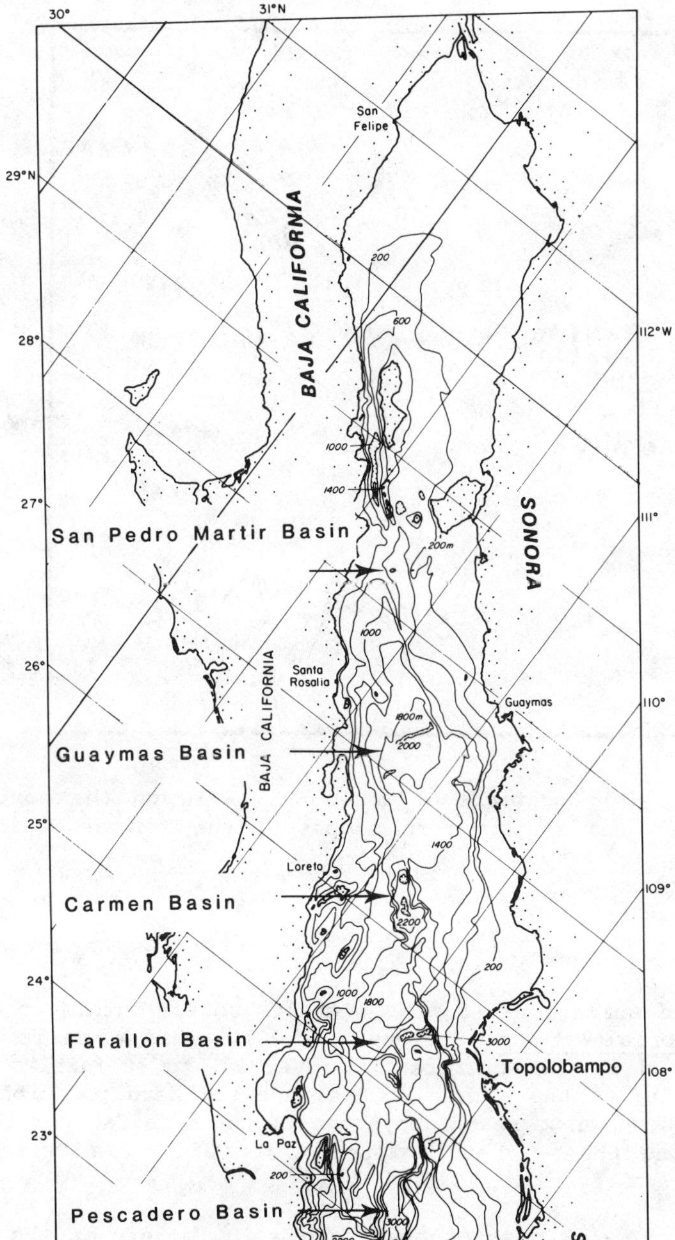

Fig. 1. (a) General bathymetry of the Gulf of California. Depth contours from Bishoff and Niemitz (1980).

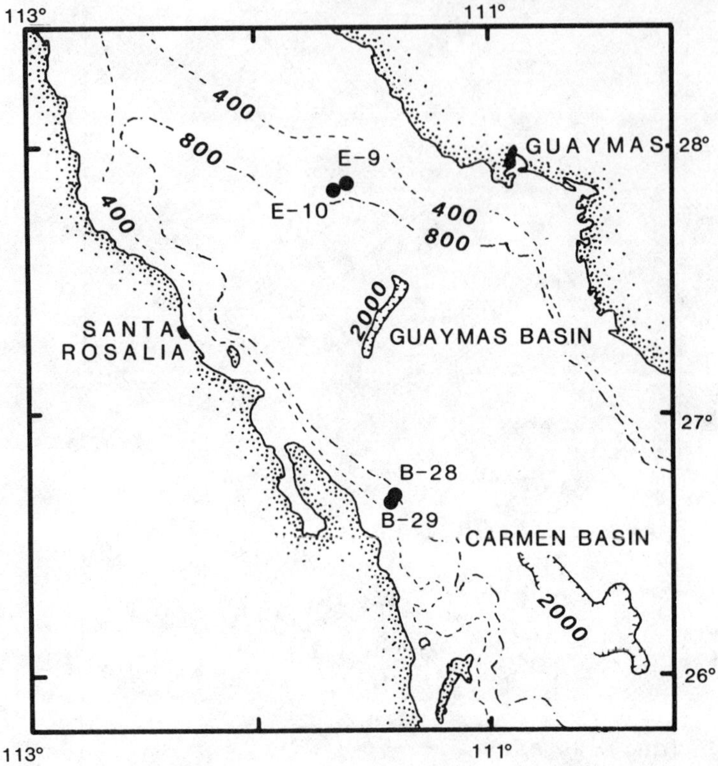

Fig. 1. (b) Locations of cores E-9, E-10 on the northern and B-28, B-29 on the southwestern slopes of the Guaymas Basin.

A second surface water mass with its source outside the Gulf is called Equatorial Water by Stevenson (1970). This water is warm (often >25°C) and is characterized by intermediate salinities (34.6 to 34.9°/₀₀). During the summer months, the northern position of this water mass is shown to reach above the tip of Baja California, apparently limiting the southern influence of California Current Water (Wyrtki, 1967; Stevenson, 1970).

The third water mass observed above 200 meters at the mouth of the Gulf is called Gulf Water. This water has its source within the Gulf, probably north of 25°N latitude, and is formed by evaporation of Equatorial Pacific Water (Stevenson, 1970). Within the Gulf, north of 25°N latitude, the only water mass observed above the thermocline is Gulf Water (Roden and Groves, 1959), with the possible exception of eddies of California Current Water, which may persist north of Topolobampo (Fig. 1a).

Climatology

The Gulf of California is a marginal sea of the subtropical Pacific Ocean lying between the arid peninsula of Baja California and the arid mainland regions of the Mexican states of Sonora and Sinaloa. The climate is continental in character with a large annual sea and air temperature change (compare Robinson, 1973). The climatology, and in turn a substantial number of oceanographic features, are controlled by four principal meteorological factors (Hastings and Turner, 1965): 1) A semipermanent and stable high pressure center building over the North Pacific reaching furthest north during the summer, which then weakens and migrates southward during the fall and winter just as a semi-permanent low is deepening over the Gulf of Alaska, 2) the subtropical anti-cyclonic high over the Atlantic which also moves northward off the east coast of North America during the spring and expands during the summer, 3) west coast tropical cyclones spawned in the intertropical convergence principally during the fall, and 4) easterly flow during the fall and retreat of the high pressure systems over both the Atlantic and Pacific. These global atmospheric circulation systems determine the regional wind patterns important to upwelling duration and intensity which control phytoplankton productivity and fluxes of organic particles to the sea floor in the southern Gulf. A simplification of the two extreme circulation patterns obtained over the Gulf is presented in Fig. 2. Condition A prevails during the summer and produces a northerly surface circulation in the Gulf, which may enhance transport of north Equatorial Pacific faunas and floras into the Gulf. Condition B, mostly during winter and spring, enhances primary productivity and exclusion of non-Gulf elements except when the California Current flow pattern is intensified.

The general surface circulation in the Gulf is strongly linked to the regional winds (Fig. 2) which blow from the northwest during winter and early spring, and change to a southeasterly direction during the summer months (Roden, 1964). Offshore transport of surface water by wind produces significant upwelling of waters along both coasts of the Gulf (Roden, 1972a). Winds with a southeasterly component develop in the summer (Hastings and Turner, 1965); these winds bring rain to Sonora and produce upwelling along the western shores of the southern Gulf. The northwesterly winds, most steady during winter and spring, are the driving force behind the winter/spring upwelling and related high primary productivity on the eastern side of the Gulf and particularly over the Guaymas shelf and slope (Soutar, Johnson and Baumgartner, 1981). The principal upwelling centers occur in the lees of capes, points and islands (Roden and Groves, 1959). A summary of the few available measurements of primary productivity in the Gulf is given by Zeitzschel (1969). The central Gulf is characterized by production rates similar to those off Peru and Southwest Africa. Averaged primary productivity values are: 0.27 g C m^{-2} day^{-1} for the area south of 25°N; 0.38 g C m^{-2} day^{-1} for the area between 25°-27°N (the Carmen Basin area); and 0.53 g C m^{-2} day^{-1} for the area between 27°-29°N (the Guaymas Basin area).

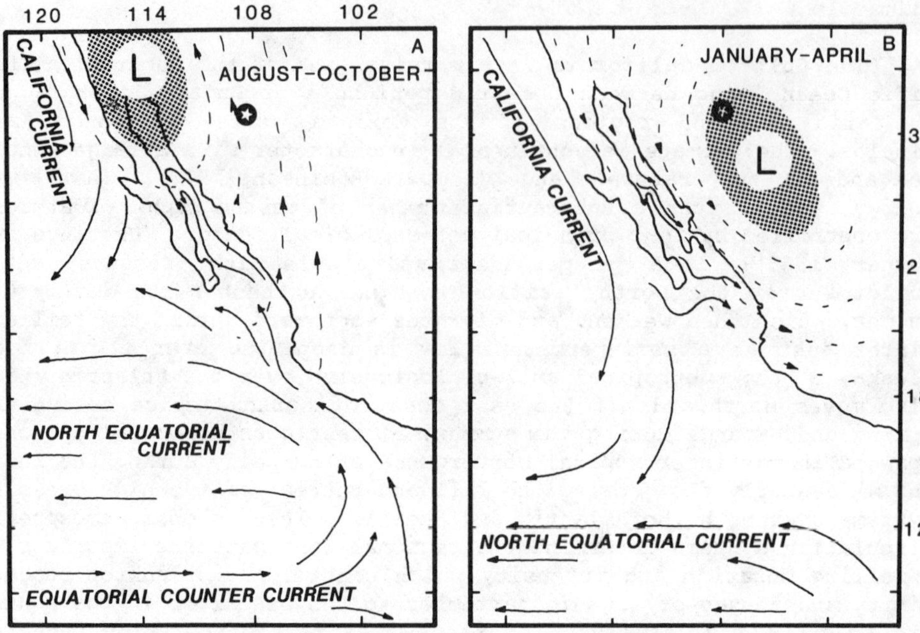

Fig. 2. Generalized wind and surface water circulation in the subtropical and tropical Eastern Pacific. Location of low pressure cells and prevailing wind systems over the Gulf from Roden (1958), circulation in the Gulf from Wyllie (1966), eastern North Pacific circulation from Namias (1971) and Wyrtki (1965; 1967). The star is the location of the Casas Grandes tree-ring site (Sierra Madre Occidental, Fritts 1965).

The regional rainfall which presumably controls terrigenous sedimentation in the Gulf falls mainly in Sonora and Sinaloa where river discharge increases toward the south (Calvert, 1966b). Very little rainfall occurs over the eastern margin of the Baja California peninsula (except at its southernmost tip) and no major rivers drain into the Gulf from it (Markham, 1972). Rainfall in Sonora and Sinaloa comes primarily in the form of a summer "monsoon" during July and August (Hastings and Turner, 1965). September rainfall over Sonora is sometimes as high as in July/August and is usually associated with tropical cyclones. Rainfall over the mainland coast and Baja California is usually extreme during such storms.

In the spring, as the North Pacific high begins its movement northward and is strengthening (Fig. 2), the anticyclonic geostrophic wind tracks begin to parallel the axis of the Gulf between this high pressure ridge and a low pressure trough over the now dry central Sonoran desert (Roden, 1958). It is this event which brings the most regular strong winds to the Gulf and the transport of surface water to the south, resulting in upwelling and the extensive winter-spring phytoplankton blooms along the eastern coast of the Gulf, normally from February to April (Roden, 1972a).

Southern Oscillation and El Niño

The existence of a relationship between variation in oceanic climate and sea level, suspected for some time (Reid, 1960), is now being detailed by more recent investigations (Wyrtki, 1975; 1977; Barnett, 1977). The El Niño event of the eastern South Pacific has been shown to affect the entire eastern tropical Pacific both north and south of the equator (Miller and Laurs, 1975). Time series of monthly, non-seasonal anomalies of sea level and shore temperatures in the Gulf exhibit variations with periods of three years or longer and amplitudes over 15 cm and nearly 2°C, respectively. Strong spatial coherence on the scale of several years is evident in comparing Gulf sea level with sea level of southern California and southern Mexico. These observed time and space scales suggest that interannual variability of sea level and shore temperatures in the Gulf is a response to large scale coupling of ocean and atmosphere changes (Baumgartner et al., 1979).

Short-term fluctuations in north Equatorial Pacific biota in the central Gulf have been associated with El Niño type events (Baumgartner and Soutar, in press). The anti-El Niño conditions, which precede El Niño events, are characterized by a strong South Equatorial Current and a weak Equatorial Counter Current (Wyrtki, 1977 and 1979, compare his figure 3 for the year 1971/1972). Under these conditions, a stronger influence of California Current Water in the central Gulf is expected (Fig. 2B). Preliminary results from a seasonal study of individual laminae over the last 70 years (Core E-13) do not yet conclusively follow this model. Here most events indicating California Current Water in the central Gulf follow indicators of strong Equatorial Pacific Water influence rather than precede them.

Laminated Sediment Deposition

Two environmental controls are principally responsible for the formation and preservation of varved sediments along the slopes of the southern Gulf and extending southward into the eastern Equatorial Pacific: a) the existence of a stable oxygen minimum layer within the water column and b) the seasonal difference in local and regional meteorology (Calvert, 1964; Soutar, Johnson and Baumgartner, 1981; Donegan and Schrader, 1982).

The oxygen minimum occurring at intermediate depth over the eastern tropical Pacific penetrates into the Gulf and is maintained and strengthened there through the oxidation of large quantities of settling organic particles (Calvert, 1964). High levels of primary productivity and possibly reduced lateral circulation due to topographic confinement of intermediate water inside the Gulf result in extremely low oxygen values (less than 0.1 ml O_2/ℓ). Slope sediments of the southern Gulf are in contact with oxygen depleted waters between 400 to 800 meters. The center of the oxygen minimum lies near 600 meters depth. Benthic fauna throughout the oxygen minimum is markedly impoverished (Calvert, 1964), and towards the center of the oxygen minimum the mixing effects of bioturbation are minimal.

The alternating light and dark bands within the laminated diatomaceous sediments were interpreted to represent seasonal fluctuations in plankton productivity (Revelle, 1950; Byrne and Emery, 1960). Later, Calvert (1966b) stressed the role of seasonal terrigenous pulses in formation of the rhythmic structure and provided evidence that the laminae pairs are true varves. The use of the term varve denotes that each lamina pair represents an annual cycle (Baumgartner et al., 1981).

METHOD AND TAXONOMY

The suite of slides used to count silicoflagellates was sampled and cleaned by Matherne (1982) as part of a diatom study. Four Kasten cores were selected for this study (Table 1; Fig. 1) and were collected during the BAV 79 cruise in September 1979 (Schrader, 1979). Around 150 to 200 silicoflagellates were counted in each slide; in most cases more than half of the slide area was scanned. Only complete skeletons or fragments greater than one half of the original skeleton were counted. Even though diatoms form the major component of these diatomaceous muds, silicoflagellate abundances are still around 10^6 specimens per gram dry sediment (Donegan and Schrader, 1982). These abundances provided sufficient silicoflagellate specimens on each slide to provide meaningful quantitative estimates.

The taxonomy strictly followed that of Poelchau (1976) for the following species: *Dictyocha messanensis*, *Dictyocha calida*, *Dic-*

Table 1. Logistics of the four Kasten cores studied (Cruise BAV 79, RV H-1 Mariano Matamoros)

Core	Location		Depth
BAV 79 - E 9	27° 53.2'N	111° 37.2'W	660 m
BAV 79 - E 10	27° 52.2'N	111° 39.7'W	644 m
BAV 79 - B 28	26° 42.5'N	111° 24.5'W	712 m
BAV 79 - B 29	26° 42.0'N	111° 25.0'W	635 m

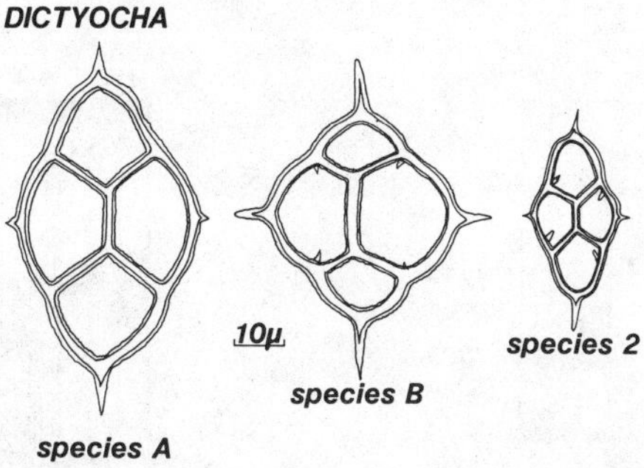

Fig. 3. *Dictyocha* species A, species B and species 2 as used throughout this paper. *Dictyocha* species A has no supporting spines at the basal ring, whereas *D.* species B and 2 have four supporting spines.

tyocha epiodon, *Octactis pulchra* and *Distephanus speculum*. In addition to these species, three others were distinguished and placed into an open classification scheme (Fig. 3): *Dictyocha* species A, *Dictyocha* species B, and *Dictyocha* species 2. Silicoflagellates, from a large surface water phytoplankton collection from inside the Gulf, from the mouth of the Gulf and along the western side of the Baja California peninsula, taken in November 1980, combined with data presented in Poelchau (1974 and 1976), provide the necessary information about species distribution and their relation to water masses. These studies resolve the silicoflagellate species distribution and their association to water masses and are discussed below.

The lithologic descriptions of the four cores presented in Figs. 4 and 5 include the distinction between lighter and darker colored intervals. These color changes are so minimal that they cannot be distinguished using the Munsell color chart; generally, the cores from area E are olive to olive-gray (5Y 5/4; 5Y 4/3) whereas the cores from area B are olive to grayish-brown (5Y 3/3). Structural distinctions (Figs. 4 and 5; Table 3) were made between intervals which are well and regularly laminated with well-defined laminae boundaries and those intervals with less well-defined laminae boundaries. The laminae over these latter intervals are generally thick, and the color differences between the light and dark laminae are minimal. Intervals where no obvious lamina structure was present as determined visually and/or by x-radiography, were termed homogenous. Disturbances with angular discontinuities, sudden wedging over 2-10 cm thick intervals, and offset of laminae are illustrated in column 6 on Figs. 4 and 5; occurrences of thick individual laminae, generally more than 3 mm thick, are documented in column 3.

Numbers of laminae per stratigraphic distance were determined by visually counting through 2.5 cm long increments (intervals identi-

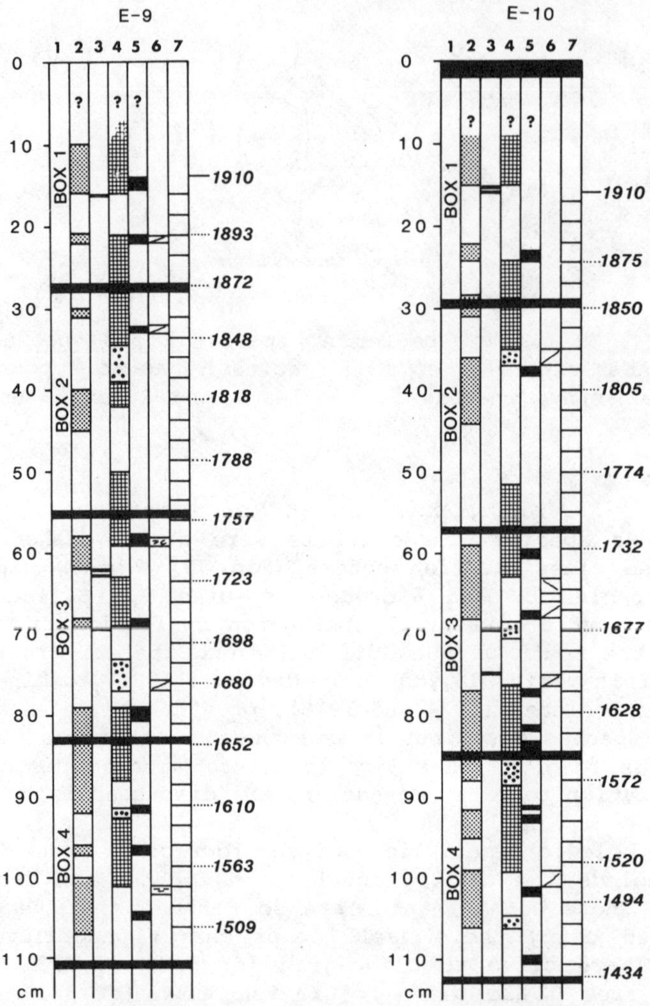

Fig. 4. Lithologies of the two cores from area E. Due to recovery and curation procedures the recovered core surface in E-10 is placed at 2 cm. Black horizontal lines identify levels where material was lost during the boxing process. Due to watery material above 15 cm, no correct interpretations are possible. The individual columns represent (2) intervals darker in color, stippled, (3) very distinct thick light laminae, dark bars, (4) intervals very distinctly laminated, hatched; intervals homogenous and/or with very indistinct structures, irregularly stippled; other less well-laminated intervals, blank, (5) intervals darker in color and very finely laminated, dark bars, (6) intervals with major disturbances dark bars, (7) intervals used for varve chronology. Ages listed on the right based on varve chronologies. Note that bedding in core E-9 is not horizontal below 40 cm depth.

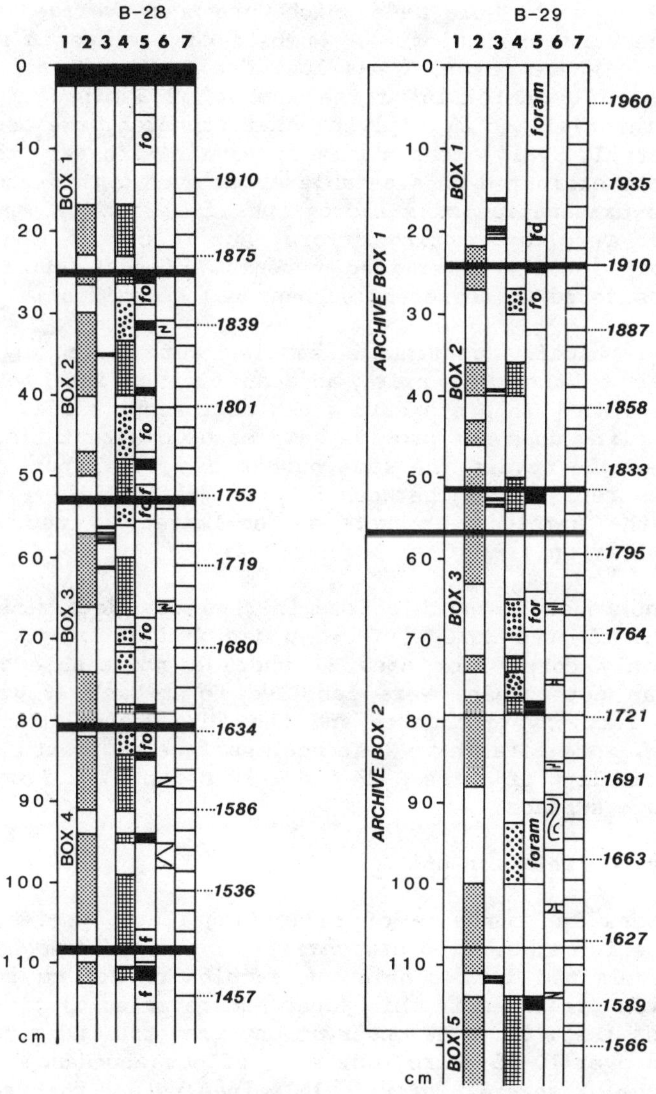

Fig. 5. Lithologies of the two cores from area B. For explanation see caption of Fig. 4. Differing from other cores, a detailed description above 15 cm in core B-29 was possible. Intervals in column (5) labelled with "fo" contain increased numbers of benthic foraminifers. Samples for silicoflagellate studies were taken at successive 2.5 cm intervals starting at 2.5-5 cm in core B-28 and 0-2.5 cm in core B-29. Bedding in core B-28 horizontal throughout, horizontal to 80 cm in core B-29 and slightly tilted (5°) below. Note the decrease in distinctly laminated intervals [column (4)] as compared with Fig. 4.

fied on Figs. 4 and 5, column 7); within each of these intervals several replicate counts were made which were then averaged and used to establish the varve chronologies. Numbers of laminae in missing intervals (those where material was lost due to the curation process) were calculated by determining the number of laminae/cm above and below that interval and multiplying that number by the length of the missing interval. Varve counts from intervals with very thin lamination sequences exhibited considerable error even with 5 and more replicates. Re-examination of x-radiographs from some of these intervals did not resolve counting errors, but results fell well within the range of visually determined numbers. These latter intervals were not considered to represent subseasonal deposition.

Stratigraphically continuous samples were taken by scraping, perpendicular to the core axis, adjacent 2.5 cm intervals over the entire core width with a standard glass microscope slide. This method of subsampling does not provide samples of constant time intervals where each sample covers the same number of years, but rather collects samples representing between 5 to 25 years at the extreme. The average of the sampled intervals is estimated to represent 10-15 years of deposition.

The occurrence of benthic foraminifers was determined visually by inspecting cleaned core surfaces under oblique light. Only those intervals (only cores from area B) where a noticeable increase in abundance was detectable, were labelled foram-rich (Fig. 5, column 5). Due to recovery procedures and the subsequent curating of the Kasten cores, some cores have lost near-surface sediments (0-2.5 cm). Therefore, sampling of cores B-28 and E-10 began at 2.5 cm below the recovered core surface.

Stratigraphy, Correlation and Age

Variations in abundance of silicoflagellate species, primarily *Dictyocha* species A, B, 2, *Dictyocha epiodon* and *Dictyocha messanensis* were determined from continuous samples of 2.5 cm intervals in all four cores presented in this paper and revealed distinct correlative horizons (Fig. 6). The observations from the individual samples are averaged over 10-15 years, but may reflect abundance changes occurring in one or several individual laminae rather than representing a large number of years during which distinct environmental conditions persisted. Independent controls were established by using distinct packages of laminae, major disturbances and mass occurrences of other microfossils (diatoms) for correlation.

By plotting these horizons according to their "varve age", determined by lamina counts, it can be seen that most events fall to within 10-20 years along the same age horizons (Fig. 7). This concordance supports the usefulness of these stratigraphic horizons as time horizons. In addition, Fig. 7 supports the assumption that

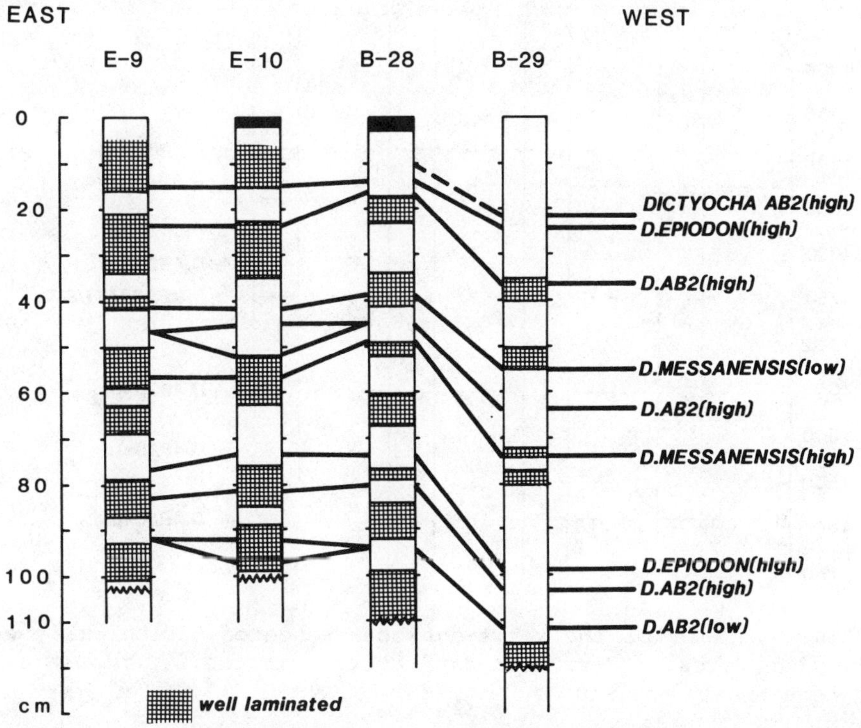

Fig. 6. Correlation between cores and identification of correlation lines. Cores are lined up from east to west. Black top intervals represent empty and/or nonrecovered intervals. Stippled correlation line tentative. Correlation of some events tentative in core E-10. Cross-hatched intervals in cores indicate distinctly laminated horizons.

little if any material has been added to or removed from the sediment column. Correlations of the biological markers shown in Fig. 7, as well as the 10-15 years resolution of the subsampling, indicate that age estimates of the samples determined by varve chronologies of the four cores examined here should fall within a range of ±20 years. A detailed Gulf-wide biological stratigraphy is in preparation and will present further evidence supporting the assumption that the sedimentary record within the laminated facies is fairly complete. Due to disturbances and loss of near-surface sediment material, it was necessary to establish a deeper reference horizon in all four cores which was used to adjust all derived age calculations. The horizon selected was the first common occurrence of *Dictyocha epiodon* (a California Current inhabitant) with *Dictyocha* species A, B, 2 (tropical Pacific coastal inhabitants). Results from a laminae by laminae study of frozen boxcore slabs from area E (Core BAM 80-E 13) revealed that this event lasted over approximately 6 laminae=3 years. Varve

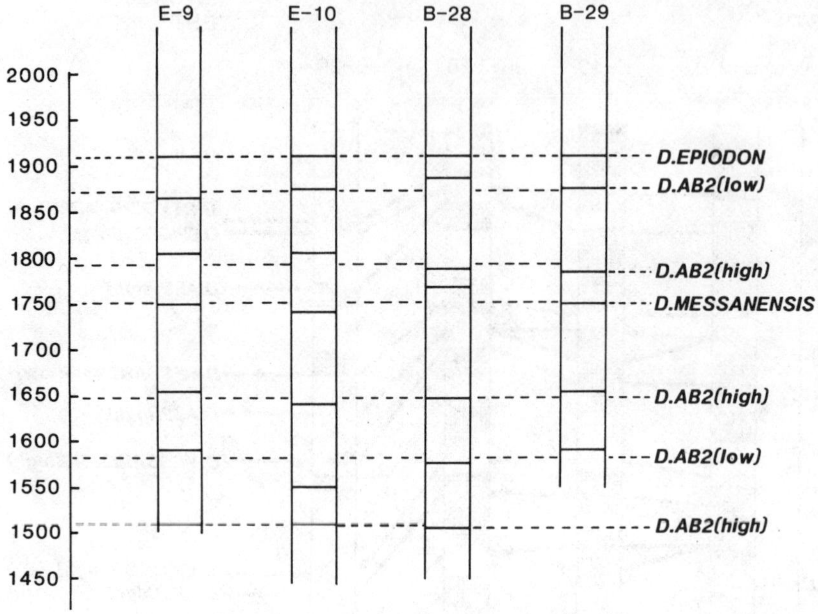

Fig. 7. Plot of the varve-age occurrence of biological events in the four cores. "Zero" standard horizon is the *D. epiodon* event dated tentatively as 1910.

chronology of that core suggest that this event occurred around 1910. Historical records listed in Quinn et al. (1978, Table 1) document a strong El Niño event occurring in 1911 with an associated strong anti-El Niño preceding it which peaked around 1910. It is assumed that during anti-El Niños, oceanic conditions permit California Current waters to enter the lower Gulf. Reconstructed summer sea surface temperatures at 22.5°N/110.5°W show a strong temperature anomaly between 1900-1910 (Douglas, 1976; 1980).

Unfortunately the strong anti-El Niño of 1910 is not a unique event. Documentation by Quinn et al. (1978) of the El Niño phenomenon also indicates a sequence of anti-El Niño conditions between 1886 and 1917, each of which is as strong or stronger than that at 1910. Because the age estimate of the *D. epiodon* appearance is based on one boxcore for which the uncertainty of the varve chronology is difficult to establish at present, the 1910 date is considered only tentative.

Bottom ages at selected horizons and resulting sedimentation rates are listed in Table 2. These ages and sedimentation rates are slightly different from the ^{14}C ages. The discrepancy between the rates derived from ^{14}C ages noted here is similar to that found by DeMaster (1979) in varved sediments from the Carmen Basin, Gulf of California, where the ^{14}C dating disagreed with ^{210}Pb and varve

Table 2. Ages and resulting sedimentation rates

Core	Depth Interval Top - Bottom (A) cm (B)	Number of Laminae Between A-B	Varve Age at B	Varve Sedimentation Rate Between A-B in cm/Year	^{14}C Sedimentation Rates in cm/Year*
E-9	14 - 106	802	1509	0.23	0.14
E-10	16 - 112	952	1434	0.20	--
B-28	14 - 114	908	1457	0.22	--
B-29	24 - 120	694	1566	0.28	0.20

* Matherne, 1982

counts of near-surface material. DeMaster suggests that the ^{14}C surface activity of this region may have varied significantly through the past, possibly due to changes in rates of upwelling or air-sea gas exchange. Such phenomena combined with the paucity of the ^{14}C samples in the dated cores are likely to have produced the observed discrepancies. Sedimentation rates determined by varve counts in the four cores examined here agree with rates determined by ^{210}Pb and varve counts on a boxcore from the Guaymas slope (Ferreira-Bartrina, Baumgartner and Schrader, in press).

RESULTS

The geographic distribution of silicoflagellate species used in this study and their associations with water masses is summarized below. For a detailed description of these relationships, see Murray (1982).

(1) *Dictyocha epiodon* occurs in the Alaskan Gyre and in the western North Pacific and extends south along the Eastern Pacific associated with the California Current (Poelchau, 1974; 1976).

(2) *Dictyocha* species A, B increase steadily toward the south along the Mexican mainland. They are also found in sediments from the Peru upper slope "mud lens" and in the Panama Basin. These species have a tropical to subtropical coastal distribution and seem to be associated with equatorial Pacific waters moving northward along the coast of Central America (Murray, 1982).

(3) *Dictyocha calida* is found in small numbers in sediments underlying the equatorial Pacific countercurrents (Poelchau, 1976). Its presence near the mouth of the Gulf in phytoplankton samples collected in November, 1980, might be an indication of mixing of equato-

rial Pacific waters and California Current waters, with the latter indicated by the presence of *D. epiodon.*

(4) *Octactis pulchra* is found in the eastern tropical-subtropical Pacific (Poelchau, 1976) associated with high primary productivity and upwelling phenomena (Murray, 1982). *Octactis pulchra* is found off Peru and in the Gulf associated with diatom florules previously classified by Schuette and Schrader (1979; 1981a; 1981b) to reflect coastal upwelling. This species had been incorrectly identified as *Distephanus speculum* by DeVries and Schrader (1981).

(5) *Dictyocha messanensis* is interpreted as an indicator of oceanic influence in the southern Gulf of California during times of reduction in upwelling.

The association of *Octactis pulchra* with diatom species indicative of high productivity supports the interpretation. Phytoplankton assemblages collected in November 1980 inside and outside the Gulf and which are dominated by *Chaetoceros radicans* and *Thalassiothrix pseudonitzschioides* contain an almost monospecific silicoflagellate flora with *Octactis pulchra*, whereas phytoplankton assemblages which are dominated by a dinoflagellate flora indicative of less productive oceanic or stratified conditions are characterized by high abundances of *Dictyocha messanensis*. Some laminae from the hydraulic piston core site at IPOD-DSDP 480 (Core 3, Section 2, individual light laminae at 145 cm), for which the diatom flora is nearly monospecific with *Chaetoceros radicans* tests and resting spores indicating high productivity (Schuette and Schrader, 1979; 1981a; 1981b), contain *Octactis pulchra* at a concentration of more than 98% of the total silicoflagellate population. Other laminae from IPOD-DSDP Site 480 (Core Catcher 30) contain an almost monospecific diatom assemblage with either *Coscinodiscus nodulifer* and/or *Coscinodiscus asterom-*

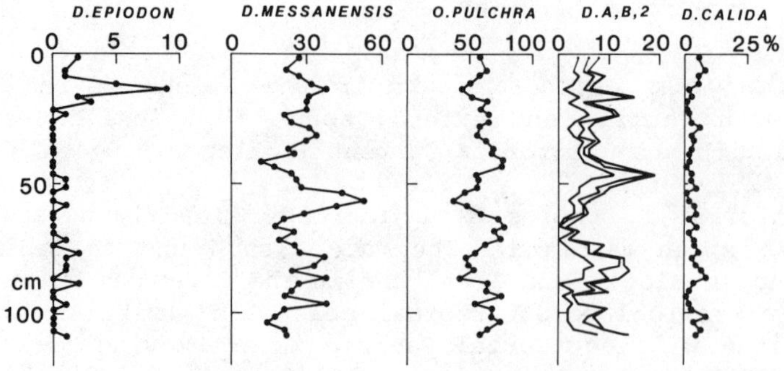

Fig. 8. Percent abundance of silicoflagellates in core E-9 (2.5 cm composite samples). The values for *Dictyocha* species A, B and 2 are combined in one column, the first curve in this column is *D.* A, the second added percent of *D.* B and the third curve added percent of *D.* 2.

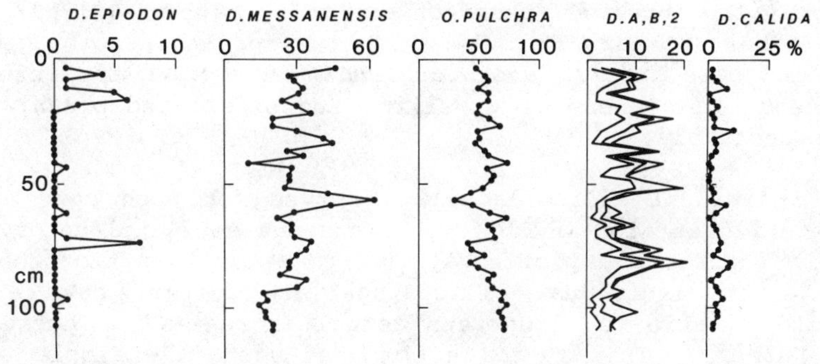

Fig. 9. Percent abundance of silicoflagellates in core E-10 (2.5 cm composite samples); for explanation see caption of Fig. 8.

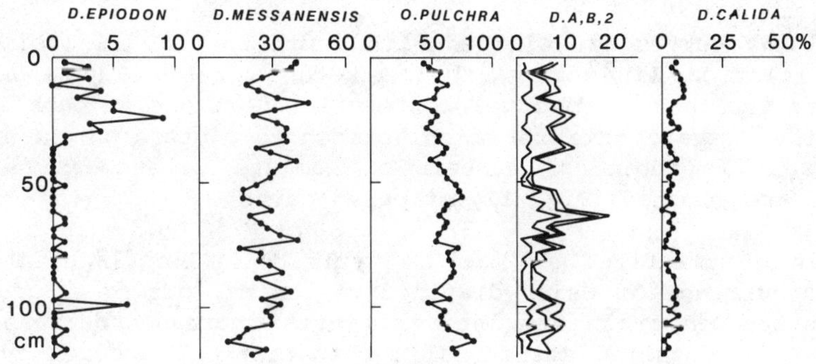

Fig. 10. Percent abundance of silicoflagellates in core B-28 (2.5 cm composite samples); for explanation see caption of Fig. 8.

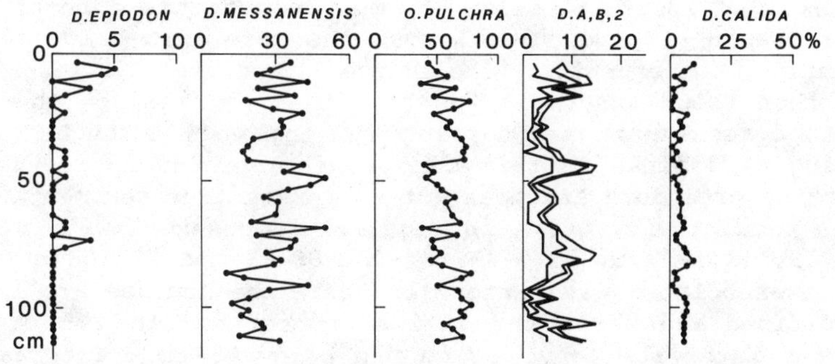

Fig. 11. Percent abundance of silicoflagellates in core B-29 (2.5 cm composite samples); for explanation see caption of Fig. 8.

phalus in which the silicoflagellate assemblage is dominated by *Dictyocha messanensis*. Similar conditions are found over the *Coscinodiscus nodulifer* interval (based on counts by Matherne, 1982) in this suite of cores (compare Fig. 14). Thus, separate assemblages dominated by *Octactis pulchra* and *Dictyocha messanensis* are interpreted to represent end members of upwelling and stratified oceanic conditions, respectively.

The individual silicoflagellate curves for each core and each species are illustrated in Figs. 8-11 and are self-explanatory. Values in all figures are plotted at the mid-point depth of the sample intervals. The individual silicoflagellate percent data are converted into a ratio to illustrate actual increases and decreases of *Octactis pulchra*. This ratio is computed using the simple equation (P = primary productivity index).

$$P = \frac{\textit{Octactis pulchra}}{\textit{Octactis pulchra} + \Sigma \text{ [of all silicoflagellates]}}$$

The resulting curves are plotted with depth on Fig. 12 (solid line) with correlation lines drawn between the cores. The approximate varve ages for these individual correlation horizons in each core are listed; the average date for each horizon is plotted on the right of the figure. In addition, numbers of laminae in successive 2.5 cm intervals are plotted (Fig. 12, stippled curve).

Table 3 summarizes all major events in silicoflagellate abundances, occurrence of major distinctive laminae packages, occurrence of homogenous intervals and intervals with increased/decreased productivity. A detailed structural interpretation in cores E-9, E-10 and B-28 above 15 cm was not possible due to too soupy and watery surfaces.

Kasten core E-9 (Figs. 4, 8 and 12): The laminae in this core start dipping at around 40 cm and further downcore and reach a maximum dip angle of about 16° below 65 cm. Due to this effect, samples which were taken perpendicular across the core axis below 45 cm cut across time. Consequently, the curves generated for E-9 appear smoother than those generated for the other cores since the subsampling method results in a two-point running mean through the data. The dipping of laminae in this core is interpreted to be the result of the coring procedure and does not represent true sedimentary dipping. The productivity curve is regularly sinusoidal with highs between 5-7.5, 17.5-27.5, 32.5-45, 62-75, 88-93 and 98-108 cm. These levels are associated with intervals where the laminae are less distinctly defined and where the laminae number within 2.5 cm is low except over intervals above 30 cm and below 86 cm. Intervals with very indistinct lamination and/or homogenous intervals occur at 34.5-39, 73-77 and 91.5-93 cm. Major angular discontinuities were found between 21-22 cm, 32-33 cm, 75.5-77 and two minor disturbances at

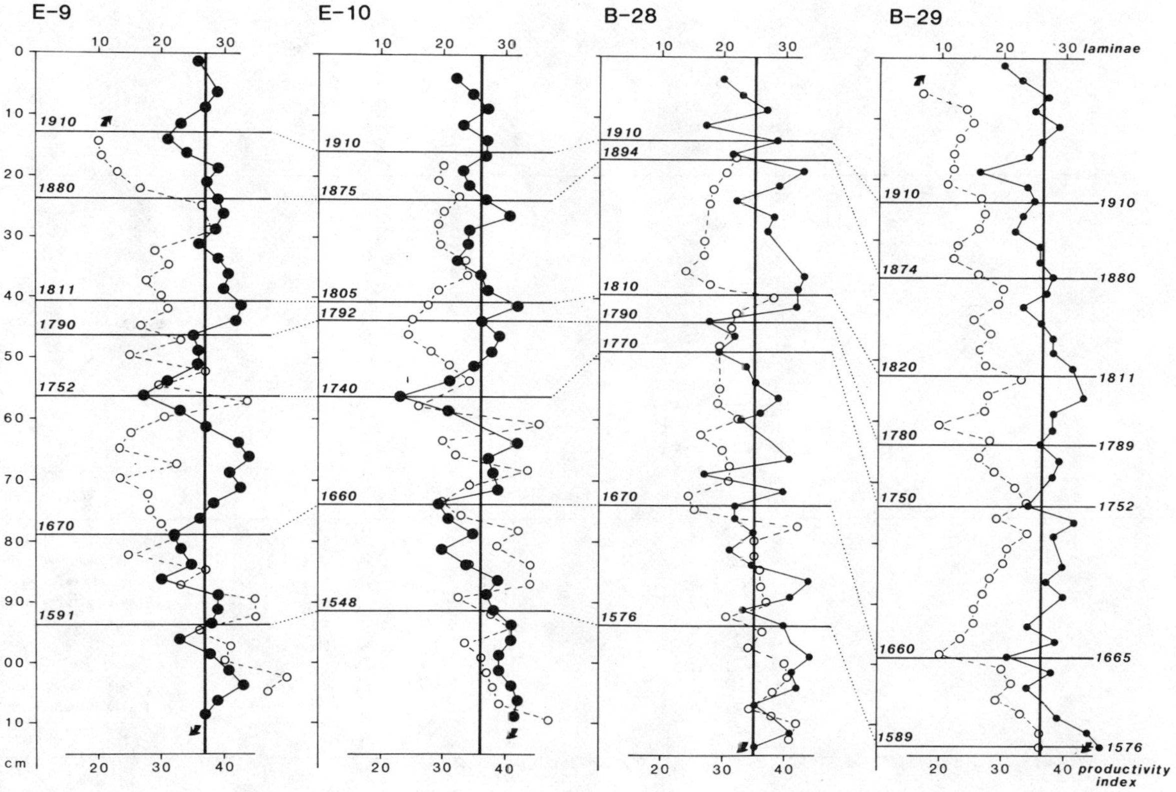

Fig. 12. Plots of productivity index (solid lines, scales at bottom) and the number of laminae (stippled lines, scales at top) per 2.5 cm intervals for all cores. Solid horizontal lines represent stratigraphic horizons which are traced between cores by punctuated lines. Individual core varve ages are listed within each core. Averaged ages of correlation horizons are plotted to the far right.

Table 3. List of major silicoflagellate occurrences in all four cores and their relation to productivity levels and structure of laminae. List of major structural horizons and relation to productivity levels and list of major productivity levels and associated structures. (Correlative horizons determined by varve chronology and silicoflagellate correlation.)

EVENTS	AREA E E-9	E-10	Laminae structure distinct	moderate	poor	AREA B B-28	B-29	Laminae structure distinct	moderate	poor	Productivity levels low	medium	high
Bio-correlation horizons	cm					cm							
Dictyocha epiodon, peak 1	12.5-15	15-17.5	+			12.5-15	22.5-25		+			+	
Dictyocha epiodon, peak 2	75-77.5	72.5-75		+	+	72.5-75	97.5-100			+	+		
Dictyocha A, B, 2, peak 1	15-17.5	22.5-25	+			12.5-17.5	35-37.5	+	+			+	
Dictyocha A, B, 2, peak 2	45-47.5	50-52.5		+		45-47.5	62.5-65			+	+		
Dictyocha A, B, 2, peak 3	77.5-87.5	80-82.5	+			77.5-80	97.5-100		+	+	+		
Dictyocha A, B, 2, peak 4	---	---				---	---						
Dictyocha messanensis, low 1	40-42.5	40-42.5	+	+		37.5-40	52.5-57.5	+					+
Dictyocha messanensis, high 1	55-57.5	55-57.5	+			47.5-50	72.5-75	+			+		
Coscinodiscus nodulifer high	55-57.5	57.5-60	+			52.5-55	72.5-75			+	+		
Structural horizons													
1. Homogenous-very faintly laminated	34.5-39	35-37				28.5-36.5	?				+		
	?	?				41.5-48	65-70				+		
	?	?				51.5-56	74-77				+		
	73-77	68.5-70.5				68.5-74	92.5-100					+	
	91.5-93	85-88.5				81-84	?					+	
	---	104.5-106.5				---	---						
2. Distinctly laminated	to 16	to 15				?	?					+	
	21-34	24-35				17-23	36-40				+		+
	39-44.5	?				33.5-40	50-54						+
	50-59	51.5-63				48-51.5	72-74				+		
	63-69	?				60-67	?						+
	79-88	76-85				84-91	?				+		
	93-101	87.5-99				99-108	114-125				+		+
Productivity horizons													
1. High	40-44	40-42		+		35-41	50-57	+	+				+
	62-72	63-65	+	+		55-57	75-77		+				+
	100-105	?		+		95-105	?	+	+				+
2. Low	52-60	52-59	+			43-50	71-74		+	+	+		
	78-87	72-84	+	+		74-82	98-100		+	+	+		

58-59 and 101-102 cm. Laminae are bent and/or offset at these intervals.

Kasten core E-10 (Figs. 4, 9 and 12): The *Dictyocha* species A, B and 2 show more variation than seen in core E-9. Four *Dictyocha calida* horizons are recognizable at 10-12.5, 27.5-30, 57.5-60, and 80-82.5 cm. The productivity curve is very similar to the previous one (E-9). Intervals of increased productivity are found between the intervals 7.5-10, 12.5-16, 23-38, 36-50, 61-72, 85 to 110 cm (and possibly below). The averaged values of the index are slightly lower than found at E-9 and follow a very similar trend observed in cores from area B. In both cases the averaged values are slightly higher in the shallower cores (E-9, B-29). The levels of increased primary productivity index coincide with levels over which the laminae number is decreased except the intervals above 30 cm and below 70 cm; over these intervals the laminae thickness curves parallel the productivity curves. Intervals with indistinct lamination and/or homogenous intervals occur at 35-37, 68.5-70.5, 85-88.5, and 104.5-106.5 cm. Disturbances in the record were found at 35-37 (angular), 63-65 and 66-67 cm (angular opposing), 75-76.5 and 99-101 cm (angular discontinuities).

Kasten core B-28 (Figs. 5, 10 and 12): The cores from area B are generally less well-laminated using the color change and show less structural detail compared with those from area E. The lamination of this core becomes gradually more distinct below 53 cm. Peaks with *Dictyocha calida* occur at 2.5-5, 10-12.5, 22.5-25, 45-47.5, and 80-82.5 cm and values below 100 cm are generally above 5%. The mean productivity ratio is slightly lower than in E-10. Highs in this ratio are found between the following levels: 7.5-10, 12.5-15, 17-22.5, 25-43, 54-59, 62-67, 70-73, 84-90, and 92-113 cm. These levels of increased values correspond to intervals with decreased laminae numbers over 2.5 cm intervals except above around 25 cm and below 75 cm. Distinct intervals with visible amounts of benthic foraminifers are generally associated with less well-laminated intervals. These are found at 6-12 (foram.), 26.5-29 (foram.) and 28.5-33.5 cm (very faintly laminated), 42-47 (foram.) and 41.5-48 (faintly laminated), 51-55.5 cm (foram.) and 52.5-56 (faintly laminated), 67.5-71 cm (foram.) and 68-70.5 and 71.5-74 (faintly laminated) and 81.5-84 cm (forams) and 81-84 cm (faintly laminated). Disturbances occurred at 94-98 cm with two opposing angular discontinuities, at 87-88 cm with an angular discontinuity and two minor disturbances with bent laminae and/or slight offsets in the normal parallel laminae at 66-66.5 and 32-34 cm.

Kasten core B-29 (Figs. 5, 11 and 12): The primary productivity index curve is less variable than in any other core, and is slightly higher in its arithmetic mean than the value for the deeper core B-28. Intervals characterized by higher primary productivity levels occur at 5-15, 31-40, 43-60, and 75-92 cm. The number of laminae per

2.5 cm is inversely correlated with the productivity index values except above 50 cm and below 80 cm. The occurrence of visible traces of benthic foraminifers is associated with intervals showing homogenous and/or faintly laminated structures. They occur at 0-12 cm (foram.), 19-21, 25-34 cm (foram.) and 27-30 cm (faintly laminated), 63-69 cm (foram.) and 65-70 cm (faintly laminated), 74-77 cm (faintly laminated), and 90-100 cm (foram.) and 92.5-100 cm (faintly laminated). A major "z"-like disturbance occurs between 89 to 96 cm. This core contains the most complete recent record from 1910 to approximately >1960 of all four Kasten cores, and may reach as close to the collection date (1979) as a couple of years.

The productivity indices of each core are plotted against time in Fig. 13. Chronological markers at 1910, 1880, 1811, 1789, 1752, 1665, 1576 indicated in Figs. 12 and 13, were used to establish reference points to anchor the productivity index values in time.

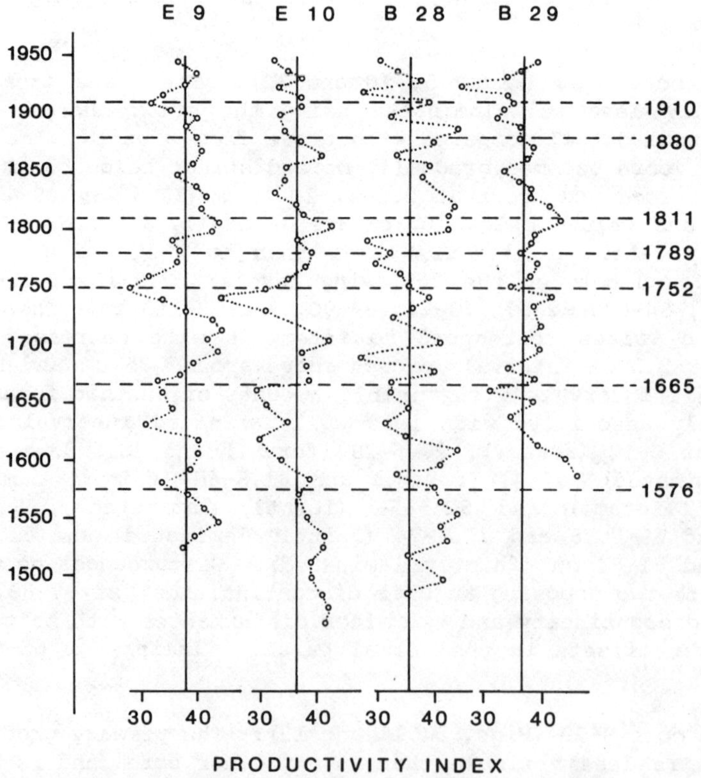

Fig. 13. Plot of productivity indices from Fig. 12 versus varve age. Horizontal lines are identical to those in Fig. 12 with average ages of correlation horizons to the far right.

Chronological positions of each value between these reference points were obtained by interpolation. The chronological references used were the mean positions of the correlation lines on Fig. 12.

Figure 13 demonstrates that, after allowing for the inherent chronological error associated with each core, the observed variation in productivity over time periods of 50 to 100 years is visually coherent among the cores. This is especially true for cores E-9 and E-10 from the eastern side of the Gulf. The somewhat diminished coherence observed in the comparison of B-28 and B-29 with E-9 and E-10 may be due to greater chronological uncertainty in the former two cores, or perhaps to some underlying difference in productivity between the eastern and western margins of the Gulf of California. The short-period fluctuations over intervals of 20 to 50 years exhibit much less coherence than do the 50 to 100-year periods. This is expected as a result of sampling resolution and chronological uncertainty.

DISCUSSION

All four cores analyzed show very similar sedimentological and microplankton variation through time. The two cores from area E contain more structural detail and carry indications of slightly increased productivity levels compared with cores from area B. In both areas, the shallower cores have higher sedimentation rates than the deeper ones which can be attributed to a closer geographic position to the actual upwelling localities and a closer position to the terrigenous source area. The lithologies in cores from area B are more uniform. Core B-29 which is the shallowest core in this suite, shows the least distinct facies changes. Four major structural units can be distinguished: homogenous, very finely laminated, faintly laminated and distinctly laminated intervals. The individual lamina thicknesses vary through time. During prevailing humid conditions, the dark lamina are thicker than the light ones since they are the repositories for the terrigenous load during runoff. Generally, the number of lamina per constant length, which represents sedimentation rates, are congruent (Fig. 12, core E-10, E-9 and B-28) with the primary productivity curves after 1850 to the present and prior to 1650 to around 1450 (the oldest date revealed by varve chronology is 1434 at 112 cm in E-10). Over the interval from 1650 to 1850 the two curves vary inversely. During periods of increased biological productivity, the sedimentation rates increase due to the increased biogenic flux. The relationship between sedimentation rate and productivity might be an indication that the increase in biologial flux does not always increase sedimentation rates, but rather terrigenous input has controlled sedimentation rates during the periods after 1850 and prior to 1650.

Fig. 14 is a comparison of the average productivity curve for the central Gulf of California with tree ring width indices from the

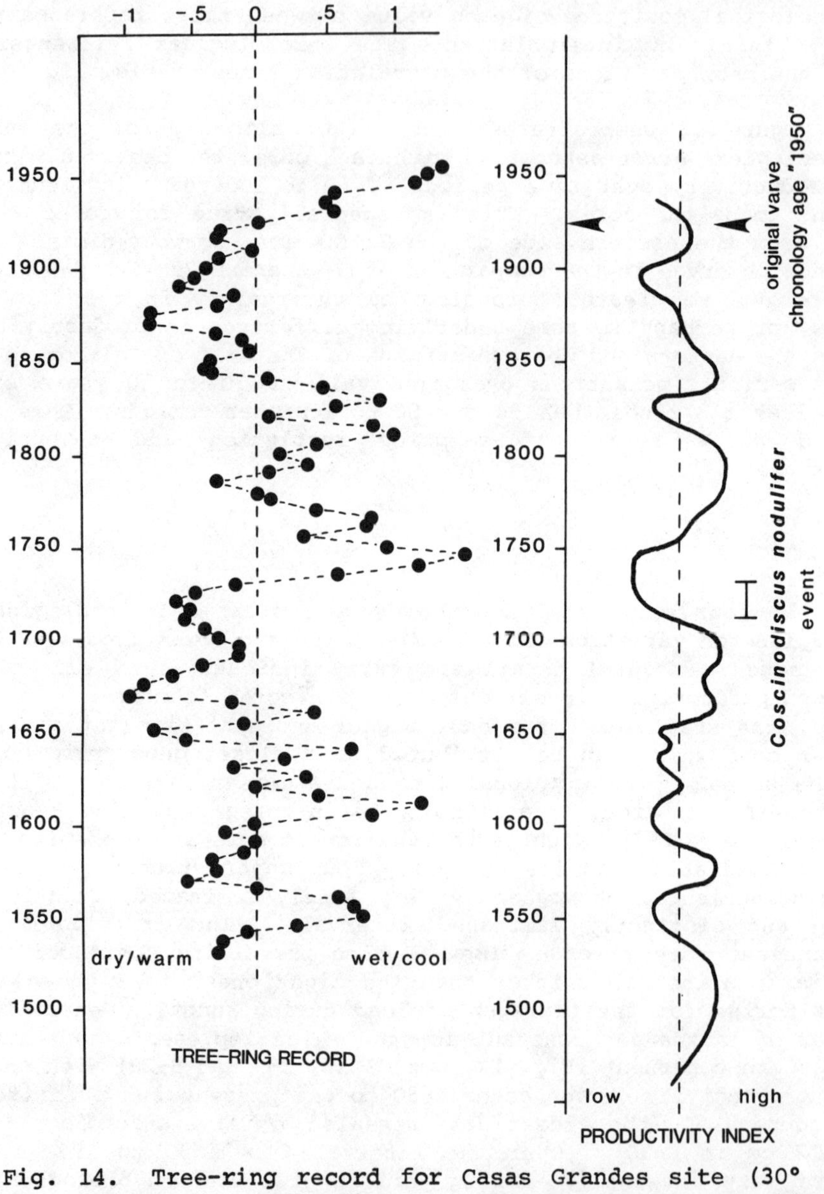

Fig. 14. Tree-ring record for Casas Grandes site (30° 07'N-108°22'W, elevation 2,520 to 1,950 m) from the Sierra Madre Occidental (Fritts, 1965). Departures to the left indicate dry conditions and departures to the right indicate moist conditions. Productivity index curve is based on a smoothed average curve of the four curves represented on Fig. 13. The original varve chronology scale is shifted by minus 25 years so that the original 1950 date is now 1925. The *Coscinodiscus nodulifer* event occurs at approximately 1720-1730.

Casas Grandes, Chihauhua, site in the Sierra Madre Occidental (Fritts, 1965). The average productivity was obtained by visually estimating a smoothed curve through superimposed plots of the productivity estimates of Fig. 13. This comparison indicates that the tentative date of 1910 used to fix the varve chronology is in error by approximately 25 years. The best fit of the productivity index to the tree ring data occurs by shifting the date of the *D. epiodon* abundance event to around 1885. It appears more likely that this event is associated with the sequence of strong anti-El Niños between 1885 and 1897 (see previous discussion).

An inverse relationship between productivity in the central Gulf and tree ring widths is now apparent. Narrow tree rings are generally correlated with high productivity over periods of 50 to 100 years. Increases in productivity occur over times when the continental climate was drier; during periods of decreased productivity, humid continental climate prevailed. Using today's model of atmospheric circulation (Fig. 2) and applying this to the sedimentary record, intervals with decreased productivity can be associated with the circulation system prevailing during the wet summer season, where the low pressure system moves northward and is positioned over the northern Gulf. The resulting air flow is responsible for bringing southeasterly winds with moisture into the Gulf. Laminae deposited over such an atmospheric condition do have higher terrigenous influence as documented in the sometimes thicker dark laminae.

The structure of the sediments also appears to vary with the productivity indices. During periods of decreased productivity, the sediments are more homogenous and/or indistinctly laminated. This is most clearly seen in the cores from area B (Table 3) where these intervals also contain increased numbers of benthic foraminifers (Fig. 5) indicating higher oxygen levels at the sediment water interface, possibly due to a decreased flux of organic matter. However, intervals indicative of high levels of productivity do not always show very distinctive laminations, most clearly visible in core E-10, and some intervals characterized by diminished productivity levels are associated with very distinct and well-laminated intervals (core E-9). A possible explanation could be the presence of different subsurface water masses during El Niño type conditions. Baumgartner et al. (1979) have argued that during these times a larger flow of Equatorial Pacific Water enters the Gulf and that this contains less dissolved oxygen. Persistence of these conditions should favor preservation of laminae (Soutar et al., 1981).

Increased abundances of *Dictyocha* species A, B, 2 and *D. calida* are associated with El Niño-type conditions. Since the sampling strategy used here does not permit exact designation of these events to individual laminae, only rough comparison to the historical record documented by Quinn et al. (1978, his Table 1) is possible.

The period associated with the Little Ice Age (1430 to 1850, National Research Council, 1975; Robock, 1979; and others) is well represented in the sedimentary record examined. During this period a steepening of the latitudinal temperature gradients and an intensification of the Equatorial Pacific circulation has been proposed (Gribbin and Lamb, 1978; Matherne, 1982) which would lead to an intensified wind circulation system and could induce intensified upwelling in the subtropical regions of the eastern North Pacific (Soutar et al., 1981). Fig. 13 indicates (Cores E-9, E-10 and B-28) an increase in the amplitude variation from high to low productivity levels between 1550 and 1800, although frequency of productivity changes and Equatorial Pacific influences during this time were similar to those of the recent past.

At present the Gulf is experiencing a period of decreased productivity and increased mid-water dissolved oxygen. Comparable conditions were present around 1550, 1600, 1740, 1830, and 1900 (Fig. 14). Coastal upwelling indices computed by Bakun (1973) covering 1946 to 1971, together with his yearly averages at a station 24°N-113°W, show a slight decrease in upwelling intensity after 1955; other stations along the western Baja Peninsula follow similar trends.

The potential of varved marine sediments from the Gulf of California to resolve decadal variations in upwelling through periods of 50 to 100 years, and variations in both terrigenous and biogenous fluxes, and their association with climatic forces controlling upwelling intensities is documented here. The most useful areas are the northeastern and southwestern slopes of the Guaymas Basin due to the high sedimentation rates, rather undisturbed sediment record and simultaneous response to outside forces.

ACKNOWLEDGEMENT

This work is supported by the Climate Dynamics Division of NSF through grants ATM 79 19458 02 and ATM 81 21775, by Julius Babisak, ARCO, and by grant CONACYT-BID PCMABNA-005321 from the Mexican government. Nick Pisias and Ken Scheidegger reviewed the first draft. Dr. Gustavo Calderon and the crew of the Mexican Research vessel H-1 Mariano Matamoros with its captain Pompeyo Leon Herrera have made this study possible through outstanding cooperation and help during two field seasons.

REFERENCES

Alvarez-Borrego, S., in press, Gulf of California, Chapter XVI, in: "Estuaries and Enclosed Seas", Ketchum, ed.

Alvarez-Sanchez, L.G., 1974, "Currents and Water Masses at the Entrance to the Gulf of California, Spring 1970," M.S. Thesis, Oregon State University, Corvallis, 75 pp.

Ayala-Castañares, A., Phleger, F.B. and Schwartzlose, R., eds., in press, "The Gulf of California: Origin, Evolution, Waters, Marine Life and Resources", Proceedings, Gulf of California Symposium, November 1979, Universidad Nacional Autonoma de México, Mazatlan, Sinaloa.

Bakun, A., 1973, Coastal upwelling indices, west coast of North America, 1946-1971, National Oceanic and Atmospheric Administration, Technical Report NMFS SSRF-671, 103 pp.

Baumgartner, T.R. and Soutar, A., in press, Preservation of the recent climatic record in laminated sediments of the Gulf of California, in: "The Gulf of California: Origin, Evolution, Waters, Marine Life and Resources", A. Ayala-Castañares et al., eds., Universidad National Autonoma de México.

Baumgartner, T., Christensen, N., Fok-Pun, L. and Quinn, W.H., 1979, Abstract, Source of interannual climatic variation in the Gulf of California and evidence for the biological response, CalCOFI Conference, Idylleville.

Baumgartner, T.R., Soutar, A., Cowen, J., Morena, P., Michaelson, J., and Bruland, K.W., 1981, Abstract, Detailed chronologies in the laminated sediments of the Guaymas slope, Geological Society of America, Cordilleran Section Meeting, Hermosillo, Sonora, Mexico, 44.

Barnett, T.P., 1977, An attempt to verify some theories of El Niño, Journal of Physical Oceanography, 7(5):633-647.

Bischoff, J.L. and Niemitz, J.W., 1980, Bathymetric maps of the Gulf of California, United States Geological Survey, Miscellaneous Investigations Series, Map 1-1244.

Byrne, J.V. and Emery, K.O., 1960, Sediments of the Gulf of California, Geological Society of America, Bulletin, 71:983-1010.

Calvert, S.E., 1964, Factors affecting the distribution of laminated diatomaceous sediments in the Gulf of California, in: "Marine Geology of the Gulf of California," Tj. H. van Andel and G.G. Shor, eds., American Association of Petroleum Geologists, Memoir 3, 311-330.

Calvert, S.E., 1966a, Accumulation of diatomaceous silica in the sediments of the Gulf of California, Geological Society of America, Bulletin, 77, 569-596.

Calvert, S.E., 1966b, Origin of diatom-rich, varved sediments from the Gulf of California, Journal of Geology, 76:546-565.

DeMaster, D.J., 1979, The Marine Budgets of Silica and ^{32}Si," Ph.D. Thesis, Yale University, 308 pp.

DeVries, T. and Schrader, H., 1981, Variation of upwelling/oceanic conditions during the latest Pleistocene through Holocene off the central Peruvian coast: A diatom record, Marine Micropaleontology, 6:157-167.

Donegan, D. and Schrader, H., 1981, Modern analogues of the Miocene Monterey Shale of California: Evidence from sedimentologic and micropaleontologic study, in: "The Monterey Formation and Related Siliceous Rocks of California", R.E. Garrison and R.G. Douglas, eds., Pacific Section, Los Angeles, 149-157.

Donegan, D. and Schrader, H., 1982, Biogenic and abiogenic components of laminated sediments in the central Gulf of California, Marine Geology, 48:215-237.

Douglas, A.V., 1976, "Past Air-sea Interactions over the Eastern North Pacific Ocean as Revealed by Tree-ring Data," Ph.D. Dissertation, University of Arizona, Tucson, 196 pp.

Douglas, A.V., 1980, Geophysical estimates of sea-surface temperatures off western North America since 1671, CalCOFI, Report 21, 102-112.

Ferreira-Bartrina, V., Baumgartner, T.R. and Schrader, H., in press, Size and abundance variations of selected opal phytoplankton species in laminated sediments of the Gulf of California: Possible ecologic response to interannual change in ocean climate, Ciencias Marinas.

Fritts, H.C., 1965, Tree-ring evidence for climatic changes in western North America, Monthly Weather Review, 93:421-443.

Gribbin, F. and Lamb, H.H., 1978, Climate change in historical time, in: "Climatic Change", F. Gribbin, ed., Cambridge University Press, London, 68-82.

Griffiths, R.C., 1965, A study of the ocean fronts off Cape San Lucas, Special Science Report, U.S. Fish and Wildlife Service 499, 54 pp.

Griffiths, R.C., 1968, Physical, chemical, and biological oceanography of the entrance to the Gulf of California, Spring 1960, Special Science Report, U.S. Fish and Wildlife Service, Fisheries 573, 47 pp.

Hastings, J.R. and Turner, R.M., 1965, Seasonal precipitation regimes in Baja California, Mexico, Geografiska Annaler, 47:204-223.

Markham, C.G., 1972, Baja California's climate, Weatherwise, April 1972, 64-76.

Matherne, A.M., 1982, "Paleoceanography of the Gulf of California: A 350-year Diatom Record," M.S. Thesis, Oregon State University, Corvallis, 111 pp.

Miller, F.R. and Laurs, R.M., 1975, The El Niño of 1972-1973 in the Eastern Tropical Pacific Ocean, Inter-American Tropical Tuna Commission Bulletin, 16(5):403-448.

Murray, D., 1982, "Paleo-oceanography of the Gulf of California Based on Silicoflagellates from Marine Varved Sediments," M.S. Thesis, Oregon State University, Corvallis, 129 pp.

Namias, F., 1971, Temporal coherence in North Pacific sea-surface temperature pattern, Journal of Geophysical Research, 75(30): 5952-5955.

National Research Council, 1975, Understanding climatic change, a program for action, National Academy of Sciences, Washington, 239 pp.

Poelchau, H.S., 1974, "Holocene Silicoflagellates of the North Pacific," Ph.D. Dissertation, University of California, San Diego, 165 pp.

Poelchau, H.S, 1976, Distribution of Holocene silicoflagellates in North Pacific sediments, Micropaleontology, 22:164-193.

Quinn, W.H., Zopf, D.O., Short, K.S., and Kuo Yang, R.T.W., 1978, Historical trends and statistics of the southern oscillation, El Niño, and Indonesian draughts, Fishery Bulletin, 76(3):663-678.

Reid, J.L., 1960, Oceanography of the Eastern North Pacific in the last ten years, California Cooperative Oceanic Fisheries Investigative Report, 7:77-90.

Revelle, R.R., 1950, Sedimentation and oceanography-A survey of field observations, in: "The 1940 E.W. Scripps Cruise to the Gulf of California," C.A. Anderson et al., eds., Geological Society of America, Memoir, 43, 6 pp.

Robinson, M.K., 1973, Atlas of monthly mean sea surface and subsurface temperatures in the Gulf of California, Mexico, Society of Natural History, San Diego, Memoir 5, 90 figs., 19 pp.

Robock, A., 1979, The "Little Ice Age": Northern hemisphere average observations and model calculations, Science, 206:1402-1404.

Roden, G.I., 1958, Oceanographic and meteorological aspects of the Gulf of California, Pacific Science, 12:21-45.

Roden, G.I., 1964, Oceanographic aspects of the Gulf of California, in: "Marine Geology of the Gulf of California," Tj. H. van Andel and G.G. Shor, eds., American Association of Petroleum Geologists, Memoir 3, 30-58.

Roden, G.I., 1972a, Large scale upwelling off northwestern Mexico, Journal of Physical Oceanography, 2:184-189.

Roden, G.I., 1972b, Thermohaline structure and baroclinic flow across the Gulf of California entrance and the Revilla Gigedo islands region, Journal of Physical Oceanography, 2(2):177-183.

Roden, G.I. and Groves, G.W., 1959, Recent oceanographic investigations in the Gulf of California, Journal of Marine Research, 18:10-35.

Schrader, H., 1979, Cruise Report Baja Vamonos 79, September 7-September 30, 1979, Data Report 78, School of Oceanography, Oregon State University, Reference 79-15, 67 pp.

Schuette, G. and Schrader, H., 1979, Diatom taphocoenoses in the coastal upwelling area off western South America, Nova Hedwigia, 64:359-378.

Schuette, G. and Schrader, H., 1981a, Diatoms in surface sediments: A reflection of coastal upwelling, in: "Coastal Upwelling," F.A. Richards, ed., 372-380.

Schuette, G. and Schrader, H., 1981b, Diatom taphocoenoses in the coastal upwelling areas off southwest Africa, Marine Micropaleontology, 6:131-155.

Soutar, A., Johnson, S.R. and Baumgartner, T.R., 1981, In search of modern depositional analogs to the Monterey Formation, in: "The Monterey Formation and Related Siliceous Rocks of California," G.E. Garrison and R.G. Douglas, eds., Society of Economic Paleontologists and Mineralogists, 123-147.

Stevenson, M.R, 1970, On the physical and biological oceanography near the entrance to the Gulf of California, October 1966-August 1967, Inter-American Tuna Commission Bulletin, 4:389-504.

Warsh, C.E. and Warsh, K.L., 1971, Water exchange at the mouth of the Gulf of California, Journal of Geophysical Research, 76:8098-8116.

Warsh, C.E., Warsh, K.L. and Stanley, R.C, 1973, Nutrients and water masses at the mouth of the Gulf of California, Deep-Sea Research, 20:561-570.

Wyllie, J.G., 1966, Geostrophic flow of the California Current at the surface and 200 m, CalCOFI, Atlas No. 4, 288 pp.

Wyrtki, K., 1965, Surface currents in the eastern equatorial Pacific Ocean, Inter-American Tropical Tuna Commission Bulletin, 9:270-304.

Wyrtki, K., 1967, Circulation and water masses in the eastern equatorial Pacific, International Journal of Oceanology and Limnology, 1:117-147.

Wyrtki, K., 1975, El Niño - The dynamic response of the Equatorial Pacific Ocean to atmospheric forcing, Journal of Physical Oceanography, 5:572-584.

Wyrtki, K., 1977, Sea level during the 1972 El Niño, Journal of Physical Oceanography, 7:779-787.

Wyrtki, K., 1979, Sea level variations: Monitoring the breath of the Pacific, EOS, Transactions, American Geophysical Union, 25-27.

Zeitzschel, B., 1969, Primary productivity in the Gulf of California, Marine Biology, 3:201-207.

OXYGEN ISOTOPE COMPOSITION OF DIATOM SILICA AND SILICOFLAGELLATE ASSEMBLAGE CHANGES IN THE GULF OF CALIFORNIA: A 700-YEAR UPWELLING STUDY

Anne Juillet and Laurent D. Labeyrie

Centre des Faibles Radioactivités
Laboratoire mixte C.N.R.S.-C.E.A.
Domaine du C.N.R.S., 91190 Gif-sur-Yvette, France

Hans Schrader
School of Oceanography
Oregon State University
Corvallis, Oregon 97331, U.S.A.

ABSTRACT

Two varved, diatomaceous cores from the northeastern slope of the Guaymas Basin and from the eastern slope of the Carmen Basin in the central Gulf of California were analyzed for oxygen isotopic composition of diatom silica and for silicoflagellate assemblages. Both indices followed the same variation pattern and are thought to represent variation in upwelling intensities and/or more Pacific influence into the central Gulf. The varve chronology provided the necessary time framework in addition to biostratigraphic horizons. Based on this time scale the following long-term, low frequency changes in the upwelling variables were observed: around 1400 A.D. a strong increase in temperature and decrease in productivity, between 1580-1450 an increase in productivity values, between 1800-1580 decreased productivity, between 1900-1800 a decrease in surface water temperature and increase in primary productivity, and between 1950-1900 a decrease in primary productivity and increase in temperature. The two areas differ in both their floral and isotopic records, indicating higher intensities of upwelling in area E and more Pacific water influence in area A.

INTRODUCTION

The Guaymas and Carmen Basin areas located in the southern Gulf of California (compare Fig. la in Schrader and Baumgartner, this volume) are characterized by year-round high upwelling and high primary productivity levels (Zeitzschel, 1969; Baumgartner, pers. comm.). The phytoplankton in these areas consist primarily of diatoms. Since their habitat is limited to the topmost part of the photic zone (Schrader and Schuette, 1981), diatom assemblages and the oxygen isotope composition of their silica frustules should be particularly suitable for information about physical oceanographic parameters of the uppermost oceanic layer through time. A method to measure the oxygen isotope composition from the opaline frustules of diatoms has recently been developed. It has been demonstrated that the isotopic composition of the diatom-opal depends upon the temperatures of the surrounding waters and their isotope contents (Labeyrie, 1974; 1979; Juillet, 1980a; 1980b).

The diatomaceous, varved sediment facies along the slopes of the Guaymas and Carmen Basin between 400-800 meters water depth (Baumgartner et al., 1981) are well suited for this study because these sediments are deposited undisturbed (Calvert, 1966), are extremely rich in opaline skeletons (DeMaster, 1979) and are deposited at very high sedimentation rates which allow high resolution studies for historical times. We document in this paper the general correspondence of variations in silicoflagellate assemblages (used as substitutes for diatom assemblage analysis), their relationship to water masses and to the oxygen isotopic composition of diatom opal.

The Gulf of California is located between 18° and 32°N latitude and is subjected to two contrasting climatic regimes: (a) dry and cool winters with northwesterly wind regimes and (b) wet and warm summers with southeasterly wind regimes. Three water masses have been distinguished in the upper 200 meters of the southern Gulf (Warsh, Warsh and Staley, 1973):

(a) California Current Water (salinity $\leq$ 34.6°/₀₀ and temperature <22°C) (Wyrtki, 1967),
(b) Equatorial Pacific Water (salinity 34.6-34.9°/₀₀ and temperatures >25°C), and
(c) Gulf Water (salinity $\geq$ 34.9°/₀₀ and temperatures 15-30°C).

The general surface circulation is affected by the regional and seasonal wind systems. Seasonal upwelling occurs preferentially in the lees of capes, points or islands (Roden and Groves, 1959). Due to these year-round upwelling phenomena which seem to be present on both sides of the Gulf, the surficial waters in the Gulf are rich in nutrients. Since most of the upwelled water has its origin from Pacific intermediate water rich in nutrients, the surface waters in the Gulf are capable of sustaining a large population of phytoplankton (mostly diatoms) (DeMaster, 1979). Diatom frustules also form the major sediment component (40-80%) of the hemipelagic sediments (Donegan and Schrader, 1982).

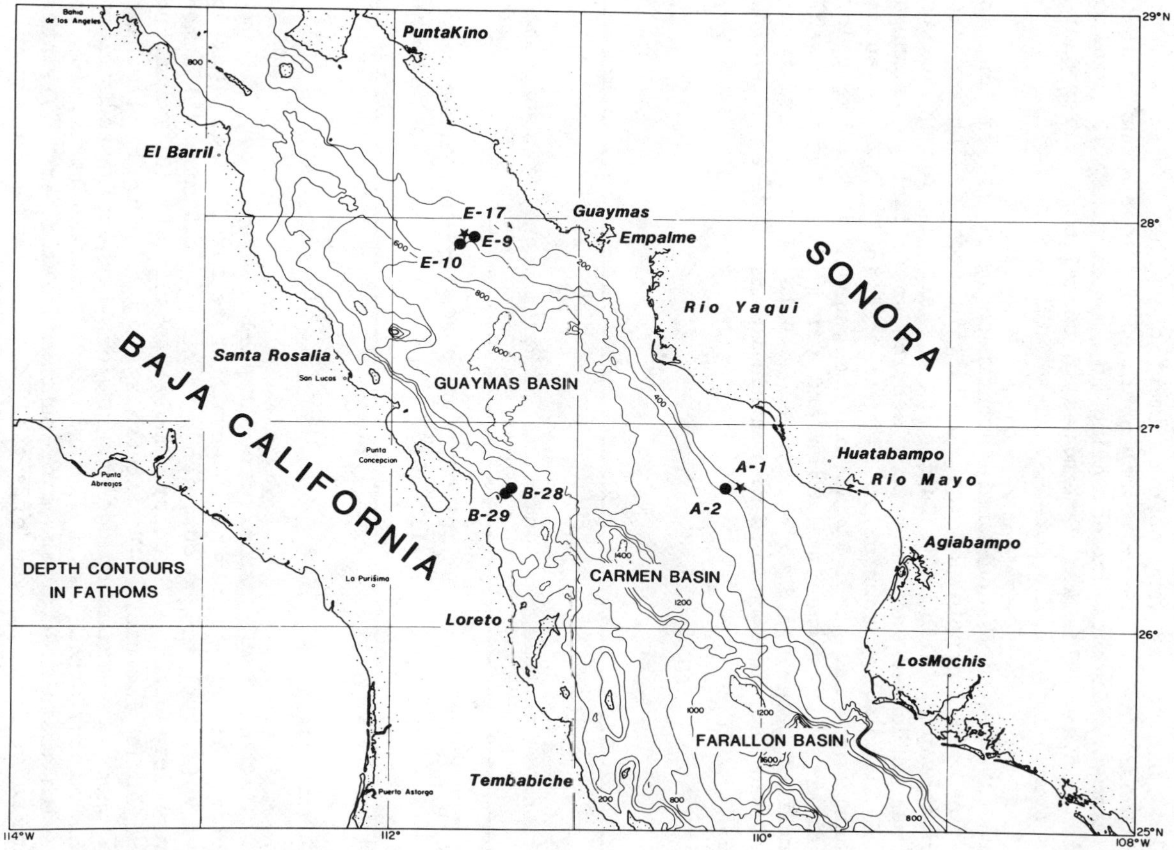

Fig. 1. Location of cores studied and discussed. Cores E-17 and A-1 marked with stars. Note depth contours in fathoms. Bathymetry from an unpublished Scripps Institution of Oceanography map.

MATERIAL

Two cores were selected for this study (Fig. 1). These cores were retrieved during two cruises in 1979 (Schrader, 1979) and 1980 on board the Mexican Navy Research Vessel H-1 "Mariano Matamoros" (Table 1). Due to the very soft sediment character (water contents up to 95%) box core BAV 79-A-1 overpenetrated and lost about 10-15 cm of the topmost sediments corresponding to approximately 40 years of deposition. This value was determined in correlating corresponding horizons from A-1 to nearby core BAV 79-A-2 (Fig. 1).

Core BAV 79-A-1 was subsampled continuously at 1.5 cm long intervals for both isotopic and floral analysis from a slab prepared for x-radiography (slab 2). During this slabbing process the top section was condensed and the bottom section was slightly disturbed. A general lithologic description is presented in Fig. 2. A sudden color change occurred at 10 cm within the general dark olive color. Two intervals with faint lamination and partially homogenous sediments occurred between 3.8 to 7.1 cm and 36.4 to 44 cm. These intervals also had increased numbers of benthic foraminifers as determined visually.

Table 1

Sample Designation	Latitude	Longitude	Water Depth	Instrument
BAV 79-A-1	26° 43.5'	110° 06.6'	640 m	Box-core
BAM 80-E-17	27° 55.2'	111° 36.6'	620 m	Kasten-core

Figure 2 (cont.)

2B. Lithology of Kasten core BAM 80-E-17 (only top 150 cm). Black horizontal bars represent intervals where material was lost due to recovery (at top) and to curation procedures. Columns: (1) number of archive tray boxes, (2) color (Munsell standard), (3) occurrence of thick light laminae, (4) intervals very distinctly laminated cross-hatched, intervals homogenous and/or very indistinctly laminated thick dots, other less well-laminated intervals blank, (6) intervals with visible occurrences of benthic foraminifers, (7) intervals with major disturbances (offsets of laminae, angular discontinuities, overturn, etc.). Ages to the far left are based on varve chronologies as determined by counting laminae over continuous 2.5 cm long intervals downcore. *Coscinodiscus* horizon at 66.5 cm.

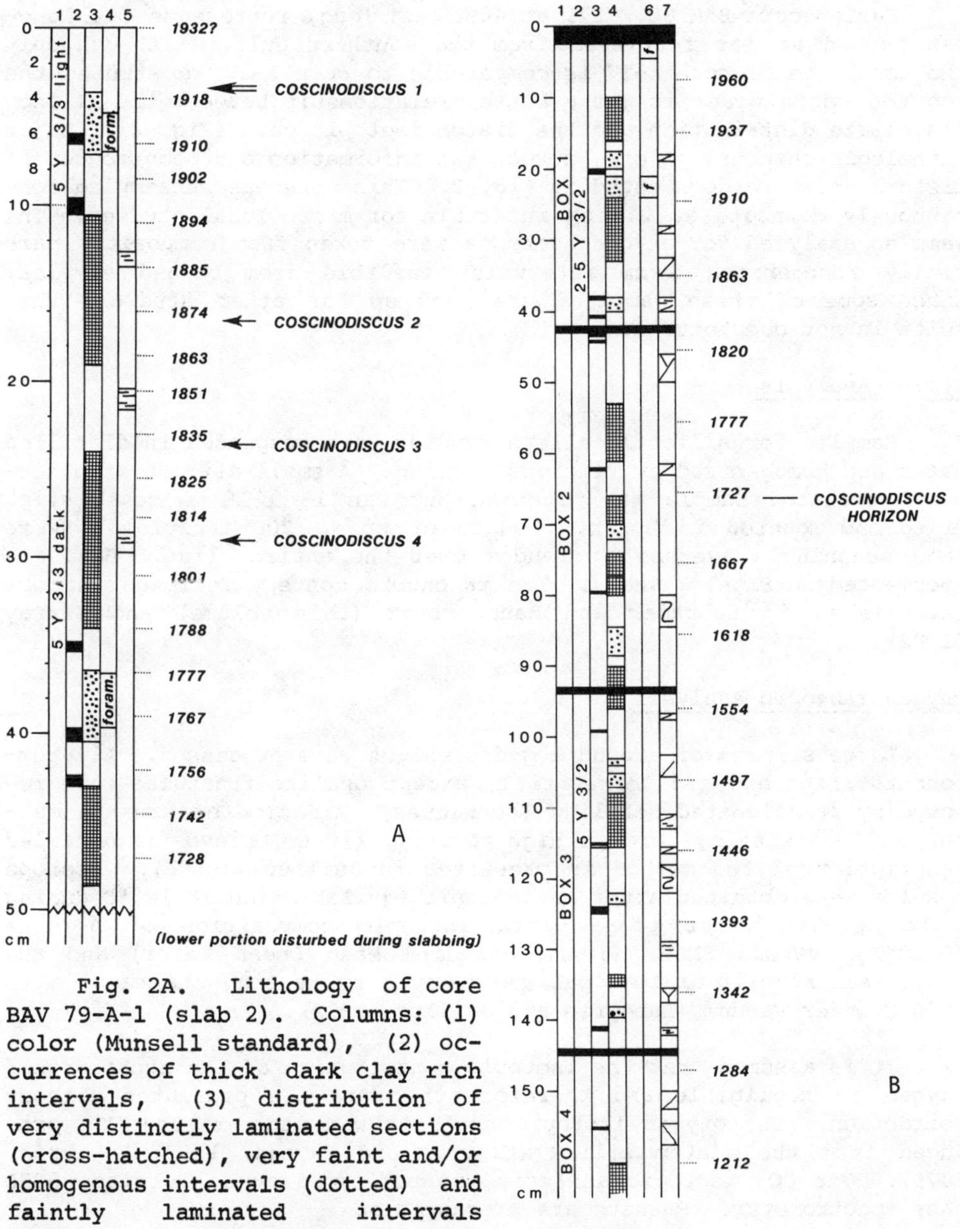

Fig. 2A. Lithology of core BAV 79-A-1 (slab 2). Columns: (1) color (Munsell standard), (2) occurrences of thick dark clay rich intervals, (3) distribution of very distinctly laminated sections (cross-hatched), very faint and/or homogenous intervals (dotted) and faintly laminated intervals (blank), (4) occurrences with visually detected intervals enriched in benthic foraminifers, (5) intervals with disturbances (mostly offsets of laminae). Ages on the right are based on uncorrected varve counts over 2.5 cm core intervals; arrows indicate position of light laminae which contain increased abundances of large *Coscinodiscus* species.

Kasten core BAM 80-E-17, at 449.5 cm long, represents our longest record so far retrieved from the southern Gulf. Although only the top 70 cm of core E-17 is comparable to core A-1, we studied the top 150 cm in order to get a better relationship between the silicoflagellate distribution and the diatom isotopic data (Fig. 2). Major lithologic changes, colors, structural information and occurrences of disturbances are documented in Fig. 2. This core was subsampled continuously downcore at 2.5 cm intervals for microfloral studies. The samples analyzed for oxygen isotopes were taken from composite, partially homogenized 5 cm intervals available from bagged samples. Since some of these samples were used up for other studies, this suite is not complete.

Floral Analysis

Samples for silicoflagellate studies were suspended in distilled water and homogenized by vigorous shaking. A small aliquot of otherwise untreated sample was pipetted onto an 18 x 18 mm cover slip, dried and mounted in Hyrax. Counts of up to 200 individuals were made scanning traverses at random over the entire slide. Data are documented in Figs. 3 and 4. The taxonomic concept followed the one established in Schrader and Baumgartner (this volume) and Murray (1982).

Oxygen Isotopic Analysis

Large samples of around 1 g dry weight were processed. All components, biogenic and terrestrial, except opaline frustules were removed by fractionated settling techniques, oxidation of organic matter and wet sieving through high porosity (10 μm) sieves (a detailed description of the method is presented in Juillet, 1980b). Isotope results were obtained after an isotopic equilibration at 180°C during 6 hours with water of a given isotopic composition of $\delta^{18}O$ = 29.20°/$_{oo}$ versus SMOW (Vienna Standard Mean Ocean Water) and the water remaining in diatom opal was removed by heating the samples to 1000°C under vacuum (Labeyrie and Juillet, 1982).

It is assumed that the isotopic contribution of the exchangeable oxygen is negligible and therefore the data are presented without correction. The oxygen isotopic analysis was made on the CO_2 produced from the diatom silica (Clayton and Mayeda, 1963; Labeyrie, 1979). The CO_2 isotopic analysis was performed on a Micromass 602D mass spectrometer. Results are presented as

$$\delta^{18}O°/_{oo} = \left\{ \frac{^{18}O/^{16}O \text{ sample}}{^{18}O/^{16}O \text{ standard}} - 1 \right\} \cdot 10^3$$

References to SMOW were obtained by calibration of our laboratory standard. The reproducibility is ±0.12°/$_{oo}$.

RESULTS

Previous oxygen isotope measurements indicate a strong correlation between the isotopic composition of diatom silica, temperature and isotopic composition of the waters in which the diatoms lived. A temperature decrease of 3°C resulted in a 1‰ isotopic increase. Values of 37‰ $\delta^{18}O$ (SMOW) were found for equatorial diatoms and 44‰ $\delta^{18}O$ (SMOW) for Antarctic diatoms. We demonstrated that there is no isotopic differentiation between different diatom species living under the same conditions (Juillet, 1980b).

The isotopic results for core BAV 79-A-1 are plotted in Fig. 5; they range from 37 to 40‰. The curve shows low values at the bottom and at the top and a steady increase over the interval 10 to 25 cm. Values between 25 to 40 cm remained approximately constant around 39-40‰. The isotopic values for core BAM 80-E-17 are plotted on Fig. 6. Low values were found above 25 cm, with a sudden increase at 30 cm to 41.21‰. The interval between 30 cm to 70 cm varied between 41.21‰ to 40.54‰; below, the values decrease steadily to 38.5‰ at 130 cm and fluctuate below that level. The isotopic values on Fig. 7 are plotted at midpoints of sample intervals and we assume that these values are "sample" representatives.

Silicoflagellates

Core BAV 79-A-1 (Fig. 3): Unlike previous documentations (Schrader and Baumgartner, this volume), here the warm water Eastern Equatorial Pacific group has been combined to consist of *Dictyocha* spec. A, B, 2 and *D. calida*. A gradual decrease in *D. messanensis* and an increase in *Octactis pulchra* downcore to 35 cm is well documented (Fig. 3). There is a strong *D. epiodon* peak between 6-10 cm with an amplitude larger than previously detected in more northern cores (E-10, E-9, B-28, B-29, Fig. 1).

Core BAM 80-E-17 (Fig. 4): Three major *D. epiodon* horizons occurred at 22-25 cm, 72.5-75 cm and 132.5-135 cm. *D. messanensis* occurred in low abundances between 10-50 cm and between 80-110 cm. *Dictyocha* A + B + 2 and *D. calida* generally occurred at lower abundances than in core A-1, but still revealed major intervals with more Pacific influence. The *Coscinodiscus nodulifer* horizon occurred at 66.5 cm.

STRATIGRAPHY AND CHRONOLOGY

A "zero" horizon was established by determining the thick (in E-17) and the thin clayey (in A-1) homogenous horizon positioned directly above the couplets of laminae containing *Dictyocha epiodon* in great abundances. This horizon was dated around 1910 (Schrader and Baumgartner, this volume) and the position and date were recently

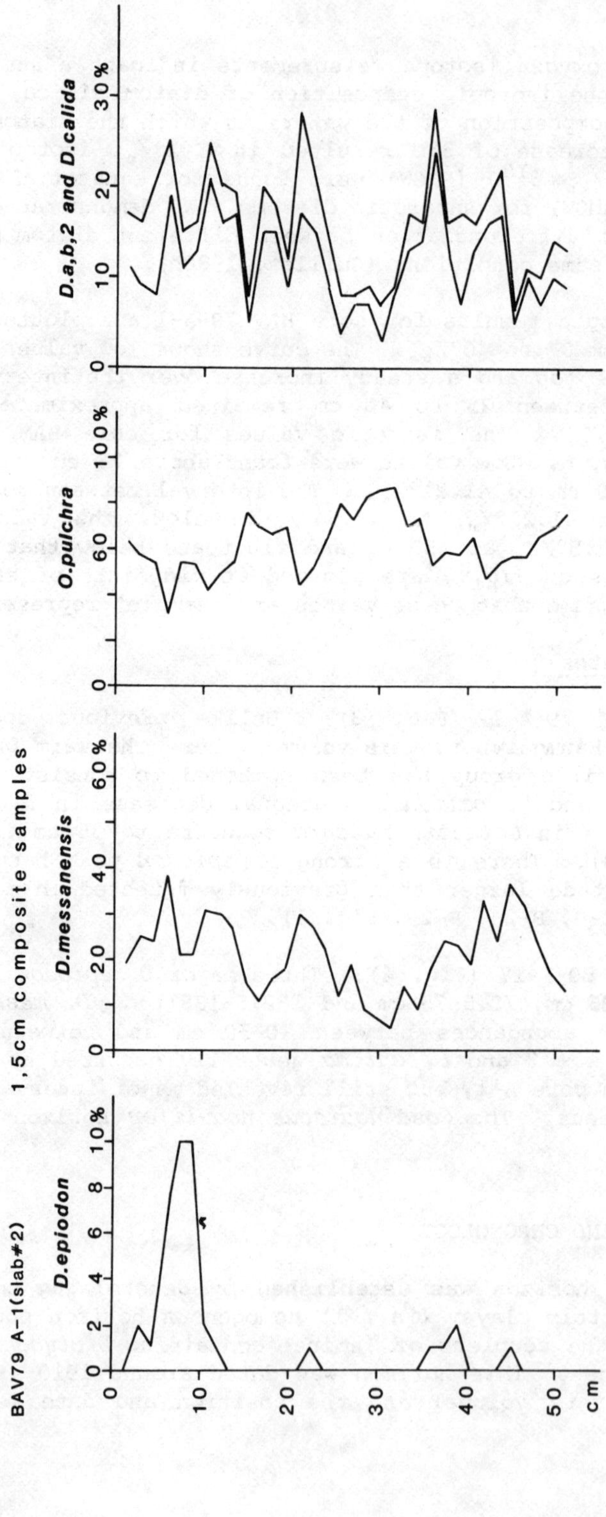

Fig. 3. Abundance of silicoflagellate species in core A-1 (%).

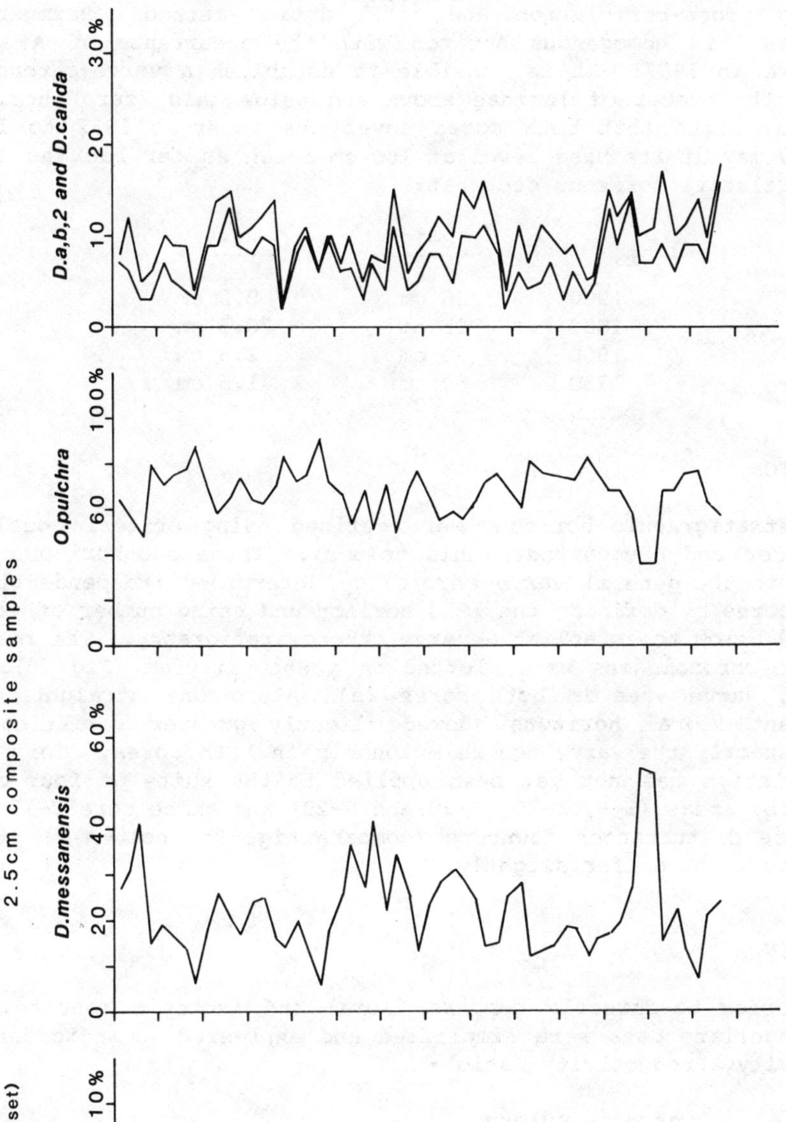

Fig. 4. Abundance of silicoflagellate species in core E-17 (%).

confirmed by Baumgartner (pers. comm.). Using a different approach including cross-correlation and ^{210}Pb dating methods, Baumgartner associates this homogenous horizon with the occurrence of a major earthquake in 1907. It is possible to establish a varve-chronology counting the number of laminae above and below this "zero" horizon. Results indicate that both cores cover the interval 1932 to 1730; core E-17 may at its base level at 160 cm reach as far back as 1200. Time correlative horizons occur at:

	E-17	A-1
1900	26 cm	8.5 cm
1850	35 cm	20.5 cm
1800	50 cm	32.5 cm
1750	60 cm	41.5 cm

CORRELATION

Biostratigraphic horizons were defined using criteria outlined in Schrader and Baumgartner (this volume). These bio-horizons were placed into the general varve chronology determined independently in the two cores by defining the 1910 horizon and using number of varves above and below to establish a varve chronostratigraphy. The resulting varve chronologies were plotted on a Shaw-diagram (Fig. 5). Interpreted varve ages in both cores fall along one straight line. Independent floral horizons showed slightly greater variation but matched exactly the varve-age relationship in both cores. This type of correlation has not yet been applied to the suite of four cores from nearby areas (E-9, E-10, B-28 and B-29) and since core E-17 contains more disturbances downcore (compare Fig. 2b, column 7), their varve ages might differ slightly.

DISCUSSION

In order to directly compare floral and isotopic results, the silicoflagellate data were simplified and expressed as indicators of productivity (Productivity Ratio):

$$P = \frac{\textit{Octactis pulchra}}{\textit{Octactis pulchra} + S \text{ silicoflagellates}} \cdot 100$$

and as a ratio indicating variations in Pacific Water influence (Pacific/Gulf Ratio)

$$P/G = \frac{\textit{Octactis pulchra}}{\textit{Dictyocha calida} + \textit{Dictyocha}\ \text{A} + \text{B}} \cdot 100$$

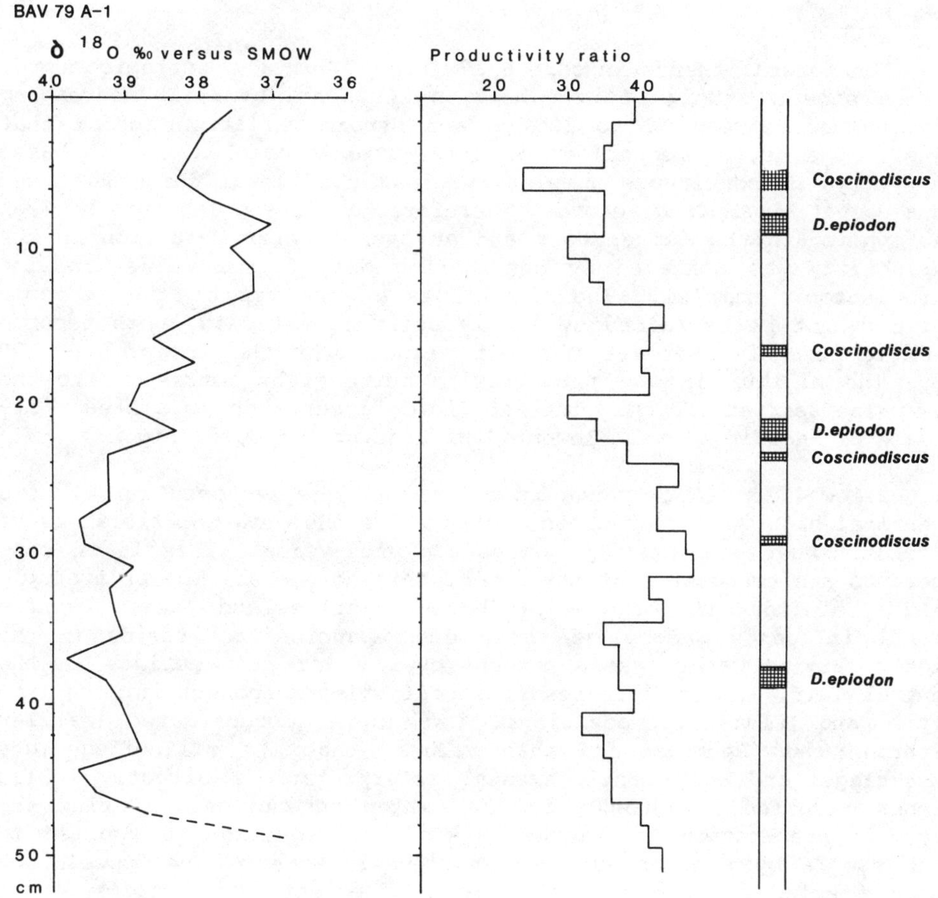

Fig. 5. Oxygen isotopic composition of 1.5 cm composite samples, sample intervals indicated as small dots along the depth bar. To the right silicoflagellate productivity ratio and occurrences of non-Gulf diatoms and silicoflagellates.

Core A-1

The general trend in the isotopic ratio in core A-1 (Fig. 6) shows a steady increase over the 0 to 40 cm interval. The productivity curve follows this general pattern but does not correlate well when comparing individual samples. In some instances the decrease in the isotopic ratio mirrors influences of outside Gulf floras with *Dictyocha epiodon* (a California current indicator) at 8 cm, 22 cm, 38 cm; other intervals do not follow this pattern. The decrease in productivity at 5 cm, 20 cm and 42.5 cm is not "seen" by the isotopic analysis.

Core E-17

The general trends in core E-17 (Fig. 7) in the isotopic composition show a strong minimum between 10-25 cm, steadily decreasing high values between 30 to 120 cm and strong variations below that level. The direct comparison of this curve is complex. It is based on spotty, discontinuous samples, whereas the floral curve is based on a complete sampling record. Therefore, a three-point running mean was applied to the floral data and an even greater reduction in the variability was achieved by considering only floral values falling into isotopic samples. The low values at the top of the isotopic curve cannot be explained by simply using floral data, both records seem to steadily decrease in their values over the interval 40-120 cm. The minimum in the productivity curve might correspond to the low value seen at 130 cm. The Pacific/Gulf curve shows a steady increase of Pacific floral elements influencing the Gulf (Fig. 7).

Below 40 cm depth range it seems that the isotopic curve shows identical high values at those levels where the original floral curve records higher productivity values and vice-versa. This is in contrast to the trend seen at the top of this core where lowest isotopic values coincide with highest productivity values and lowest Pacific floral influence. Some of these discrepancies may reside in the highly varying water masses present over short time periods in the central Gulf. These will result in differing responses in both isotopic and floral compositions. We have demonstrated earlier (Schrader and Baumgartner, this volume) that the silicoflagellate assemblages and their environmental interpretations directly follow trends revealed by changes in the diatom assemblages. During the isotopic preparation techniques, a size fractionation is applied to each sample by sieving with 10 µm sieves. Some of the "upwelling" index species may be removed (broken fragments of *Thalassiothrix pseudonitzschioides*, small *Chaetoceros*, small *Synedra indica*, small *Thalassiosira* species), leading to bias towards the larger "oceanic" species which are brought into our study areas by warmer Pacific water masses. The mean productivity ratio over time equivalent horizons (A-1 0-50 cm and E-17 15-65 cm) are in core A-1 P = 37.85 and in E-17 P = 40.80 corresponding to a 6% difference in magnitude.

Comparison Between the Two Curves

During the period jointly represented in both cores, isotopic curves show a similar trend: low values at the top followed by an important decrease. The oxygen isotopic signal has a similar per mil amplitude difference over these intervals with a 2°/₀₀ difference. At the present time, we do not have the water samples necessary to evaluate the effect of salinity on the isotope fractionation. Assuming that the freshwater input is a very local phenomenon and that the evaporation effects in the northern part of the Gulf are negligible, we can compare the isotopic importance of the salinity inside the

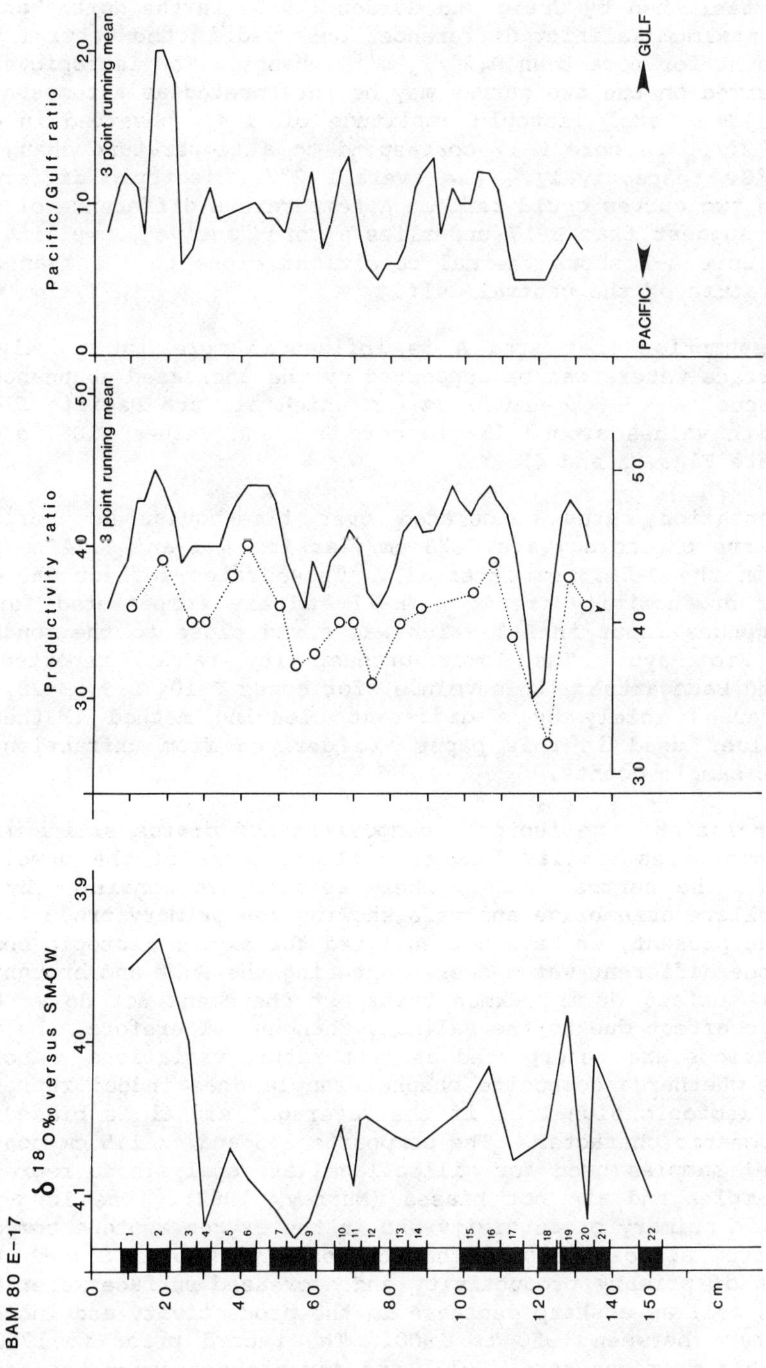

Fig. 6. Oxygen isotopic composition of 20 samples, numbered and labelled alongside the depth scale. Silicoflagellate productivity ratio (3-point running mean) and the same curve only using data over isotopic samples (note scale at bottom offset to top scale); left, the silicoflagellate Pacific/Gulf ratio (3-point running mean).

Gulf and in the Pacific Ocean. According to the $\delta^{18}O$ salinity relationship established by Craig and Gordon (1965) in the North Pacific Water, the maximum salinity differences observed in the Central Gulf cannot account for more than 0.2°/₀₀ $\delta^{18}O$ change. The isotopic variations observed on the two curves may be interpreted as a temperature variation. The total isotopic amplitude of 3°/₀₀ observed in core A-1 and of 2°/₀₀ in core E-17 correspond to a temperature change of 9°C and 6°C, respectively. The overall 2°/₀₀ isotopic difference between the two curves could reflect a temperature difference of 6°C. This would suggest that E-17 underlies a more "active" upwelling regime while core A-1 shows thermal conditions close to the mean surface temperature of the central Gulf.

The assumption that area A is influenced more intensively by Pacific surface waters can be supported by the increased abundance of *Dictyocha* spec. A + B + 2 and *D. calida* which all are Pacific floral elements with values around 15% in core A-1 and values <10% in core E-17 (compare Figs. 3 and 4).

Sedimentation rates calculated over time-equivalent horizons based on varve chronology are 0.23 cm/year for A-1 and 0.24 cm/year for E-17 (in the 0-56.5 cm interval). These rates reflect the general higher productivity trends in E-17 and are compensated for by more terrigenous input in A-1 which was cored close to the mouth of the river Rio Mayo. The lower productivity ratios reported by Schrader and Baumgartner (this volume) for cores E-10, E-9, B-28, and B-29 are caused solely by a different cleaning method (Matherne, 1982). Values used in this paper are derived from unfractionated smear slide sample counts.

In conclusion, the isotopic composition of diatom silica indicates in both cores similar long-term fluctuations of the upwelling intensity in the central Gulf. These results are confirmed by the silicoflagellate assemblage analysis showing the primary productivity changes. At present, we have not analyzed the oxygen isotopic composition of the different water masses entering the Gulf and brought up to the sea surface during Ekman transport phenomena nor do we know the isotopic effect due to the salinity changes. Therefore, the isotopic variations are interpreted as temperature variations. Another question is whether a composite channel sample does indeed represent an average isotopic signal or if the "average" signal is biased toward more oceanic character. The composite 2.5 and/or 1.5 cm continuous channel samples used for silicoflagellate analysis do represent averaged samples and are not biased (Murray, 1982). The long-term variations in primary productivity and in the oxygen isotope composition in diatom silica indicate that between 1900 to 1800 both have high levels of primary productivity and decreased surface water temperature as well as a sharp decrease in the productivity and increase in temperature between 1950 to 1900. The record prior to 1750 is only available from one core (E-17) and reveals decreased productiv-

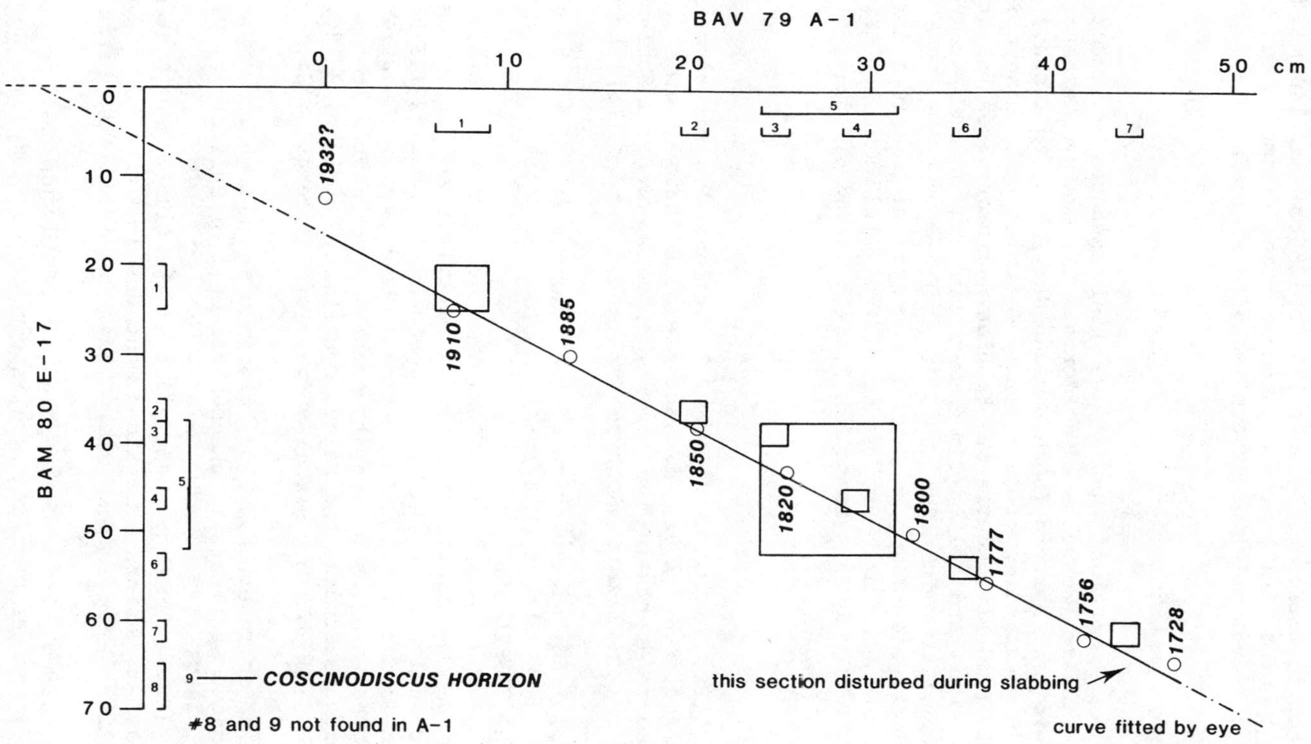

Fig. 7. Shaw-diagram of core E-17 and A-1 over correlative intervals. Varve ages are plotted at their determined downcore depth. Bio-horizons show error bars due to length of sample and/or uncertainties in their exact position.

ity values between 50 to 90 cm (1800 to 1580), increased values between 90 to 115 cm (1580 to 1450) and a sharp decrease in productivity and in Pacific influence between 115 to 125 cm (ca. 1400). Only one of the two cores (E-17) provided near-surface sediments but these samples were not available for oxygen isotopic analysis.

ACKNOWLEDGEMENT

Samples were obtained during two field seasons on board the Mexican Navy Research Vessel H-1 Matamoros. Dr. Gustavo Calderon and Captain Pompeio Leon Herrera (both Mexican Navy) made this study possible through outstanding cooperation. We thank J.J. Labeyrie, C. Lalou and J.C. Duplessy, Centre des Faibles Radioactivités for stimulating discussions and help. Tim Baumgartner contributed with providing the supportive data for the 1910 horizons and compiling much of the oceanographic data. Financial support was received through C.N.R.S. and C.E.A., completed by the A.T.P. Evolution des Climats (CNRS-DGRST), ARCO and NSF grants ATM 7919458-01 and ATM 8121775.

REFERENCES

Baumgartner, T.R., Soutar, A., Cowen, J., Morena, P., Michalson, J., and Bruland, K.W., 1981, Abstract, Detailed chronologies in the laminated sediments of the Guaymas slope, Geological Society of America, Cordilleran Section Meeting, Hermosillo, Sonora, Mexico, 44.

Calvert, S.E., 1966, Accumulation of diatomaceous silica in the sediments of the Gulf of California, Geological Society of America, Bulletin, 77:569-596.

Clayton, R.N. and Mayeda, T.K., 1963, The use of bromine pentafluoride in the extraction of oxygen from oxides and silicates for isotopic analysis, Geochimica et Cosmochimica Acta, 27:43-52.

Craig, A. and Gordon, L.I., 1965, Deuterium and oxygen 18 variations in the ocean and the marine atmosphere, in: "Stable Isotopes in Oceanographic Studies and Paleotemperature," Spoleto, 9-130.

DeMaster, D.J., 1979, "The Marine Budgets of Silica and ^{32}Si," Ph.D. Thesis, Yale University, 308 pp.

Donegan, D. and Schrader, H., 1982, Biogenic and abiogenic components of laminated sediments in the central Gulf of California, Marine Geology, 48:215-237.

Juillet, A., 1980a, "Analyse Isotopique de la Silice des Diatomées Lacustres et Marines: Fractionnement des Isotopes de l'Oxygène en Fonction de la Température," Thèse de spécialité, Paris XI, 71 pp.

Juillet, A., 1980b, Structure de la silice biogénique: nouvelles données apportées par l'analyse isotopique de l'oxygène, Compte-Rendu à l'Académie des Sciences de Paris, 290(Série D): 1237-1239.

Labeyrie, L.D., 1974, New approach to surface seawater paleotemperatures using $^{18}O/^{16}O$ ratio in silica of diatom frustules, Nature, 248:40-42.

Labeyrie, L.D., 1979, "La Composition Isotopique de l'Oxygène de la Silice des Valves de Diatomées. Mise au Point d'une Nouvelle Méthode de Paléoclimatologie Quantitative," Thèse d'Etat, Université de Paris XI, 171 pp.

Labeyrie, L.D. and Juillet, A., 1982, Oxygen isotopic exchangeability of diatom valve silica: interpretation and consequences for paleoclimatic studies, Geochimica et Cosmochimica Acta, 46: 967-975.

Matherne, A., 1982, "Paleoceanography of the Gulf of California: A 350-year Diatom Record," M.S. Thesis, Oregon State University, Corvallis, 111 pp.

Murray, D., 1982, "Paleo-oceanography of the Gulf of California Based on Silicoflagellates from Marine Varved Sediments," M.S. Thesis, Oregon State University, Corvallis, 129 pp.

Roden, G.I. and Groves, G.W., 1959, Recent oceanographic investigations in the Gulf of California, Journal of Marine Research, 18(1):10-35.

Schrader, H., 1979, Cruise report Baja Vamonos 79, September 7-September 30, 1979, Data Report 78, School of Oceanography, Oregon State University, Reference 79-15, 67 pp.

Schrader, H. and Schuette, G., 1981, Marine diatoms, in: THE SEA, Vol. 7, "The Oceanic Lithosphere," C. Emiliani, ed., 1179-1232.

Warsh, C.E., Warsh, K.L. and Staley, R.C., 1973, Nutrients and water masses at the mouth of the Gulf of California, Deep-Sea Research, 20:561-570.

Wyrtki, K., 1967, Circulation and water masses in the Eastern Equatorial Pacific Ocean, International Journal of Oceanology and Limnology, 1(2):117-147.

Zeitzschel, B., 1969, Primary productivity in the Gulf of California, Marine Biology, 3:201-207.

STABLE ISOTOPES OF FORAMINIFERS OFF PERU RECORDING HIGH FERTILITY AND CHANGES IN UPWELLING HISTORY

Gerold Wefer

Geologisches-Paläontologisches Institut
Olshausenstrasse 40-60
D-2300 Kiel, Federal Republic of Germany

Robert B. Dunbar

Scripps Institution of Oceanography
La Jolla, California 92093, U.S.A.
Present address:
Rice University
Houston, Texas 77251, U.S.A.

Erwin Suess

School of Oceanography
Oregon State University
Corvallis, Oregon 97331, U.S.A.

ABSTRACT

Oxygen and carbon isotope analyses of benthic and planktonic foraminifers from the upwelling region off Peru showed that *Neogloboquadrina dutertrei* shells from high fertility areas are depleted in ^{18}O in the larger size classes. This may reflect specific coastal upwelling characteristics because of the unique temperature pattern in the upwelling habitat. Downcore isotopic variations in *Uvigerina peregrina, Cancric oblongus* and *N. dutertrei* yielded information on the upwelling history for three late Quaternary time intervals. There are indications that about 15,000 years ago off Peru worldwide cooling of the surface waters may have been offset by warming due to weaker upwelling. From 8,000 to 5,000 years B.P. $\delta^{18}O$ values in *N. dutertrei* increase slightly, which in conjection with the benthic $\delta^{18}O$ curve, suggests a gradual deterioration of the temperature-stratified water column. There are also indications for warmer bottom waters (by about 2°C) during the past 200 years.

INTRODUCTION

Several attempts have been made to identify signals of coastal upwelling in sediments and to trace their variability in space and time. A high abundance of organic carbon and opal in continental margin sediments has been suggested as an aid to locating upwelling centers (Diester-Haass, 1978; Koopmann, Sarnthein and Schrader, 1978). Such sedimentological features, combined with faunal data on radiolarians and foraminifers and carbon isotope data of *Globigerinoides ruber*, were proposed to reflect intensity changes in upwelling regimes (Berger, Diester-Haass and Killingley, 1978a). Usually, though, these parameters indicate high fertility and not necessarily coastal upwelling; moreover there is generally a strong preservational component (Müller and Suess, 1979; Thiede, Suess and Müller, 1982). Other biogenous skeletal indicators of coastal upwelling are the floral remains of *Chaetoceros* spores and *Delphineis karstenii* (Schuette and Schrader, 1981) and high abundances of the foraminiferal species *Globigerina bulloides* and *Neogloboquadrina dutertrei* (Thiede, 1975; Duplessy, Bé and Blanc, 1981a). Both of these species generally prefer habitats at intermediate depth between 50 and 100 m in subtropical waters, but in coastal upwelling regimes they appear unusually frequent close to the sea surface (Fig. 1; Thiede, this

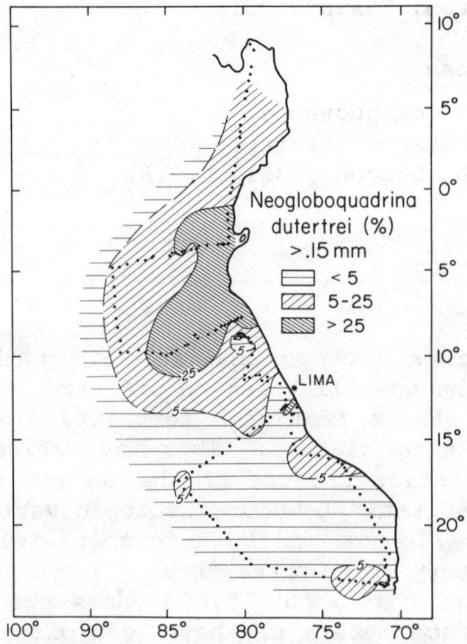

Fig. 1. Frequency distribution of *Neogloboquadrina dutertrei* in surface plankton off the Peru continental margin (Thiede, this volume); *N. dutertrei* is also one of the very few calcareous species preserved in sediments of the upper slope mud lens facies.

volume). Previous work by Deuser et al. (1981) and Erez and Honjo (1981) showed that *N. dutertrei* appear to incorporate oxygen isotopes into their tests close to equilibrium with seawater.

The distribution of *G. bulloides* and its oxygen isotope characteristics were used by Prell and Curry (1981) to delineate sea surface temperature anomalies related to variability of monsoonal upwelling in the western Arabian Sea during the last glacial/interglacial transition. Also, Dunbar (this volume) relies heavily on the stable isotope record of *G. bulloides* to track upwelling and climate over the last 250 years of the continental borderland off Southern California. By combining the stable isotope characteristics of different planktonic and benthic species, Ganssen and Sarnthein (this volume) were further able to detect coastal upwelling off northwest Africa and its seasonal variability with latitude. For the sediments of the Peruvian coastal upwelling regime, *N. dutertrei* is the only planktonic foraminiferal species sufficiently frequent to possibly provide any information. The scarcity of calcareous remains in these sediments is attributed to the well-known phenomenon that the lysocline rises toward continental margins (Berger, 1978), where $CaCO_3$ is dissolved by CO_2 released from the decomposition of organic matter at the sediment/water interface (Emerson and Bender, 1981).

Our present studies on *N. dutertrei* are preliminary in nature and attempt to discover (a) what kinds of fertility signals may be contained in the stable isotope make-up of this species along the continental margin off Peru and Ecuador where upwelling and river input from the Guayaquil River cause high fertility; (b) how these signals might have changed during the last glacial/interglacial transition; and (c) if bottom water characteristics could also be deduced from the isotopic composition of the benthic species *Uvigerina peregrina* and *Cancris oblongus*.

MATERIAL AND METHODS

The samples are from cores collected in June of 1977 during Oregon State University's WELOC Cruise to the continental margin of Peru and in March of 1973 and September of 1971 during the YALOC Cruises to the Peru Basin (Fig. 2). Cores 7706-76 and -69 are box cores from areas of high primary production ranging between 0.2-0.5 g C-org m^{-2} day^{-1}, due to nutrients from rivers emptying into the Gulf of Guayaquil and upwelling in the Eastern Equatorial Pacific. Core 7706-43 is a box core and cores 7706-41, -44 are Kasten cores from a recent deposit of an upper slope mud lens facies (Reimers and Suess, this volume) forming in the region of very high primary production from coastal upwelling (> 1.0 g C-org m^{-2} day^{-1}). Core 7706-33 is located at the eastern-most edge of the Nazca Ridge seaward of a region of strong upwelling at 15°S. The cores Y 73-3-23 and Y 71-9-89 are piston cores taken about 2,000 km offshore where the primary production

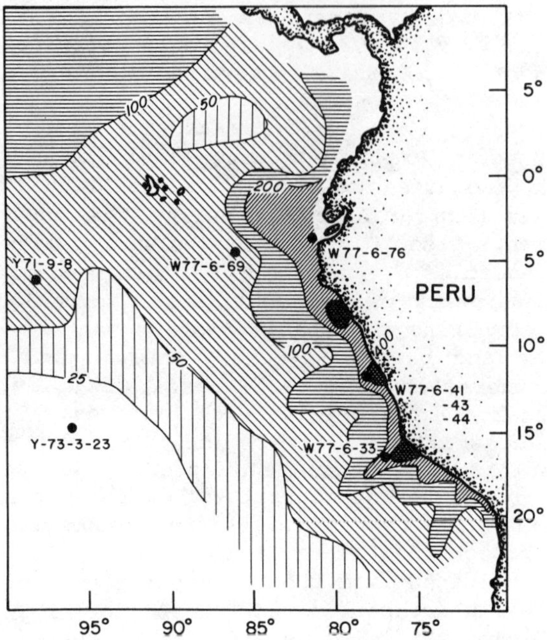

Fig. 2. Core locations and integrated biological productivity from open ocean and high fertility areas on the Peru continental margin and adjoining Nazca Plate region; the contour intervals are in units of g C-org • m^{-2} • y^{-1}; productivity at upwelling centers off Peru between 6°-15°S is >400g C-org • m^{-2} • y^{-1}.

is less than 0.1 g C-org m^{-2} day^{-1}. The data on primary production are compiled from Koblentz-Mishke, Volkovinsky and Kabanova (1970); Zuta and Guillen (1970); and Love and Allen (1975).

The upper one or two centimeters of the box cores and 2-5 cm intervals at depths in the Kasten cores were sampled at Oregon State University and shipped to Scripps Institution of Oceanography for further treatment. The fine fraction was removed by sieving with calgon solution through a 62 µm screen. After the sample was separated into six size fractions by dry sieving, specimens were picked and placed in a small 150 µm mesh enclosure and ultra-sonicated in a buffered water bath for final cleaning. Only whole specimens showing no traces of contamination inside or outside were used for isotopic analysis. The mean number of individual shells per isotope analysis was 18. After reaction in 100% phosphoric acid at 50°C, the liberated CO_2 gas was analyzed with an "on line" V.G. Micromass 602C mass spectrometer. Analytical procedures are described by Berger and Killingley (1977). Results are reported in the standard δ notation relative to PDB. Precision of our determinations expressed at ± one standard deviation is based on analyses of NBS-20 and was 0.10°/$_{\circ\circ}$

for $\delta^{18}O$ and 0.07°/₀₀ for $\delta^{13}C$. When duplicate samples of foraminiferal species were run, reproducability averaged 0.12°/₀₀ for $\delta^{18}O$ and 0.10°/₀₀ for $\delta^{13}C$.

RESULTS AND DISCUSSION

Oxygen Isotope Variability

The variation of oxygen isotope composition with shell size for the planktonic foraminifer *N. dutertrei* from high fertility and open ocean (low fertility) areas is illustrated in Fig. 3. Shells from high fertility areas show a marked depletion in ^{18}O in the larger size classes, whereas the shells from the low fertility area show either no change or slight enrichment in ^{18}O with increasing size. No change or enrichment of ^{18}O agrees with previous observations from other oceanic areas, e.g., *N. dutertrei* (=*Globoquadrina eggeri*, Emiliani, 1971) from the Caribbean and Mediterranean Seas show no

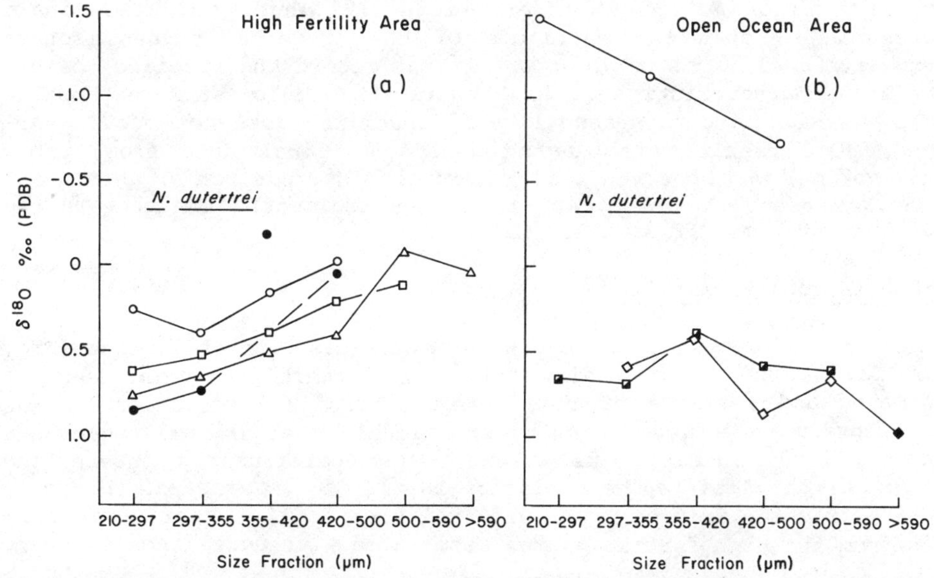

Fig. 3. Oxygen isotopic composition of the planktonic foraminifers *Neogloboquadrina dutertrei* as a function of shell size.

(a) coastal high fertility environment: ○ = 7706-43 ● = 7706-33
□ = 7706-76 △ = 7706-69

(b) open ocean environment: ◇ = Y7303-23 ◪ = Y7109-89
○ = western equatorial Pacific (Berger, Killingley and Vincent, 1978)

trend in $^{18}O/^{16}O$ ratios for different sizes while larger shells of *N. dutertrei* from the western Equatorial Pacific are enriched in ^{18}O (Berger, Killingley and Vincent, 1978). Such a trend is considered 'normal' and probably reflects a deeper (and colder) habitat of adult forms or other species living in tropical waters (Vincent and Berger, 1981).

The 'reverse' trend of depletion of ^{18}O with increasing shell size, as found off Peru for *N. dutertrei*, has hitherto only been reported by Emiliani (1971) from shells of *Orbulina universa* and *Globigerinoides conglobatus* in open ocean areas. But these changes were so small, -0.2 to -0.4°/$_{\circ\circ}$, that they might easily be attributed to seasonal temperature changes. For *N. dutertrei* in high fertility areas, as off Peru, it is likely that the size distribution of the foraminiferal population changes in response to fertility and temperature during upwelling. When low surface temperatures and high nutrients prevail, reproduction occurs early producing large numbers of smaller forms, thus preferentially imprinting the isotopic temperature signature of the small size classes in the sediment record. During non-upwelling periods, when higher temperatures prevail, the larger forms of *N. dutertrei* are produced. Thus, our preliminary suggestion is that the magnitude of the 'reverse' oxygen isotope trend with shell size in *N. dutertrei* may record the specific coastal upwelling characteristics of high nutrients and low surface temperatures, perhaps tied to seasonality if upwelling does not occur year-round. Our results are supported by additional data from Dunbar (this volume) who observed similar 'reverse' trends for *N. dutertrei*, *O. universa* and *G. bulloides* in surface sediments from beneath the upwelling systems off Southern California.

Carbon Isotope Variability

In *N. dutertrei*, the ^{13}C content increases with shell size (Fig. 4). This trend is generally found in foraminifers from deep-sea sediments and reflects greater disequilibrium fractionation by the small, presumably faster growing individuals (Berger, Killingley and Vincent, 1978; Dunbar, this volume). One particularly interesting feature of the data is the depletion in ^{13}C of *N. dutertrei* in sample 7706-76 from the Gulf of Guayaquil, relative to the other three samples from the high fertility area associated with upwelling.

Several possibilities have to be taken into account when explaining this depletion in ^{13}C. It is likely that temperature is not responsible for the observed difference as the mean temperature varies by only about 3°C between station 7706-76 and the other stations. Temperature exerts only a small control on $\delta^{13}C$ in marine carbonates, the dependency being 0.035°/$_{\circ\circ}$ per °C (Emrich, Ehhalt and Vogel, 1970). We suggest that the light $\delta^{13}C$-values in the sample from the Gulf of Guayaquil are not the result of increased metabolic CO_2 in the regional oxygen minimum layer. The oxygen minimum layer

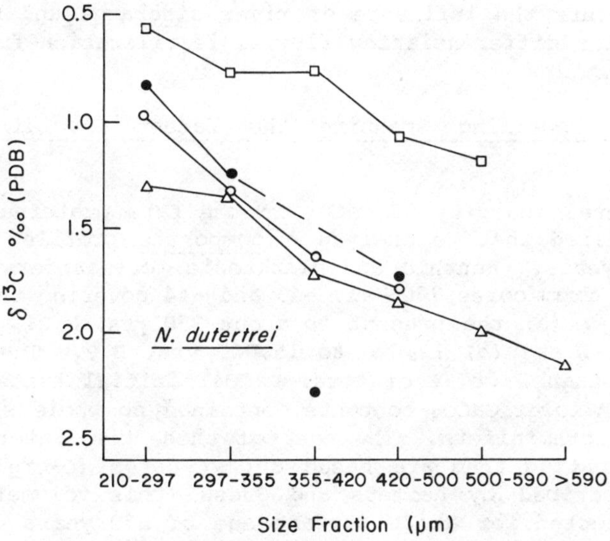

Fig. 4. Carbon isotopic composition of the planktonic foraminifers *Neogloboquadrina dutertrei* as a function of shell size; note 'light' $\delta^{13}C$ of sample from the Gulf of Guayaquil (□ = 7706-76); other symbols as in Fig. 3a.

is located between about 200 and 300 m water depth at all sites north of approximately 12°S (Reimers and Suess, this volume), well below the habitat of *N. dutertrei* in the high fertility regimes. Other parameters such as light transmission and primary production do not show differences large enough to account for the observed differences in carbon isotope make-up of *N. dutertrei* in core 7706-76 and the other cores.

A possible explanation for this is a depletion in the initial carbon isotope composition of the seawater from the Gulf of Guayaquil caused by input of fresh water. The total inorganic carbon species in terrestrial run-off are enriched in ^{12}C relative to seawater, as suggested by Pastouret et al. (1978). This is partly the effect of fractionation among the carbonate species at the different pH values of fresh water and seawater, but is also due to the increased load of terrigenous organic carbon which contributes ^{13}C depleted CO_2 to the river water (Degens, 1969). In addition, if rivers emptying into the Gulf of Guayaquil transport a substantial amount of land-derived organic material which decomposes in the Gulf, the $\delta^{13}C$ may be further depleted in comparison to the other areas. Unfortunately, data on the stable isotope composition of ΣCO_2 in the water are not available. The lower salinity values show that river water is mixed with seawater at the mouth of the Gulf of Guayaquil (Pak, Menzies and Zaneveld, 1979). Thus, *N. dutertrei* may record in its stable carbon isotope signature the influence of river discharge and thereby provide a tool for differentiating fluvial fertilization from upwelling fertilization.

Variation of Upwelling Spanning the Latest Glacial/Interglacial Transition

The general scarcity of $CaCO_3$ in the Late Quaternary sediments off Peru required that we analyze a composite profile spanning the last 15,000 years. Benthic and planktonic foraminifers were therefore selected from cores 7706-41, -43 and -44 covering three discrete time intervals: (a) the present to about 250 yrs. B.P., (b) 5,000 to 9,000 yrs. B.P. and (c) 12,000 to 15,000 yrs. B.P. These intervals all had more than 2 wt.-% of total $CaCO_3$; initial tests showed that sediments with lower $CaCO_3$-contents contained no whole shells of any identifiable foraminifers. The ages of these time intervals and the hiatuses separating them are based on ^{14}C-dates (C-org) and unconformities described by Reimers and Suess (this volume). The time scale is corrected for a ^{14}C surface age of 570 years which is assumed to be due to incorporation of older carbon into the surface sediments.

Downcore isotopic variations in *Uvigerina peregrina, Cancris oblongus* and *N. dutertrei* are shown in Fig. 5. With a few exceptions, all foraminiferal tests were taken from the size class 355-420 µm for the analyses. Resig (1981), in a recent treatment of the biogeography of benthic foraminifers of the Nazca Plate and adjacent continental margin, identifies the genera *Cancris sp.* and *Uvigerina sp.* as members of an upper to middle slope bathyal assemblage ranging in water depth from 150-1000 m.

Period: 250-0 Years Ago

Only shells of *Cancris oblongus* were available here for analyses. The isotope ratios show a significant depletion in ^{18}O with age. This change is partly due to a 2°C temperature difference as a function of water depth between sites 7706-41 and 7706-43 and therefore has to be reduced by ∿0.4°/₀₀. The temperature profiling at the core sites is reported by Pak et al. (1979). In spite of this adjustment and further assuming a constant oxygen isotope make-up of the water during the last several hundreds of years, the difference is still significant and indicates that the bottom water was about 2°C warmer 200 years ago.

There are other independent indications of a general warming trend, or more correctly, for frequent warm surface water incursions, during this time off Peru when elsewhere a general cooling trend was recorded (Salinger, 1981). DeVries and Schrader (1981) report from floral analyses of the same core sequence, that during the third Neoglacial Period, i.e., 200-400 yrs. B.P., tropical diatom species were

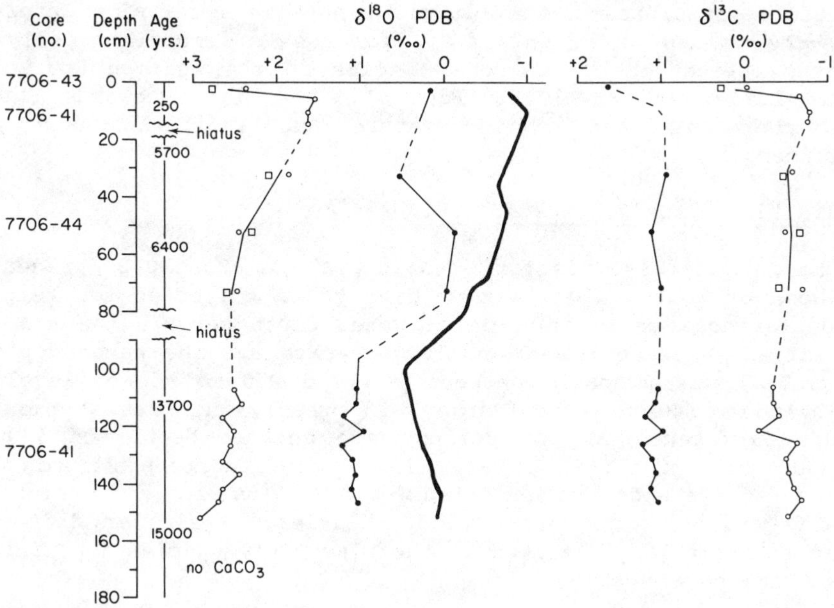

Fig. 5. Downcore oxygen and carbon isotopic variations in the planktonic foraminifera *Neogloboquadrina dutertrei* (●) and the benthic foraminifers *Cancris oblongus* (○) and *Uvigerina peregrina* (□) from the coastal upwelling region off Peru. The solid line shows the oxygen isotope signal of *N. dutertrei* from the western equatorial Pacific (Berger, Killingley and Vincent, 1978). The composite profile is based on data from three closely spaced cores spanning a water depth from 411-583 m; chronology is from Reimers and Suess (this volume).

more common than subtropical, upwelling related ones. The same interpretation was reached by Reimers and Suess (this volume) from examining characteristics of the proto-kerogen record of this and neighboring cores.

Period: 8,000 to 5,000 Yrs. B.P.

At about 8,000 years ago $\delta^{18}O$ values in benthic foraminifers were as 'heavy' as values found in surface sediments of today (Fig. 5). The mean difference of 0.6°/$_{oo}$ between the datum level at 5,000 years B.P. and both the surface and the 8,000 year level could be an indication of bottom water temperature changes of about 2.5°C. It is not very likely that the isotope difference was caused by sea level or ice-volume changes, because during that period sea level remained rather constant (Berger, 1982). From 8,000 to 5,000 yrs. B.P. $^{18}O/^{16}O$ ratios in *N. dutertrei* increase slightly, which in conjunction with the benthic $\delta^{18}O$ curve suggests the gradual deterioration of a temperature stratified water column during this interval. Interestingly, the values of *N. dutertrei* from the Peru margin closely reflect those recorded by the same species in the western Equatorial Pacific (Berger et al., 1978b; Fig. 5). The offset towards lighter values there is due to temperature and regional water mass differences.

Period: 15,000 to 12,000 Yrs. B.P.

The mean $\delta^{18}O$ values for *C. oblongus* are about 0.2°/₀₀ heavier than those of today. Again, they have to be corrected for temperature differences as a function of water depth between the two sampling sites. Between 15,000 and 12,000 years ago the water depth at site 7706-41 was probably between 330 and 270 m, as sea level was lower by about 140 to 80 m (Berger, in press). At present the temperature difference at our coring site between depths of 580 and about 300 m is 4°C (Pak et al., 1979) which corresponds to about -0.8°/₀₀ difference in expected $\delta^{18}O$. Instead $\delta^{18}O$-values of shells from *C. oblongus* are +0.2°/₀₀ heavier. A similar difference in $\delta^{18}O$ of about 1°/₀₀ is seen in the planktonic species *N. dutertrei* of that time period.

Differences in $\delta^{18}O$ between glacial and interglacial periods are caused mainly by two effects: an ice effect due to the locking of ^{16}O-rich water in polar ice caps and a corresponding enrichment of ^{18}O in seawater, and a temperature effect related to growth in colder water during glacial times. Generally, at comparable geographical latitudes an oxygen isotope difference of about 1.6°/₀₀ is observed between glacial and interglacial periods (Shackleton, 1977). Dunbar (this volume) reported from the Santa Barbara Basin that *G. bulloides* became depleted in ^{18}O by 2.2°/₀₀ and *O. universa* by 1.6°/₀₀. We assume that about 15,000 years ago in the study area off Peru only the ice-effect is seen in the oxygen isotopes and that the worldwide cooling of the surface waters was offset by a warming due to weaker upwelling. Between 13,000-12,000 yrs. B.P. and today, the difference in $\delta^{18}O$ is similar to values observed elsewhere (e.g., Duplessy et al., 1981b).

Stable Carbon Isotope Characteristics

The calcareous shells of the planktonic and the benthic species from the pre-recent sediments are depleted in ^{13}C relative to today's samples (Fig. 5). The recent carbon and oxygen values shown here for core 7706-43 were, however, also observed in other cores deposited at similar water depths (*N. dutertrei*, see Figs. 3 and 4). Therefore, we believe the data from core 7706-43 to be typical for a recent foraminiferal fauna and that no glacial tests were included in the material analyzed.

Again, as discussed above for planktonic species, several factors may be considered responsible for the carbon shift seen in the benthic species: (a) temperature, (b) a change in the isotopic composition of the water, and (c) a change in the fractionation of carbon isotopes during carbonate precipitation. The temperature effect is small as shown above, and moreover it changes the $\delta^{13}C$ values towards heavier ratios. Also, the decomposition of organic matter cannot be responsible for the carbon shift, which increases ^{12}C relative to ^{13}C. Measurements of the $\delta^{13}C$ of the dissolved inorganic carbon species in the Pacific Ocean show a decrease from about 2°/₀₀ at the surface to -0.5°/₀₀ within the oxygen minimum layer at 900 m water depth (Kroopnick, Deuser and Craig, 1970). The coring sites and the surface sample (7706-43) are located within the present day oxygen minimum layer where oxygen contents are less than 0.2 mlO_2/ℓ. Consumption of the remaining 0.2 ml/ℓ would lead to a $\delta^{13}C$ change of about only 0.1°/₀₀.

Another effect which changes the long-term isotopic composition of the inorganic carbon species in the water is proposed by Shackleton (1977). He suggests that the difference in $\delta^{13}C$ between glacial and interglacial times, universally observed in benthic species, is related to fluctuations in the size of the terrestrial plant biomass reservoir which is enriched in ^{12}C. During interglacial times, more ^{12}C is fixed in this reservoir on land causing heavier $\delta^{13}C$ values in the ocean. This mechanism would, however, only explain the lighter carbon isotope values during 15,000-12,000 yrs. B.P. but not the especially light values of about 200 years ago. For these we favor fractionation resulting from environmental conditions. In general, small planktonic foraminifers are enriched in ^{12}C compared with the larger ones due to rapid $CaCO_3$ production of the initial structure of the test (Berger et al., 1978b). Similarly, an increase in bottom water temperature of about 2°C might have stimulated faster growth and thus caused preferential uptake of ^{12}C by *C. oblongus*.

At this stage of our investigations no simple explanation can be given for the observed carbon isotope differences in benthic as well as for planktonic foraminifers of recent and late glacial ages. One conclusion seems safe in that a combination of effects rather than any one single effect is likely responsible for the observed differences. The possibilities of environmental control of stable carbon isotopes on continental margins are numerous and are increased by still another factor in the fresh water effect suggested here for the Gulf of Guayaquil samples.

ACKNOWLEDGEMENT

This work was supported by Deutscher Akademischer Austauschdienst through a grant to G. Wefer. Isotope measurements were made in the Sediment Isotope Laboratory of Scripps Institution of Oceanography. We owe thanks to J.S. Killingley (La Jolla) for assistance with the micro mass spectrometer and to W.H. Berger and M. Sarnthein for a critical reading of the manuscript.

REFERENCES

Berger, W.H., 1978, Sedimentation of deep-sea carbonates: maps and models of variations and fluctuations, Journal of Foraminiferal Research, 8:286-302.

Berger, W.H., 1982, On the definition of the Pleistocene-Holocene boundary in deep-sea carbonates, Sveriges geologiska undersökning, ser. C794:270-280 (Uppsala).

Berger, W.H. and Killingley, J.S., 1977, Glacial-Holocene transitions in deep-sea carbonates: selective dissolution and the stable isotope signal, Science, 197:563-566.

Berger, W.H., Diester-Haass, L. and Killingley, J.S., 1978, Upwelling off Northwest Africa: The Holocene decrease as seen in carbon isotopes and sedimentological indicators, Oceanologica Acta, 1:3-7.

Berger, W.H., Killingley, J.S. and Vincent, E., 1978b, Stable isotopes in deep-sea carbonates: Box Core ERDC-92, West equatorial Pacific, Oceanologica Acta, 1:203-216.

Degens, E.T., 1969, Biogeochemistry of stable carbon isotopes, in: "Organic Geochemistry", G. Eglinton and M.T.J. Murphy, eds., Springer-Verlag, Berlin, 304-329.

Deuser, W.H., Ross, E.H., Hemleben, C. and Spindler, M., 1981, Seasonal changes in species composition, numbers, mass, size and isotopic composition of planktonic foraminifera settling into the deep Sargasso Sea, Palaeogeography, Palaeoclimatology, Palaeoecology, 33:103-128.

DeVries, T.J. and Schrader, H.J., 1981, Variations of upwelling/oceanic conditions during the latest Pleistocene through Holocene off the central Peruvian coast: a diatom record, Marine Micropaleontology, 6:157-167.

Diester-Haass, L., 1978, Sediments as indicators of upwelling, in: "Upwelling Ecosystems", R. Boje and M. Tomczak, eds., Springer-Verlag, Berlin, 261-281.

Duplessy, J.C, Bé, A.W.H. and Blanc, P.L., 1981, Oxygen and carbon isotopic composition and biogeographic distribution of planktonic foraminifera in the Indian Ocean, Palaeogeography, Palaeoclimatology, Palaeoecology, 33:9-46.

Duplessy, J.C., Delibrias, G., Turon, J.L., Pujol, C., and Duprat, J., 1981b, Deglacial warming of the northeastern Atlantic Ocean. Correlation with the paleoclimatic evolution in the European Continent, Palaeogeography, Palaeoclimatology, Palaeoecology, 35:121-144.

Emerson, S.E. and Bender, M., 1981, Carbon fluxes at the sediment-water interface of the deep-sea: Calcium carbonate preservation. Journal of Marine Research, 39:139-162.

Emiliani, C., 1971, Depth habitats of growth stages of pelagic foraminifera, Science, 173:1122-1124.

Emrich, K., Ehhalt, H.D., and Vogel, J.C., 1970, Carbon isotope fractionation during the precipitation of calcium carbonate, Earth and Planetary Science Letters, 8:363-371.

Erez, J. and Honjo, S., 1981, Comparison of isotopic composition of planktonic foraminifera in plankton tows, sediment traps and sediments, Palaeogeography, Palaeoclimatology, Palaeoecology, 33:129-156.

Koblentz-Mishke, O.J., Volkovinsky, V.V. and Kabanova, J.G., 1970, Plankton primary production of the world ocean, in: "Scientific Exploration of the South Pacific", W.S. Wooster, ed., National Academy of Science, Washington, 183-193.

Koopmann, B., Sarnthein, M. and Schrader, H.J., 1978, Sedimentation influenced by upwelling in the subtropical Baie du Levrier (West Arica), in: "Upwelling Ecosystems", R. Boje and M. Tomczak, eds., Springer-Verlag, Berlin, 282-300.

Kroopnick, P., Deuser, W.G. and Craig, H., 1970, Carbon 13 measurements on dissolved inorganic carbon at the North Pacific (1969) Geosecs Station, Journal of Geophysical Research, 75:7668-7671.

Love, C.M. and Allen, R.M., 1975, Eastropac Atlas 10. U.S. Government Printing Office, Washington, Circulation #330.

Müller, P.J. and Suess, E., 1979, Productivity, sedimentation rate and sedimentary organic matter in the oceans. I. Organic carbon preservation, Deep-Sea Research, 26A:1347-1362.

Pak, H., Menzies, D. and Zaneveld, J.R.V., 1979, Optical and hydrographical observations off the coast of Peru during May-June, 1977. Data Report 77, School of Oceanography, Oregon State University, Corvallis, 93 pp.

Pastouret, L., Chamley, H., Delibrias, G., Duplessy, J.C. and Thiede, J., 1978, Late Quaternary climatic changes in western tropical Africa deduced from deep-sea sedimentation off the Niger delta, Oceanologica Acta, 1:217-232.

Prell, W.L. and Curry, W.B., 1981, Faunal and isotopic indices of monsoonal upwelling: Western Arabian Sea, Oceanologica Acta, 4:91-98.

Resig, J., 1981, Biogeography of benthic foraminifera of the northern Nazca Plate and adjacent continental margin, in: "Nazca Plate: Crustal Formation and Andean Convergence", L.D. Kulm, J. Dymond, E.J. Dasch and D.M. Hussong, eds., Geologial Society of America Memoir 54, 619-666.

Salinger, J.M., 1981, Paleoclimates north and south, Nature, 291: 106-107.

Schuette, G. and Schrader, H.J., 1981, Diatom taphocoenosis: The reflection of coastal upwelling, in: "Coastal Upwelling," F.A. Richards, ed., Coastal and Estuarine Sciences, American Geophysical Union, Washington, 372-380.

Shackleton, N.J., 1977, Carbon-13 in *Uvigerina*: tropical rainforest history and the equatorial Pacific carbonate dissolution cycles, in: "The Fate of Fossil Fuel CO_2 in the Oceans", N.R. Andersen and A. Malahoff, eds., Plenum Press, New York, 401-427.

Thiede, J., 1975, Distribution of foraminifera in surface waters of a coastal upwelling area, Nature, 253:712-714.
Thiede, J., Suess, E. and Müller, P.J., 1982, Late Quaternary fluxes of major sediment components to the sea floor at the Northwest African continental slope, in: "Geology of the Northwest African Continental Margin", U. von Rad, K. Hinz, M. Sarnthein and E. Seibold, eds., Springer-Verlag, Berlin, 605-631.
Vincent, E. and Berger, W.H., 1981, Planktonic foraminifera and their use in paleoceanography, in: "The Oceanic Lithosphere," C. Emiliani, ed., THE SEA, Vol. 7, Wiley and Sons, New York, 1025-1119.
Zuta, S. and Guillen, O., 1970, Oceanografia de las aguas costeras del Peru, Boletin del Instituto del Mar Peru, 2:157-324.

PLEISTOCENE TIME SCALES

SPATIAL AND TEMPORAL PATTERNS OF ORGANIC MATTER ACCUMULATION ON THE PERU CONTINENTAL MARGIN

Clare E. Reimers* and Erwin Suess

School of Oceanography
Oregon State University
Corvallis, Oregon 97331, U.S.A.
*Present address:
Marine Biology Research Division
Scripps Institution of Oceanography
La Jolla, California 92093, U.S.A.

ABSTRACT

The present-day regional pattern of organic carbon on the Peru continental margin is characterized by two areas of preferential accumulation: the outer shelf-upper slope at about 100-450 m of water depth between about 11°S and 16°S, and the lower continental slope (>2000 m). The middle slope lacks significant recent organic-rich sediments. Deposition on the upper slope originates predominately from high biological production in response to persistent coastal upwelling. Two organic carbon concentration maxima (>10% dry wt.) are found on the slope between 11°S and 14°S, where the shelf is only ~15 km wide below or on the fringes of centers of maximum annual primary production (~1000 $gC \cdot m^{-2} \cdot y^{-1}$). At 7°-10°S, where a similarly productive upwelling center is located, the wide shelf (~30 km) and shallow water depths promote continuous reworking by bottom currents. Interaction of the Peru Current system with shelf-slope morphology is the controlling factor for the present distribution of organic matter on the continental margin. Between about 11° and 14°S, bottom currents that fluctuate in both strength and direction appear responsible for the non-deposition on the mid- slope and the subsequent downslope accumulation of resuspended particulate organic matter. At 15°S, the subsurface current that flows predominantly poleward relaxes at slope depths, and, as a consequence, organic carbon and fine-grained clays accumulate here preferentially (>9 $gC \cdot cm^{-2} \cdot 1000\ y^{-1}$). This accumulation maximum is not reflected in the carbon content of the sediments (<10%) due to dilution by fine-grained terrigenous debris. Organic carbon accumulation is further enhanced here by intensification of the O_2-minimum from north to south.

The temporal pattern of organic carbon accumulation between 11° and 14°S is recorded in a series of sediment sequences separated by hiatuses. The present mode and magnitude of organic carbon accumulation is higher at the northern end of the upper-slope mud lens facies (11°S) than in the south (13°S), and has been active for no longer than 500 years. A second record (1,500-3,000 years ago) is similar in magnitude to the present and separated from a third period of organic carbon accumulation by a major hiatus lasting from about 4,000 to 10,000 years ago. Prior to that time organic carbon accumulation rates were about 1/6 of the present accumulation at the 11°S sites, but nearly equal to the present rates at the 13°S sites. The timing of the hiatuses and the regional changes in organic carbon accumulation suggest a climatic control whereby both the intensities and the positions of currents and primary productivity centers must have responded to global warming and sea level changes which have occurred since the Late Pleistocene.

INTRODUCTION

The Peru continental margin represents one of five major and persistent coastal upwelling regions of the world's oceans. In these regions nutrient-rich subsurface waters are brought to the euphotic zone as a mass balance for offshore surface Ekman flow in response to coastal winds. The subsurface water may also carry the "seeds" of phytoplankton communities which develop into large blooms as upwelled waters move offshore. Therefore, it is not surprising that a multiscalar interplay of physical, biological and chemical controls determines the chemical character of the underlying sediment and its organic matter content. The purpose of this paper is to identify the most significant of these controlling factors for the Peru margin and to trace their history in the sediment record for the last 15,000 years.

METHODS AND SAMPLES

We have relied heavily on data collected as part of an interdisciplinary study at Oregon State University on the formation, variability and stability of hemipelagic sediments (Pak, Codispoti and Zaneveld, 1980; Krissek, Scheidegger and Kulm, 1980; Suess, 1981; Busch and Keller, 1981; DeVries and Schrader, 1981; Reimers, 1982). These were supplemented where appropriate, by data and samples obtained by the Nazca Plate Project (Kulm et al., 1981) and data from the concurrent CUEA Program (Rowe, 1979). Since not all areas of the margin were equally well sampled, most of our discussion addresses organic matter sedimentation in a few locations between 6° and 16°S latitude where dense sampling and other supportive information were available.

All of the carbon analyses were made by the standard LECO induction combustion technique; i.e.: organic carbon was determined as the difference in the CO_2 liberated between two high temperature combustions, one with and one without prior heating of the sample to 450°C for two hours to oxidize organic carbon (Heath, Moore and Dauphin, 1977). This is a rapid technique suitable for large numbers of samples, but its precision varies widely between 1-10% depending on the amount and mineralogy of carbonates present. Therefore, it imposes certain limitations on estimates of organic carbon budgeting but is well suited for mapping intra-regional distribution patterns. Total nitrogen, inorganic phosphorous and quartz analyses were determined by micro-Kjeldahl digestion, HCℓ extraction-spectrophotometry, and x-ray diffractometry techniques, respectively (Bremner, 1960; Ellis, 1972; Müller, 1977; Suess, 1981). Textural compositions (sand, silt, and clay percentages) are the results of standard siev-

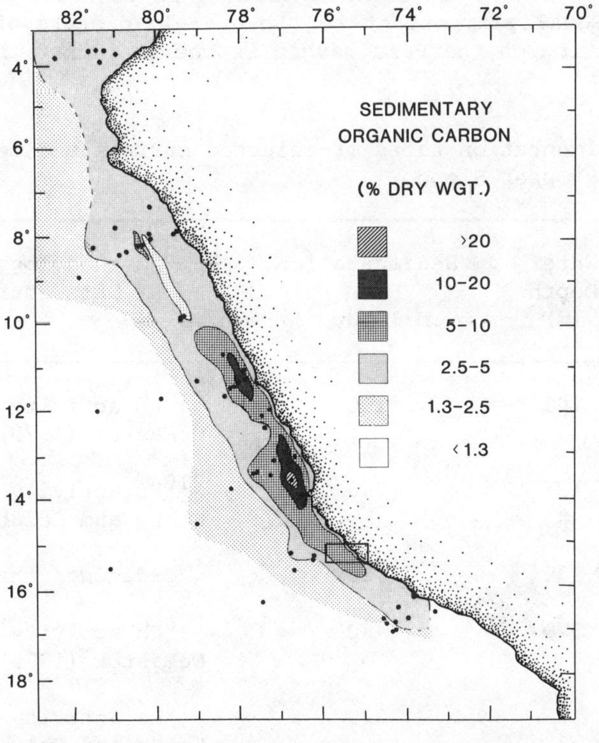

Fig. 1. The distribution of organic carbon (% dry weight) in surface sediments on the Peru continental margin. Two maxima with concentrations >10% are located within an upper slope mud lens facies between approximately 11°S and 14°S. To the north of 11°S a minimum of organic carbon contents is observed at the center shelf edge. High contents of organic matter are also found at the lower slope at 8°S and 14°S.

ing and particle-settling techniques (Busch and Keller, 1981). Supplemental CUEA data were determined according to methods described by Rowe (1979). The results are listed in the Appendix.

SPATIAL PATTERNS OF ORGANIC MATTER ACCUMULATION ON THE PERU CONTINENTAL MARGIN

In many ways the meso-scale surface distribution of organic carbon on the Peru margin illustrates the variability of sedimentation conditions in this particular coastal upwelling region (Fig. 1). Based on the results of organic carbon analyses of surface sediments from 88 cores, the outstanding feature of the sedimentary organic carbon distribution is the distinctive region in the vicinity of the upper slope between 11°S and 16°S with organic carbon contents >5% dry weight. Sediments in this area are presently accumulating at rates >30 cm/1000 y as determined from ^{210}Pb-activity depth profiles (Table 1). Equally important to the overall picture of organic matter sedimentation on the Peru margin is the fact that there is little

Table 1. Sedimentation rates of selected surface sediments from the Peru Upper Slope

Station Location	Water Depth (m)	Sedimentation Rate ($cm \cdot 1000y^{-1}$)	Methodology and Data Source
11°15'S 77°57'W	186	160	^{210}Pb activity profiles; DeMaster (1979)
12°02'S 77°43'W	194	340	^{210}Pb activity profiles; Koide and Goldberg (1982)
12°59'S 76°58'W	325	30	^{14}C-dating; this study
13°37'S 76°51'W	370	50	^{210}Pb activity profiles; DeMaster (1979)
14°39'S 76°10'W	183	320	^{210}Pb activity profiles; Koide and Goldberg (1982)
15°04'S not listed	92	1100	^{210}Pb activity profiles; Henrichs (1980)
15°09'S not listed	268	1200	^{210}Pb activity profiles; Henrichs (1980)

or no contemporary sedimentation on the upper slope between approximately 6°S and 10°S. Five box cores taken in this region recovered partially cemented biogenous sands with phosphorites and fish debris from water depths ranging from 192 m to 486 m. Further to the north at approximately 3°45'S, a well-compacted olive-gray clay, intensely drilled by boring macro-organisms, was recovered from 713 m water depth indicating an erosional surface. The organic carbon contents of these older surface sediments are at most 3 wt-%. On the lower slope two regions stand out with preferred organic matter accumulations; one located off 8°S at depths 3000 m with a faint cross-trench extension to bathyal depths of 5000 m and a second one located at 13°-15°S. There it is restricted to the area of the ancient structural Lima Basin, and no cross-trench extension of high organic matter accumulations is observed.

To decipher the causes of these and other changes in organic matter distribution and sedimentary facies over the Peru slope and

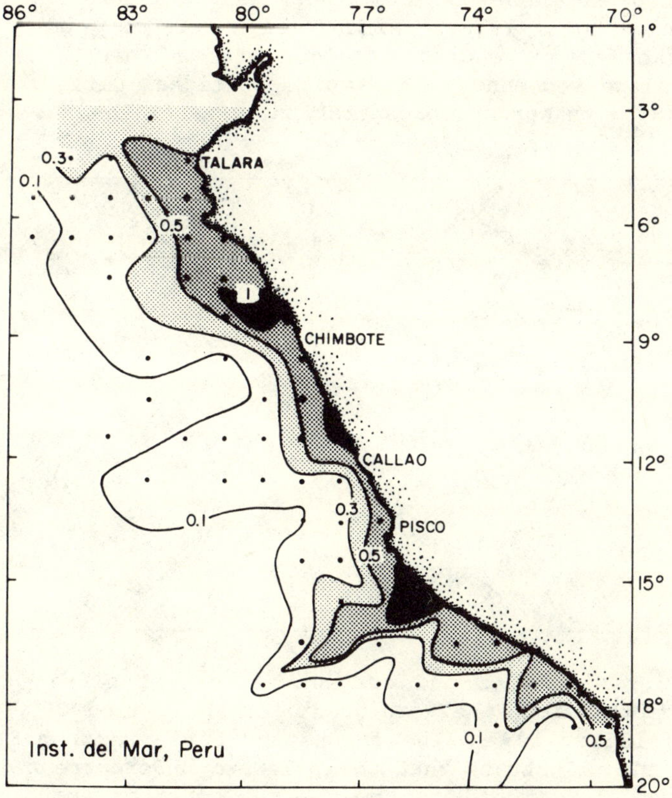

Fig. 2. The distribution of total integrated, long-term mean primary production in $g \cdot m^{-2} \cdot day^{-1}$ over the Peru continental margin as compiled by Zuta and Guillén (1970) from measurements during the years 1960-1970 by IMARPE; Instituto del Mar, Peru.

adjoining shelf, we will consider a typical distribution of primary productivity in Peruvian surface waters (Fig. 2); the mean wind and subsurface current vectors along the Peru coast (Fig. 3); and five cross-margin profiles illustrating from north to south the changes in bathymetry, dissolved oxygen in slope waters, $CaCO_3$ contents and organic carbon contents (Fig. 4).

The primary productivity data are adopted from observations during the period 1960-1970 (Zuta and Guillén, 1970). The wind and current data are reproduced from Brockmann et al. (1980), and Barber and Smith (1981), whereas the cross-margin profiles represent a compilation of new data and data from CUEA reports by Rowe (1979) and Hafferty, Codispoti and Huyer (1978), and from STEP-1 reports by Wooster (1961).

Areas of high primary production along the Peru coast are generally centered within latitudes 7°-8°S, 11°-12°S, and 14°-16°S. This would be expected from present day prevailing patterns of newly upwelled waters in these regions (Zuta and Guillén, 1970; Guillén, Mendiola and Rondán, 1973). Since a high rate of primary production is the first factor leading to high rates of carbon input to the shelf and slope sediments, the two depositional sites found between 11° and 14°S with carbon concentrations greater than 10% may be ex-

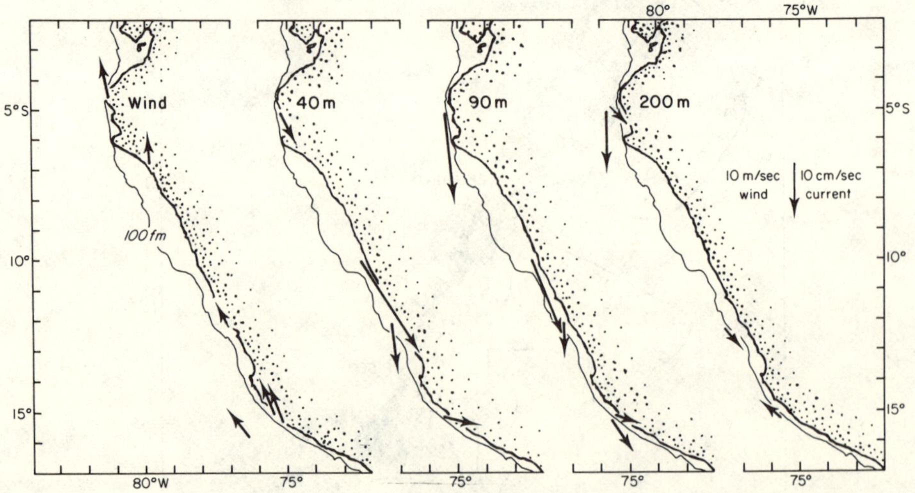

Fig. 3. Mean wind and current vectors at different locations and water depths along the Peru coast for the period 0000 Z, 2 April to 0600 Z, 10 May 1977. The figure and data are from Brockmann et al. (1980) and illustrate that the poleward undercurrent is a prominent feature along the Peru upper slope north of about 14°S. Less information is available on currents for the middle slope, but at these depths there is also evidence for intermittently strong poleward flow. By permission from Deep-Sea Res., Copyright (c) 1980, Pergamon Press, Inc.

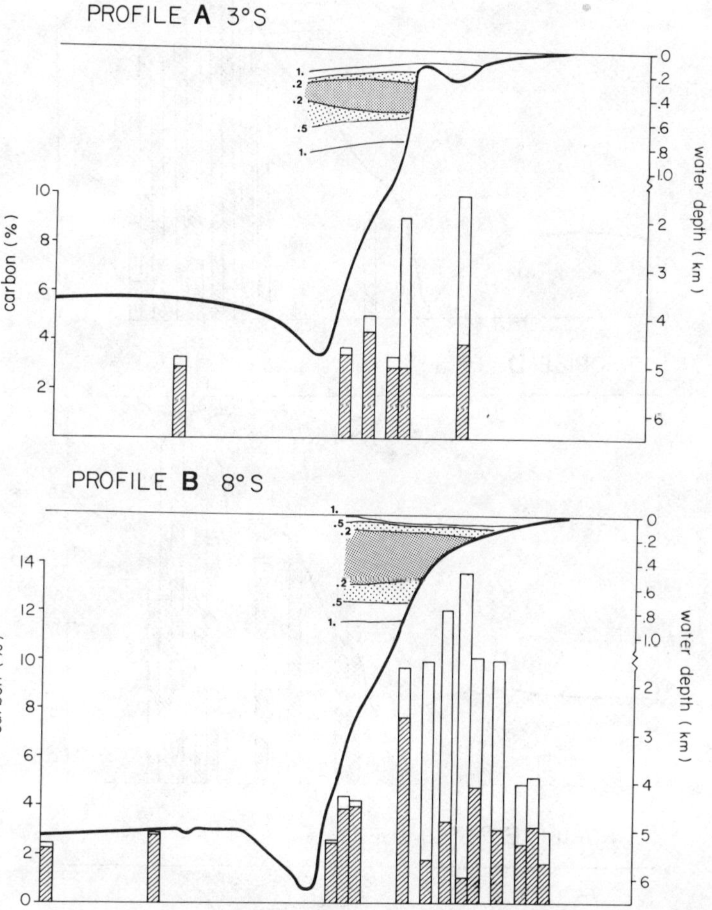

Fig. 4. The vertical offshore distribution of dissolved oxygen and sedimentary organic carbon (hatched histograms) and $CaCO_3$-carbon contents (open histograms) in relationship to the bathymetry of the Peru Margin at approximately 3°-4°S (A), 7°-10°S (B), 11°-12°S (C), 12°-14°S (D), and 15°S (E). The oxygen distributions are contoured from hydrocast data collected during STEP-1 and CUEA sampling programs. Profile letters correspond to data listed in Appendix Table 1. (continued overleaf)

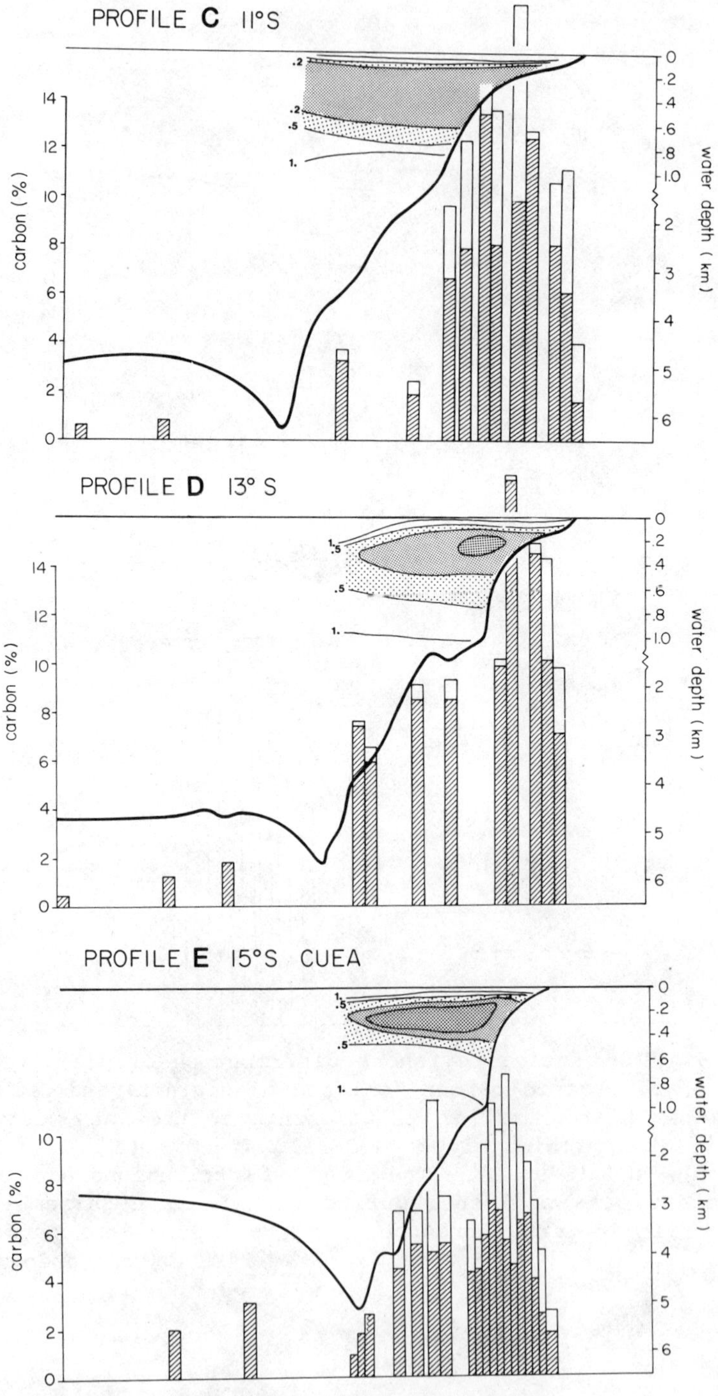

Fig. 4 (continued)

plained as direct consequences of present patterns of biological production. Within latitudes 7°-8°S and 15°-16°S, the surface sediment organic carbon contents are lower in spite of highly fertile waters there. This is related to the interaction between bottom morphology and the Peru Current system at these latitudes.

Barber and Smith (1981) have elegantly developed this latter idea as a point of comparison between the northwest African and Peruvian upwelling ecosystems and it is further refined in Smith (this volume). They argue that off northwest Africa strong onshore and alongshore currents (10-15 $cm \cdot sec^{-1}$) are present next to a shallow bottom, and that the resulting interaction prevents the accumulation of recent fine-grained, organic-carbon-rich sediments across the shelf and upper slope. Off Peru they confine their observations on current to the upper slope at approximately 15°S. There they observe that maximal onshore and alongshore current velocities are at mid-water depths and that bottom currents are "weak or nonexistent," and hence organic debris may accumulate. Applying this same thesis to the entire Peru margin demonstrates why patterns of organic matter concentration differ from north to south. Below a thin surface water layer less than 25 m deep, a strong poleward undercurrent exists off Peru. During times of upwelling this current appears to be strongest in the north and to have its maximum velocity at about 100 m (Brockmann et al., 1980), i.e. near the sea bottom. Therefore, from 7°-10°S (and perhaps further north) the consequences of these flow patterns appear to create a situation similar to those off northwest Africa. That is, a wide shelf (30 km) and shallow water depths result in a bottom that is almost continuously swept clean of new sediments. These currents rework the sediments so that at water depths less than about 400 m they are coarse and well oxidized; the conditions are ideal for carbonates to be concentrated as a type of lag deposit and for phosphorite formation. At greater depths the currents are less intense and this, combined with redeposition of fine sediments from upper slope depths, results in a narrow lens of recent organic-carbon-rich sediments along the lower slope.

In the south, from about 11°-14°S, the shelf is deeper than to the north and the shelf-slope transition is gentle. Here the bottom is not subjected to currents, hence accumulation of fine-grained, organic-carbon-rich debris is facilitated. In the water column two clear water minima centered at approximately 90 m and 350 m suggest the occurrence of two flow centers well off the bottom (Fig. 5 after Kullenberg, 1981) although it is not quite clear why low light scattering (i.e., low particle concentrations) is confined to the apparent core waters of maximum flow. In addition, the synclinal Lima Basin, that apparently began to subside in Late Miocene time (Thornburg and Kulm, 1981), has probably contributed to the accumulation of the materials here, and so may account in part for the organic-carbon-rich, low-carbonate, upper slope mud lens that has been delineated by Krissek et al. (1980).

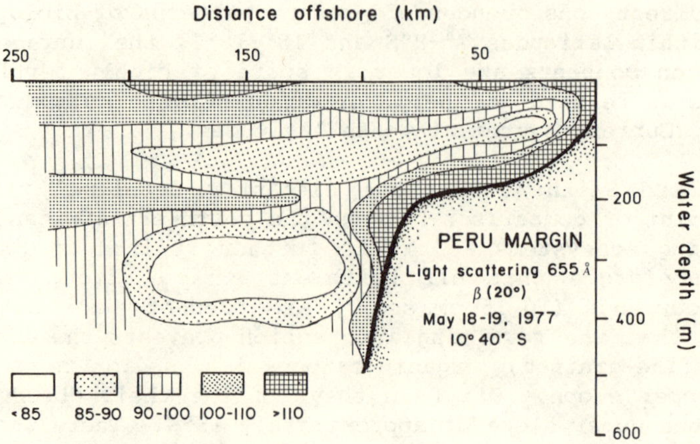

Fig. 5. Light scattering of 655 nm at 10°40'S off Peru, May 18-19, 1977. The clear water minima are thought to correspond to two cores of predominantly poleward alongshore flow. From Kullenberg (1981).

The third center of high biological production of the Peru margin is the region around 15°S. Here also bottom currents are weak and more variable (Brockmann et al., 1980; Friederich and Codispoti, 1981), and the shelf is narrow. Thus, our observation that organic carbon contents are lower and $CaCO_3$-concentrations more erratic than to the immediate north is at first surprising, but finds its explanation in a dilution effect and not low organic carbon input (Ibach, 1982). Utilizing information on actual bulk sedimentation rates along the length of the margin (Table 2), the burial rate of organic carbon in upper slope sediments at approximately 15°S is 9-12 $gC \cdot cm^{-2} \cdot 1000\ y^{-1}$ (Henrichs, 1980; Rowe, in press). These rates are higher by factors of 3-9 than what we estimate for sediments at comparable water depths within the two organic carbon concentration maxima further north. Thus, in units of flux, the true carbon accumulation maximum on the Peru margin may be in the vicinity of 15°S, whereas the carbon concentration maximum is between 11° and 14°S.

The dilution effect we attribute to increased fluxes of fine-grained mineral detritus from local river input or ephemeral runoff, and mineral matter that is advected south and then settles out where the Peru undercurrent weakens. The latter process has also been suggested by Koide and Goldberg (1982) as an explanation for higher fluxes of plutonium nuclides to sediments at 14°39'S compared to 12°02'S.

Particulate settling and poleward flow have two other coupled effects on organic matter accumulation, and these are related to preservation. The preservation of organic matter in marine sediments

Table 2. Accumulation pattern of sediment constituents from the Peru upper slope mud lens facies during selected time intervals ($g \cdot cm^{-2} \cdot [1000y]^{-1}$)

Cores		North 7706-39/-40	7706-41	7706-04	South 7706-36/-37
	Bulk	28. ±5	33. ±5	9. ±3	11. ±3
Present	C-org*	3.3±0.5	6.3±0.5	1.3±0.5	1.6±0.5
± 500 yrs.	$CaCO_3$	0.6±0.1	3.4±0.1	0.2±0.1	0.2±0.05
	Quartz	4.4±0.8	3.7±0.5	1.4±0.5	1.7±0.5
	Bulk	52. ±12	no record	15. ±4	no record
2,500	C-org	4.8±1.		2.0±0.5	
± 500 yrs.	$CaCO_3$	1.0±0.2		0.3±0.2	
	Quartz	11. ±2		2.4±1	
	Bulk	12. ±3	13. ±2	11. ±4	17. ±2
12,000	C-org	0.5±0.1	1.4±0.2	1.1±0.3	0.9±0.1
±1000 yrs.	$CaCO_3$	0.1±0.05	0.3±0.05	0.2±0.05	0.3±0.05
	Quartz	5.0±1	2.9±0.5	1.9±0.5	2.8±0.5
	Bulk	no core	10. ±1	17. ±2	17. ±2
14,500	C-org		1.1±0.1	1.0±0.1	0.8±0.1
± 500 yrs.	$CaCO_3$		1.3±0.1	1.0±0.1	0.4±0.05
	Quartz		3.4±0.5	2.1±0.5	2.6±0.5

* Surface organic carbon accumulation rates are not corrected for further losses via diagenetic regeneration reactions.

is enhanced by low oxygen contents of bottom waters and rapid sedimentation (Müller and Suess, 1979; Demaison and Moore, 1980). Fig. 4 shows the vertical offshore distribution of dissolved oxygen from the north to the south along the Peru coast in relation to the carbon and carbonate profiles in the underlying sediments. At all latitudes a mid-water oxygen minimum is present, but it shoals and becomes more intensified to the south. This condition is the consequence of *in situ* oxygen utilization by bacteria and animals as organic matter settles, and of the poleward two-layered flow (Barber and Smith, 1981). Its effect on sediment composition is seen in the latitudinal organic carbon maxima of the cross-slope surface sediment profiles, as previously discussed, which also tend to intensify and shoal from north to south.

The effect of bulk sedimentation rate on organic carbon preservation can be inferred from Fig. 6. Here the initial dissolved sulfate gradients and sedimentation rates of three of the Peru cores (7706-39, -36, and -04) are plotted with data from other marine localities. Steeper sulfate gradients signify faster rates of sulfate-reduction and more labile organic matter escaping consumption prior to burial and, thus, enhanced preservation. Presumably, sediments from 15°S would plot in the far upper right-hand corner of Fig. 6. This relationship was first observed by Berner (1978) and results from the interaction of SO_4-diffusion, SO_4-reduction by microbial

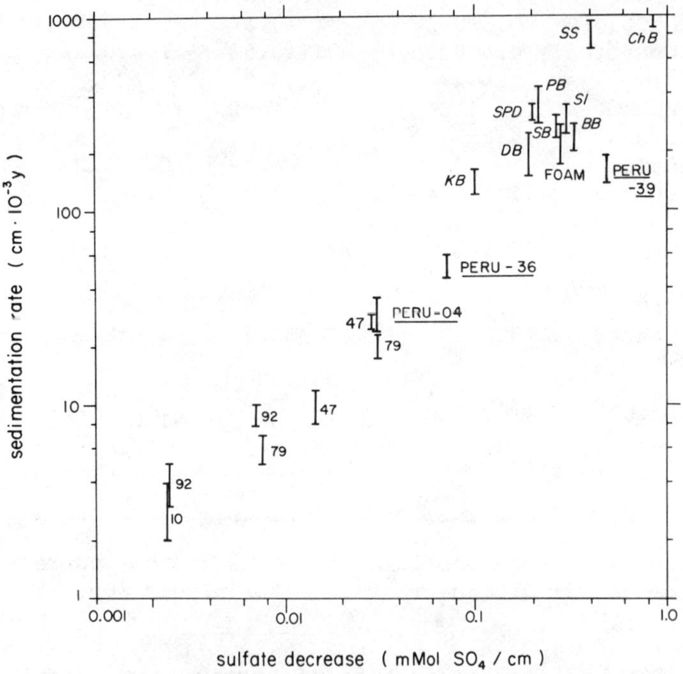

Fig. 6. Dissolved interstitial sulfate gradients and the bulk sedimentation rates of several marine sediments including three Peru sites (7706-39, -04, -36). This empirical relationship, first described by Berner (1978), illustrates the reducing chemical environment of the Peru cores and the influence of sedimentation rate on the preservation of organic matter. Bar lengths represent the uncertainties in the sedimentation rates. The data 10, 92, 47, and 79 are for sediments from northwest Africa (Hartmann et al., 1973, 1976). Two bars for one core represent Holocene (lower sedimentation rate) and Pleistocene (higher sedimentation rate) sections. The other data are for sediments from Kieler Bucht (KB) (Hartmann and Nielsen, 1969); Danziger Bucht (DB) and Bornholm Basin (BB) (Suess, 1976); Long Island Sound (FOAM), Saanich Inlet (SI), Chesapeake Bay (ChB), Somes Sound (SS), Santa Barbara Basin (SB), Pescadero Basin (PB), and San Pedro Basin (SPD) (Berner, 1978 and references cited therein).

decomposition, and bulk sedimentation rates. Independent of these considerations, Müller and Suess (1979) showed that a ten-fold increase in bulk sedimentation rate roughly increases the preservation rate of organic matter by a factor of 20; the preservation rate here being the percentage of primary produced organic matter which escapes recycling and is buried. Thus, oxygen contents of the water column and sedimentation rates, both of which are responses to primary particle fluxes, continental runoff and currents, account for the maximum in organic carbon accumulation located beneath the persistent center of primary production at about 15°S, and not below either of the other two centers at 7°-8°S and 11°-12°S.

In summary then, the present-day regional pattern of sedimentary organic carbon along the Peru continental margin is characterized by two areas of preferential accumulation: the outer shelf-upper slope region at about 100-400 m of water depth between 11°S and 16°S, and the lower continental slope (>2000 m). The middle slope of the margin lacks significant recent sediment accumulation. The high organic carbon contents of recent deposits of the upper slope originate predominantly from high biological production in response to persistent coastal upwelling centers at fixed localities. Two of these centers

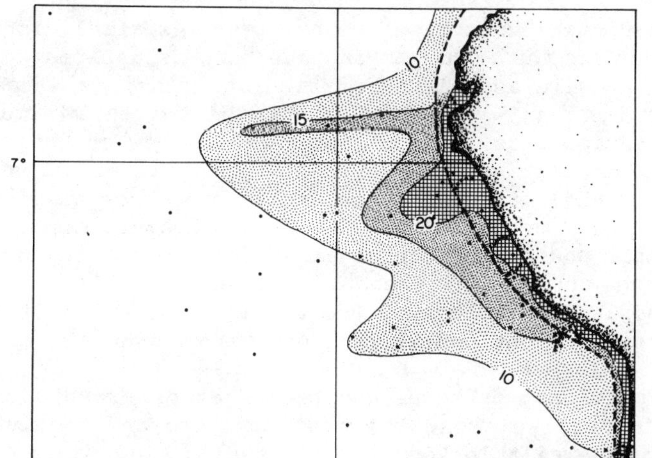

Fig. 7. Quartz distribution in surface sediments of the Peru continental margin and adjoining Nazca Plate; the contours are in wt-% quartz on a carbonate- and opal-free basis. The general pattern reflects eolian and riverine input from the South American continent and distribution by prevailing surface circulation. However, the prominent westward maximum of quartz between 11°-7°S may reflect winnowing by the undercurrent flow from the shallow shelf of the margin and subsequent cross-trench transport.

and resulting organic carbon concentrations (>10% dry weight) are found between 11°S and 14°S. At 15°S, the third site of persistent upwelling, subsurface flow that is predominantly poleward relaxes at slope depths, and the maximum organic carbon accumulation rates result from concurrent input of fine-grained terrigenous debris. At 6°-10°S, where a similarly high productivity center from coastal upwelling is located, the wide shelf (30 km) and shallow water depths cause the bottom to be almost continuously reworked by strong onshore and poleward current flow, thereby preventing the accumulation of fine-grained, organic-carbon-rich deposits. In fact, this reworking mechanism might preferentially transport organic-carbon-rich material to the lower slope region of the margin and be responsible as well for a significant cross-trench transport of terrigenous quartz at this latitude (Fig. 7). Essentially, the interaction of the Peru Current system with shelf-slope morphology is a very important controlling factor for the present distribution of organic matter and most inorganic sediment components as well.

PALEOPRODUCTIVITY AND ORGANIC MATTER ACCUMULATION OFF PERU DURING THE LATE QUATERNARY

Deciphering Temporal Patterns on the Upper Slope

Three essential controls for the present spatial distribution of organic matter on the Peru margin have been delineated: (1) primary productivity and its associated particulate flux, (2) organic matter decomposition and its relationship to the oxygen minimum zone and bulk sedimentation rate, and (3) interaction between shelf-slope morphology and the Peru current system. These interactions are local and fall under the category of meso-scale processes as used by Barber and Smith (1981) in describing upwelling ecosystems. On a global scale, climatic phenomena such as are reflected in sea level fluctuations, warming and cooling trends, or changes in the strength of meridional winds, also influence chemical cycling and sediment distribution patterns in continental margin environments.

To decipher past patterns and processes of organic carbon sedimentation off Peru we chose the most complete and highly resolvable sediment record available to us. Depicted in Fig. 8, it is the age/facies pattern of the upper slope mud lens between approximately 11°S and 14°S. Erosional surfaces and non-deposition to the north and inadequate sample coverage to the south excluded these areas from detailed consideration.

The three time-stratigraphic sections are based on cores collected using a 15 cm-diameter square barrel Kasten corer; Section 8A is located NW-SE along the present day depositional axis of the mud lens, and Sections 8B and 8C are located NE-SW at approximately 11°20'S. At two of the seven sites represented by these sections,

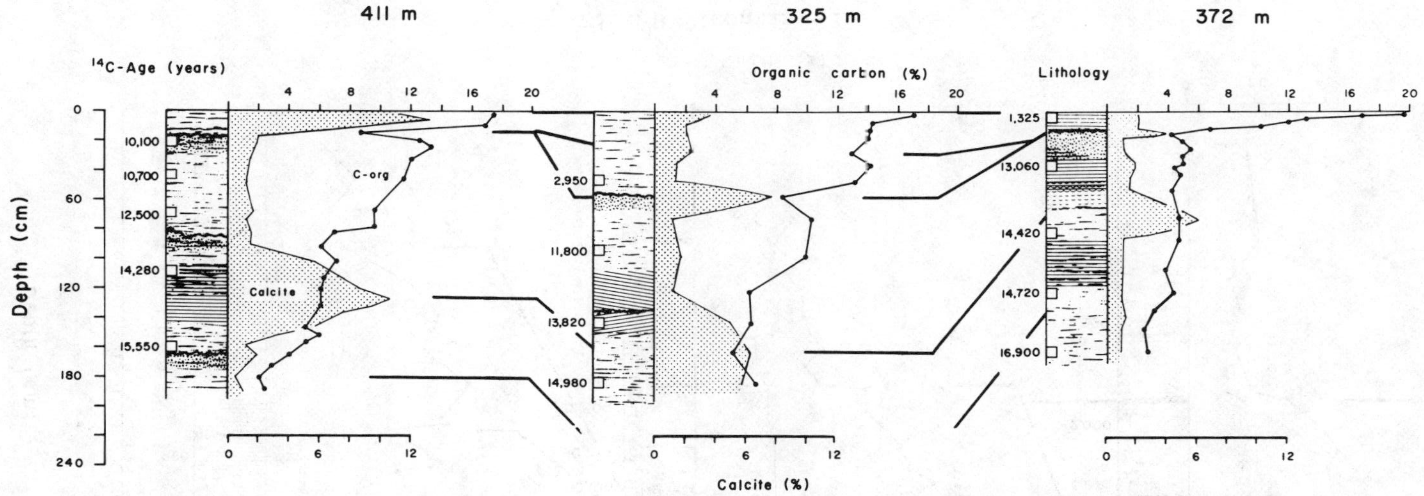

Fig. 8. The stratigraphies and lithologies of cores from the Peru continental margin. (A) represents sedimentation NW-SE along approximately 300 km of the present day depositional axis of the upper slope mud lens facies. (B) extends NE-SW at approximately 11°20'S and illustrates the foreset bed configuration of the upper slope. (C) completes the NE-SW profile downslope and across the trench. The mid-slope is without any recent sediment cover.

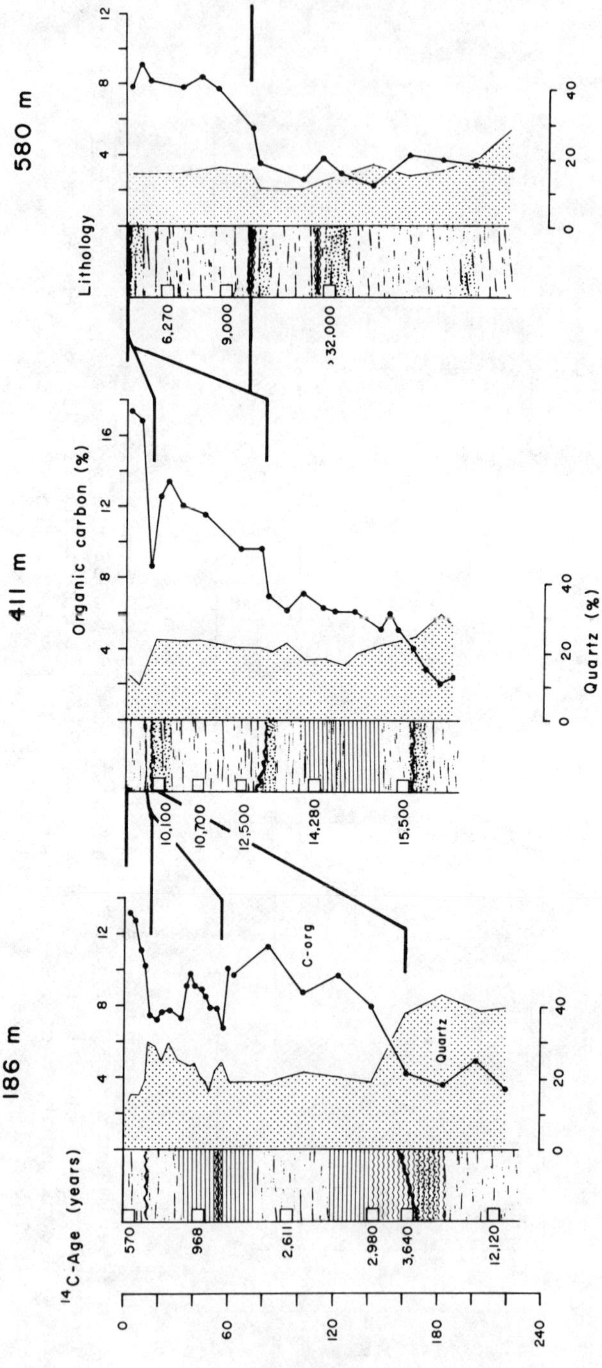

Fig. 8 (continued)

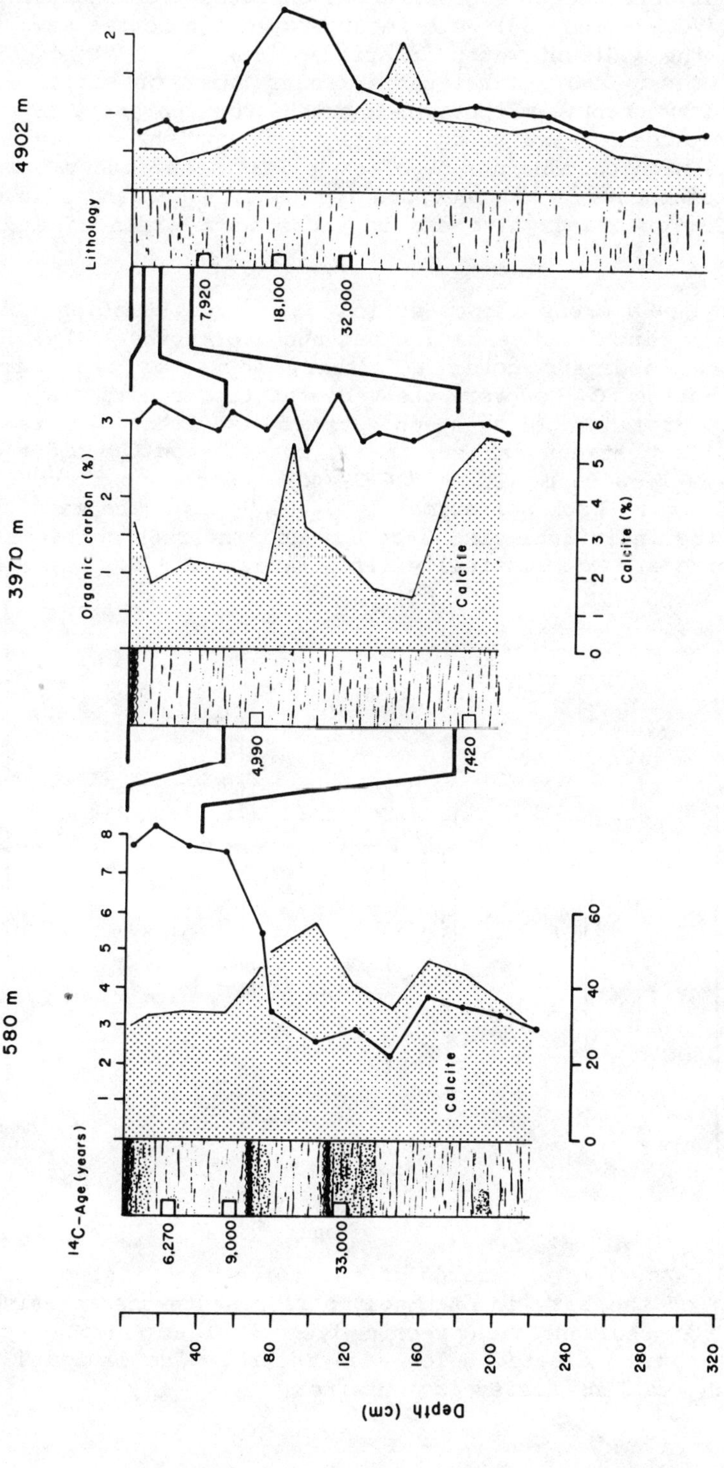

Fig. 8 (continued)

compositional data and ^{210}Pb sedimentation rates from shorter Reineck box cores (7706-39 and -36) were incorporated for better sample resolution near the sediment-water interface. The NW-SE alongshore section shows a number of correlatable facies units of variable thickness bounded by unconformities which extend for approximately 300 km. These unconformities are recognizable by abrupt changes in lithologies such as high concentrations of fish debris and other coarse sedimentary components. ^{14}C-ages and compositional changes in organic carbon, $CaCO_3$ or quartz also aid in cross-correlation of individual units.

In the NE-SW cross slope Section 8B, several facies units are present which pinch out seaward along unconformities. The mid-slope has no recent sediment cover. Further, downslope and across the trench in Section 8C, however, the sedimentation record again becomes continuous. From the ^{14}C-ages and estimated sedimentation rates, the time resolution between samples in all cores is on the order of 500-1,000 years. The 186 m site (7706-39/40) has the most complete recent record, going back approximately 3,600 years. The magnitude and quality of the individual time sections and intervening hiatuses for all the cores are schematically illustrated in Fig. 9. Note core

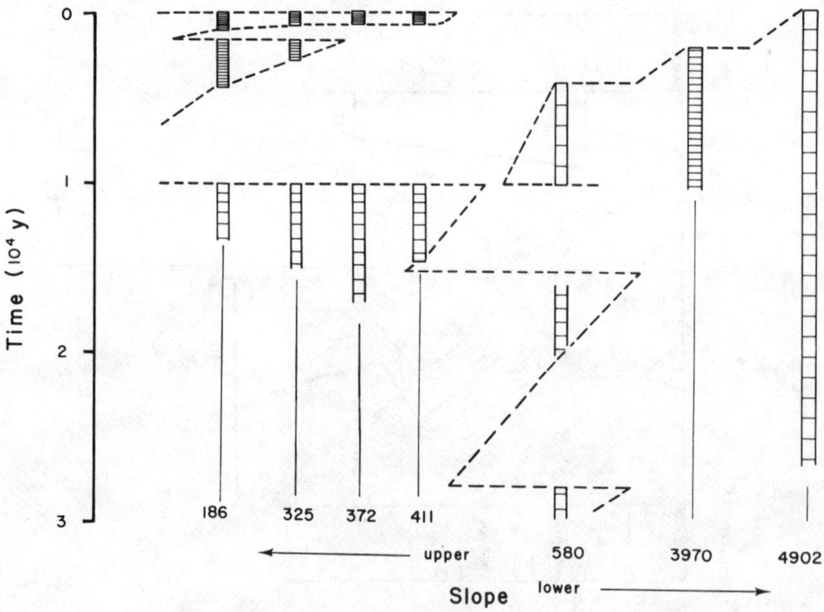

Fig. 9. Age/facies records of the cores across slope transect. The spacing of the bars in each column represents the relative resolution of its sedimentation record from 5-10 cm sample spacings. Water depths are indicated below each record. The landward boundaries of each mud lens facies are inferred.

7706-44, at 580 m of water depth and with a surface age of approximately 6,000 y B.P., is not part of the upper slope mud lens facies, but rather contains the time-stratigraphic record from between 6,000 and 10,000 years ago which is missing from the shallower sections but is generally contained in the lower slope record. Its non-depositional surface is probably generated by high-velocity current pulses related to the poleward flow (>10 cm/sec) at these depths (Brink, Allen and Smith, 1978; Brockmann et al, 1980). Resuspension of particles along the middle slope, combined with offshore advection and sinking, may further explain the observed increases in organic carbon contents and preferred sedimentation on parts of the lower slope; e.g., at 3970 m.

Interpreting the Temporal Record: Accumulation Rates During Selected Time Intervals

Based on ^{210}Pb and ^{14}C-sedimentation rates, compositional data and physical properties, the bulk accumulation rates and individual component accumulation rates of organic carbon, $CaCO_3$ and quartz in cores 7706-39/40, -41, -04, and -36/37 have been calculated and are averaged over select time slices as listed in Table 2. Estimated uncertainties which reflect variabilities in both real fluxes and methodology are also listed.

Having shown that the present accumulation pattern on the slope is related to surface distribution processes, it remains an intriguing question to determine why certain sections of the sediment record are preserved and others not, and why component accumulation rates fluctuate at these sites.

It seems that the hiatuses in the sedimentation record off Peru are caused by currents. Two independent observations lead to this conclusion. First, there is a similarity in texture and composition between downcore sediment intervals that mark discontinuity boundaries (Figs. 8A, 8B, 8C) and surface sediments in regions that are being swept by bottom currents today. Table 3 compares the textures (% sand, silt, clay), organic carbon contents, $CaCO_3$-contents and inorganic (HCℓ extractable) phosphorus contents of sediments from reworked, current-swept surfaces or discontinuities to undisturbed surface sediments and similar downcore intervals. As stated by DeVries (1980) in his study of the nekton remains in these cores:

> "The coarsest sediments tend to have lower values of organic carbon. The coarseness of these intervals can often be attributed to greater concentrations of fish debris."

We would add that phosphorite nodules and benthic foraminifers may also contribute to the sand and silt fractions of these reworked deposits.

Table 3. The textural and compositional data of selected sediment samples that show evidence of (1) bottom scouring by currents and (2) deposition in "quiet" environments. The scope of this table was limited by the availability of textural data

Core, Sample Interval	Water Depth (m)	Texture % Sand	% Silt	% Clay	Bulk Composition % Org-C	% $CaCO_3$	% Inorg-P
Reworked surface sediments and discontinuities							
7706-53, 0-3	414	81	17	2	3.3	72	1.11
7706-57, 0-5	192	70	14	16	3.2	67	0.42
7706-44, 0-5	580	24	49	27	7.8	31	1.20
7706-45, 0-1	810	44	34	22	6.7	25	0.66
7706-40, 180-185	186	17	67	16	3.6	2.4	0.56
7706-41, 185-191	411	11	54	35	2.5	1.2	0.27
Undisturbed surface and downcore sediments							
7706-03, 0-3	304	2	38	60	14.5	2.9	0.36
7706-37, 0-5	370	5	35	60	13.2	2.8	0.61
7706-40, 0-5	186	4	28	68	13.7	1.7	0.13
7706-40, 100-105	186	2	28	70	8.4	0.5	0.14
7706-41, 4-8	411	12	37	51	18.6	11	0.52
7706-41, 125-130	411	4	19	77	6.0	10	0.21

Under the influence of bottom currents, organic carbon contents of sediments are reduced due to resuspension that reintroduces fine particulate organic matter into the benthic boundary layer thereby increasing its residence time and the efficiency of benthic decomposition processes, or by causing material to be physically redistributed and winnowed. Inorganic phosphorous contents reflect the abundances of phosphorites and fish debris (Suess, 1981). To be in high abundance these require a significant local fish population and selective winnowing.

The concentrations of $CaCO_3$ in these deposits are less predictable. Surface sediments from the current-swept region between 8° and 10°S contain high concentrations of $CaCO_3$ due most often to high standing stocks of benthic foraminifers species belonging to the genus *Uvigerina* (Hutson, pers. comm.; Wefer, Dunbar and Suess, this volume). In the coarse sediments from the area of the mud lens, *Uvigerina* are also present but in lower numbers. They occur together with several species of another genus, *Bolivina* (Hutson, pers. comm.). The interesting sidelight of this mixed assemblage is that *Bolivina* are usually associated with bottom waters of extremely low oxygen concentrations whereas *Uvigerina* cannot tolerate low oxygen contents. Thus these mixed assemblages of *Bolivina* and *Uvigerina* may be the result of fluctuating oxygen conditions at sediment surfaces to the south. Oxygen fluctuations could be generated by changes in current strength and bottom turbulence, or changes in the intensity of upwelling.

The second observation that provides evidence for past episodes of bottom current scouring, is the time continuity of over 300 km NW-SE of the hiatuses in cores 7706-41, -04, and -37 (Figs. 8A, 8B, 8C). Hiatuses that occur on account of slumping, another possible explanation, would not be expected to be time transgressive and to cover such a large areal extent, whereas periods of non-deposition triggered by strong alongshore (or on-offshore) currents would be.

Accordingly, then, if currents impinging on certain parts of the slope have caused hiatuses, this would also imply that during periods when there is an extensive sediment record preserved such bottom currents were either not present or were located somewhere off the slope bottom in the water column. Before speculating on potential oceanographic and climatic controls which may have led to these conditions, we will more closely examine the accumulation rate changes of bulk and individual sediment components with time (Table 2).

The accumulation record of sediment constituents in Table 2 demonstrates that the sedimentation conditions in this region have varied latitudinally throughout the past 15,000 years. Highest bulk accumulations occur in the north (at ∿11°S) today and did so 2,500 ± 500 years ago but diminish to the south (∿13°S). Quartz accumulation patterns exhibit the same trend and suggest that the control may be precipitation and river discharge which at present show similar north-south gradients (Zuta and Guillén, 1970; Krissek et al., 1980). Alternatively, primary productivity could be the major mechanism influencing accumulation in this region because the present day centers of primary production occur between these latitudes. If, as suggested by McCave (1975), Chesselet (1980) and Scheidegger and Krissek (this volume), the process of zooplankton and other marine grazers feeding and forming feces accelerates the transport of mineral as well as organic matter to the sediments, then the detrital flux should increase as well where these organisms have an abundant food source and constitute dense populations, i.e., in areas of high primary production.

At 12,000 ± 1,000 years and 14,500 ± 500 years ago bulk accumulation rates appear to have been more uniform over the latitudes of the mud lens facies and significantly lower in the north than they are today. Organic carbon accumulation follows the same pattern (Table 2). It seems that at least some of the numerous factors that enhance sediment and carbon accumulation in the vicinity of 11°S today were not at work before about 11,000 years B.P., but which ones?

The paleo-climate of the subantarctic and equatorial Pacific Ocean underwent a major change between about 14,000 and 11,000 years ago. Earlier, at the end of the last Pleistocene glacial advance, the earth was extremely cold and intensified atmospheric circulation is indicated by quartz distributions off several coasts and carbonate distributions in the equatorial region (Emiliani, 1971; Molina-Cruz, 1978; Thiede, 1979). DeVries and Schrader (1981) suggest that these circulation changes may have been accompanied by a shift in the location of the Intertropical Convergence southward of its present day position, and Salinger (1981) argues for a weakening of the Southern Oscillation. Both suggestions imply a decreased rate of primary productivity off Peru during Pleistocene glacial stages, although the reasoning is not exactly the same.

First, DeVries and Schrader (1981) reason that if the Intertropical Convergence Zone was shifted southward, this would have lead to more frequent intrusions of equatorial waters along the South American coast. These intrusions, called El Ninõs today, lead to the appearance of warm, less nutrient-rich water, lower primary and secondary production, and torrential rains off Peru. Whether upwelling actually decreases during these times or is only masked by the thickness of the warm surface layer is controversial (Wyrtki, 1975). The second argument, a weaker Southern Oscillation, however, would mean less intense coastal upwelling, warmer associated surface waters and again lower productivity (Quinn, 1971).

As a consequence of either of these two climatic developments, we speculate that the productivity and sedimentation maxima associated with cold water upwelling was restricted to waters south of about 13°S prior to 11,000 y B.P. The added importance of changes in precipitation and subsequent changes in input of terrigenous matter causing these accumulation variations is uncertain. From palynological studies it appears that the Northern Andes were much drier during the latter part of the last glacial (van der Hammen, 1974).

Between 14,000 and 11,000 B.P. the southern climate responded to a global warming trend accompanied by major glacial recessions and a rise in worldwide sea level (Mercer, 1976). Table 2 and Fig. 8 show that this transition had a pronounced effect on sedimentation off Peru. If we speculate that with global warming came the establishment of intense *cold* water upwelling and present day current systems off Peru, then it could be concluded that after about 11,000 y B.P. a relatively persistent poleward undercurrent system was present and actively removed sediment along a relatively shallow shelf. This configuration of the poleward undercurrent system, analogous to that off northwest Africa today, then produced the major hiatus between approximately 11,000 and 4,000 years ago. But when sea level rose to near its maximum height around 6,000 years ago (Fairbridge, 1976), the centers of poleward and onshore flows must have gradually become detached from the bottom allowing renewed sediment accumulation along the present-day upper slope. This interpretation accounts for the long time concordant hiatuses in sedimentation in the upper slope cores (7706-39/40, -41, -04, and -36/37), the non-deposition of new sediment after about 6,000 y B.P. in core 7706-44 located at greater water depth, and recent patterns of higher bulk and carbon accumulation in the north.

SUMMARY

Parallels between spatial patterns of bulk sediment and organic carbon accumulation today, and past records from the Peru upper slope mud lens facies provide a rationale for linking local environmental controls to larger scale global climatic conditions. Maximum bulk sediment and organic carbon accumulation on the Peru margin occur today at approximately 15°S on a narrow shelf (<5 km) and upper slope. Here at extremely high rates of sedimentation, dilution of organic matter by terrigenous constituents in the sediment record is effective. The high rates of sedimentation result from high surface productivity, relaxation of the Peru undercurrent at bottom depth, an intense oxygen minimum, and high local fluvial discharge. Higher organic carbon *concentrations* but lower *accumulation rates* occur to the north in a region between 11°S and 14°S where the margin is much broader (∿15 km) and the upper slope mud lens facies most prominently developed. Throughout this region surface productivity is high, but the oxygen minimum becomes shallower and more depleted to the south. Along the middle slope, in the region of the mud lens facies, and further north little or no sediment is presently accumulating. This can be attributed to the effects of bottom current activity in dominantly poleward and onshore directions. In the past, the evidence shows that the average strength and/or position of this flow pattern may have fluctuated relative to sea bottom depth in response to changing sea-level and global wind patterns. Another feature of the past sediment record is that bulk sediment and organic carbon accumulation rates were lower north of 13°S at the end of the last Pleistocene glacial than they are today. We attribute this primarily to lower primary productivity due to warmer surface water, but changes in precipitation/runoff and the chemical environment of the water may have also been significant.

ACKNOWLEDGEMENTS

We would like to thank the officers and crew of the RV "Wecoma" who contributed to the success of the sampling portion of this study. Sediment analyses were performed in part by C.A. Ungerer and P. Price. ^{14}C ages were determined by Radiocarbon, Ltd. (Lampasas, Texas). This research was supported by the Office of Naval Research, Grant N00014-76C-0067.

REFERENCES

Barber, R.T. and Smith, R.L., 1981, Coastal upwelling ecosystems, in: "Analysis of Marine Ecosystems," A.R. Longhurst, ed., Academic Press, 31-68.

Berner, R.A., 1978, Sulfate reduction and the rate of deposition of marine sediments, Earth and Planetary Science Letters, 37:492-498.

Bremner, J.M., 1960, Determination of nitrogen in soil by the Kjeldahl method, Journal of Agricultural Science, 55:11-33.

Brink, K.H., Allen, J.S. and Smith, R.L., 1978, A study of low frequency fluctuations near the Peru coast, Journal of Physical Oceanography, 8:1025-1041.

Brockmann, C., Fahrbach, E., Huyer, A., and Smith, R.L., 1980, The poleward undercurrent along the Peru coast: 5-15°S, Deep-Sea Research, 27:847-856.

Busch, W.H. and Keller, G.H., 1981, The physical properties of Peru-Chile continental margin sediments--The influence of coastal upwelling on sediment properties, Journal of Sedimentary Petrology, 51:705-719.

Chesselet, R., 1980, Modes of settling and organic input to the sediment seawater interface: a review, Collogues Internationaux du C.N.R.S., Centre National de la Recherche Scientifique, Paris, 293:27-34

Demaison, G.J. and Moore, G.T., 1980, Anoxic environments and oil source bed genesis, Organic Geochemistry, 2:9-31.

DeMaster, D.J., 1979, "The Marine Budget of Silica and ^{32}Si," Ph.D. Dissertation, Yale University, New Haven, 308 pp.

DeVries, T.J., 1980, "Nekton Remains, Diatoms, and Holocene Upwelling off Peru," M.S. Thesis, Oregon State University, 85 pp.

DeVries, T.J. and Schrader, H., 1981, Variation of upwelling/oceanic conditions during the latest Pleistocene through Holocene off the central Peruvian coast: A diatom record, Marine Micropaleontology, 6:157-167.

Ellis, D.B., 1972, "Holocene Sediments of the South Atlantic Ocean: The Calcite Compensation Depth and Concentrations of Calcite, Opal and Quartz," M.S. Thesis, Oregon State University, 77 pp.

Emiliani, C., 1971, The amplitude of Pleistocene climatic cycles at low latitudes and the isotopic composition of glacial ice, in: "The Late Cenozoic Glacial Ages," K.K. Turekian, ed., Yale University Press, 183-198.

Fairbridge, R.W., 1976, Effects of Holocene climatic change on some tropical geomorphic processes, Quaternary Research, 6:529-556.

Friederich, G.E. and Codispoti, L.A., 1981, The effects of mixing and regeneration on the nutrient content of upwelling waters off Peru, in: "Coastal Upwelling," F.A. Richards, ed., Coastal and Estuarine Sciences 1, American Geophysical Union, Washington, 221-227.

Guillén, O., de Mendiola, B.R. and de Rondán, R.I., 1973, Primary productivity and phytoplankton in the coastal Peruvian waters, in: "Oceanography of the South Pacific," 1972 New Zealand National Commission for UNESCO, Wellington, 405-418.

Hafferty, A.J., Codispoti, L.A. and Huyer, A., 1978, CUEA Data Report 45, Joint II Ref. M78-48, University of Washington, 779 pp.

Hartmann, M. and Nielsen, H., 1969, δ^{34}S-werte in rezenten Meeressedimenten und ihre Deutung am Beispiel einiger Sedimentprofile aus der westlichen Ostsee, Geologische Rundschau, 58:621-655.

Hartmann, M., Müller, P., Suess, E., and Van der Weijden, C.H., 1973, Oxidation of organic matter in recent marine sediments, "Meteor" Forschungs-Ergebnisse, C12:74-86.

Hartmann, M., Müller, P., Suess, E., and Van der Weijden, C.H., 1976, Chemistry of late Quaternary sediments and their interstitial waters from the NW African continental margin, "Meteor"Forschungs-Ergebnisse, 24:1-67.

Heath, G.R., Moore, T.C. and Dauphin, J.P., 1977, Organic carbon in deep-sea sediments, in: "The Fate of Fossil Fuel CO_2 in the Oceans," N.R. Andersen and A. Malahoff, eds., Plenum Press, New York, 605-625.

Henrichs, S.M., 1980, "Biogeochemistry of Dissolved Free Amino Acids in Marine Sediments," Ph.D. Dissertation, Woods Hole Oceanographic Institution, WHOI 80-39, 253 pp.

Ibach, L.E.J., 1982, Relationship between sedimentation rate and total organic carbon content in ancient marine sediments, American Association of Petroleum Geologists, Bulletin, 66:170-188.

Koide, M. and Goldberg, E.D., 1982, Transuranic nuclides in two coastal marine sediments off Peru, Earth and Planetary Science Letters, 57:263-277.

Krissek, L.A., Scheidegger, K.F. and Kulm, L.D., 1980, Surface sediments of the Peru-Chile continental margin and the Nazca plate, Geological Society of America, Bulletin, 91:321-330.

Kullenberg, G., 1981, A comparison between distributions of suspended matter in the Peru and Northwest African upwelling areas, in: "Coastal Upwelling", F.A. Richards, ed., Coastal and Estuarine Sciences 1, American Geophysical Union, Washington, 282-290.

Kulm, L.D., Dymond, J., Dasch, E.J., and Hussong, D.M., eds., 1981, "Nazca Plate: Crustal Formation and Andean Convergence," Geological Society of America, Memoir 154, 824 pp.

McCave, I.N., 1975, Vertical flux of particles in the ocean, Deep-Sea Research, 22:491-502.

Mercer, J.H., 1976, Glacial history of southernmost South America, Quaternary Research, 6:125-166.

Molina-Cruz, A., 1978, "Late Quaternary Oceanic Circulation Along the Pacific Coast of South America," Ph.D. Thesis, Oregon State University, Corvallis, 246 pp.

Müller, P.J., 1977, C/N ratios in Pacific deep-sea sediments: Effect of inorganic ammonium and organic nitrogen compounds sorbed by clays, Geochimica et Cosmochimica Acta, 41:765-776.

Müller, P.J. and Suess, E., 1979, Productivity, sedimentation rate, and sedimentary organic matter in the oceans - I. Organic carbon preservation, Deep-Sea Research, 26:1347-1362.

Pak, H., Codispoti, L.A. and Zaneveld, J.R.V., 1980, On the intermediate particle minima associated with oxygen-poor water off western South America, Deep-Sea Research, 27:783-798.

Quinn, W.H., 1971, Late Quaternary meteorological and oceanographic developments in the equatorial Pacific, Nature, 299:330-331.

Reimers, C.E., 1982, Organic matter in anoxic sediments off central Peru: relations of porosity, microbial decomposition and deformation properties, Marine Geology, 46:175-197.

Rowe, G.T., 1979, Coastal Upwelling Ecosystems Analysis (CUEA) Data Report 65. Sediment data from short cores during JOINT II off Peru, 57 pp.

Rowe, G.T., in press, Benthic production and processes off Baja California, Northwest Africa and Peru, in: "Symposium on Bio-productivity of Upwelling Ecosystems," Moscow.

Salinger, M.J., 1981, Paleoclimates north and south, Nature, 291: 106-107.

Suess, E., 1976, Nutrients near the depositional interface, in: "The Benthic Boundary Layer", I.N. McCave, ed., Plenum Press, New York, 57-80.

Suess, E., 1981, Phosphate regeneration from sediments of the Peru continental margin by dissolution of fish debris, Geochimica et Cosmochimica Acta, 45:577-588.

Thiede, J., 1979, Wind regimes over the late Quaternary southwest Pacific Ocean, Geology, 7:259-262.

Thornburg, T. and Kulm, L.D., 1981, Sedimentary basins of the Peru continental margin: structure, stratigraphy, and Cenozoic tectonics from 6°S to 16°S latitude, Geological Society of America, Memoir, 154:393-422.

Van der Hammen, T., 1974, The Pleistocene changes of vegetation and climate in tropical South America, Journal of Biogeography, 1:3-26.

Wooster, W.S., 1961, STEP-1 Expedition, Part 1 Physical and Chemical Data, SIO Ref. 61-9.

Wyrtki, K., 1975, El Niño - the dynamic response of the Equatorial Pacific Ocean to atmospheric forcing, Journal of Physical Oceanography, 5:572-584.

Zuta, S. and Guillén, O., 1970, Oceanografia de las aguas costeras del Peru, Instituto del Mar del Peru, Boletin, 2:161-323.

Appendix I: Locations and characteristics of surface sediment samples from the Peru continental margin and adjacent Nazca Plate.

Core or Station #	Latitude	Longitude	Water Depth (m)	Type[2] Sampler	Sediment Description	Surface Composition (%) Org-C	Total-N	$CaCO_3$	Qtz	Data Source
Profile A										
7706-76	3°34.7'S	81°00.1'W	365	RB	olive gray silty clay	3.92	0.45	9.7	21	This study
7706-78	3°29.0'S	81°17.2'W	539	RB	grayish brown foram sand	2.99	0.34	50.6	12	This study
7706-74	3°45.4'S	81°24.3'W	713	RB	olive gray well compacted clay	2.93	0.31	3.0	22	This study
7706-73	3°29.4'S	81°29.0'W	2116	K	dark olive gray silty sand	4.14	0.49	5.6	13	This study
7706-72	3°31.2'S	81°38.5'W	3601	K	olive gray silty clay	3.35	0.48	2.9	13	This study
7706-71	3°40.9'S	82°30.3'W	3600	G	gray clay	2.88	0.33	3.1	11	This study
Profile B										
30928	7°39.7'S	79°31.9'W	33	G	--	1.67	0.22	11.4	--	Rowe (1979)
30933	7°45.2'S	79°42.8'W	95	G	--	3.17	0.40	16.4	--	Rowe (1979)
30947	7°07.7'S	80°16.4'W	108	G	--	2.36	0.29	21.2	--	Rowe (1979)
7706-57	7°55.7'S	80°10.9'W	192	RB	green foram sand with shell fragments	3.20	0.44	67.5	8	This study
30938	7°51.7'S	80°17.9'W	208	G	--	4.86	0.67	44.0	--	Rowe (1979)
7706-51	9°44.6'S	79°24.3'W	259	RB	Green foram sand	1.11	0.15	86.8	2	This study
7706-52	9°46.0'S	79°25.0'W	352	RB	Green foram sand with fish debris and phosphorites	--	--	--	--	This study
7706-53	9°47.0'S	79°26.5'W	414	RB	Green foram sand with fish debris and phosphorites	3.32	0.49	72.4	6	This study
7706-58	8°03.4'S	80°24.3'W	486	RB	Green foram sand	1.77	0.23	67.1	2	This study
7706-60	8°03.7'S	80°25.9'W	830	RB	olive gray silty clay	7.75	0.92	16.7	17	This study
7706-61	8°03.7'S	80°25.9'W	838	K	olive gray silty clay	6.02	0.77	19.6	21	This study
7706-62	8°13.0'S	80°47.7'W	2670	K	olive gray silty clay	3.93	0.50	1.8	10	This study

Appendix I (cont.)

Core or Station #	Latitude	Longitude	Water Depth (m)	Type Sampler	Sediment Description	Org-C	Total-N	$CaCO_3$	Qtz	Data Source
Y71-8-83	7°40.0'S	81°02.2'W	3103	MG	olive gray clay	3.94	0.54	1.1	20	This study
7706-63	8°16.9'S	80°55.4'W	4513	K	olive gray silty clay	2.51	0.39	1.3	19	This study
Y71-8-76	8°07.0'S	81°36.0'W	5122	MG	olive gray silty clay	2.81	0.39	0.9	25	This study
7701-64	8°49.0'S	81°56.6'W	4404	K	dark grayish brown	2.24	0.29	1.6	18	This study
Profile C										
30883	11°58.5'S	77°11.3'W	32	G	--	1.82	0.23	15.7	--	Rowe (1979)
30889	11°58.2'S	77°20.3'W	110	G	--	6.09	0.81	42.6	--	Rowe (1979)
7611	~10°58.0'S	~78°00.0'W	~150	--	--	8.13	--	13.9	--	*Delgado (pers. commun.)
7706-39	11°15.1'S	77°57.4'W	186	RB	dark olive gray organic rich silty clay	13.2	1.60	2.4	14	This study
7706-40[1]	11°15.3'S	77°57.8'W	186	K	same as above	13.7	1.56	1.7	13	This study
20914	10°29.9'S	78°33.2'W	290	G	--	9.68	1.13	67.9	--	Rowe (1979)
30892	12°08.5'S	77°37.4'W	340	G	--	7.85	0.97	47.1	--	Rowe (1979)
7706-42	11°20.6'S	78°07.0'W	411	RB	over penetrated olive silty clay with fish debris	13.5	1.54	4.2	18	This study
7706-41[1]	11°20.6'S	78°07.0'W	411	K	black silty clay with fish debris	19.6	2.29	9.8	11	This study
7706-43	11°24.6'S	78°13.8'W	584	RB	black (benthic) foram silty clay	7.94	0.90	37.8	13	This study
7706-44[1]	11°24.6'S	78°13.8'W	580	K	black (benthic) foram sand with pteropods and fish debris	7.83	1.01	30.6	15	This study
7706-45	11°26.6'S	78°17.2'W	810	RB	black foram sand with fish debris and Mn nodules	6.69	0.81	25.3	17	This study
7706-47	11°40.2'S	78°25.8'W	1500	RB	black silty clay with forams, fish debris and worm tubes	1.93	0.24	3.1	29	This study

Appendix I (cont.)

Core or Station #	Latitude	Longitude	Water Depth (m)	Type Sampler	Sediment Description	Surface Composition (%) Org-C	Total-N	$CaCO_3$	Qtz	Data Source
7706-49	11°16.6'S	79°05.9'W	3970	K	dark olive gray silty clay	3.23	0.44	1.7	21	This study
7706-50	11°40.7'S	79°57.9'W	4902	K	brown clay	0.78	0.18	0.6	14	This study
V19-36	11°59.0'S	81°31.0'W	4731	P	--	0.57	0.11	0.1	14	This study
Profile D										
30878	13°00.0'S	76°47.0'W	155	G	--	7.11	0.88	22.9	--	Rowe (1979)
30873	13°51.8'S	76°34.0'W	184	G	--	10.1	1.12	35.1	--	Rowe (1979)
Y73-7-98	12°22.0'S	77°28.2'W	300	MG	black organic rich clay	9.83	1.17	2.7	24	This study
7706-03	12°58.3'S	76°57.4'W	304	RB	black organic rich silty clay with fish debris	14.5	1.66	2.9	18	This study
7706-04[1]	12°58.9'S	76°58.0'W	325	K	black organic rich silty clay with fish debris	17.3	--	3.1	15	This study
7706-36	13°37.3'S	76°50.5'W	370	RB	black organic rich silty clay with fish debris	21.2	2.46	2.0	9	This study
7706-37[1]	13°37.8'S	76°50.9'W	370	K	black organic rich silty clay with fish debris	13.2	1.61	2.8	14	This study
7706-08[1]	13°08.4'S	77°32.6'W	464	K	compacted dark olive gray silty clay					
7706-05[1]	13°02.4'S	77°04.9'W	625	K	silty sand with sedimentary rock fragments					
7706-06[1]	13°02.4'S	77°04.9'W	701	K	silty sand with sedimentary rock fragments					
7706-07[1]	13°16.0'S	77°27.6'W	1774	G	dark gray black mudstone					

Appendix I (cont.)

Core or Station #	Latitude	Longitude	Water Depth (m)	Type Sampler	Sediment Description	Surface Composition (%) Org-C	Total-N	$CaCO_3$	Qtz	Data Source
7706-11	13°09.0'S	77°14.0'W	1380	K	silty clay	8.53	1.06	6.6	21	This study
Y71-6-5	13°30.2'S	77°19.9'W	2286	MG	olive gray silty clay	8.55	--	--	16	This study
7706-13	13°26.0'S	77°40.0'W	3470	G	silty clay	5.90	0.69	4.5	17	This study
7706-14	13°27.3'S	77°42.9'W	3621	K	clay	7.45	0.89	3.0	13	This study
7706-15	13°48.4'S	78°18.3'W	4581	K	dark grayish brown clay	1.79	0.28	1.0	15	This study
Y71-6-4	14°36.0'S.	79°05.5'W	4310	MG	olive gray clay	1.02	0.11	0.0	15	This study
Profile E										
TGT 237	15°03.1'S	75°25.2'W	60	G	--	1.80	0.24	7.5	--	Rowe (1979)
KN 73-3-C1	15°05.4'S	75°29.6'W	60	G	--	2.59	0.46	21.7	--	Rowe (1979)
KN 73-3-C3	15°06.0'S	75°30.4'W	120	G	--	4.01	0.59	26.7	--	Rowe (1979)
TGT 190	15°06.1'S	75°31.0'W	110	G	--	6.69	0.85	17.8	--	Rowe (1979)
TGT 258	15°13.2'S	75°29.6'W	240	G	--	6.29	0.73	24.7	--	Rowe (1979)
TGT 243	15°10.4'S	75°33.9'W	388	G	--	4.62	1.02	48.4	--	Rowe (1979)
30870	15°10.8'S	75°35.2'W	430	G	--	5.64	0.72	56.7	--	Rowe (1979)
30854	15°10.2'S	75°37.0'W	580	G	--	6.84	0.93	27.5	--	Rowe (1979)
TGT 267	15°10.2'S	75°37.3'W	600	G	--	7.18	0.88	34.3	--	Rowe (1979)
30858	15°11.5'S	75°37.9'W	820	G	--	5.77	0.71	30.6	--	Rowe (1979)
TGT 302	15°12.1'S	75°42.1'W	1140	G	--	4.41	0.52	10.0	--	Rowe (1979)
30848	15°14.5'S	75°40.7'W	1240	G	--	4.32	0.49	17.3	--	Rowe (1979)
TGT 197	15°15.7'S	75°49.0'W	2050	G	--	5.50	0.66	15.9	--	Rowe (1979)
30832	15°17.9'S	75°49.7'W	2250	G	--	5.15	0.54	51.9	--	Rowe (1979)
TGT 238	15°20.2'S	75°55.7'W	3000	G	--	5.44	0.65	15.5	--	Rowe (1979)
Y71-6-23	15°11.9'S	76°14.7'W	3930	MG	dark olive gray silty clay	4.45	0.59	1.6	14	This study
TGT 245	15°21.9'S	75°58.0'W	4110	G	--	4.45	0.54	19.9	--	Rowe (1979)
Y71-6-25	15°15.5'S	76°18.9'W	4614	P	grayish olive clay	2.59	0.29	0.7	13	Rowe (1979)
Y71-6-24	15°16.1'S	76°18.9'W	4899	MG	moduled olive gray clay	1.78	0.21	0.9	15	Rowe (1979)

Appendix I (cont.)

Core or Station #	Latitude	Longitude	Water Depth (m)	Type Sampler	Sediment Description	Surface Composition (%) Org-C	Total-N	$CaCO_3$	Qtz	Data Source
Y71-6-26	15°17.9'S	76°21.3'W	4813	MG	olive gray silty clay	0.95	0.13	4.4	14	Rowe (1979)
7706-34	15°19.2'S	76°51.0'W	3316	K	dark olive gray silty clay	3.11	0.37	0.8	14	This study
7706-33	15°43.6'S	76°47.8'W	2967	K	olive gray silty clay	1.95	0.26	3.3	15	This study
Additional Peru Margin Cores[1]										
7706-31	16°17.6'S	73°53.6'W	645	RB	silty clay with Mn micronodules	5.79	0.72	2.6	13	This study
7706-30	16°42.0'S	73°24.2'W	2154	RB	sand	0.04	0.02	3.1	13	This study
Y73-6-95	16°25.1'S	74°13.9'W	2509	G	dark olive gray clayey silt	2.99	0.37	1.4	18	This study
Y71-6-12	16°26.6'S	77°33.8'W	2685	MG	light olive gray calcareous ocze	0.82	0.16	31.6	11	This study
Y71-6-22	16°48.6'S	74°03.4'W	5301	MG	olive gray clay	3.49	0.44	1.4	10	This study
Y71-6-19	17°02.9'S	74°24.5'W	5791	MG	olive brown clay	0.61	0.10	0.2	11	This study
Y73-6-94	16°50.7'S	74°33.3'W	5920	MG	olive brown silty clay	0.92	0.12	0.6	12	This study
Y73-6-93	16°46.1'S	74°34.6'W	6285	G	dark olive gray silt	1.74	0.22	1.0	14	This study
Y71-6-21	16°50.8'S	74°21.7'W	6505	MG	olive gray clay	2.25	0.27	1.0	15	This study
Y71-6-20	16°57.9'S	74°20.8'W	7293	FF	dark grayish brown silty clay	1.44	0.17	0.4	15	This study
7706-16	15°38.2'S	81°14.8'W	4521	K	light brown clay	0.37	0.10	0.9	13	This study

[1] These data are not illustrated in the across-margin profiles (Figure 6).

[2] RB = Reineck Box, K = Kasten, G = Gravity, MG = Multigravity, FF = Free fall.

Appendix II: Downcore profiles of ^{14}C-ages and sediment constituents from nine cores off the Peru coast. The cores are listed in order of increasing water depth. ^{14}C-age and compositional variations indicate facies changes and the resolvability of the sediment record.

Core, Sample Interval (cm)	^{14}C-Age (y.B.P.)	Org-C	$CaCO_3$	Quartz	Inorg-P
			(% dry weight)		
7706-39 R.B.					
0-3		13.2	3.1	14	0.16
3-6		12.6	3.4	15	0.18
6-9		11.7	3.4	16	0.16
9-12		10.3	3.3	18	0.19
12-15		7.3	1.1	30	0.22
15-18		7.1	0.9	29	0.21
18-21		7.6	1.3	25	0.20
21-24		7.8	1.2	30	0.20
24-27		7.6	1.1	25	0.20
27-30		7.5	1.2	24	0.17
30-33		7.3	0.8	24	0.21
33-36		9.2	0.9	24	0.24
36-39		9.8	1.1	21	0.24
39-42		9.1	0.8	19	0.16
42-45		8.9	1.4	16	0.17
45-48		8.5	1.6	22	0.20
48-51		8.0	2.1	25	0.23
51-54		7.9	2.9	22	0.21
54-57		6.6	1.7	26	0.23
57-59		10.0	3.2	20	0.19
7706-40 K					
0-5	571 ± 82	13.7	1.7	13	0.13
15-20		6.8	0.5	n.d.[1]	n.d.
20-25		8.3	1.8	28	0.19
40-45		9.6	2.1	26	0.20
45-50	967 ± 63	n.d.	n.d.	n.d.	n.d.
50-55		9.8	1.4	n.d.	n.d.
60-65		9.6	2.1	22	0.17
75-80		9.7	1.7	n.d.	0.18
80-85		10.7	1.9	19	0.16
90-95	2,611 ± 65	n.d.	n.d.	n.d.	n.d.
100-105		8.4	0.5	21	0.14
110-115		7.8	1.2	n.d.	0.18
120-125		9.6	2.4	21	0.20
135-140		8.3	3.3	n.d.	n.d.
140-145	2,980 ± 140	7.9	2.4	19	0.16
160-165	3,640 ± 130	5.1	1.4	39	0.33
180-185		3.4	2.2	43	0.56
200-205		4.9	1.1	38	0.21
205-210		3.7	0.9	n.d.	0.20
210-215	12,114 ± 412	n.d.	n.d.	n.d.	n.d.
215-222		3.3	0.9	40	0.20
7706-41					
0-4		19.6	9.8	11	0.63
4-8*	840 ± 100	18.6	11.0	11	0.52
10-15		8.4	6.0	n.d.	0.56
15-20		12.6	1.4	22	0.68
20-25	10,100 ± 250	13.3	2.0	n.d.	0.43
30-35		12.0	1.4	24	
40-45	10,700 ± 270	11.4	1.8	n.d.	0.52
45-50		11.5	1.6	24	0.40
65-70	12,500 ± 340	9.7	1.5	n.d.	0.36
75-80		9.7	1.4	20	0.30
80-85		7.1	1.4	19	0.23

* 8-10 cm actual depth of ^{14}C sample

Appendix II (cont.)

Core, Sample Interval (cm)	^{14}C-Age (y.B.P.)	Org-C	$CaCO_3$ (% dry weight)	Quartz	Inorg-P
90-95		6.2	7.6	21	0.21
105-110	14,280 ± 420	n.d.	n.d.		
110-115		6.4	9.4	17	0.20
120-125		6.1	8.7	n.d.	0.26
130-135		5.5	1.8	16	0.20
150-155		6.2	2.8	n.d.	0.23
155-160	15,500 ± 550	5.2	1.0	n.d.	0.18
165-170		4.2	1.0	23	0.24
170-175		3.0	0.8	n.d.	0.27
180-185		2.2	0.8	28	0.28
185-191		2.5	1.3	27	0.27
7706-44 K					
0-5		7.8	30.6	15	1.20
5-10		9.3	30.3	n.d.	0.95
10-15		8.3	33.5	15	0.95
20-25	6,270 ± 160	n.d.	n.d.	n.d.	n.d.
30-35		7.8	33.9	15	0.62
42-47		8.3	32.5	n.d.	0.58
50-55		7.7	33.1	16	0.57
55-60	9,000 ± 220y	n.d.	n.d.	n.d.	n.d.
70-75		5.5	42.3	14	0.81
75-80		3.4	49.4	12	1.08
100-105	* > 34,000	2.6	57.0	12	0.44
110-115		3.8	50.9	n.d.	0.62
115-120	> 34,000	n.d.	n.d.	n.d.	n.d.
120-125		5.9	41.1	13	0.74
140-145		2.2	36.6	17	0.69
160-165		3.9	47.4	16	0.31
180-185		3.7	44.4	16	0.43
200-205		3.4	37.9	18	0.43
220-225		2.9	27.4	28	0.30
*90-95 actual depth of ^{14}C sample					
7706-04 K					
0-1		17.3	3.1	15	0.35
6-9		14.3	2.9	17	0.39
9-13		14.1	2.0	15	0.41
15-18		13.9	2.0	16	0.45
25-28		13.0	2.3	16	0.88
35-38		14.2	1.7	16	0.71
45-48	2,950 ± 130	13.3	1.7	16	0.81
55-58		8.6	9.4	19	1.36
70-73		10.3	1.2	19	0.44
95-98		9.5	1.8	16	0.51
120-123		6.4	1.4	17	0.26
140-143	13,820 ± 440	6.4	5.0	16	0.26
160-163		5.8	6.6	14	0.25
180-183		6.8	5.7	12	0.24
7706-36 R.B.					
0-2		20.0	2.2	10	0.31
2-5		16.8	2.1	15	0.49
5-8		12.0	2.2	17	0.96
8-12		9.1	2.2	18	1.98
14-17		4.4	3.7	11	3.95
17-23		5.0	0.8	17	0.29
23-28		5.3	1.0	17	0.26
28-33		4.9	0.9	16	0.24
33-39		5.1	1.3	17	0.24
39-47		4.5	1.4	18	0.25

Appendix II (cont.)

Core, Sample Interval (cm)	^{14}C-Age (y.B.P.)	Org-C	$CaCO_3$ (% dry weight)	Quartz	Inorg-P
7706-37 K					
0-5	1,328 ± 72	13.2	2.8	14	0.61
15-20		6.8	2.4	16	0.60
30-35		5.9	2.5	16	0.42
35-40	13,060 ± 200	n.d.	n.d.	n.d.	n.d.
40-45		4.9	1.7	17	0.23
50-55		4.3	1.4	15	0.19
70-75		4.8	4.5	14	0.22
80-85	14,420 ± 370	n.d.	n.d.	n.d.	n.d.
85-90		4.8	1.2	15	0.14
105-110		4.0	1.1	18	0.14
120-125	14,720 ± 490	4.5	1.1	14	0.16
133-138		3.0	1.1	18	0.17
145-150		2.6	1.1	20	0.37
160-165	16,900 ± 630	2.7	1.0	21	0.21
7706-49 K					
0-5		2.9	3.3	13	0.15
10-15		3.2	1.7	21	0.18
30-35		2.9	2.3	22	0.16
45-50		2.8	2.3	n.d.	0.19
50-55		3.1	2.1	23	0.18
65-70	4,990 ± 160	n.d.	n.d.	n.d.	n.d.
70-75		2.8	1.9	25	0.20
85-90		4.0	10.1	25	0.26
90-95		2.6	3.2	26	0.20
110-115		3.3	2.6	23	0.20
120-125		2.7	2.8	n.d.	0.19
130-135		2.8	1.5	22	0.17
150-155		2.7	1.4	22	0.16
170-175		3.0	4.7	21	0.18
180-185	7,420 ± 220	n.d.	n.d.	n.d.	n.d.
190-195		3.0	5.7	21	0.18
200-205		2.9	5.3	22	0.18
7706-50 K					
0-5		0.8	0.6	14	0.08
5-15		0.9	0.7	17	0.08
15-20		0.9	0.4	17	0.09
20-25		0.8	0.6	19	0.08
35-40	7,920 ± 480	n.d.	n.d.	n.d.	n.d.
40-45		0.8	0.7	18	0.08
45-50		0.8	1.4	n.d.	0.08
58-65		1.7	0.9	19	0.06
75-80	18,100 ± 750	n.d.	n.d.	n.d.	n.d.
80-85		2.4	1.1	16	0.05
100-105		2.1	1.0	17	0.05
120-125		1.4	0.7	16	0.06
130-135		1.1	2.0	n.d.	0.06
140-145		1.2	1.9	16	0.06
160-165		1.1	1.1	18	0.06
180-185		1.2	0.9	18	0.06
200-205		1.1	0.8	18	0.06
220-225		1.0	1.0	16	0.07
240-245		0.9	0.9	17	0.07
260-265		0.8	0.6	16	0.08
272-278		0.9	0.7	17	0.08
290-295		0.8	0.4	18	0.08
302-330		0.9	0.5	16	0.08

[1] n.d. = not determined

VARIABILITY OF UPWELLING REGIMES (NORTHWEST AFRICA AND SOUTH ARABIA) DURING THE LATEST PLEISTOCENE: A COMPARISON

Monique Labracherie, Marie-France Barde,
Jean Moyes, and Annick Pujos-Lamy

Département de Géologie et Océanographie
Université de Bordeaux I, Avenue des Facultés
F-33405 Talence Cedex, France

ABSTRACT

The distribution of diatoms, radiolarians and coccoliths was analyzed and compared in two cores from two different recent upwelling areas (Mauritania and south Arabia). Off south Arabia the high abundance of radiolarian skeletons associated with rather low abundance of diatoms is probably a signal recording the high level of nutrients produced by a divergence along current boundaries rather than by wind-induced upwelling. The scarcity of neritic and meroplanktonic diatoms may support the former hypothesis of nutrient enrichment of surface waters at a front between two opposing currents. Off the Mauritanian coast, on the other hand, distinctive siliceous skeletal assemblages, rather strongly diluted by diatom frustules, might have been deposited under nutrient-rich upwelled waters driven by winds.

Upwelling regimes both present and past appear diversified and highly variable. Off the Oman coast (south Arabia) upwelling intensity at 18,000 years B.P. might have been much weaker than today and during the Holocene period. Off Cap Blanc (Mauritania) changes in the planktonic assemblages during the last glacial time (youngest glacial maximum and latest deglaciation) may have resulted in more productive coastal upwelling than today. The distribution of some species (*Phormospyris stabilis scaphipes*, *Gephyrocapsa oceanica*) could support the conclusion that shortly after 18,000 years B.P. the penetration of the South Atlantic Central Water flowing northward might have been better marked than today and might have increased the nutrient enrichment of surface waters. However, during the last deglaciation, the flux of the South Atlantic Central Water might have been weakened. Some species (*Coccolithus pelagicus*) could indicate a stronger yet shorter influence of the North Atlantic Central Water.

INTRODUCTION

In recent years, investigations on the biogenic material in surface waters and sediment assemblages from the upwelling areas have been numerous and varied. Works on the distribution of phytoplankton in the surface sediments show that some species of diatoms may record the signal of newly upwelled water (Schuette and Schrader, 1981). From studies of deep-sea sediment cores other authors have described upwelling related changes in several parts of the ocean and over certain time intervals, especially during glacial and interglacial times (Diester-Haass, 1978; Prell, 1978; Labracherie, 1980a; DeVries and Schrader, 1981).

Our present study of the Holocene and latest glacial sediments from two well-known regions of coastal upwelling, northwest Africa and south Arabia, has three distinct objectives:

1. The first is to evaluate the usefulness but also the limits of utilization of opaline skeletons of radiolarians and diatoms as upwelling indicators. In order to approach this problem, we compared their abundance to the distribution of some selected species groups because there are variables which are more independent of sedimentation rate. The few species we have selected have been related to the present day dynamics of upwelling; they can also reflect hydrographic differences in the surface and subsurface waters (Margalef, 1978; Blasco, Estrada and Jones, 1980; Schuette and Schrader, 1981).

For the upwelling areas off northwest Africa and south Arabia, planktonic foraminifers have been the most studied organisms (Thiede, 1975; McIntyre et al., 1976; Prell, 1978; Pflaumann, 1980). They provided a useful tool (cool-water assemblages) for the interpretation of upwelling activity. For example, "sea-surface temperatures during the glacial maximum are interpreted as responses to....an increased upwelling off Africa during February" (Gardner and Hays, 1976). "On the northeastern Atlantic continental margin the highest radiolarian/planktonic foraminifera ratios are found in the region of most intense upwelling" (Diester-Haass, 1977). The distribution patterns of other fossils (nannofossils, diatoms) seemed to be less useful. However, Diester-Haass, Schrader and Thiede (1973) wrote "upwelling is characterized by underlying opal-rich sediments, and by high production of cold water preferring phytoplankton assemblages". Preferred biotopes of large centric planktonic diatoms are upwelled cool waters (Thiede, 1975). In this study we considered upwelling as "an ecological and hydrological event" which is not only characterized by generation of cold water but can be defined by the distribution of major water masses, which controls the distribution patterns of planktonic organisms in the surface and subsurface waters.

2. A second objective is to compare the distribution of organisms in different regimes of upwelling, each having different characteristics (climate, influence of winds upon hydrography, water mass stratification, etc.).

3. The final objective is to interpret not only the distribution patterns of fossils as reflecting major water masses or hydrological features but also to reconstruct the paleo-history of upwelling areas during a period extending from the last glacial maximum to present times.

OCEANOGRAPHIC SETTING

The upwelling processes off the Mauritanian coast are connected with the intensity of trade winds, the deplacement of the intertropical convergence zone, and the source of upwelled water. There are two different upwelling systems. To the north of the 21°N latitude the upwelling source water is the North Atlantic Central Water. Coastal upwelling occurs year round. South of 21°N the upwelling moves during winter along a north-south axis so that its intensity decreases seasonally. The upwelling source water is from the South Atlantic Central Water. A mixing zone separates these two systems. Warm tropical surface water is found offshore which may advance mainly in summer into the region close to Cap Blanc when the cooler waters of the Canary Current have shifted seaward (Fraga, 1974; Tomczak, 1978).

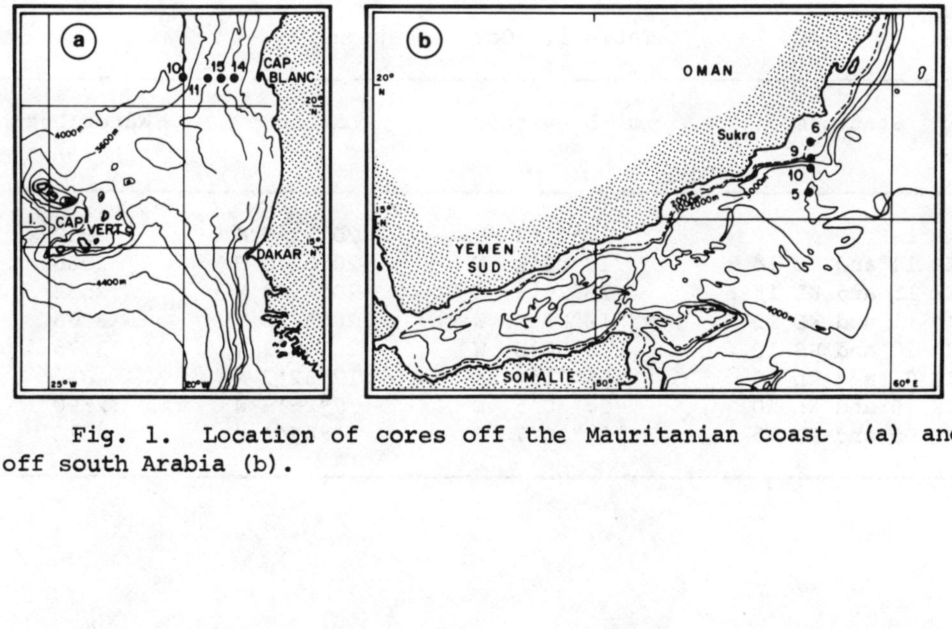

Fig. 1. Location of cores off the Mauritanian coast (a) and off south Arabia (b).

The upwelling off the south Arabian coast is unique because it does not occur along a western continental margin. In this area the surface currents are exposed to the strong variations of the monsoon winds. The upwelling processes are initiated during the southwest monsoon which blows northward parallel to the coast, but there are different explanations given about the circulation pattern. Some authors (Ryther and Menzel, 1965) think vertical deep water movements could be produced at a front between two opposite currents rather than being produced directly by wind stress.

SAMPLE MATERIAL

Off the Mauritanian coast we have selected four cores (Fig. 1) located along a transect at 21°N which are distributed between about 1000 and 4000 meters water depth. Off the south Arabian coast, at the northeast end of the Muscat-Oman coast, we chose four cores (Fig. 1) extending from 200 to 4000 meters of depth. The sediments in the shallowest core were barren or very poor in siliceous microfossils and have therefore been eliminated from consideration.

The cores --8 Reineck box cores (KR) and 7 Kullenberg piston cores (KS and KL)-- were taken during ORGON III and ORGON IV cruises (Table 1). Previous investigations contain lithologic descriptions and textural analyses (Moyes et al., 1981).

The downcore $\delta^{18}O$ fluctuations were used as a stratigraphic base. The oxygen isotope curve was determined from tests of *Globigerinoides ruber* from core KL 10 of the Sukra profile. The correlations between the other cores are based on the distribution of assemblages of planktonic foraminifers and on the Moyes' ecostratigraphy (Moyes et al., 1979). For the cores off Cap Blanc the approach is

Table 1. Core locations

Stations	Longitude	Latitude	Water Depth (m)
KR 13	17°45'2 W	20°53'0 N	910
KR 14 and KL 14 b	18°02'9 W	20°59'0 N	2030
KR 15 and KL 15 a	18°29'3 W	20°59'2 N	2552
KR 11 and KL 11	19°07'06 W	20°58'6 N	3263
KR 10 and KS 10	19°56'8 W	20°59'9 N	3775
KR 09 and KL 09	57°31'6 E	17°52'7 N	830
KR 10 and KL 10	57°30'5 E	17°26'6 N	2390
KR 05 and KL 05 a	57°30'7 E	16°45'9 N	4010

similar. The *Globigerina bulloides* records in core 15 are used as a correlation base. The oxygen isotopic stratigraphy shows that the Y/X boundary, previously established by planktonic foraminifers (Moyes et al., 1979) and used by Labracherie (1980a) does not coincide with the oxygen isotope stage boundary 1/2. The new Holocene/ Pleistocene boundary has shifted slightly downward.

METHODS

Samples were taken from Reineck box cores for the surface sediments study. Each Kullenberg core was generally sampled at 20 cm intervals. Methods for the study of radiolarians and nannofossils are well known (Pujos-Lamy, 1977; Riedel and Sanfilippo, 1977). Preparation of samples for diatoms followed Riedel's procedure with some modifications (Barde, 1980). Radiolarian abundance was expressed in two size fractions: estimated values related to volume in the coarse fraction >125 μm, and counts per gram of dried sediment in the fraction 45-125 μm. We have used previous data on abundance of radiolarian fauna which showed good correlations between the two kinds of data (Labracherie, 1980a). For diatoms we have taken into account the results from tests by Schrader and Gersonde (1978). Diatoms were counted on two slides and on 120 fields each measuring 280 μm in diameter.

VARIATIONS OF UPWELLING INFLUENCE. A COMPARISON OF DIATOM AND RADIOLARIAN RECORD

Results

The radiolarian/planktonic foraminifer ratio has been considered useful to characterize upwelling areas (Diester-Haass, 1977) and radiolarian amounts are "a valuable indicator of fertility changes because they reflect the opal input by diatoms" (Berger, Diester-Haass and Killingley, 1978). However, differences appear between the fluctuations of radiolarian and diatom contents. Off Cap Blanc both diatoms and radiolarians can be found in recent sediments from the upper continental slope. The associations are never high in abundance (5000 to 8000 radiolarians and 5×10^6 to 8×10^6 diatoms per gram of dried sediment). Diatoms disappear offshore and radiolarians become less numerous.

The downcore distributions (Fig. 2) show that radiolarians are always more abundant in oxygen isotope stage 2 than in the recent period. Diatoms also increase, but from east to west there is a sudden rise in the number of cells, with a maximum near the lower continental slope. There are also greater variations within stage 2. This means that at the end of the Pleistocene, the upwelling processes could have been more intense than today and during the Holo-

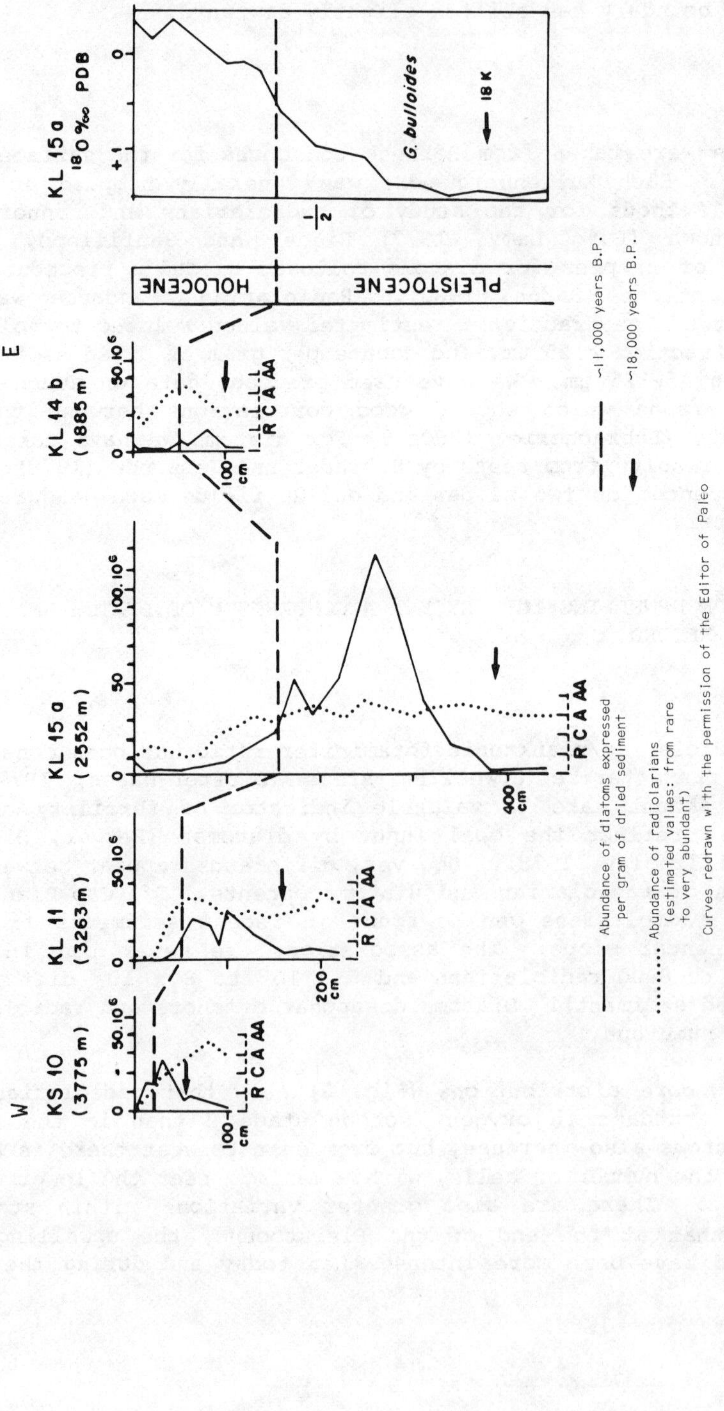

Fig. 2. Transect of Cap Blanc (northwest Africa). Downcore abundance of diatoms and radiolarians

cene. However, during stage 2 the quantitative fluctuations of the diatom flora, and consequently the variations of primary productivity, seem to have been rather important. They could indicate temporal changes of upwelling influence that have not been discerned with the radiolarian fauna.

In the Sukra transect (Fig. 3), a similar study shows that high concentrations today and during Holocene times are located in the farthest offshore stations at around 3000 meters of depth. Diatoms are hardly more abundant than in the recent sediments off Cap Blanc. They are always well preserved. On the other hand, radiolarian associations are very abundant (10 times as high). The sedimentation rates calculated for the Holocene are ∿10 cm/1000 yrs. for KL 10-KL 05a (Sukra profile) and KL 15a (Cap Blanc). During the last glacial maximum, abundance of diatom frustules and radiolarian skeletons is very low which means a decrease in productivity and a weakening of upwelling. An interesting feature in the Sukra transect is a very high recent radiolarian content associated with a rather low amount of diatoms. Such a distribution is comparable to what has been observed around the Pacific equatorial divergence (Labracherie, unpublished data, 1981). Along this profile, biosiliceous sedimentation far offshore could mean the presence of a front between two opposing currents resulting from the special hydrology of the Arabian Sea during the southwest monsoon. These characteristics are found again during the last glacial off Cap Blanc (Fig. 2).

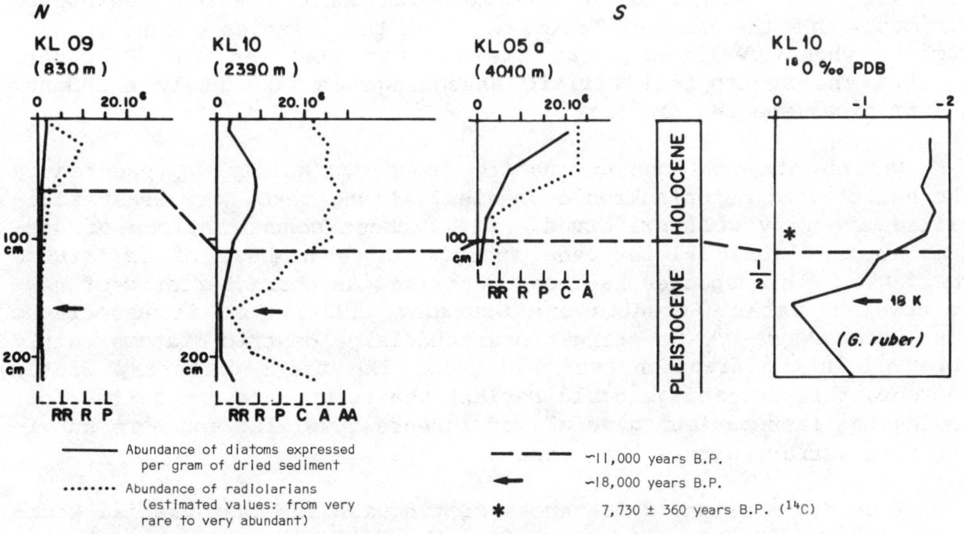

Fig. 3. Transect of Sukra (south Arabia). Downcore abundance of diatoms and radiolarians.

PHYTOPLANKTON COMPOSITION AND RELATION TO HYDROLOGY. VARIABILITY IN OCEANOGRAPHIC CONDITIONS SINCE 18,000 YEARS B.P.

The quantitative observations on sedimentary components do not give more than an average value which combines numerous signals of more or less strong upwelling conditions and which is controlled by dilution. On the other hand, species distribution in the sediments can well account for the hydrological events in surface water (Schuette and Schrader, 1981). The species distribution of phytoplankton during the last 18,000 years has been examined in relation to the diatom and radiolarian concentrations in the two cores from the same water depth, which presents a detailed stratigraphy and the strongest fluctuations of abundance of opaline skeletons in each transect. Nannoplankton distributions based on a duality between warm and cold species are discussed as a consequence of the distribution of some major surface water masses and of their nutrient contents.

Off the Mauritanian Coast

In core KL 15a of the Cap Blanc transect, the most frequent diatoms have been reported in other upwelling regions as well. They have been distributed among three group categories (DeVries and Schrader, 1981): subtropical neritic species, "tropical" oceanic species and meroplanktonic species (Fig. 4). According to Schuette and Schrader (1981), the first group is "defined" as characteristic of cooler upwelled waters. The "tropical" oceanic species are warm water species and could be related to a weakening of upwelling processes. In Recent and latest Holocene sediments, the diatom association has been dominated by the meroplanktonic species *Cyclotella striata*. Off the central Peruvian coast this species occurs in high numbers when upwelling is at its weakest (DeVries and Schrader, 1981). The subtropical neritic assemblage is moderately abundant; the tropical species are rare.

During stage 2, the dominating group was mainly represented by the neritic and meroplanktonic species. Among them, the least silicified are very well preserved. The richest concentrations of diatoms in the latest Pleistocene include large numbers of *Delphineis karstenii*. This species has been described as characteristic of newly upwelled water (Schuette and Schrader, 1981). It is associated with *Chaetoceros spp.* resting spores and large centric diatoms mainly observed in the fraction over 125 μm. Like the sedimentary diatom content, this assemblage could reflect the recurrence, at a time following the last maximum glacial, of intense upwelling and more nutrient-rich surface waters than today.

Indeed, important differences distinguish the last glacial sediments (around 18,000 years B.P.) from the following ones, but changes in the composition do not give a good idea of modifications of inten-

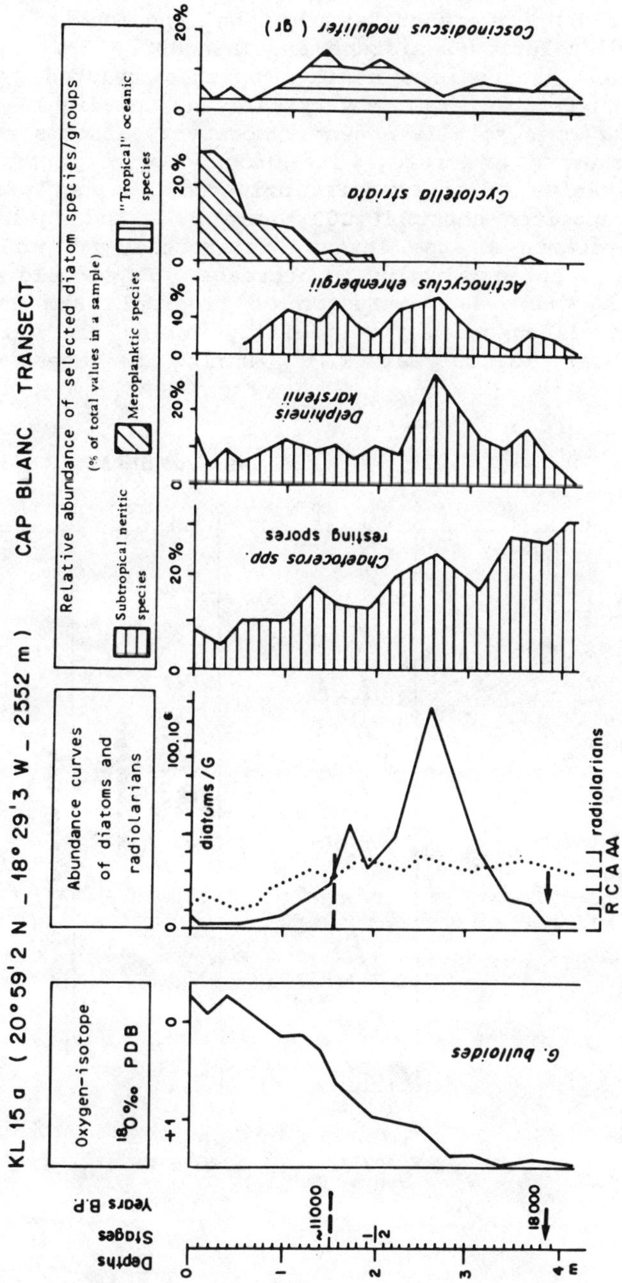

Fig. 4. Downcore distribution of some diatom species.

sity of upwelling. During the last glacial maximum, diatom concentrations are less. Diatom assemblages are dominated by *Chaetoceros spp.* resting spores, which are dissolution-resistant forms. DeVries and Schrader (1981) write that "strong upwelling conditions favor the dominance of dissolution-resistant assemblages". However, off Cap Blanc the other subtropical neritic species related to the most nutrient-rich surface waters are very rare and the highly silicified large centric species totally absent. Moreover, diatoms representing an oceanic influence are rare. The group *Coscinodiscus nodulifer/radiatus*, indicative of an oceanic influence, is poorly represented in sediments deposited about 18,000 years B.P. and thereafter this group becomes common in the latest Pleistocene and early Holocene when the diatom abundance begins to decrease. This could be taken as a signal for an increased intrusion of tropical warm water and an increase in upwelling/oceanic variability during the warming phase which starts about 13,500 years B.P. During the latest Holocene as

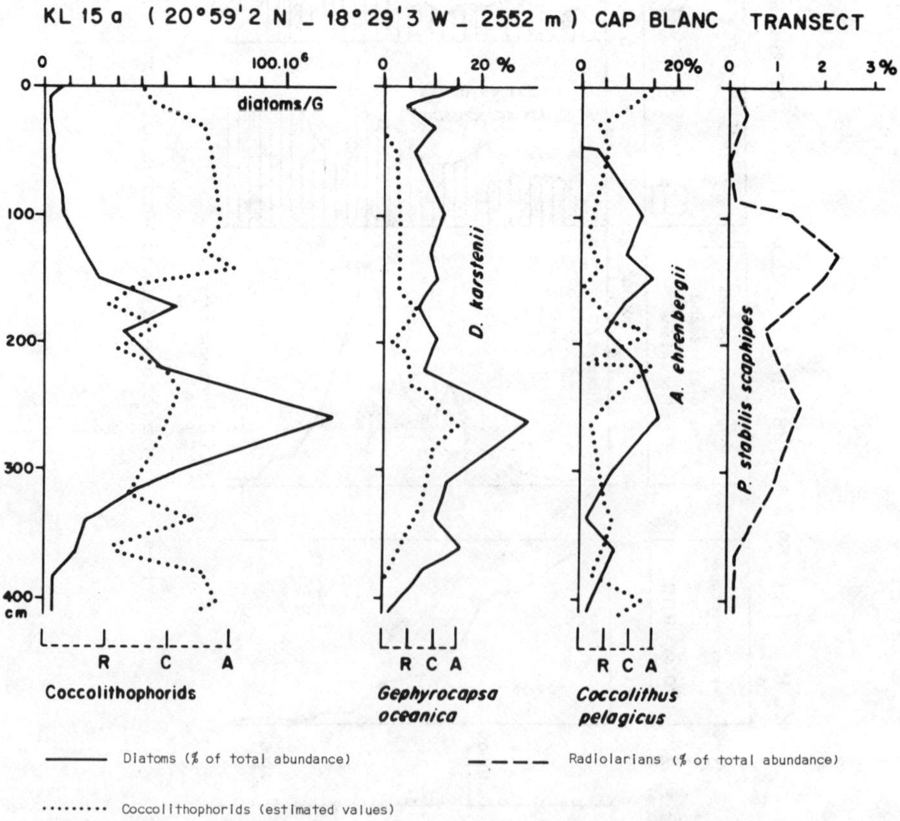

Fig. 5. Comparison of the distribution of some selected species (diatoms, coccolithophorids and radiolarians).

today, the tropical water invades the upwelling zone only during short periods and high productivity indicator species are more rarely found.

Downcore abundance and compositon of nannoflora do not change markedly in core KL 15a (Fig. 5). However, the distribution of two species seems strongly related or opposite to the high nutrient level indicator distribution. The downcore distribution of *Gephyrocapsa oceanica* (large form) and the downcore distribution of the diatoms are rather similar. This oceanic species is frequent in the surface sediments underlying the inner Cap Vert Basin (Müller and Rothe, 1975) and could be characteristic of warm water. The contribution of *G. oceanica* is important during the deglaciation in core KL 15a. The location of this species in the plankton off the Mauritanian coast could be related to the proportion of South Atlantic Central Water in the upwelling source waters (Blasco et al., 1980). We think that its high abundance at the end of the Pleistocene could reflect an increase of nutrients and a lower salinity related to a higher intensity of the upwelling system including southern water masses.

The recent distribution and downcore abundance of the radiolarian *Phormospyris scaphides* in the upwelling region off Cape Blanc have been used as indicating the presence and movement of the South Atlantic Central Water (Labracherie, 1980b). It has been proposed that this surface water mass from the south could spread northward beyond latitude 21°N and its advance further to the north sector off Cape Blanc could contribute to an increase of nutrients. This event is well marked and occurred twice during the glacial/interglacial transition (Fig. 5).

The species *Coccolithus pelagicus* (nannoflora) is present in the surface sediments off Cape Blanc. It is an unusual species at this latitude. Müller and Rothe (1975) consider that this cold-water species might indicate the course of the Canary Current. But its presence in the Recent sediments of the core KL 15a can not be directly related to this water since it is shifted more offshore. This species could, however, follow the nutrient-poorer North Atlantic Central Water. Strong concentrations appear when the influence of the South Atlantic Central Water seems to be lowest. It could be hypothesized that the maximum abundance of *C. pelagicus* represents the signal of the water coming from the north and supplying the upwelling system to the north of 21°N. The signal might be weakened or interrupted by an increase of the proportion of South Atlantic Central Water surfacing during a stronger upwelling.

Finally, it could have been immediately after the last glacial maximum that productivity was the highest off Cap Blanc. The influence of the water mass from the south, flowing northward might have been better marked than today and could have increased the nutrient-enrichment of surface waters when upwelling processes were highest.

If upwelling activity was then in conjunction with winds blowing parallel to the coast, wind stress could have reached a maximum intensity. During the deglaciation, modifications in the composition of planktonic assemblage could have reflected a changing influence of central water masses and recorded the advance of warm waters. Alternations could therefore be related to a higher variability in the intensity of upwelling.

Off South Arabia

The downcore distribution of selected species in the core KL 10 of the Sukra transect (Fig. 6) is less complex and their significance more readily seen than in core KL 15a off Cap Blanc. Recent and Holocene sediments contain a high proportion of warm oceanic species. *Coscinodiscus nodulifer* is the dominant species of the oceanic diatom assemblage. Subtropical neritic species (*Chaetoceros spp.* resting spores) or meroplanktonic, strictly related to the shelf species (*Cyclotella striata*), are poorly represented. The different associations are dominated by tropical elements. The species *Gephyrocapsa oceanica* (large form) is common; its distribution parallels that of the diatoms.

During the last glacial maximum , the abundance of planktonic diatoms and radiolarians is very low. Diatoms are rare and represented only by dissolution-resistant species. Upwelling processes could have been reduced and the oceanic circulation pattern, generally induced by the southwest monsoon, weakened. The species composition off the Oman coasts distinctly differs from that off Cap Blanc. The mechanism of upwelling could explain the differences since off the Mauritanian coast, the offshore transport in the surface layer is important. Neritic species indicative of high productivity may then extend further offshore and their abundance could be related to the wind stress. If the surface offshore flow is weaker and if the dynamic processes are induced by an offshore oceanic front at current boundaries, the oceanic species may have been the most important component of the diatom flora, as they are off the Arabian coast.

CONCLUSIONS

From this comparison of the variations of sedimentary biosiliceous contents observed in two upwelling regions, it can be concluded that:

Quantitative and qualitative diatom fluctuations could give a better gradation on the temporal evolution of coastal upwelling than can radiolarian abundance changes.

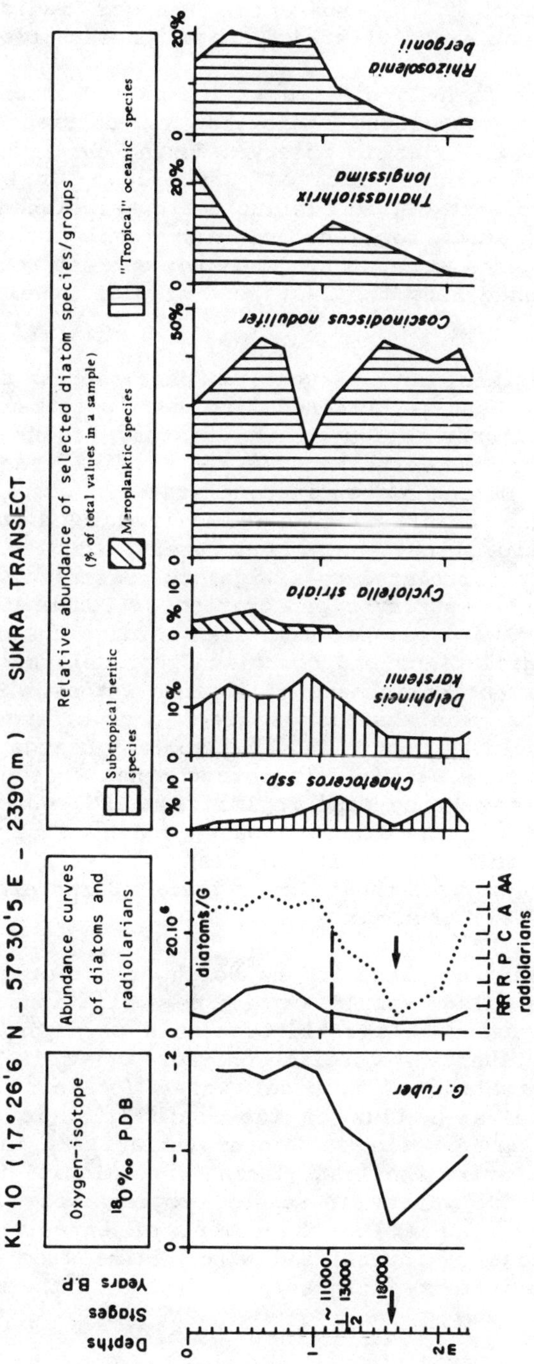

Fig. 6. Downcore distribution of some diatom species.

In fertile regions and fertile periods both diatoms and radiolarians are found in the sediments. However, radiolarian preservation does not exactly reflect diatom preservation.

An assemblage rich in opaline skeletons and strongly diluted by diatoms, could be a specific indication of nutrient-rich upwelled waters driven by winds, as is observed during the latest glacial upwelling on the continental slope off Cap Blanc. On the other hand, the very high abundance of radiolarian tests associated with a smaller abundance of diatoms could record high levels of nutrients produced at a divergence zone at two opposing current boundaries, as is observed today and during the last post-glacial upwelling off south Arabia.

Upwelling regimes, both present and past, appear diversified and highly variable. Results obtained from two localized regions allow us a tentative interpretation of the history of upwelling regimes during the last 18,000 years both in the tropical east Atlantic and the Arabian Sea. During oxygen-isotope stage 2 (glacial period) the high proportion of neritic diatoms, some of which have been considered characteristic of newly upwelled waters, suggests an important transport offshore correlated to a vigorous coastal upwelling off the Mauritanian coast. A stronger penetration of South Atlantic Central Water flowing northward, as has been seen through the distribution of some species (radiolarians and coccolithophorids), might be able to increase the nutrient enrichment of surface waters. Simultaneously, in the Arabian Sea, the absence of glacial upwelling processes off the Oman coast could be controlled by important changes in the circulation of the surface waters. In both oceans, these changes should be induced by the weakened wind regime from SSW, which controls upwelling processes off the Arabian coast, and by a strongly intensified wind regime from NNE as it has been inferred by other methods, more specifically off northwestern Africa (Sarnthein, 1978; 1979; Diester-Haass, 1979; Sarnthein et al., 1981).

During the warming phase in the North Atlantic, the composition of planktonic assemblages could explain new variations in the distribution of water masses. It could suggest a stronger North Atlantic Central Water influence (coccolithophorids and radiolarians) and a more marked penetration of tropical water (oceanic diatoms) which appear with a weakened flux of the South Atlantic Central Water. These variations could reflect a higher variability in the upwelling off Mauritania. During the last glacial-interglacial transition, the fertility of surface waters increased progressively in the Arabian Sea. It could be related to the coming of an alternating monsoon regime and a gradually intensified wind regime from the southwest, which would correspond to a higher variability in the northwest African Trade Winds.

ACKNOWLEDGMENTS

The authors wish to thank M. Arnould (Comité d'Etudes Géochimiques Marines) and R. Pelet (Institut Francais du Pétrole) for allowing them to work on the cores from ORGON III and ORGON IV cruises. $\delta^{18}O$ curves were kindly provided by J. C. Duplessy (Centre des Faibles Radioactivités, Laboratoire Centre National Recherche Scientifique-C.E.A., Gif-sur-Yvette).

REFERENCES

Barde, M. F., 1980, Les diatomées des sédiments actuels et du Quaternaire supérieur de l'Atlantique nord-oriental. Intérêt hydrologique et climatique, Bulletin Institut de Géologie du Bassin d'Aquitaine, 29:85-111.

Berger, W. H., Diester-Haass, L. and Killingley, J. S., 1978, Upwelling off North-West Africa: the Holocene decrease as seen in carbon isotopes and sedimentological indicators, Oceanological Acta, 1:3-7.

Blasco, D., Estrada, M. and Jones, B., 1980, Relationship between the phytoplankton distribution and composition and the hydrography in the northwest Africa upwelling region near Cabo Corbeiro, Deep-Sea Research, 27A:799-821.

DeVries, T. J. and Schrader, H., 1981, Variation of upwelling/oceanic conditions during the latest Pleistocene through Holocene off the Central Peruvian coast: a diatom record, Marine Micropaleontology, 6:157-167.

Diester-Haass, L., 1977, Radiolarian/planktonic foraminiferal ratios in a coastal upwelling region, Journal of Foraminiferal Research, 7:26-33.

Diester-Haass, L., 1978, Sediments as indicators of upwelling, in: "Upwelling Ecosystems," R. Boje and M. Tomczak, eds., Springer Verlag, 261-281.

Diester-Haass, L., 1979, Indicators of continental climates in marine sediments: A reply, "Meteor"Forschungs-Ergebnisse, 31:53-58.

Diester-Haass, L., Schrader, H. and Thiede, J., 1973, Sedimentological and paleoclimatological investigations of two pelagic ooze cores off Cape Barbas, North-West Africa, "Meteor"Forschungs-Ergebnisse, 16:19-66.

Fraga, F., 1974, Distribution des masses d'eau dans l'upwelling de Mauritanie, in: Analyse de l'écosystème des "upwellings," deuxième conférence, Marseille, 20-30 Mai 1973, Tethys, 6:5-10.

Gardner, J. V. and Hays, J. D., 1976, Responses of sea-surface temperature and circulation to global climatic change during the past 200,000 years in the eastern equatorial Atlantic ocean, Geological Society of America, Memoir, 145:221-246.

Labracherie, M., 1980a, Modifications de la circulation océanique au large du Cap Blanc (Afrique du North-Ouest) entre le dernier maximum glaciaire et l'époque actuelle. Apport des diatomées et des radiolaires, Comptes - Rendus des séances de l'Académie des Sciences, Paris, 291:601-604.

Labracherie, M., 1980b, Les radiolaires témoins de l'évolution hydrologique depuis le dernier maximum glaciaire au large du Cap Blanc (Afrique du Nord-Ouest), Palaeogeography, Palaeoclimatology, Palaeoecology, 32:163-184.

MacIntyre, A. and Kipp, N.G., with A.W.H. Bé, T. Crowley, T. Kellog, J. V. Gardner, W. Prell, and W. F. Ruddiman, 1976, Glacial North Atlantic 18,000 years ago: a CLIMAP reconstruction, Geological Society of America, Memoir, 145:43-75.

Margalef, R., 1978, Phytoplankton communities in upwelling areas, the example of NW Africa, Oecologia Aquatica, 3:97-132.

Moyes, J., Duplantier, F., Duprat, J., Faugères, J.C., Pujol, C., Pujos-Lamy, A., and Tastet, J.P., 1979, Etude stratigraphique et sédimentologique, in: "Géochimie organique des sédiments marins profonds," ORGON III - Mauritanie, Sénégal, Îles du Cap Vert, Centre National Recherche Scientifique, Paris, 121-213.

Moyes, J., Duprat, J. Faugères, J.C., Gonthier, E., and Pujol, C., 1981, Etude stratigraphique et sédimentologique, in: "Géochimie organique des sédiments marins profonds, ORGON IV, Golfe d'Aden, Mer d'Oman, Centre National Recherche Scientifique, Paris, 189-264.

Müller, C. and Rothe, P., 1975, Nannoplankton contents in regard to petrological properties of deep-sea sediments in the Canary and Cape Verde area, Marine Geology, 19:259-273.

Pflauman, U., 1980, Variations of the surface water temperatures along the Eastern North Atlantic continental margin (sediment surface samples, Holocene climatic optimum, and Last Glacial maximum), in: "Palaeoecology of Africa and the Surrounding Islands," E.M. van Zinderen Bakker and J.A. Coetzee, eds., 12: 191-212.

Prell, W.L., 1978, Glacial/interglacial variability of monsoonal upwelling: western Arabian sea, in: "Evolution of Planetary Atmospheres and Climatology of the Earth," 1978 International Conference, Centre National d'Etudes Spatiales, Nice, 149-156.

Pujos-Lamy, A., 1977, *Emiliania* et *Gephyrocapsa* (nannoplancton calcaire): biométrie et intérêt biostratigraphique dans le Pléistocéne supérieur marin des Açores, Revista Española de Micropaleontología, IX, 1:69-84.

Riedel, W.R. and Sanfilippo, A., 1977, Cenozoic radiolaria, in: "Oceanic micropaleontology," A.T.S. Ramsay, ed., Academic Press, London, 847-912.

Ryther, J.H. and Menzel, D.W., 1965, On the production, composition and distribution of organic matter in the Western Arabian Sea, Deep-Sea Research, 12:199-209.

Sarnthein, M., 1978, Sand deserts during glacial maximum and climatic optimum, Nature, 271(5648):45-51.

Sarnthein, M., 1979, Indicators of continental climates in marine sediments. A discussion, "Meteor"Forschungs-Ergebnisse, 31: 49-51.

Sarnthein, M., Tetzlaff, G., Koopmann, B., Wolter, K., and Pflaumann, U., 1981, Glacial and interglacial wind regimes over the eastern subtropical Atlantic and North-West Africa, Nature, 293(5829): 193-196.

Schrader, H. and Gersonde, R., 1978, Diatoms and silicoflagellates, Utrecht Micropaleontological Bulletin, 17:129-176.

Schuette, G. and Schrader, H., 1981, Diatom taphocoenoses in the coastal upwelling area off southwest Africa, Marine Micropaleontology, 6:131-155.

Thiede, J., 1975, Shell- and skeleton-producing plankton and nekton in the eastern North Atlantic Ocean, "Meteor" Forschungs-Ergebnisse, 20:33-79.

Tomczak, M., Jr., 1978, De l'origine et la distribution de l'eau remontée à la surface au large de la côte nord-ouest africaine, Annales Hydrographiques, 6(748):5-15.

GLACIAL-INTERGLACIAL CYCLES IN OCEANIC PRODUCTIVITY INFERRED FROM ORGANIC CARBON CONTENTS IN EASTERN NORTH ATLANTIC SEDIMENT CORES

Peter J. Müller, Helmut Erlenkeuser*
and Rita von Grafenstein

Geologisch-Paläontologisches Institut der Universität
Olshausenstrasse 40/60
D-23 Kiel, Federal Republic of Germany

*Institut für Reine und Angewandte Kernphysik der
Universität, Olshausenstrasse 40/60
D-23 Kiel, Federal Republic of Germany

ABSTRACT

Sediments from the continental margin off South Morocco, Mauritania, Gambia and from the Sierra Leone Rise reveal pronounced cycles with generally higher levels of organic carbon in sediments deposited during glacial stages of the Quaternary Period. The cycles can be traced back to oxygen isotope stage 13, i.e., to about 500,000 years B.P. The organic matter preserved in the cores is dominantly of marine origin as verified by stable carbon isotope ratios in one core and deduced from sedimentological and climatic evidence for the others. A previously developed empirical relationship between sedimentary organic carbon, sedimentation rate and primary production rate was used to estimate paleo-productivities.

The results suggest that oceanic productivity was approximately the same as today during interglacial stages but higher by a factor of 2-3 during glacial stages. The higher productivity of the glacial ocean is attributed to increased coastal and equatorial upwelling, as well as to a generally stronger mixing in response to stronger temperature gradients, winds, and currents. The results point to the possible significance of productivity changes of the ocean as a cause for a CO_2-depleted glacial atmosphere.

INTRODUCTION

Upwelling results in an increased production rate of organic matter and a higher organic matter flux to the sea bottom which in turn may leave some imprint in the sediments in terms of organic carbon content, accumulation rate, or chemical environment at the sea floor (Müller, Hartmann and Suess, in press). Recent sediment trap studies, as compiled by Suess (1980), have shown that the flux of organic matter to the sea bottom can roughly be predicted from primary production rate at the sea surface and water depth.

Prediction becomes much more complicated, however, when attempting to go the opposite way, i.e., to reconstruct paleo-productivity from organic carbon accumulation rate patterns in sediments, since numerous processes alter the original signal. For instance, waves and currents may prevent the deposition of organic material. This is typically the case off northwest Africa where strong undercurrents (e.g., Mittelstaedt, 1976; Smith, this volume) prevent fine-grained material and hydraulic equivalents, such as organic detritus, from accumulating in water depths of less than about 300 m. The bulk of organic matter, therefore, is deposited somewhere offshore at the continental slope (Diester-Haass and Müller,1979) and not directly below the waters of highest productivity (Schemainda, Nehring and Schulz, 1975). The same holds true for other components equally susceptible to winnowing by bottom currents (Fütterer, this volume).

Another complicating factor in certain regions of upwelling is additional nutrient supply by rivers, which also stimulates productivity (Diester-Haass, this volume). This is presently so off northwest Africa just south of about 16°N. There upwelling occurs only a few months per year but the annual primary production is comparable to that off Cape Blanc where upwelling prevails year-round (Schemainda et al., 1975). Hence, additional evidence is required in order to distinguish between upwelling and river-induced productivity.

Finally, the possibility has to be considered that sedimentary organic matter may have originated from terrestrial sources and then been carried to the site of deposition by winds and rivers. Presently, this contribution may be roughly quantified by the stable carbon isotope signature of preserved carbon compounds (Sackett and Thompson, 1963; Hedges and Parker, 1976; Erlenkeuser, 1978; Sweeney and Kaplan, 1980) and by their chemical composition (Gardner and Menzel, 1974; Pocklington, 1976; Palmer and Baker, 1978; Simoneit, 1978).

Only a small portion of the organic matter settling to the sea floor becomes fossilized, while its major part is recycled in benthic biological consumption and microbial oxidation. In environments with oxygenated bottom water the fraction that will eventually escape sur-

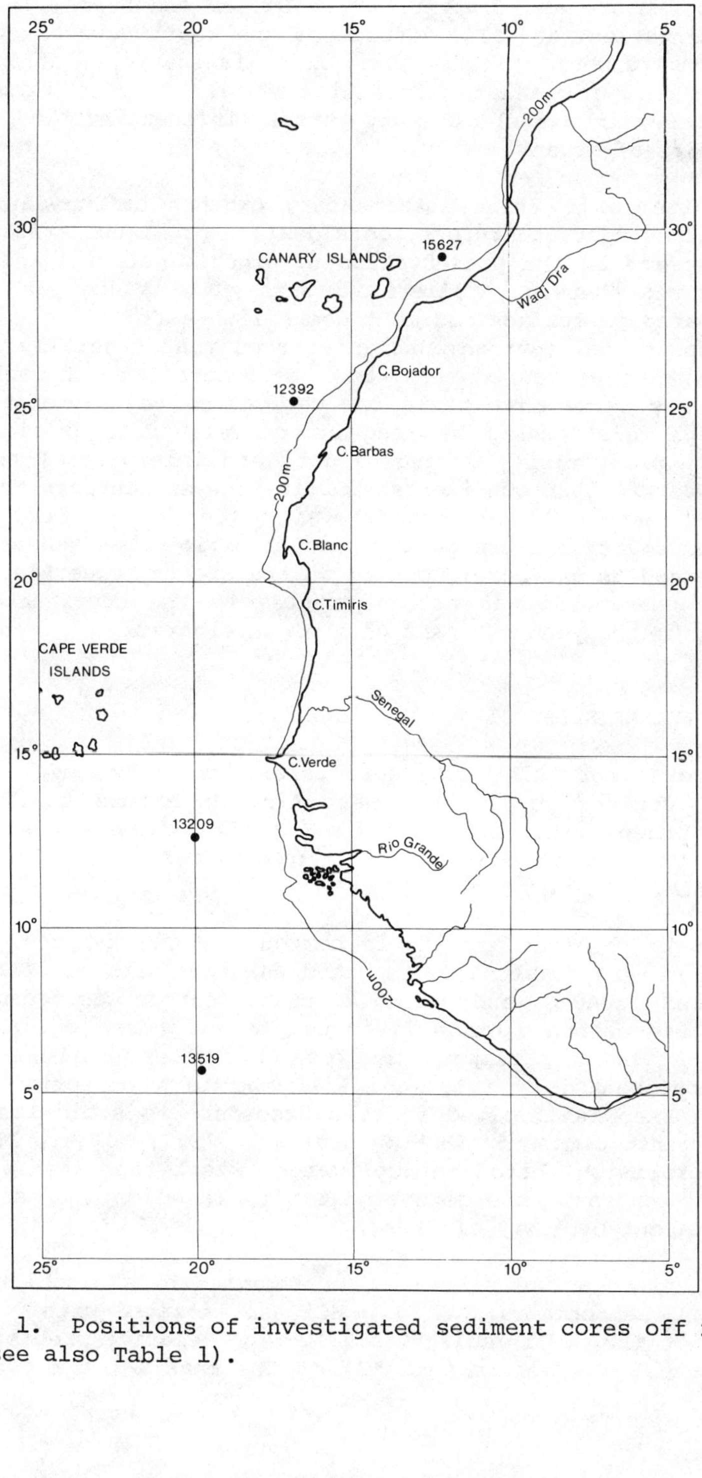

Fig. 1. Positions of investigated sediment cores off northwest Africa (see also Table 1).

face utilization seems to depend primarily on the bulk sedimentation rate (Heath, Moore and Dauphin, 1977; Müller and Suess, 1979). The bulk sedimentation rate also determines the depth downcore where the chemical environment switches from oxic diagenetic reaction to sub-oxic and anoxic mechanisms (Froehlich et al., 1979; Lerman, 1979; Müller et al., in press), thereby further influencing the post-depositional fate of organic material.

Any attempt to relate sedimentary organic carbon data to past productivities must, therefore, consider the prominent effect of sedimentation rate on the preservation of organic material and the signals contained therein. Müller and Suess (1979) used an empirical approach based on surface sediment compositions from widely differing marine depositional environments and showed that generally the sedimentary organic carbon content, bulk sedimentation rate and production rate may serve as a basis for estimating paleo-productivities. We use this relationship here again to evaluate the regional fluctuations in productivity during glacial and interglacial periods on the northwest African continental margin between southern Morocco and Gambia where coastal upwelling prevails today. We find that the paleo-productivity pattern during glacial periods was generally higher and one additional core, located on the Sierra Leone Rise, reveals that this phenomenon was not restricted to the continental margin region nor to the youngest part of the Quaternary.

MATERIALS AND METHODS

The cores for this study were taken during RV "Meteor" and RV "Valdivia" cruises to the northwest African continental margin and the Sierra Leone Rise (Table 1 and Fig. 1), where cores 13519 and 13209 are located at the same positions as DSDP Sites 366 and 367, respectively.

Prior to chemical and stable carbon isotope analyses, all sediment samples were desalted by repeated washings with distilled water, vacuum dried at 40°C, and ground. Organic carbon was measured volumetrically as carbon dioxide liberated by sulphuric acid/dichromate oxidation at boiling temperature for 20 minutes; carbonate-CO_2 was expelled beforehand by treatment with dilute phosphoric acid. No filtration step is involved in this procedure to avoid loss of acid soluble organic compounds (Hartmann et al., 1971; Müller, 1977). The relative precision, based on duplicates, was better than ±3% of the mean C-org contents; the detection limit, based on 2 g sample aliquots, is about 0.01 wt% of C-org.

Total nitrogen was liberated as ammonium by Kjeldahl digestion, distilled as ammonia into 0.01 N HCl and titrated with 0.01 N NaOH (Hartmann et al., 1971; Müller, 1977). The relative precision, based on duplicates, was better than ±3% of the mean N-total contents but

Table 1. Geographic positions and water depths of sediment cores investigated in this study

Cruise	Core	Latitude	Longitude	Water Depth	Reference
METEOR 25-1971	12392-1	25°10'N	16°51'W	2575 m	Seibold (1972)
VALDIVIA 10-1975	13209-2	12°29'N	20°03'W	4713 m	Seibold and Hinz (1976)
METEOR 51-1979	13519-2	05°40'N	19°51'W	2862 m	------
METEOR 53 C/D-1980	15627-3	29°10'N	12°05'W	1058 m	------

up to ±7% in case of sediments with low (<300 ppm) nitrogen contents. The detection limit, based on 250 mg aliquots, is about 50 ppm of nitrogen.

Carbonate-CO_2 was measured by infra-red absorption after dissolution of carbonates in dilute phosphoric acid using a closed circulation system. The results are expressed as $CaCO_3$ assuming calcite as the predominant carbonate phase. The relative precision, based on duplicates, was better than ±2% of the mean value in samples with more than 5 wt% $CaCO_3$. At sample weights of a few mg and appropriate gain settings of the IR-analyzer, the detection limit was about 0.5 wt% $CaCO_3$.

For carbon isotope analysis on the organic matter, aliquots of about 450 mg of desalted, dry sediment were treated with dilute phosphoric acid at about 95°C for several hours in order to remove the carbonates. To avoid losses of solubilized organic substances, the phosphoric acid/sediment mixture as a whole was subjected to sulphuric acid (70%)/dichromate oxidation under boiling conditions (150°C, 20 min) at reduced pressure in a closed reaction system (Schmidt, 1981). The CO_2 evolved was dried at -80°C, the yield measured volumetrically, and the gas transferred to a Micromass 602 D mass-spectrometer. $^{13}C/^{12}C$ ratios are expressed in the usual δ-notation and refer to the PDB-scale. The total isotopic reproducibility of the CO_2 preparation and measuring procedure was better than 0.1°/₀₀ on the δ-scale as determined by replicate analyses.

The oxygen isotope stratigraphy of core 13519 is based on the planktonic foraminifer *Globigerinoides sacculifer* (Sarnthein et al., in press). Carefully selected shells (0.1 to 0.5 mg per sample) were cleaned ultrasonically in methanol, dried, reacted with ∿100% phosphoric acid under vacuum at 50 ± 0.2°C, and analyzed by mass-spectrometry for stable carbon and oxygen isotope composition of the CO_2 gas thus obtained. Reference to the PDB scale was established via the NBS-20 isotope standard.

GLACIAL-INTERGLACIAL ORGANIC CARBON CYCLES

Core 15627

The $CaCO_3$ and organic carbon distributions in core 15627 from the upper continental slope off southern Morocco are illustrated in Fig. 2. The core was recovered from a water depth of 1058 m and appears to resolve eight oxygen isotope stages. The stage boundaries predicted in Fig. 2 are based on the organic carbon cycles which were calibrated primarily on oxygen isotope distributions in cores 13519 and 12392; these boundaries are in accordance with preliminary biostratigraphic results (U. Pflaumann, pers. comm.). The carbonate distribution provides an independent control, at least for the upper

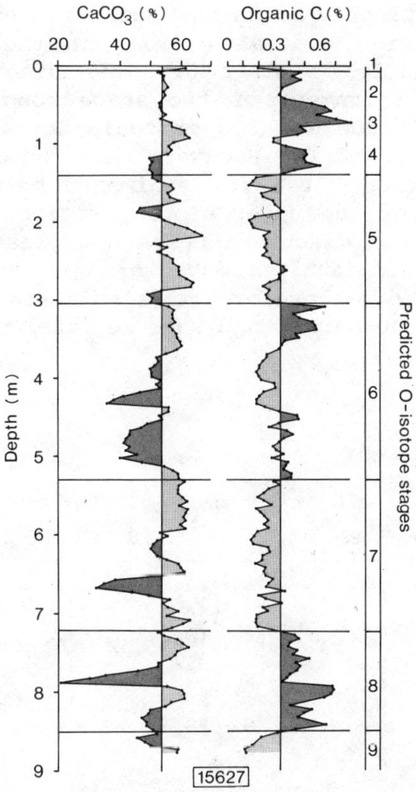

Fig. 2. Distribution of $CaCO_3$ and organic carbon in core 15627-3 from the continental slope off South Morocco. The stratigraphy is "predicted" from the organic carbon cycles as outlined in the text. Approximately 30 cm of surface sediment were lost during coring as suggested by the age/depth curve in Fig. 6. See Table 2 (Appendix) for original carbon and nitrogen data.

3 m of the core, in that three prominent carbonate maxima, typical for stage 5 in many Atlantic sediment cores, are readily apparent (e.g., Gardner, 1975; Bé et al., 1976; Sarnthein et al., 1982). Thus we feel strongly that the predicted stratigraphy is a good approximation although the detailed oxygen isotope record will certainly modify the position of the boundaries.

Sediments from glacial stages 2-4 and 8 (Fig. 2) have organic carbon contents higher by a factor of 2-4 than those from interglacial stages 1, 5, 7, and 9, although overall organic carbon concentrations are relatively low. In addition, close sample spacing at 5 cm intervals, corresponding to a time resolution of 1000 to 3000 years, reveals distinct fluctuations within single stages. Stage 6 sediments do not perfectly match the general trend of higher organic carbon contents during glacial periods in that they show a broad minimum for the middle of stage 6. This discrepancy might simply reflect incorrect assignment of the stage boundaries such that the core section between 300 and 520 cm could actually represent oxygen isotope stages 6, 7, and 8. However, this would result in unreasonably low sedimentation rates for sediments between 300 and 520 cm which is unlikely for the upper slope position of this core. Moreover, two other cores, discussed below and presented in Figs. 3 and 4, also reveal significantly reduced organic carbon contents during the middle of stage 6. Therefore, this suggests correct time assignment of the stage boundaries and some regional oceanographic control on carbon distribution.

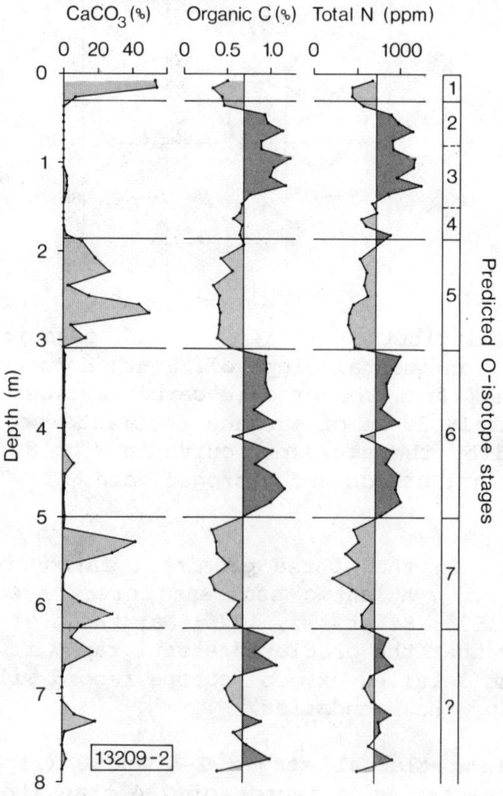

Fig. 3. Distribution of $CaCO_3$, organic carbon and total nitrogen in core 13209-2 from the abyssal plain off Gambia. The stratigraphy is "predicted" from the organic carbon cycles and the $CaCO_3$ distribution as outlined in the text. See Table 3 (Appendix) for analytical data.

Core 13209

The distributions of $CaCO_3$, organic carbon and total nitrogen in core 13209 are shown in Fig. 3. Again, the cyclicity in organic matter distribution is readily apparent. This core was recovered from the abyssal plain off Gambia from a water depth of 4713 m. As for core 15627, the stratigraphy was inferred from the organic carbon and $CaCO_3$ distributions calibrated by the oxygen isotope stratigraphy of cores 13519 and 12392.

The $CaCO_3$ contents are a sensitive reflection of changes in the level of the carbonate compensation depth (CCD) during the past eight oxygen isotope stages. Presently the CCD is at about 5000 m in this part of the Atlantic Ocean (Berger, 1978) and was obviously at the same depth during the interglacial stages 1, 5, and 7. Also the prominent $CaCO_3$ peaks during stage 5 are again apparent, supporting the assigned stratigraphy. The CCD must have shifted upwards considerably during glacial stages 2-4, 6, and 8. This is probably the result of two factors:

a) enhanced northward flow of Antarctic Bottom Water, which is enriched in CO_2 and thus chemically more aggressive towards biogenous carbonates; and

b) higher surface production rate of organic matter, as suggested by the organic carbon distribution, which increases the carbon dioxide content of the bottom water and thereby favors the dissolution of calcareous tests (Berger, 1976, Bender and Keigwin, 1979, Diester-Haass and Müller, 1979, Emerson and Bender, 1981).

The increased fraction of carbon dioxide from isotopically light organic matter in the glacial bottom water is evidenced by the carbon isotope composition of benthic foraminiferal tests, which is significantly lighter in glacial samples (Shackleton, 1977; Broecker, 1981; or Fig. 4 in Thiede, Suess and Müller, 1982).

Core 13519

This core (Fig. 4) had been recovered from the Sierra Leone Rise at a water depth of 2862 m and is dominantly composed of calcium carbonate with concentrations of 60 ± 30%. The oxygen isotope and calcium carbonate results will be discussed in detail by Sarnthein et al. (in press), but for the purpose of this study we extract only the stratigraphic information.

Core 13519 resolves the last 21 oxygen isotope stages corresponding to the past 800,000 years. The Brunhes/Matuyama boundary (730,000 yrs. B.P.; Mankinen and Dalrymple, 1979) occurs close to the oxygen isotope boundary 19/20 at a depth of 990 cm downcore (Bleil, in press).

The organic carbon content is very low ranging from 0.1 to 0.5 wt%. Total organic matter, as reflected by organic carbon and total

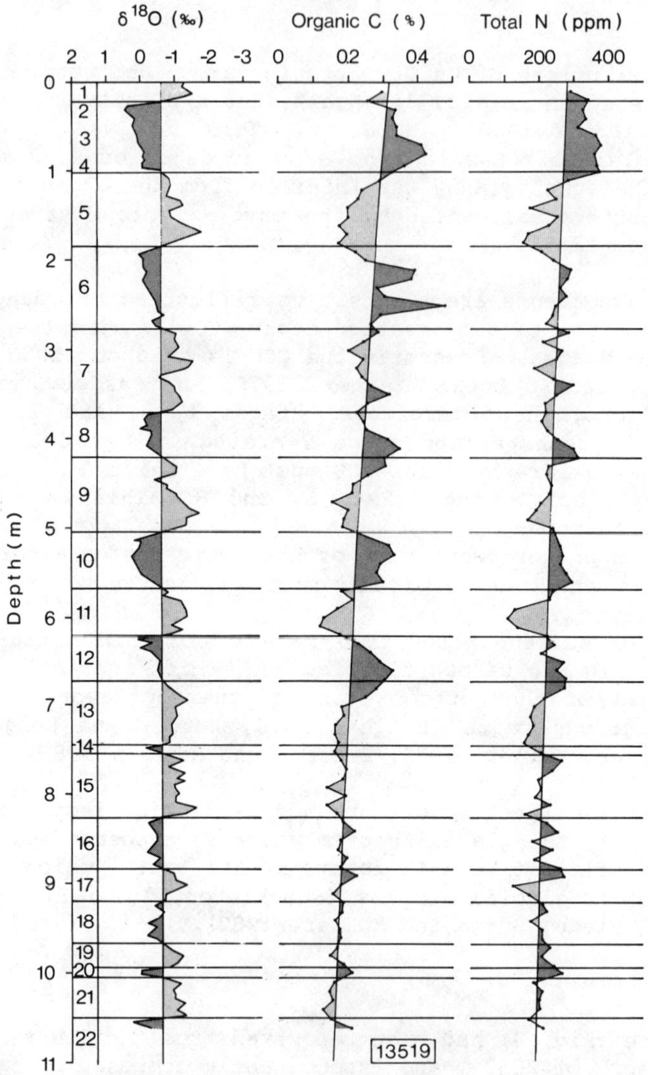

Fig. 4. Distribution of organic carbon and total nitrogen in core 13519 from the Sierra Leone Rise. Oxygen isotope stratigraphy after Sarnthein et al. (in press). The core penetrates approximately the past 800,000 year record. The vertical curves represent exponential fits through all data points and indicate significant diagenetic breakdown of organic compounds:

$$\text{C-org } (\%) = 0.317 \exp(-6.70 \times 10^{-4} z),\ r = 0.67$$

$$\text{N-tot (ppm)} = 281 \exp(-3.83 \times 10^{-4} z),\ r = 0.49$$

z = depth in core (cm); r = correlation coefficient. See Table 4 (Appendix) for carbon and nitrogen data.

nitrogen contents, decreases exponentially with depth in the core; for equations see caption of Fig. 4. Notwithstanding the low organic matter contents, core 13519 also reveals a distinct cyclicity through oxygen isotope stage 13, with glacial stages generally having more organic matter than interglacial stages. The cycles appear much less distinct in older core sections or even diminish entirely in certain intervals, perhaps signaling a smaller amplitude of the primary marine productivity signal. Interestingly, the amplitudes of the glacial-interglacial changes of the $\delta^{18}O$ record appear also to be reduced in these sediments. Applying the ages proposed by Morley and Hays (1981; see Fig. 6) to the oxygen isotope boundaries, reveals a small but significant reduction in average sedimentation rates in the lower part of this core which would tend to suppress the primary organic matter signal slightly more than in the upper part. The mechanism for suppression is increased decomposition of organic matter by oxygen consumption at low sedimentation rate. Müller and Mangini (1980) and Müller et al., (in press) have suggested a bulk sedimentation rate threshold of 2 ± 1 cm/1000 years below which the chemical environment of pelagic sediments remains oxygenated below the sediment/water interface and thereby severely reduces organic matter contents. The sedimentation rate of core 13519 essentially falls within this range. Therefore, small differences in average sedimentation rates, at generally low rates, might have had a relatively large negative effect on the preservation of organic matter during oxic diagenesis.

Bioturbation could additionally have contributed to the apparent reduction of the amplitudes of the organic carbon cycles in the deeper section of the core, for example in the short interval represented by glacial isotope stage 14.

ORIGIN OF SEDIMENTARY ORGANIC MATTER

Any attempt to interpret the organic carbon cycles in terms of paleo-productivity requires some knowledge of the relative proportions of marine and terrestrial organic matter. The continental margin sediment cores in particular may contain organic matter of terrestrial origin supplied by wind and rivers, depending on the climatic conditions on the adjacent continent. One method to distinguish between marine and terrestrial organic matter in sediments is by stable carbon isotopes. Marine planktonic organic matter formed at low and intermediate geographic latitudes has $\delta^{13}C$ values of -18 to -21°/₀₀ (Degens, 1969; Fontugne and Duplessy, 1981). Terrestrially derived organic matter is lighter by 5-8°/₀₀ as shown in several studies for the Gulf of Mexico (Newman, Parker and Behrens, 1973; Hedges and Parker, 1976; Northam et al., 1981).

We have chosen core 12392 to study the carbon isotope composition of the organic fraction in detail. This core was recovered from a water depth of 2575 m at the continental rise off Mauritania and resolves the past 140,000 years (Fig. 5). The sediments contain 50-75% $CaCO_3$, dominantly of coccoliths and foraminifers, up to 4% biogenic opal, and 5-20% quartz (Thiede et al., 1982). The grain size fraction <6 μm constitutes on average of 64% of the acid insoluble material and is about 20% higher in interglacial sediments than in glacial sediments (Chamley, Diester-Haass and Lange, 1977). The clay mineral composition is dominated by smectite and illite and enriched in so-called mixed-layer clay minerals in glacial core sections. Kaolinite occurs in about the same proportions throughout the core and chlorite is of minor importance. The interpretation of poorly defined mixed-layer clay minerals in the sedimentary record is controversial. Chamley et al. (1977) and Chamley and Diester-Haass (1979) claimed these clay minerals to be indicative of river-derived material. However, Lange (1982) and Sarnthein et al. (1982) have shown that they represent irregular clay aggregates which are difficult to disperse by conventional laboratory treatments and their occurrence appears to be confined, at the present time, to dust falls and evaporite environments of sebkhas.

Figure 5 shows the oxygen isotope stratigraphy of core 12392 (Shackleton, 1977), the organic carbon distribution, and the stable carbon isotope composition of the total organic matter. The organic carbon distribution in this core has been discussed by Müller and Suess (1979) and Thiede et al. (1982). The main conclusion was that a higher productivity of glacial surface waters as compared to the present oceanographic situation and to interglacial oxygen isotope stage 5 may be responsible for the organic carbon accumulation patterns. The stable carbon isotope results verify and support this conclusion in that they reveal higher organic matter contents of glacial sediments are <u>not</u> the result of increased terrestrial organic matter input.

Except for a few samples from stage 5c-d, sediments from the interglacial periods show $\delta^{13}C$ values of around -20.0 to -20.5 °/₀₀. Glacial sediments with high organic matter contents reveal significantly heavier $\delta^{13}C$ values which amount to -18.4 ± 0.2°/₀₀ for stage 2 deposits and appear to approach similarly high ^{13}C-contents at the base of this core (late glacial stage 6).

The fraction of land-derived organic matter in the sediment may be estimated from mixing models balancing the relative contributions from the presumed sources of sedimentary organic carbon by means of the indigenous carbon isotope composition of the respective source materials (Sackett and Thompson, 1963). Assuming the carbon isotope composition of planktonic organic matter to represent the $^{13}C/^{12}C$-ratio of the bulk marine-derived organic detritus arriving at the sea bottom (Sackett et al., 1965; Hedges and Parker, 1976) and using the model of Fontugne and Duplessy (1981) for estimating $\delta^{13}C$ of planktonic organic carbon, the $\delta^{13}C$ value should be about -20°/₀₀ for the

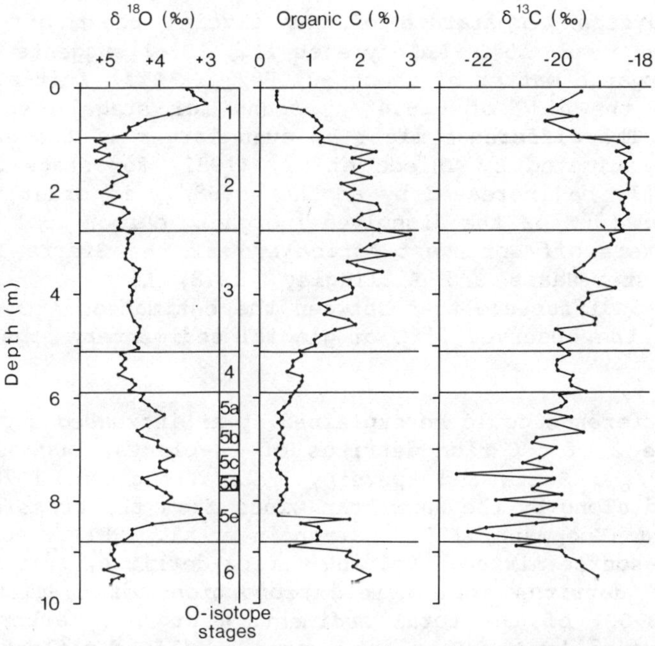

Fig. 5. Distribution and stable isotope composition of organic carbon in core 12392-1 from the continental rise off Mauritania (see also Table 1, Appendix). Oxygen isotope stratigraphy after Shackleton (1977) and Thiede et al. (1982). Note that glacial stages with high organic matter contents show the heaviest $\delta^{13}C$ values, suggesting marine planktonic organic matter as the primary organic carbon source.

marine organic fraction in the recent sediments at the coring site (surface water temperature: ca. 20°C; Ganssen and Sarnthein, this volume). This $\delta^{13}C$ value closely matches the carbon isotope composition actually found in interglacial sediments. Thus, there is little reason to consider a major contribution of terrestrial organic carbon to interglacial sediments at the site of core 12392. However, the considerable variability of $\delta^{13}C$ among the individual plankton samples analyzed by Fontugne and Duplessy (1981) renders it difficult to define a representative mean value and an extreme value such as $\delta^{13}C = -17°/_{oo}$ seems not impossible. Hence, as a worst-case estimate, we calculate a terrestrial contribution of 40% of the total sedimentary organic matter or 0.14 wt% of organic carbon, assuming $\delta^{13}C$ values of -26, -20.5 and $-17°/_{oo}$ for the terrigenous, sedimentary and planktonic organic carbon, respectively.

For the glacial sediments, particularly those of stage 2, the stable carbon isotope record is not as well understood and a major discrepancy remains unsolved: Following Fontugne and Duplessy (1981) the lower surface temperature of the glacial ocean off northwest Africa (annual mean: 15°C, McIntyre et al., 1976) suggests a $\delta^{13}C$ for planktonic organic matter of about -21.8‰. This initial value is lighter than the $\delta^{13}C$ of -18.4‰ found for stage 2 sediments in core 12392. The difference might be even larger if the sea surface temperature estimated by Thiede et al. (1982) for stage 2 is used, and must likely be increased by another 0.6‰ in order to correct for the lower $\delta^{13}C$ of the dissolved inorganic carbon pool of glacial upwelling waters off Northwest Africa and at the Sierra Leone Rise (Berger, Diester-Haass and Killingley, 1978; Lutze et al., 1979). The compounded difference then between the estimated $\delta^{13}C$ of glacial plankton and the observed $\delta^{13}C$ of glacial sedimentary organic matter is >3‰.

This difference could be explained by an increased supply during glacial stage 2 of ^{13}C-rich detritus of C_4-plants, such as grasses ($\delta^{13}C \sim -13$‰, Smith and Epstein, 1971; Troughton, 1972; Lerman, 1979) carried along by the Harmattan winds from the grasslands south of the Sahara (Koopmann, 1981; Sarnthein et al., 1981). Accordingly, simple three-source mixing involving marine detritus, terrestrial C_3- and C_4-plant detritus would yield proportions of C_4-plant organic carbon of 40-60% of the total sedimentary organic carbon. Such a high contribution by the C_4-plant source should be reflected by high C/N ratios since grass contains low amounts of proteineceous matter when compared to plankton. Typical C/N wt.-ratios are 5-8 for marine plankton, but 20-50 for grasses (Redfield, Ketchum and Richards, 1963; Allen et al., 1974; Suess and Ungerer, 1981). However, the C/N ratios of stage 2 sediments cluster around a relatively low mean value of 8.7 (Thiede et al., 1982). It would be difficult to reconcile any appreciable contribution from a terrestrial plant source with these C/N characteristics (Suess and Müller, 1980).

Fractionation during degradation of planktonic organic material appears to shift the $\delta^{13}C$ values of the residue towards lighter values (Degens et al., 1968). This may also explain the lighter values of interglacial sediments which accumulated at a slower rate than glacial sediments (Thiede et al., 1982; see also Fig. 6) thereby favoring a more intensive oxic surface diagenesis. This view is supported by amino acid studies which suggest that the organic material preserved in slowly accumulating sediments has suffered more extensive degradation. In addition, the large scattering of modern plankton $\delta^{13}C$ data (Fontugne and Duplessy, 1981) indicates that effects other than water temperature also seem to be involved in constituting the carbon isotope composition. Degens et al. (1968), Degens (1969), Eadie, Jeffrey and Sackett (1978), and Galimov (1980) report significant differences in $\delta^{13}C$ between different biochemical constituents of planktonic material. Therefore, the enrichment of ^{13}C from interglacial to glacial sedimentary organic matter may be partly related to a changing biochemical quality of the planktonic matter which may depend on species composition and surface water mass properties (Fontugne and Duplessy, 1978). Similar heavy $\delta^{13}C$ values as found in stage 2 sediments of core 12392 were reported by Degens et al. (1968) and Reimers and Suess (this volume) for modern plankton from the upwelling region off Peru.

The foregoing discussion did not include the samples between 830 and 880 cm representing the period of rapid deglaciation at the transition from stage 6 to stage 5. The very light organic carbon isotope ratios of these sediments could indeed reflect a higher amount of terrestrial organic matter, both in relative and absolute terms. The same effect appears to hold true also for the deglaciation period from stage 2 to stage 1. An increased deposition of terrestrial organic matter during the rapid deglaciation periods may be explained by reworking and downslope transport of organic material deposited on the dry parts of the shelf during times of low sea level.

Since the present and past climates at latitude 25°N remained dominantly arid during the time interval cored, interrupted by humid periods during stages 1, 3 and 5 (Sarnthein, 1978; Lutze et al., 1979), we can exclude rivers as a significant influence and conclude that the small amount of terrestrial organic matter which might be present in core 12392 represent eolian fallout. This is also in accordance with the composition of the terrigenous sediment fraction, which does not show the typical bimodal grain-size distribution with an excess of clay indicative of river load (Koopmann, 1981; Sarnthein et al., 1982). Accordingly, the other cores of this study which are located to the north and south of the region with maximum eolian fallout (Sarnthein and Koopmann, 1980; Sarnthein et al., 1981) should have received even less eolian organic matter.

River influence can *a priori* be excluded at the distant and elevated position of core 13519 from the Sierra Leone Rise. There is also no evidence that core 13209 from the abyssal plain off Gambia has received significant amounts of river load (Koopmann, 1981). Several recent studies have revealed that the material supplied by rivers south of about 16°N is transported parallel to the coast by undercurrents, rather than to the west, and deposited at the continental slope in water depths between 500 and 2500 m (Lange, 1975; Diester-Haass and Müller, 1979; Fütterer, 1980).

Moreover the sedimentological study of numerous cores from the continental margin between 10 and 30°N off northwest Africa by Koopmann (1981) lead to the conclusion that rivers were even less active during the last glacial maximum than during the Holocene throughout the region. This is in accordance with the view that subtropical and tropical climatic belts were compressed during the last glacial period whereas the arid zone extended further to the south (Sarnthein, 1978). Palynological investigations off northwest Africa (Caratini, Bellet and Tissot, 1979; Rossignol-Strick and Duzer, 1979; Agwu and Beug, 1982) and the history of the Niger River during the past 30,000 years (Pastouret et al., 1978) point in the same direction.

In summary, the isotopic composition of the organic material preserved in core 12392 reveals that interglacial-glacial organic carbon cycles are not caused by varying inputs of terrestrial organic matter. Sedimentological and climatic evidence suggests that this is also true in the other cores of this study. We may thus relate the organic carbon cycles to changing oceanic productivities in the past.

ESTIMATION OF PALEO-PRODUCTIVITY

In order to extract quantitative information about past changes of oceanic productivity from the sedimentary organic carbon record, we must account for effects of water depth, sedimentation rate, surface and post-depositional diagenesis, all of which determine the organic matter fraction which is eventually permanently buried (Reimers and Suess, this volume). Sediment trap data suggest that roughly 90% of the original production is recycled in the upper 1000 m of the water column (Suess, 1980). The fraction reaching the sea floor becomes further reduced and transformed by benthic biological processes leaving only a very small portion for burial. The absolute amount of organic matter escaping surface diagenesis appears to depend primarily on the bulk sedimentation rate which determines whether the chemical environment affecting decomposition remains oxic, and thereby relatively more efficient, or whether it switches to suboxic and anoxic conditions which are less efficient (Müller et al., in press).

In the following, we apply an empirically derived relationship (Müller and Suess, 1979) to estimate absolute productivity changes encoded in the organic carbon cycles. This relationship does not account individually for the assorted decomposition mechanisms and sites of significant reductions of organic matter as listed above, but simply expresses their overall integrated effects.

$$R = \frac{C \cdot p_s \cdot (1 - \phi)}{0.0030 \cdot S^{0.30}} \tag{1}$$

where R = estimated paleo-productivity ($gC\ m^{-2}\ y^{-1}$)
C = sedimentary organic carbon content (% dry wt.)
p_s = dry sediment density ($g\ cm^{-3}$)
ϕ = porosity/100
S = sedimentation rate ($cm\ 1000^{-1}\ y^{-1}$)

Equation (1) is based on a set of surface sediments from different oceanic regions with widely differing primary production rates, organic carbon contents, and sedimentation rates, but of similar lithologies and textures. Therefore, it should best be applied to sediments with low sand contents and those not affected by turbidites or slumping and/or containing significant portions of terrestrial organic matter.

Sedimentation rates usually are linearly extrapolated over relatively large depth (time) intervals depending on the number of absolute ages available. For core 12392, we use here the sedimentation rates as proposed by Thiede et al. (1982). Sedimentation rates of the three other cores are based on the time scale proposed by Morley and Hays (1981) although other scales (e.g., Kominz et al., 1979) could be used as well.

These time scales for the past 750,000 years are based on variations of orbital parameters and their corresponding frequency components in the oxygen isotope record of several deep-sea cores. Both scales are in perfect agreement to within 5,000 years for the last 350,000 years, but differ by up to 41,000 years in older sediments. The age/depth relationships of the four cores of this study are illustrated in Fig. 6 based on the time scale of Morley and Hays (1981). The dotted line in the lower part of core 13519 connects ages proposed by Kominz et al. (1979) for the respective oxygen isotope boundaries. Ages assigned by both authors to oxygen isotope boundary 10/11 are indicated by the question mark. They would yield

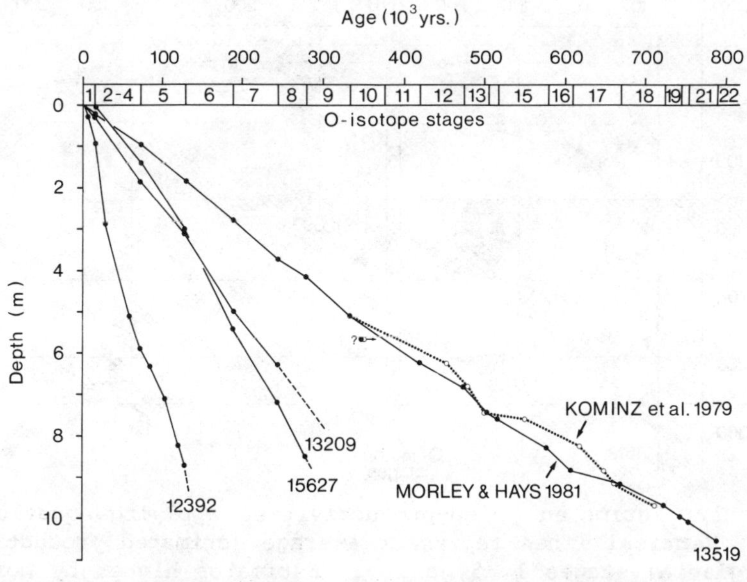

Fig. 6. Age/depth curves for the four sediment cores of this study. Heavy dots designate ages proposed by Morley and Hays (1981) for the oxygen isotope boundaries. Ages for stage boundaries 20/21 and 21/22 were extrapolated using the mean sedimentation rate of stages 13-19 (1.3 cm/1000 yr). Open circles designate ages proposed by Kominz et al. (1979). The ages proposed for oxygen isotope stage boundary 10/11 (question mark) by both authors result in unusually high sedimentation rates for stage 10 and were not considered in calculating sedimentation rates.

relatively high sedimentation rates for stage 10, not compatible with the general sedimentation rate pattern of this core. Therefore, we feel justified in using the same sedimentation rate for stages 10 and 11.

Assumptions had to be made with respect to the physical properties of the sediments because they were only available for core 12392 (Thiede et al., 1982). A mean dry density of 2.71 g cm^{-3} and a mean porosity of 70% were used with the other three cores. These values are based on unpublished data of F.-C. Kögler (Kiel, pers. comm.) and on our own measurements of water contents of several sediment cores from the northwest African continental margin. Porosities decreased by about 10% in the upper 10 cm of sediment but remained constant (at about 70%) in deeper core sections.

In applying equation (1) we did not include a correction for the post-depositional diagenetic loss of organic carbon in cores 15627,

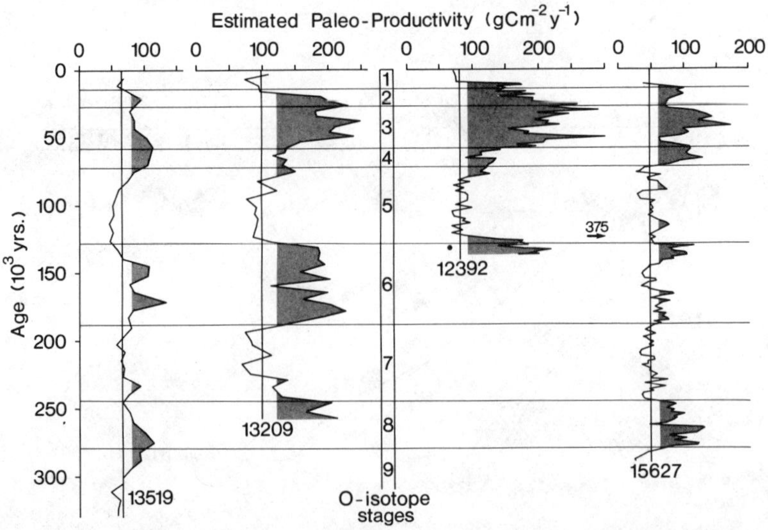

Fig. 7. Estimated paleo-productivities applying equation (1), see text. Vertical lines represent average estimated productivities for interglacial stages 1, 5, and 7. Estimates higher by more than one standard deviation of the average interglacial productivity values are accentuated by the stippled pattern. Note that the estimated productivities fall into a reasonable order of magnitude and that higher productivities consistently occur during glacial periods. Two data points close to Termination II (stage boundary 5/6) in core 12392 deviate significantly from the general pattern and were not considered. In core 13519 from the Sierra Leone Rise the glacial-to-interglacial differences are less pronounced although still significant. The complete profile of this core is shown in Fig. 8.

13209, and 12392, since the organic carbon/depth profiles (Figs. 2, 3, 5) suggest this loss to be small when compared to the pronounced differences in interglacial and glacial sediments. Core 13519, on the other hand, shows a significant organic carbon decrease with age. Before substituting organic carbon values into equation (1) they were corrected by a factor obtained for each depth interval from the exponential fit through all data (Fig. 4 and equation therein).

The estimated paleo-productivities for oxygen isotope stages 1 to 9 at all four sites are shown in Fig. 7. The vertical lines indicate average productivities for interglacial stages 1, 5 and 7 and estimates, higher by more than one standard deviation of this aver-

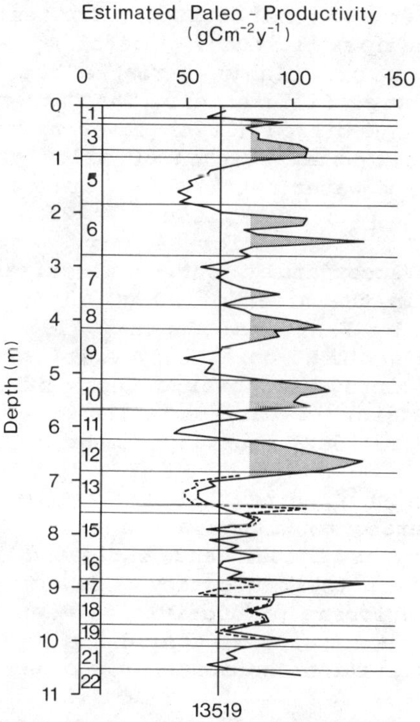

Fig. 8. Estimated paleo-productivities for Sierra Leone Rise station 13519 based on the time scale of Morley and Hays (1981) and empirical equation (1), see text. Distinct glacial-to-interglacial cycles can be traced back to oxygen isotope stage 13 corresponding to about 500,000 years B.P. Older sediments do not reveal a consistent pattern due to diagenetic breakdown of organic matter and uncertain ages of the stage boundaries (cp. Fig. 6). The dotted line below stage boundary 12/13 is based on sedimentation rates calculated from the time scale of Kominz et al. (1979) to illustrate the effect of uncertain sedimentation rates.

age, are accentuated by the stippled pattern. The decoding of sedimentary organic carbon percentages into units of productivity at the sea surface via equation (1) results in reasonable order of magnitude estimates when compared with actual measurements. Core 13519 from the Sierra Leone Rise and core 13209 from the abyssal plain off Gambia are located beneath water masses of low to moderately high primary production of 50 to 150 gC m^{-2} y^{-1} (Steemann Nielsen and Jensen, 1957; Schemainda et al., 1975; Degens and Mopper, 1976) and the average interglacial production rate estimates are 66 and 98 gC m^{-2} y^{-1}, respectively. The primary production rate today in surface waters at site 12392 is 75 gC m^{-2} y^{-1} and stage 5 sediments yield essentially the same value (Müller and Suess, 1979).

For the waters of site 15627, located over the continental slope off South Morocco, it is more difficult to assign them an average production rate. At this latitude, coastal upwelling is most pronounced during summer but highly variable both in space and time leading to large seasonal differences. The data reported by Grall et al. (1974) and Groupe Mediprod (1974) cover a range from 0.2 to 4.2 gC m^{-2} day^{-1} for shelf waters within 60 miles off the coast. Offshore waters at >200 m water depth, which are of particular interest here, were found to be much less productive with mean values ranging from 0.06 to 0.43 gC m^{-2} day^{-1} for January/February and July/August 1972, respectively. Accordingly, our best estimate for average primary production is in the order of 90 gC m^{-2} y^{-1}. Measurements of ~10 mg C m^{-3} day^{-1} by Fiala and Jacques (1974) in surface waters close to the site of sediment core 15627 yield an estimated mean rate of 110 gC m^{-2} y^{-1} if integrated over a depth of 50 m and assuming a linear decrease. Again, comparison with the average interglacial production rate of core 15627 shows agreement within a factor of 2.

We emphasize at this point that our estimates of paleo-productivity are only accurate to within a factor of 2-3. Therefore, the close agreement between actual and estimated Holocene production rates is not a proof of their accuracy. However, the fact that actual and estimated Holocene productivities meet within a reasonable range in all four cores gives some confidence as to the relative temporal and regional variations estimated for older periods.

We also attach little significance to the productivity fluctuations within single oxygen isotope stages, which merely reflect the variations of absolute organic carbon contents, because constant sedimentation rates were assumed between adjacent stage boundaries except for the 30,000 year old section of core 12392 (Thiede et al., 1982) for which ^{14}C dates were available. Actually, sedimentation rates may vary considerably, and probably do, at a depth-/time-scale not resolvable at the present time. The following interpretation therefore focusses on differences between interglacial and glacial periods rather than on short-term variations.

DISCUSSION AND CONCLUSIONS

We attribute the pronounced productivity differences between interglacial and glacial stages (Figs. 7 and 8) to changing intensities of upwelling as a response to changing global climatic and oceanographic parameters. In addition, shifting climatic belts may have resulted in a higher global nutrient discharge to the ocean, for example by erosion of high-latitude interglacial soils in glacial periods by advancing ice sheets. The low and relatively constant levels of productivity found during interglacial stages 1 through 13 indicate that the factors determining oceanic productivity in the eastern interglacial North Atlantic were of the same magnitude during the past 500,000 years.

The present current regime off northwestern Africa is characterized by the Canary Current which carries cold surface water to the south, and by two undercurrents of opposite flow direction at depths between 100 and 600 m (Smith, this volume). The southward flowing undercurrent follows the Canary Current and carries North Atlantic Central Water (NACW) and the northward flowing undercurrent transports South Atlantic Central Water (SACW) which is richer in nutrients. Both undercurrents meet and veer out to the open sea somewhere near 23°N (Mittelstaedt, 1972; Hughes and Barton, 1974; Mittelstaedt, Pillsbury and Smith, 1975; Tomczak and Hughes, 1980). Accordingly, coastal upwelling waters are fed by NACW north of 23°N and by SACW south of 23°N. The fact that the NACW is richer in oxygen and more depleted in nutrients is one reason for the generally lower productivity of the northern upwelling regime. The most productive waters are near ∿21°N off Cape Blanc, where SACW feeds coastal upwelling (Fraga, 1974; Schemainda et al., 1975; Shaffer, 1976; Huntsman and Barber, 1977).

The situation apparently was quite different during the last glacial stage. Estimates of sea surface temperatures from planktonic foraminiferal assemblages of North Atlantic sediment cores suggest that the Canary Current waters were cooler then by 6-8°C and extended farther south by ∿5° latitude and also somewhat farther offshore (McIntyre et al., 1976; Thiede, 1977; Sarnthein et al., 1982). There is also evidence for a penetration of nutrient-rich SACW farther northward based on the radiolarian species *Phormospyris stabilis scaphipes* (Labracherie, 1980) and supporting the CLIMAP reconstruction that the NACW ceased to form (McIntyre et al., 1976). Finally, there is strong evidence for an increased atmospheric circulation during the last glacial maximum based on modelling of CLIMAP data (Manabe and Hahn, 1977), on the distribution of sand dunes in the Saharan Desert (Sarnthein and Walger, 1974; Sarnthein, 1978), and on the distribution of eolian-marine deposits (Sarnthein and Koopmann, 1980; Koopmann, 1981; Sarnthein et al., 1981; 1982).

Thus, it appears well established that the increased productivities estimated here for Quaternary glacial stages are the result of intensified upwelling as a response to a general increase in atmospheric and oceanic circulation and to the tapping of nutrient-rich SACW by the poleward undercurrent to feed coastal upwelling. This conclusion is further corroborated by abundant deep and cold water species of planktonic foraminifera in glacial deposits off Senegal and Mauretania (Pflaumann, 1975; Lutze et al., 1979) and by the lighter $^{13}C/^{12}C$ ratios in the shell carbonate of glacial benthonic foraminifers off northwest Africa (Shackleton, 1977; or Fig. 4 in Thiede et al., 1982). Isotopically light shell carbonate suggests an increased contribution of metabolic CO_2 from decaying organic matter to the bottom water. The view of higher glacial productivity is also in accordance with the conclusions reached by Diester-Haass (1977; 1978; this volume), Labracherie (1980), and Labracherie et al. (this volume) from radiolarian distributions and coarse-grained component studies.

Higher glacial productivity during Quaternary times appears not to be restricted to continental margin regions as seen in core 13519 from the Sierra Leone Rise. Again, the agreement with the CLIMAP reconstruction of the glacial eastern Atlantic oceanography is striking (McIntyre et al., 1976). It shows an increased equatorial divergence between 0° and 6°N latitude and a greater mass transport of Southern Atlantic surface waters into the North Atlantic, both suggesting a higher nutrient flux to equatorial surface waters.

Equatorial and coastal upwelling was apparently also intensified in the glacial north Pacific Ocean at mid- and low-latitudes (CLIMAP Project Members, 1976; Seibold and Berger, 1982). However, to the best of our knowledge, there is no comparable organic carbon record in north Pacific pelagic sediments (refs. Heath et al., 1977). The problem may be due to poor preservation as a function of the extremely low sedimentation rates in vast parts of the Pacific which do not allow burial of enough organic material. Hemipelagic sediment cores, however, should show similar carbon preservation cycles as our cores do for the eastern North Atlantic if productivity has changed significantly during the Late Quaternary.

If it were possible to show that oceanic productivity was increased worldwide during glacial Quaternary stages, this would have important implications for the global CO_2 budget. It has been postulated from gas contents of ice cores from Greenland and Antarctica (Berner, Oeschger and Stauffer, 1980; Delmas, Ascenio and Legrand, 1980) that the carbon dioxide content of the late glacial atmosphere was lower by about 50% than at present. Broecker (1981) has discussed the apparent increase in atmospheric CO_2 content during interglacial stage 1 which "involves the loss of phosphorus from the sea to the shelf sediments during the early postglacial transgression of sea level, reducing the amount of plant matter formed per unit of upwelled water and thereby increasing the CO_2 pressure in surface water and the atmosphere". Berger (1982) has discussed the coral reef hypothesis which "calls for shelf carbonate buildup during transgression, which releases CO_2 to the upper ocean and the atmosphere".

Since the ocean contains 60 times more carbon than the atmosphere, slight changes in its inventory may have a pronounced effect on the CO_2 content of the atmosphere. According to Garrels, MacKenzie and Hunt (1975) a significant change in the ratio of photosynthesis to respiration and decay could change atmospheric levels several-fold in 100 years. The results of our study appear to give evidence for the significance of such processes in that they suggest a higher rate of both glacial primary production and burial of organic carbon in the sediments. The enhanced extraction of CO_2 by photosynthesis from the euphotic zone could have resulted in an increased flux of atmospheric CO_2 into surface waters thereby reducing the glacial atmospheric CO_2 level.

ACKNOWLEDGEMENTS

We thank H. Hensch and D. Schmidt for the careful chemical analyses and H. Cordt for his expert work with the isotope measurements. M. Sarnthein provided helpful advice on stratigraphic interpretations. E. Suess critically reviewed the manuscript. We are also indebted to captains and crews of RV "Meteor" and RV "Valdivia", and to F.-C. Kögler, N. Mühlhan, and F. Werner who recovered these outstanding sediment cores from the eastern Atlantic. The financial support by the Deutsche Forschungsgemeinschaft (DFG), Bonn is gratefully acknowledged.

REFERENCES

Agwu, C. and Beug, H.J., 1982, The history of the vegetation and climate as recorded in marine sediments off the West African coast, "Meteor"Forschungs-Ergebnisse, C36:1-30.

Allen, S.E., Grimshaw, H.M., Parkinson, J.A., and Quarmby, C., 1974, "Chemical Analysis of Ecological Materials," Blackwell Scientific Publication, Oxford, England, 565 pp.

Bé, A.W.H., Damuth, J.E., Lott, L., and Free, R., 1976, Late Quaternary climatic record in western equatorial Atlantic sediment, Geological Society of America, Memoir, 145:165-200.

Bender, M.L. and Keigwin, L.D., Jr., 1979, Speculations about the upper Miocene change in Abyssal Pacific dissolved bicarbonate $\delta^{13}C$, Earth and Planetary Science Letters, 45:383-393.

Berger, W.H., 1976, Biogenous deep-sea sediments: production, preservation and interpretation, in: "Chemical Oceanography," J.P. Riley and R. Chester, eds., Academic Press, London, 265-388.

Berger, W.H., 1978, Sedimentation of deep-sea carbonate: maps and models of variations and fluctuations, Journal of Foraminiferal Research, 8:286-302.

Berger, W.H., 1982, Increase of carbon dioxide in the atmosphere during deglaciation: the Coral Reef Hypothesis, Naturwissenschaften, 69:87-88.

Berger, W.H., Diester-Haass, L. and Killingley, J.S, 1978, Upwelling of North-West Africa: The Holocene decrease as seen in carbon isotopes and sedimentological indicators, Oceanologica Acta, 1:3-7.

Berner, W., Oeschger, H. and Stauffer, B., 1980, Information on the CO_2 cycle from ice core studies, Radiocarbon, 22:227-235.

Bleil, U., in press, Magnetostratigraphy of "Meteor" core 13519, Sierra Leone Rise, "Meteor"Forschungs-Ergebnisse, C.

Broecker, W.S., 1981, Glacial to interglacial changes in ocean and atmospheric chemistry, in: "Climatic Variations and Variability: Facts and Theories," A. Berger, ed., D. Reidel Publishing Company, Dordrecht, Holland, 111-121.

Caratini, C., Bellet, J. and Tissot, C., 1979, Etude microscopique de la matiére organique: Palynologie et palynofacies, in: "Géochemie Organique des Sédiments Marins Profonds," ORGON III, Mauritanie - Sénégal - Îles du Cap-Vert, M. Arnould and R. Pelet, eds., Editions du Centre National de la Recherche Scientifique, Paris, 215-247.

Chamley, H. and Diester-Haass, L., 1979, Upper Miocene to Pleistocene climates in NW Africa deduced from terrigenous components of Site 397 sediments, in: "Initial Reports DSDP," 47A, W.B.F. Ryan, U. von Rad, et al., U.S. Government Printing Office, Washington, 641-646.

Chamley, H., Diester-Haass, L. and Lange, H., 1977, Terrigenous material in East Atlantic sediment cores as an indicator of NW African climates, "Meteor"Forschungs-Ergebnisse, C26:44-59.

CLIMAP Project Members, 1976, The surface of ice-age earth, Science, 191:1131-1137.

Degens, E.T., 1969, Biogeochemistry of stable carbon isotopes, in: "Organic Geochemistry," G. Eglinton and M.T.J. Murphy, eds., Springer-Verlag, Berlin, 304-329.

Degens, E.T., Behrendt, M., Gotthardt, B., and Reppmann, E., 1968, Metabolic fractionation of carbon isotopes in marine plankton - II. Data on samples collected off the coasts of Peru and Ecuador, Deep-Sea Research, 15:11-20.

Degens, E.T. and Mopper, K., 1976, Factors controlling the distribution and early diagenesis of organic material in marine sediments, in: "Chemical Oceanography," Vol. 6, J.P. Riley and R. Chester, eds., Academic Press, London, 59-113.

Delmas, R.J., Ascenio, J.-M. and Legrand, M., 1980, Polar ice evidence that atmospheric CO_2 20,000 yrs B.P. was 50% of present, Nature, 284:155-157.

Deuser, W.G. and Ross, E.H., 1980, Seasonal change in the flux of organic carbon to the deep Sargasso Sea, Nature, 283:364-365.

Diester-Haass, L., 1977, Radiolarian/planktonic foraminiferal ratios in a coastal upwelling region, Journal of Foraminiferal Research, 7:26-33.
Diester-Haass, L., 1978, Sediments as indicators of upwelling, in: "Upwelling Ecosystems," R. Boje and M. Tomczak, eds., Springer-Verlag, Berlin, 261-281.
Diester-Haass, L. and Müller, P.J., 1979, Processes influencing sand fraction composition and organic matter content in surface sediments off W. Africa (12-19°N), "Meteor"Forschungs-Ergebnisse, C31:21-47.
Eadie, B.J., Jeffrey, L.M. and Sackett, W.M., 1978, Some observations on the stable carbon isotope composition of dissolved and particulate organic carbon in the marine environment, Geochimica et Cosmochimica Acta, 42:1265-1269.
Emerson, S. and Bender, M.L., 1981, Carbon fluxes at the sediment-water interface of the deep-sea: calcium carbonate preservation, Journal of Marine Research, 39:139-161.
Erlenkeuser, H., 1978, Stable carbon isotope characteristics of organic sedimentary source materials entering the estuarine zone, in: "Biogeochemistry of Estuarine Sediments," Proceedings of a UNESCO/SCOR Workshop held in Melreux, Belgium, United Nations Educational, Scientific and Cultural Organization, Paris, Publication 199-206.
Fiala, M. and Jacques, G., 1974, Relations entre ATP, chlorophylle et production dans la couche euphotique d'une zone d'upwelling (Campagne CINECA-CHARCOT II, 14 mars-30 avril 1971), Tethys, 6:261-268.
Fontugne, M.R. and Duplessy, J.-C., 1978, Carbon isotope ratio of marine plankton related to surface water masses, Earth and Planetary Science Letters, 41:365-371.
Fontugne, M.R. and Duplessy, J.-C., 1981, Organic carbon isotopic fractionation by marine plankton in the temperature range -1 to 31°C, Oceanologica Acta, 4:85-89.
Fraga, F., 1974, Distribution des masses d'eau dans l'upwelling de Mauritanie, Tethys, 6:5-10.
Froelich, P.N., Klinkhammer, G.P., Bender, M.L., Luedtke, N.A., Heath, G.R., and Maynard, V., 1979, Early oxidation of organic matter in pelagic sediments of the eastern Equatorial Atlantic: suboxic diagenesis, Geochimica et Cosmochimica Acta, 43:1075-1090.
Fütterer, D., 1980, Sedimentation am NW-afrikanischen Kontinentalrand: Quantitative Zusammensetzung und Verteilung der Siltfraktion in den Oberflächensedimenten, "Meteor" Forschungs-Ergebnisse, C33:15-60.
Galimov, E.M., 1980, C^{13}/C^{12} in kerogen, in: "Kerogen. Insoluble Organic Matter from Sedimentary Rocks," B. Durand, ed., Editions Technip, Paris, 271-299.
Gardner, J.V., 1975, Late Pleistocene carbonate dissolution cycles in the eastern Equatorial Atlantic, in: "Dissolution of Deep-Sea Carbonates," W.V. Sliter, A.W.H. Bé and W.H. Berger, Cushman

Foundation for Foraminiferal Research Special Publication No. 13, Tulsa, 129-141.

Gardner, W.S. and Menzel, D.W., 1974, Phenolic aldehydes as indicators of terrestrially derived organic matter in the sea, Geochimica et Cosmochimica Acta, 38:813-822.

Garrels, R.M., MacKenzie, F.T. and Hunt, C., 1975, "Chemical Cycles and the Global Environment," William Kaufmann, Inc., Los Altos, California, 206 pp.

Grall, J.-R., Laborde, P., Le Corre, P., Neveux, J., Traguer, P., and Thiriot, A., 1974, Caractéristiques trophiques et production planctonique dans la région sud de l'atlantique Marocain. Resultats des campagnes CINECA-CHARCOT I et III, Tethys, 6:11-28.

Groupe Mediprod, 1974, Généralités sur la campagne CINECA-CHARCOT II (15 Mars-29 Avril 1971), Tethys, 6:33-42.

Hartmann, M., Lange, H., Seibold, E., and Walger, E., 1971, Oberflächensedimente im Persischen Golf und Golf von Oman. I. Geologisch-hydrologischer Rahmen und erste sedimentologische Ergebnisse, "Meteor"Forschungs-Ergebnisse, C4:1-76.

Heath, G.R., Moore, T.C. and Dauphin, J.P., 1977, Organic carbon in deep-sea sediments, in: "The Fate of Fossil Fuel CO_2 in the Oceans," N.R. Andersen and A. Malahoff, eds., Plenum Press, New York, 749 pp.

Hedges, J.I. and Parker, P.L., 1976, Land-derived organic matter in surface sediments from the Gulf of Mexico, Geochimica et Cosmochimica Acta, 40:1019-1029.

Hughes, P. and Barton, E.D., 1974, Stratification and water mass structure in the upwelling area off northwest Africa in April/May 1969, Deep-Sea Research, 21:611-628.

Huntsman, S.A. and Barber, R.T., 1977, Primary production off northwest Africa: the relationship to wind and nutrient conditions, Deep-Sea Research, 24:25-33.

Kominz, M.A., Heath, G.R., Ku, T.-L. and Pisias, N.G., 1979, Brunhes time scales and the interpretation of climatic change, Earth and Planetary Science Letters, 45:394-410.

Koopmann, B., 1981, Sedimentation von Saharastaub im subtropischen Nordatlantik während der letzten 25,000 Jahre, "Meteor" Forschungs-Ergebnisse, C35:23-59.

Labracherie, M., 1980, Les radiolaires temoins de l'évolution hydrologique depuis le dernier maximum glaciaire au large de Cap Blanc (Afrique du Nord-Ouest), Palaeogeography, Palaeoclimatology, Palaeoecology, 32:163-184.

Lange, H., 1975, Herkunft und Verteilung von Oberflächensedimenten des westafrikanischen Schelfs und Kontinentalhanges, "Meteor" Forschungs-Ergebnisse, C22:61-84.

Lange, H., 1982, Distribution of chlorite and kaolinite in eastern Atlantic sediments off north Africa, Sedimentology, 29:427-431.

Lerman, A., 1979, Geochemical Processes, in: "Water and Sediment Environments," John Wiley and Sons, New York, 481 pp.

Lutze, G.F., Sarnthein, M., Koopmann, B., Pflaumann, U., Erlenkeuser, H., and Thiede, J., 1979, "Meteor" cores 12309: Late Pleistocene reference section for interpretation of the Neogene of Site 397, in: "Initial Reports DSDP," 47, Part 1, U. von Rad, W.B.F. Ryan et al., U.S. Government Printing Office, Washington, 727-739.

Manabe, S. and Hahn, D.G., 1977, Simulation of the tropical climate of an ice age, Journal of Geophysical Research, 82:3889-3909.

Mankinen, E.A. and Dalrymple, G.B., 1979, Revised geomagnetic polarity time scale for the interval 0-5 my B.P., Journal of Geophysical Research, 84:615-626.

McIntyre, A., Kipp, N.G., Bé, A.W.H., Crowley, T., Kellogg, T., Gardner, J.V., Prell, W., and Ruddiman, W.F., 1976, Glacial North Atlantic 18,000 years ago: A CLIMAP reconstruction, Geological Society of America, Memoir, 145:43-76.

Mittelstaedt, E., 1972, Der hydrographische Aufbau und die zeitliche Variabilität der Schichtung und Strömung im nordwestafrikanischen Auftriebsgebiet im Frühjahr 1968, "Meteor" Forschungs-Ergebnisse, A11:1-57.

Mittelstaedt, E., 1976, On the currents along the northwest African coast south of 22°N, Deutsche Hydrographische Zeitschrift, 29: 97-117.

Mittelstaedt, E., Pillsbury, D. and Smith, R.L., 1975, Flow patterns in the Northwest African upwelling area, Deutsche Hydrographische Zeitschrift, 28:145-167.

Morley, J.J. and Hays, J.D., 1981, Towards a high-resolution, global, deep-sea chronology for the last 750,000 years, Earth and Planetary Science Letters, 53:279-295.

Müller, P.J., 1977, C/N ratios in Pacific deep-sea sediments: Effect of inorganic ammonium and organic nitrogen compounds sorbed by clays, Geochimica et Cosmochimica Acta, 41:765-776.

Müller, P.J. and Suess, E., 1979, Productivity, sedimentation rate and sedimentary organic matter in the oceans. I. Organic carbon preservation, Deep-Sea Research, 26A:1347-1362.

Müller, P.J. and Mangini, A., 1980, Organic carbon decomposition rates in sediments of the Pacific Manganese Nodule Belt dated by ^{230}Th and ^{231}Pa, Earth and Planetary Science Letters, 51: 94-114.

Müller, P.J., Hartmann, M. and Suess, E., in press, The chemical environment of pelagic sediments, in: "The Manganese Nodule Belt of the Pacific Ocean. Scientific, Technical, and Economical Aspects," P. Halbach and G. Friedrich, eds., Enke-Verlag, Stuttgart.

Newman, J.W., Parker, P.L. and Behrens, E.W., 1973, Organic carbon isotope ratios in Quaternary cores from the Gulf of Mexico, Geochimica et Cosmochimia Acta, 37:225-238.

Northam, M.A., Curry, D.J., Scalan, R.S., and Parker, P.L., 1981, Stable carbon isotope ratio variations of organic matter in Orca Basin sediments, Geochimica et Cosmochimia Acta, 45:257-260.

Palmer, S.E. and Baker, E.W., 1978, Copper porphyrins in deep-sea sediments: A possible indicator of oxidized terrestrial organic matter, Science, 201:49-51.

Pastouret, L., Chamley, H., Delibrias, G., Duplessy, J.-C. and Thiede, J., 1978, Late Quaternary climatic changes in Western

Tropical Africa deduced from deep-sea sedimentation off the Niger delta, Oceanologica Acta, 1:217-232.

Pflaumann, U., 1975, Late Quaternary stratigraphy based on planktonic foraminifera off Senegal, "Meteor"Forschungs-Ergebnisse, C23: 1-46.

Pocklington, R., 1976, Terrigenous organic matter in surface sediments from the Gulf of St. Lawrence, Journal Fisheries Research Board Canada, 33:93-97.

Redfield, A.C., Ketchum, B.H. and Richards, F.A., 1963, The influence of organisms on the composition of seawater, in: THE SEA, Vol 2, M.N. Hill, ed., Interscience Publishers, New York, 26-77.

Rossignol-Strick, M. and Duzer, D., 1979, Late Quaternary pollen and dinoflagellate analysis of marine cores off West Africa, "Meteor"Forschungs-Ergebnisse, C30:1-14.

Sackett, W.M. and Thompson, R.R., 1963, Isotopic organic carbon composition of recent continental derived clastic sediments of eastern gulf coast, Gulf of Mexico, American Association of Petroleum Geologists, Bulletin, 47:525-531.

Sackett, W.M., Eckelmann, W.R., Bender, M.L., and Bé, A.W.H., 1965, Temperature dependence of carbon isotope composition in marine plankton and sediments, Science, 148:235-237.

Sarnthein, M., 1978, Sand deserts during glacial maximum and climatic optimum, Nature, 271:43-46.

Sarnthein, M. and Walger, E., 1974, Der äolische Sandstrom aus der W-Sahara zur Atlantikküste, Geologische Rundschau, 63:1065-1087.

Sarnthein, M. and Koopmann, B., 1980, Late Quaternary deep-sea record on NW African dust supply and wind circulation, in: "Palaeoecology of Africa, Vol. 12: Sahara and Surrounding Seas," M. Sarnthein, E. Seibold and P. Rognon, eds., A.A. Balkema, Rotterdam, 239-253.

Sarnthein, M., Tetzlaff, G., Koopmann, B., Wolter, K., and Pflaumann, U., 1981, Glacial and interglacial wind regimes over the eastern subtropical Atlantic and NW Africa, Nature, 293:193-196.

Sarnthein, M., Thiede, J., Pflaumann, U., Erlenkeuser, H., Fütterer, D., Koopmann, B., Lange, H., and Seibold, E., 1982, Atmospheric and oceanic circulation patterns off Northwest Africa during the past 25 million years, in: "Geology of the Northwest African Continental Margin," U. von Rad, K. Hinz, M. Sarnthein, and E. Seibold, eds., Springer-Verlag, Berlin, 584-604.

Sarnthein, M., Erlenkeuser, H., von Grafenstein, R., and Schröder, C., in press, Brunhes time sale: Isotope and carbonate stratigraphy of Sierra Leone Rise core 13519, "Meteor"Forschungs-Ergebnisse, C.

Schemainda, R., Nehring, D. and Schulz, S., 1975, Ozeanologische Untersuchungen zum Produktionspotential der nordwestafrikanischen Wasserauftriebsregion 1970-1973, Geodätische und geophysikalische Veröffentlichungen, 4, 88 pp.

Schmidt, D., 1981, "Isotopenverhältnis des organischen Kohlenstoffs aus marinen Sedimenten," Staatsexamensarbeit, University of Kiel, 125 pp.

Seibold, E., 1972, Cruise 25/1971 of R.V. "Meteor": continental margin of West Africa. General report and preliminary results, "Meteor"Forschungs-Ergebnisse, C10:17-38.

Seibold, E. and Hinz, K., 1976, German cruises to the continental margin of North West Africa in 1975: General reports and preliminary results from "Valdivia" 10 and "Meteor" 39, "Meteor" Forschungs-Ergebnisse, C25:47-80.

Seibold, E. and Berger, W.H., 1982, "The Sea Floor. An Introduction to Marine Geology," Springer-Verlag, Berlin, 288 pp.

Shackleton, N.J., 1977, Carbon-13 in Uvigerina: Tropical rainforest history and the equatorial Pacific carbonate dissolution cycles, in: "The Fate of Fossil Fuel CO_2 in the Oceans," N.R. Andersen and A. Malahoff, eds., Plenum Press, New York, 401-427.

Shaffer, G., 1976, A mesoscale study of coastal upwelling variability off NW-Africa, "Meteor"Forschungs-Ergebnisse, A17:33-70.

Simoneit, B.R.T., 1978, The organic chemistry of marine sediments, in: "Chemical Oceanography, Vol. 7," J.P. Riley and R. Chester, eds., Academic Press, London, 233-311.

Smith, B.N. and Epstein, S., 1971, Two categories of $^{13}C/^{12}C$ ratios for higher plants, Plant Physiology, 47:380-384.

Steemann Nielsen, E. and Jensen, E.A., 1957, Primary oceanic production. The autotrophic production of organic matter in the oceans, Galathea Report I, 49-135.

Suess, E., 1980, Particulate organic carbon flux in the oceans. Surface productivity and oxygen utilization, Nature, 288:260-263.

Suess, E. and Müller, P.J., 1980, Productivity, sedimentation rate and sedimentary organic matter in the oceans. II. Elemental fractionation, Colloques Internationaux du C.N.R.S., No. 293, Editions du Centre National de la Recherche Scientifique, Paris, 17-26.

Suess, E. and Ungerer, C.A., 1981, Element and phase composition of particulate matter from the circumpolar current between New Zealand and Antarctica, Oceanologica Acta, 4:151-160.

Sweeney, R.E. and Kaplan, I.R., 1980, Natural abundances of ^{15}N as a source indicator for nearshore marine sedimentary and dissolved nitrogen, Marine Chemistry, 9:81-94.

Thiede, J., 1977, Aspects of the variability of the glacial and interglacial North Atlantic eastern boundary current (last 150,000 years), "Meteor"Forschungs-Ergebnisse, C28:1-36.

Thiede, J., Suess, E. and Müller, P.J., 1982, Late Quaternary fluxes of major sediment components to the sea floor at the Northwest African Continental Slope, in: "Geology of the Northwest African Continental Margin," U. von Rad, K. Hinz, M. Sarnthein, and E. Seibold, eds., Springer-Verlag, Berlin, 605-631.

Tomczak, M., Jr. and Hughes, P., 1980, Three dimensional variability of water masses and currents in the Canary Current upwelling region, "Meteor"Forschungs-Ergebnisse, A21:1-24.

Troughton, J.H., 1972, Carbon isotope fractionation by plants, Proceedings of the 8th International Conference on Radiocarbon Dating, The Royal Society of New Zealand, Wellington, 613-624.

Table 1. Stable isotope composition of organic carbon in core 12392-1

Depth (cm)	Organic Carbon (%)	$\delta^{13}C_{PDB}$ (‰)	Depth (cm)	Organic Carbon (%)	$\delta^{13}C_{PDB}$ (‰)	Depth (cm)	Organic Carbon (%)	$\delta^{13}C_{PDB}$ (‰)
2- 10	0.35	-19.45	330-340	1.75	n.a.	663-673	0.41	-19.37
10- 20	0.34	n.a.	340-350	1.95	-19.39	673-683	0.50	-20.56
20- 29	0.35	n.a.	350-360	2.49	-19.61	683-693	0.45	-20.67
30- 40	0.37	-20.06	363-373	1.59	-19.47	693-703	0.43	-20.05
40- 50	0.89	-20.31	373-383	1.82	-19.00	703-713	0.33	n.a.
50- 60	1.07	-19.53	383-393	1.88	-18.59	713-723	0.41	-20.49
60- 70	1.13	-20.37	393-403	1.85	-19.28	723-733	0.36	-20.92
70- 80	1.24	n.a.	403-413	1.52	-19.25	733-743	0.37	-20.18
83- 90	1.11	-19.66	413-423	1.20	-19.40	743-753	0.45	-22.58
90-100	1.27	-19.51	423-433	1.40	-19.52	753-763	0.56	-20.09
100-110	1.23	-18.52	433-443	1.31	-19.42	763-773	0.55	-20.53
110-120	1.83	-18.25	443-453	1.78	-19.08	773-783	0.55	-20.69
120-130	2.33	-18.59	453-463	1.94	-19.10	783-793	0.43	-19.93
130-140	1.87	-18.23	463-473	1.52	-20.08	793-803	0.48	-21.58
140-150	2.26	-18.36	473-483	1.24	-19.93	803-813	0.42	-20.07
150-160	2.01	-18.33	483-493	1.09	-19.63	813-823	0.33	-20.56
160-170	1.85	-18.41	493-503	1.13	-20.00	823-833	0.40	-21.17
170-180	2.27	-18.24	503-513	0.81	n.a.	833-839	1.81	-19.68
180-190	1.62	n.a.	513-523	0.85	-19.72	842-848	0.83	-20.59
190-200	1.91	-18.28	523-533	0.68	-20.04	848-855	1.18	-21.82
200-210	1.72	-18.62	533-543	0.69	-20.06	855-863	1.24	-22.29
210-220	1.85	-18.33	543-553	0.56	-20.02	863-873	1.16	-21.65
220-230	2.38	-18.41	553-563	0.86	-19.63	873-883	1.17	-20.35
230-240	2.43	-18.40	563-573	0.86	-19.72	883-893	0.60	-21.09
240-250	2.05	-18.27	573-583	0.81	-19.69	893-903	1.83	-19.92
250-260	1.91	-18.49	583-593	0.74	-19.31	903-913	1.73	-19.84
260-270	2.10	-18.49	593-598	0.63	-20.01	913-923	1.98	-19.50
270-280	2.42	-18.40	603-613	0.62	-19.84	923-933	1.87	-19.42
280-290	3.03	-18.71	613-623	0.50	-19.95	933-943	1.94	n.a.
290-300	2.39	-18.64	623-633	0.39	-20.37	943-953	2.23	-19.01
300-310	2.24	-18.68	633-643	0.47	-19.67	953-963	1.84	n.a.
310-320	2.02	-19.10	643-653	0.38	-20.34			
320-330	2.71	-19.13	653-663	0.48	n.a.			

n.a. = not analysed

Table 2. Analytical results of core 15627-3. Concentrations in % or ppm of the bulk dry sediment

Depth (cm)	Calcium Carbonate (%)	Organic Carbon (%)	Total Nitrogen (ppm)	C/N Ratio	Depth (cm)	Calcium Carbonate (%)	Organic Carbon (%)	Total Nitrogen (ppm)	C/N Ratio
0- 5	61.6	0.20	440	4.6	225-230	51.2	0.26	520	5.0
5- 10	52.3	0.48	770	6.2	230-235	55.5	0.24	430	5.6
10- 15	54.3	0.42	660	6.4	235-240	52.3	0.26	460	5.7
15- 20	54.1	0.48	780	6.2	240-245	55.5	0.23	370	6.2
25- 30	55.7	0.36	590	6.1	245-250	57.5	0.28	430	6.5
30- 35	54.3	0.35	590	5.9	250-255	57.5	0.29	430	6.7
35- 40	54.9	0.34	500	6.8	255-260	59.6	0.37	500	7.4
40- 45	54.1	0.37	570	6.5	260-265	59.4	0.39	540	7.2
45- 50	53.0	0.55	780	7.1	265-270	59.6	0.34	510	6.7
50- 55	53.2	0.64	900	7.1	270-275	61.9	0.34	530	6.4
55- 60	53.9	0.68	1000	6.8	275-280	64.4	0.28	490	5.7
60- 65	55.0	0.58	930	6.2	280-285	63.4	0.23	430	5.4
65- 70	53.2	0.71	1000	7.1	285-290	53.7	0.28	530	5.3
70- 75	56.9	0.81	1030	7.9	290-295	49.1	0.25	590	4.2
75- 80	57.1	0.47	750	6.3	295-300	50.9	0.26	560	4.6
80- 85	61.2	0.51	710	7.2	300-305	50.3	0.30	570	5.3
85- 90	61.4	0.46	720	6.4	305-310	56.2	0.64	810	7.9
90- 95	63.0	0.30	570	5.3	310-315	57.8	0.55	730	7.5
95-100	57.8	0.30	560	5.4	315-320	57.3	0.43	580	7.4
100-105	55.9	0.39	660	5.9	320-325	57.5	0.47	640	7.3
105-110	57.1	0.51	810	6.3	325-330	58.4	0.56	690	8.1
110-115	55.8	0.52	790	6.6	330-335	56.4	0.57	700	8.1
115-120	49.6	0.50	770	6.5	335-340	57.5	0.58	700	8.3
120-125	50.7	0.51	760	6.7	340-345	58.7	0.45	620	7.3
125-130	49.8	0.61	820	7.4	345-350	59.1	0.47	640	7.3
130-135	49.8	0.44	680	6.5	350-355	60.0	0.37	530	7.0
135-140	50.9	0.37	570	6.5	355-360	60.5	0.36	530	6.8
140-145	50.7	0.19	330	5.8	360-365	59.1	0.34	520	6.5
145-150	52.5	0.17	300	5.7	365-370	55.3	0.26	450	5.8
150-155	56.9	0.14	270	5.2	370-375	52.1	0.26	460	5.7
155-160	57.8	0.32	540	5.9	375-380	52.8	0.24	480	5.0
160-165	55.7	0.22	350	6.3	380-385	51.6	0.22	440	5.0
165-170	56.6	0.23	430	5.4	385-390	50.0	0.21	440	4.8
170-175	60.0	0.30	500	6.0	390-395	50.7	0.20	400	5.0
175-180	53.9	0.31	570	5.4	395-400	50.7	0.21	410	5.1
180-185	46.6	0.35	670	5.2	400-405	53.5	0.22	400	5.5
185-190	45.8	0.37	700	5.3	405-410	52.1	0.31	480	6.5
190-195	52.8	0.18	460	3.9	410-415	53.3	0.32	480	6.7
195-200	54.6	0.14	420	3.3	415-420	46.7	0.24	430	5.6
200-205	58.2	0.16	320	5.0	420-425	41.5	0.21	460	4.6
205-210	62.3	0.17	360	4.7	425-430	38.3	0.19	340	5.6
210-215	64.6	0.28	400	7.0	430-435	36.1	0.19	370	5.1
215-220	67.1	0.24	430	5.6	435-440	56.3	0.25	460	5.4
220-225	58.9	0.26	500	5.2	440-445	56.3	0.34	520	6.5

Table 2 (cont.)

Depth (cm)	Calcium Carbonate (%)	Organic Carbon (%)	Total Nitrogen (ppm)	C/N Ratio	Depth (cm)	Calcium Carbonate (%)	Organic Carbon (%)	Total Nitrogen (ppm)	C/N Ratio
445-450	53.1	0.46	610	7.5	665-670	32.6	0.25	560	4.5
450-455	53.9	0.47	610	7.7	670-675	44.4	0.38	680	5.6
455-460	49.4	0.34	570	6.0	675-680	49.9	0.21	400	5.3
460-465	47.2	0.27	440	6.1	680-685	56.8	0.25	440	5.7
465-470	43.5	0.39	620	6.3	685-690	54.2	0.36	540	6.7
470-475	43.4	0.43	640	6.7	690-695	56.0	0.23	410	5.6
475-480	42.2	0.37	570	6.5	695-700	59.6	0.27	410	6.6
480-485	41.6	0.34	560	6.1	700-705	55.1	0.19	330	5.8
485-490	43.4	0.30	530	5.7	705-710	63.2	0.19	360	5.3
490-495	42.2	0.32	560	5.7	710-715	60.8	0.19	420	4.5
495-500	45.8	0.37	550	6.7	715-720	57.4	0.20	440	4.6
500-505	40.3	0.29	560	5.2	720-725	52.6	0.34	540	6.3
505-510	47.6	0.40	530	7.6	725-730	54.6	0.44	600	7.3
510-515	54.8	0.38	540	7.0	730-735	59.1	0.47	640	7.3
515-520	57.7	0.37	510	7.3	735-740	60.7	0.41	600	6.8
520-525	59.4	0.42	540	7.8	740-745	62.8	0.40	570	7.0
525-530	59.5	0.42	530	7.9	745-750	58.8	0.46	670	6.9
530-535	61.5	0.30	420	7.1	750-755	58.4	0.43	610	7.1
535-540	59.5	0.35	420	8.3	755-760	56.5	0.54	760	7.1
540-545	59.4	0.27	440	6.1	760-765	52.9	0.45	600	7.5
545-550	59.4	0.24	440	5.5	765-770	52.8	0.48	670	7.2
550-555	60.8	0.20	370	5.4	770-775	46.1	0.46	570	8.1
555-560	62.6	0.24	400	6.0	775-780	36.2	0.41	550	7.5
560-565	59.7	0.29	380	7.6	780-785	30.3	0.39	510	7.7
565-570	62.6	0.19	370	5.1	785-790	20.7	0.25	370	6.8
570-575	61.8	0.22	310	7.1	790-795	55.8	0.69	880	7.8
575-580	61.1	0.24	400	6.0	795-800	60.7	0.70	850	8.2
580-585	61.8	0.27	420	6.4	800-805	61.3	0.68	770	8.8
585-590	60.0	0.24	370	6.5	805-810	61.7	0.63	760	8.3
590-595	62.0	0.27	420	6.4	810-815	57.1	0.50	690	7.3
595-600	53.3	0.20	400	5.0	815-820	55.3	0.41	650	6.3
600-605	56.0	0.18	390	4.6	820-825	50.2	0.56	740	7.6
605-610	52.6	0.16	460	3.5	825-830	49.4	0.50	720	6.9
610-615	51.2	0.17	470	3.6	830-835	47.4	0.46	690	6.7
615-620	50.4	0.27	630	4.3	835-840	49.1	0.65	870	7.5
620-625	n.a.	0.28	520	5.4	840-845	48.9	0.65	860	7.6
625-630	52.4	0.23	520	4.4	845-850	51.1	0.45	710	6.3
630-635	55.7	0.23	490	4.7	850-855	52.4	0.33	610	5.4
635-640	59.8	0.18	400	4.5	855-860	45.6	0.22	460	4.8
640-645	60.5	0.25	430	5.8	860-865	48.4	0.15	350	4.3
645-650	61.6	0.24	420	5.7	865-870	50.5	0.16	340	4.7
650-655	53.6	0.25	500	5.0	870-875	59.2	0.11	280	3.9
655-660	39.0	0.30	600	5.0	875-878	58.9	0.12	250	4.8
660-665	35.2	0.27	550	4.9					

n.a. = not analysed

Table 3. Analytical results of core 13209-2. Concentrations in % or ppm of the bulk dry sediment

Depth (cm)	Calcium Carbonate (%)	Organic Carbon (%)	Total Nitrogen (ppm)	C/N Ratio	Depth (cm)	Calcium Carbonate (%)	Organic Carbon (%)	Total Nitrogen (ppm)	C/N Ratio
4- 10	52.6	0.51	680	7.5	355-370	0.8	1.01	870	11.6
10- 20	54.0	0.34	440	7.7	370-385	-	0.82	790	10.4
20- 30	6.8	0.44	440	10.0	385-400	0.7	1.06	1000	10.6
30- 42	- x)	0.47	570	8.3	400-415	0.5	0.58	560	10.4
42- 50	-	0.93	900	10.3	415-435	0.6	1.04	930	11.2
50- 59	-	0.97	940	10.3	435-445	5.9	0.84	850	9.9
59- 69	-	1.14	1150	9.9	445-460	1.0	1.06	990	10.7
69- 80	-	0.89	910	9.8	460-475	0.6	1.18	960	12.3
80- 90	-	0.90	920	9.8	475-490	0.6	1.01	1020	9.9
90-100	-	1.23	1170	10.5	490-507	0.8	0.71	780	9.1
100-110	-	1.04	1160	9.0	507-520	0.6	0.34	470	7.2
112-121	1.5	1.00	970	10.3	520-536	42.2	0.39	530	7.4
121-130	1.8	1.18	1260	9.4	536-546	28.4	0.40	380	10.5
130-142	0.7	0.78	860	9.1	546-562	2.6	0.53	530	10.0
142-152	-	0.68	680	10.0	562-575	-	0.32	330	9.7
152-161	-	0.66	730	9.0	575-590	-	0.38	460	8.3
161-164	-	0.57	530	10.8	590-602	10.0	0.65	690	9.4
164-178	-	0.68	600	11.3	602-617	28.8	0.54	570	9.5
178-185	1.4	0.65	890	7.3	617-628	10.7	0.61	570	10.7
185-200	10.5	0.70	730	9.6	628-645	4.7	1.04	890	11.7
200-215	18.6	0.43	540	8.0	645-660	17.4	0.85	800	10.6
215-230	27.0	0.57	620	9.2	660-675	1.0	1.09	930	11.7
230-245	2.4	0.35	590	5.9	675-685	-	0.61	670	9.1
245-255	14.9	0.42	630	6.7	685-700	-	0.50	600	8.3
255-265	43.7	0.41	470	8.7	700-715	-	0.56	690	8.1
265-275	50.1	0.43	420	10.2	715-725	7.1	0.71	910	7.8
275-290	4.1	0.42	400	10.5	725-736	19.2	0.90	760	11.8
290-305	13.0	0.40	470	8.5	736-750	0.8	0.59	790	7.5
305-311	-	0.52	480	10.8	750-765	-	0.74	650	11.4
311-325	1.3	0.96	1010	9.5	765-781	2.2	1.01	930	10.9
325-340	0.9	0.97	930	10.4	781-791	-	0.41	510	8.0
340-355	0.6	0.96	860	11.2					

x) - = less than 0.5 wt% $CaCO_3$

Table 4. Analytical results of core 13519-2. Concentrations in % or ppm of the bulk dry sediment

Depth (cm)	Calcium Carbonate (%)	Organic Carbon (%)	Total Nitrogen (ppm)	C/N Ratio
10- 12	82.3	0.30	290	10.3
20- 22	51.4	0.26	340	7.7
30- 32	50.0	0.34	330	10.3
40- 42	49.1	0.33	340	9.7
50- 52	59.4	0.34	310	11.0
60- 62	57.3	0.34	370	9.2
70- 72	55.9	0.41	380	10.8
80- 82	58.2	0.43	360	11.9
93- 95	33.0	0.36	370	9.7
100- 102	34.6	0.33	380	8.7
110- 112	59.1	0.29	290	10.0
120- 122	69.4	0.24	230	10.4
130- 132	62.5	0.23	250	9.2
140- 142	78.9	0.20	170	11.8
150- 152	57.1	0.20	260	7.7
160- 162	56.4	0.18	230	7.8
170- 172	76.9	0.20	170	11.8
180- 182	64.1	0.18	160	11.3
190- 192	45.0	0.22	210	10.5
200- 202	48.4	0.25	240	10.4
210- 212	55.3	0.39	290	13.5
220- 222	45.9	0.38	280	13.6
230- 232	43.4	0.28	260	10.8
240- 242	39.3	0.30	280	10.8
250- 252	76.6	0.48	250	19.2
260- 262	70.5	0.29	240	12.1
270- 272	50.7	0.27	220	12.3
280- 282	54.1	0.29	290	10.0
290- 292	65.7	0.24	210	11.4
300- 302	57.1	0.20	270	7.4
310- 312	64.8	0.24	250	9.6
320- 322	77.8	0.22	190	11.6
330- 332	51.2	0.24	240	10.0
340- 342	37.1	0.26	300	8.7
350- 352	68.7	0.32	250	12.8
360- 362	78.5	0.26	220	11.8
370- 372	58.2	0.22	220	10.0
380- 382	65.5	0.24	210	11.4
390- 392	78.5	0.25	220	11.4
400- 402	58.2	0.31	250	12.4
410- 412	43.2	0.35	300	11.7
420- 422	41.8	0.30	320	9.4
430- 432	79.6	0.31	230	13.5
440- 442	67.3	0.26	220	11.8
450- 452	58.0	0.21	240	8.8
460- 462	68.2	0.21	230	9.1
470- 472	60.0	0.16	230	7.0
480- 482	77.8	0.20	190	10.5
490- 492	78.2	0.19	170	11.2
498- 500	53.7	0.18	240	7.5
520- 522	60.0	0.32	260	12.3
530- 532	58.9	0.33	270	12.2
540- 542	54.3	0.29	260	11.2
550- 552	62.1	0.28	280	10.0
560- 562	55.7	0.30	300	10.0
570- 572	59.6	0.18	240	7.5
580- 582	57.5	0.21	230	9.1
590- 592	83.7	0.17	130	13.1
600- 602	90.5	0.13	110	11.8
610- 612	84.8	0.12	140	8.6
620- 622	52.1	0.17	220	7.7
630- 632	41.8	0.23	240	9.6
640- 642	61.6	0.26	200	13.0
650- 652	40.9	0.29	270	10.7
660- 662	56.2	0.33	250	13.2
670- 672	46.2	0.30	280	10.7
680- 682	35.9	0.28	270	10.4
690- 692	72.8	0.25	210	11.9
700- 702	70.0	0.18	180	10.0
710- 712	79.1	0.18	170	10.6
720- 722	72.1	0.16	180	8.9
730- 732	80.7	0.16	170	9.4
740- 742	76.6	0.17	150	11.3
750- 752	59.6	0.16	200	8.0
760- 762	51.6	0.17	260	6.5
770- 772	55.9	0.19	240	7.9
780- 782	67.3	0.18	200	9.0
790- 792	54.6	0.14	210	6.7
800- 802	78.2	0.19	190	10.0
810- 812	58.0	0.14	240	5.8
820- 822	74.6	0.17	160	10.6
830- 832	49.1	0.18	220	8.2
840- 842	57.3	0.22	260	8.5
850- 852	69.4	0.18	180	10.0
860- 862	38.7	0.17	240	7.1
870- 872	71.9	0.19	180	10.6
880- 882	35.5	0.18	260	6.9
890- 892	30.9	0.22	270	8.2
900- 902	87.5	0.17	120	14.2
910- 912	78.5	0.15	170	8.8
920- 922	52.1	0.18	210	8.6
930- 932	61.9	0.18	220	8.2
940- 942	80.5	0.16	170	9.4
950- 952	58.2	0.17	210	8.1
960- 962	58.2	0.17	200	8.5
970- 972	55.0	0.17	230	7.4
980- 982	61.6	0.14	200	7.0
990- 992	41.8	0.18	240	7.5
998-1000	42.3	0.21	260	8.1
1010-1012	66.2	0.16	180	8.9
1020-1022	56.9	0.14	210	6.7
1030-1032	63.2	0.15	200	7.5
1040-1042	61.2	0.12	200	6.0
1050-1052	61.6	0.14	170	8.2
1062-1064	56.6	0.21	210	10.0

DIFFERENTIATION OF HIGH OCEANIC FERTILITY IN MARINE SEDIMENTS CAUSED BY COASTAL UPWELLING AND/OR RIVER DISCHARGE OFF NORTHWEST AFRICA DURING THE LATE QUATERNARY

Liselotte Diester-Haass

Geographisches Institut der Universität
Schloss, Postfach 2428
D-6800 Mannheim, Federal Republic of Germany

ABSTRACT

Increased fertility in continental margin sediments off northwestern Africa can be detected by means of increased accumulation rates of planktonic and benthic foraminifers, biogenous opal and organic carbon as well as by increased radiolarian/planktonic foraminiferal ratios and benthos/plankton foraminiferal ratios. The attribution of this increased fertility to either upwelled water or to river discharge in subtropical areas is possible by means of two methods: 1) a micropaleontological investigation: the recognition of cold water assemblages of planktonic foraminifers and of radiolarians favors upwelling influence; 2) a sedimentological investigation of terrigenous particles. Grain size data of terrigenous matter, as well as clay mineral compositions, and mica and wood fiber contents makes it possible to distinguish fluvial from eolian sediment supply. Eolian sediment supply indicates that fertilization is most probably due to upwelling, whereas river influenced sediments indicate a fertilization by dissolved river load.

Preservation of organic matter and biogenous opal is enhanced seaward of river mouths where biogenous carbonate is easily dissolved. In arid, upwelling influenced areas, however, organic matter and biogenous opal are less well preserved, and in contrast, biogenous carbonate suffers less from dissolution. The application of these parameters to northwest African continental slope sediments, to help find increased fertility and attribute it to upwelling or river discharge, made possible a reconstruction of Late Quaternary fluctuations in climate and fertility. The Sahara desert shifted from its actual position at 18-27°N to 13-19°N during the last Glacial period.

INTRODUCTION

Coastal upwelling may lead to a highly increased production of marine organisms because nutrient-rich subsurface waters are supplied to the ocean surface where plants and animals proliferate. Consequently, in underlying sediments accumulation rates of organic matter (Müller, 1975) and skeletal debris of various benthic and planktonic organisms may increase significantly (Diester-Haass, 1976b).

Regions of prevailing coastal upwelling are concentrated on the eastern margins of the ocean basins adjacent to the arid climatic belt on land. Some are subject to seasonal migrations as off northwest Africa where, during winter, upwelling occurs as far south as 10°N (Schemainda, Nehring and Schulz, 1975), i.e., at a geographical latitude outside the arid belt, where rivers can influence marine sedimentation. The mean annual duration of upwelling decreases from 20 to 12°N (Fig. 1). The mean annual precipitation over northwest Africa increases from 29 mm at 21°N to 579 mm at Dakar and to 1236 mm at Bathurst, Gambia (Deutsches Hydrographisches Institut, 1973; Fig. 1). As a consequence, dissolved nutrients fertilize the surface waters off river mouths (Lafond and Lafond, 1971; Milliman and Boyle, 1975; Schemainda et al., 1975). As a result, the mean annual primary production offshore from these northwest African rivers is nearly as high (>300 g C/m^2/y) as in the center of the coastal upwelling area at about 20°N (>325 g C/m^2/y) (Fig. 1). The effect on the sediments underlying these fluvially-influenced zones of enhanced productivity may be similar to that in upwelling areas with increased accumulation rates of organic matter and skeletal debris of various marine organisms.

It is the aim of this contribution to attempt to find parameters that make possible the differentiation of fertilization by rivers from that by coastal upwelling. Such parameters cannot yet be found in the organic particles themselves, but only indirectly by the recognition of an arid climate on the neighboring continent. If sedimentological parameters indicate an arid climate, increased fertility can be attributed to upwelling. If sedimentological parameters point to river discharge, fertilization by rivers will be responsible for the increased fertility, although an additional seasonal upwelling effect cannot be excluded.

Another problem considered here is the effect of preservation on the productivity signal in these two environments. Sediments accumulating off rivers differ greatly from those off arid regions and allow a better preservation of, for example, opal and organic matter due to rapid burial leading to increased dissolution of calcium carbonate (Müller and Suess, 1979; Diester-Haass and Müller, 1979). Finally, a distinction between humid and arid climatic conditions on the neighboring continent, and thereby between fertilization of the

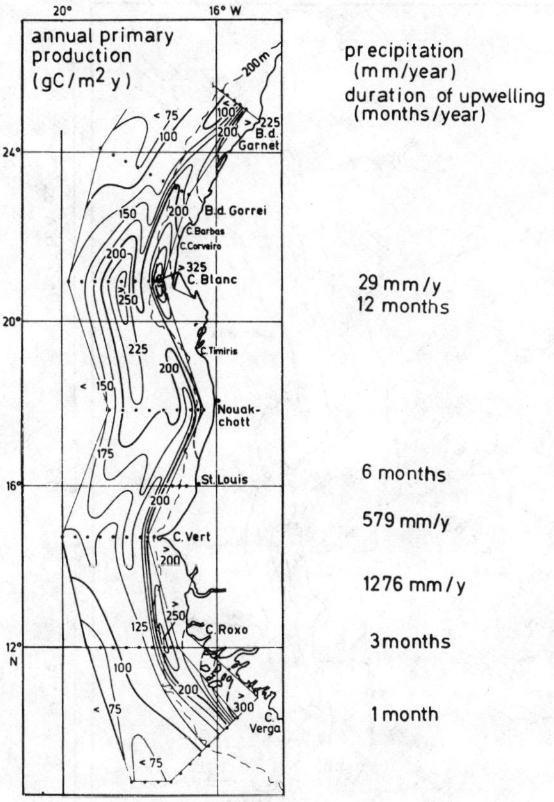

Fig. 1. Mean annual primary production (g/m^2·y) off northwest Africa (from Schemainda et al., 1975), precipitation (mm/y) and duration of upwelling (months/year).

near coastal area by rivers or by upwelling, will be used in an attempt to reconstruct the Late Quaternary history of upwelling and climate on the northwest African margin.

MATERIALS AND METHODS

The results presented here summarize the data from 28 cores taken between 12 and 17°N on the northwest African continental slope below 1000 m water depth (Fig. 2) ("Meteor" cruises 25 and 39, "Valdivia" cruise 10, Seibold, 1972; Seibold and Hinz, 1976) (for original data see Diester-Haass, Schrader and Thiede, 1973; Diester-Haass, 1975; 1976a and 1976b; 1977; 1978; 1981a; 1981b; in press; Chamley, Diester-Haass and Lange, 1977; Diester-Haass and Müller, 1979). Data were obtained by means of a quantitative coarse fraction analysis (Sarnthein, 1971). About 30 biogenic, terrigenous

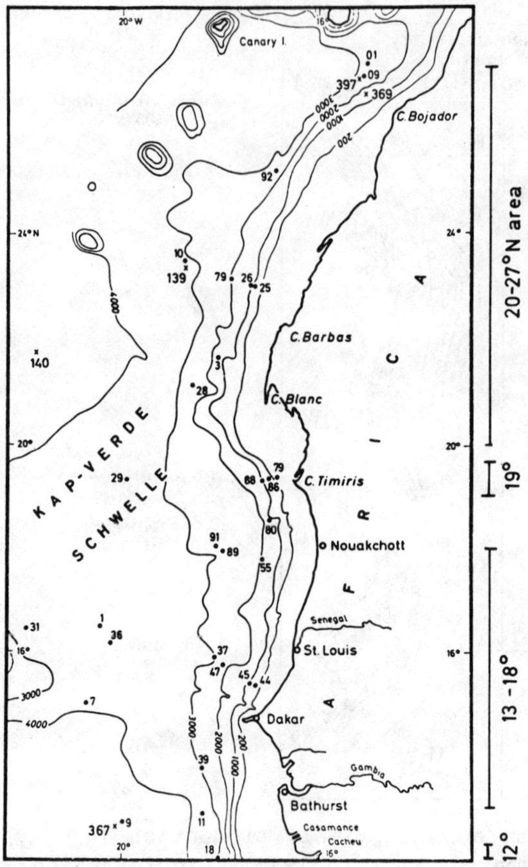

Fig. 2. Position of investigated "Meteor" and "Valdivia" cores; 1 and 3 are TAG/72 cores (dots). Crosses indicate position of Glomar Challenger drill sites.

and authigenic components have been distinguished in six coarse fractions (40-63, 63-125, 125-250, 250-500, 500-1000, >1000 µm) and the composition of the coarse fraction (40-63 and >63 µm fraction) has been calculated. Where available, organic carbon contents (Müller, 1975), clay mineralogy and grain size distributions of the total terrigenous component (Lange, 1975, and unpublished results) have also been used.

PARAMETERS FOR THE RECOGNITION OF INCREASED FERTILITY IN MARINE SEDIMENTS OFF NORTHWEST AFRICA

Figure 3 summarizes schematically the relationship between nutrient input, productivity and water properties in upwelling and river discharge areas off northwest Africa and their influence on the underlying sediments. It is obvious that the fertilization by upwelling and by rivers leads to similar characteristics: organic carbon, biogenous opal and calcareous benthos are enriched in the sediments compared to those from neighboring less fertile areas. If calculation of accumulation rates of the individual particles is not possible, the ratio of radiolarians/planktonic foraminifers and of benthic/planktonic foraminifers can be used as a rough indicator of increased fertility (Diester-Haass, 1976b; 1977; 1978).

Fig. 4 shows as an example Meteor core 12392-1 (for position see Fig. 2: 92) which reveals strong glacial (stage 2-4) and weak interglacial upwelling influence. The increase in accumulation rates of radiolarians and benthic foraminifers (Figs. 4D and 4E) is clearly reflected by increased ratios of radiolarians/planktonic foraminifers (Fig. 4B) and benthos/plankton foraminifers (Fig. 4C) in oxygen isotope stages 2-4 compared to stages 1 and 5. The strong upwelling influence in stages 2-4 is also reflected by increased contents and accumulation rates of organic carbon (Figs. 4G and 4H).

Fish production is usually highly increased in the surface waters, but neither an enrichment of fish debris in the sediments nor phosphorite formation could be detected off northwest Africa (cf. Suess, 1981).

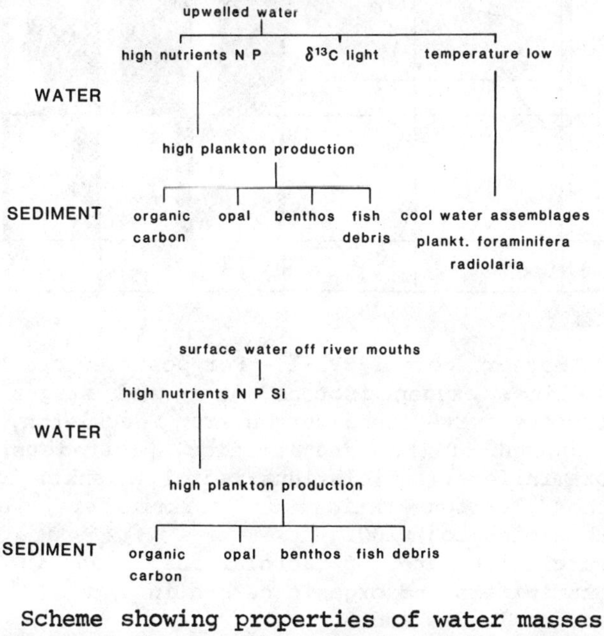

Fig. 3. Scheme showing properties of water masses in upwelling areas and in front of rivers off northwest Africa and their influence on the underlying sediments.

The main difference between fertile areas in upwelling regions and off river mouths can be seen in the following (see Fig. 3):

a) cool water temperatures in upwelling areas, which are reflected in cold water planktonic foraminiferal assemblages, especially in the presence of *Globigerina bulloides* (Prell and Curry, 1981; Thiede, 1975) and in cold water species of radiolarians (Labracherie, 1980);

b) $\delta^{13}C$ values which were found in *Globigerinoides ruber* by Berger, Diester-Haass and Killingley (1978) to be isotopically lighter during periods of Late Glacial upwelling off northwest Africa.

The light carbon isotope is preferentially fixed during photosynthesis and subsequently released at the time of nutrient remineraliza-

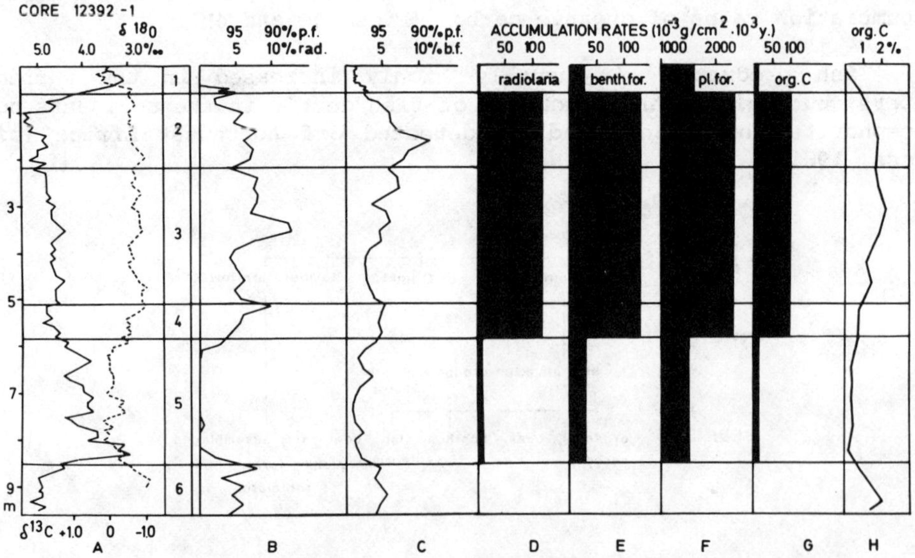

Fig. 4. "Meteor" core 12392-1. For position see Fig 2.

A. Heavy line: oxygen isotope curve and stages 1-6. Dotted line: carbon isotope curve (both curves from Shackleton, 1977).

B. Opal content of the sand fraction, plotted as a ratio with planktonic foraminifers: (radiolarians/radiol.+plankt. foram.)x100.

C. Benthos/plankton ratios of foraminifers, calculated as (benthos/benthos+plankton)x100.

D-G. Accumulation rates of radiolarians, benthic foraminifers, planktonic foraminifers and organic carbon in $g/cm^2 \cdot 10^3$ years in oxygen isotope stages 1, 2-4 and 5.

H. Percentage of organic carbon in total sediment (values concerning organic carbon from Müller, 1975).

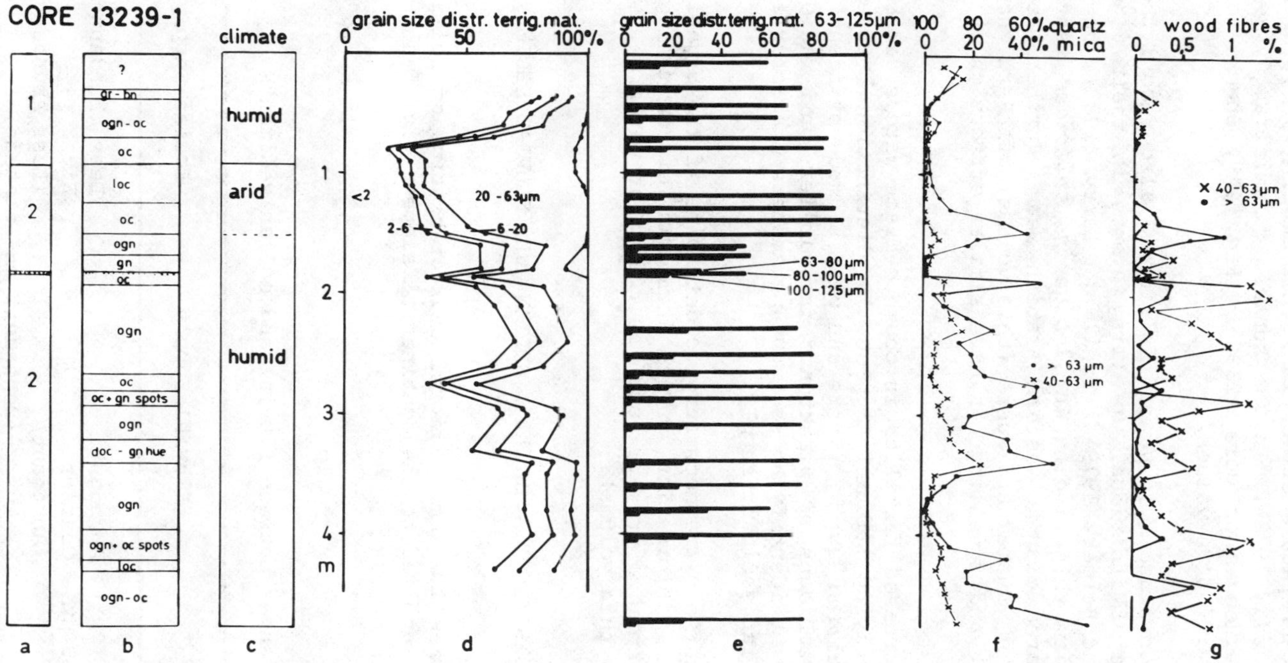

Fig. 5. "Valdivia" core 13239-1. For position see Fig. 2.
a. Stratigraphy, oxygen isotope stages.
b. Sediment colors: gr=grey, gn=green, bn=brown, oc=ocre, ogn=olive-green, l=light, d=dark.
c. Climatic interpretation.
d. Grain size distribution of terrigenous matter (unpublished data, kindly provided by H. Lange, Kiel.
e. Grain size distribution of terrigenous matter in the 63-125 µm fraction.
f. Mica content in the coarse fractions, plotted as (mica/mica+quartz)x100.
g. Content of wood fibers in the coarse fractions.

tion. With the upwelled water it comes back to the surface and becomes enriched in the shells; thus $\delta^{13}C$ is lower than in areas without upwelling. The temperature effect is rather small (~0.2% for a 5°C range) (Berger et al., 1978). In the benthic foraminifer *Uvigerina* from glacial sequences of core 12392-1 (Fig. 4A), Shackleton (1977) also found lowered $\delta^{13}C$ values.

Shells from the mollusc *Mytilus* off California also record the onset and termination of individual upwelling events in their $\delta^{13}C$ signal (Killingley and Berger, 1979). Prell and Curry (1981), however, could not confirm this finding in *Globigerina bulloides* tests from the upwelling area of the western Arabian Sea. Furthermore, investigations by Wefer, Dunbar and Suess (this volume) off Peru show that both river input and upwelling can have similar effects on $\delta^{13}C$ values.

As a result, increased fertility in ocean surface waters can be detected in the underlying sediments. But the distinction between fertilization by upwelling and by rivers can only be made by identifying cool water assemblages of planktonic foraminifers and radiolarians or indirectly by identifying sediment parameters characteristic of fluvial or eolian sediment supply.

Further investigations are necessary to determine if the organic matter itself contains information on its formation either in upwelled water or in front of rivers (such as molecular markers, see Brassel and Eglinton, this volume).

PARAMETERS FOR THE DISTINCTION BETWEEN EOLIAN AND FLUVIAL SEDIMENT SUPPLY

Table 1 summarizes the compositional characteristics of eolian and fluvial sediment supplied to the continental slope. Fig. 5 shows

Table 1. Differences between fluvial and eolian sediments

	Fluvial Supply	Eolian Supply
% <2 µm fraction	Abundant	Less abundant
Terrigenous matter >63 µm	Coarser	Finer
(Mica)	(Abundant)	(Less Abundant)
Clay minerals	(Dependent on latitude)	
Fluviatile diatoms	Present	Absent
Plant debris	Present	Absent

as an example core 13239-1, situated at 14°N (Fig. 2) off the mouth of the Gambia River. This core contains the essential parameters for climatic interpretations which are found in the other cores as well.

The <2 µm fraction is much more abundant in front of rivers draining tropical weathering areas than in arid areas (H. Lange, pers. comm.). In core 39 (Fig. 5d) the <2 µm fraction is abundant in the humid Holocene and lower stage 2, when the Gambia River transported its load to the sea. In the upper stage 2 the 20-63 µm fraction--a typical grain size of eolian matter--increases from 5 up to 60% and leads to a strong reduction in clay content. Parallel to the grain size variations, there is an obvious change in sediment colors: green in the river mud sections and ocher in the sections with eolian silt (Fig. 5b).

The grain size distribution of terrigenous matter within the sand fraction is coarser in fluvial sediments than in wind-derived dusts. The 100-125 µm and the >125 µm fractions are noticeably absent in eolian sediments on the continental slope (Diester-Haass, 1980; Diester-Haass and Chamley, 1982). In core 39 (Fig. 5e) the sediments from the arid interval in the upper stage 2 contain no quartz >100 µm, whereas those in these core sections where river mud is present (higher <2 µm fraction percentages), the 100-125 µm fraction contains 3-15% quartz. Table 2 summarizes grain size data of terrigenous matter in the 63-125 µm fraction of cores from the northwest African continental slope. It reveals that those core sections that suggest an arid climate on the northwest African continent contain <1 or 0% terrigenous matter >100 µm, whereas in sections from humid periods the 100-125 µm and 80-100 µm sized terrigenous fraction is more abundant.

Table 2. Mean values of grain size distributions of terrigenous particles within the 63-125 µm fraction, given for humid intervals (fluviatile sediment input) and arid intervals (eolian sediment supply) in cores taken off northwestern Africa

	Eolian Supply (weight-%)			Fluviatile Supply		
Core No.	63-80	80-100	100-125 µm	63-80	80-100	100-125 µm
13289	83.4	16.1	0.2	73.9	24.9	1.0
13291	83.8	15.7	0.1	73.0	26.5	0.4
13239	85.0	13.7	0.9	67.0	28.2	4.7
13211				26.5	44.6	28.8
13218				64.3	28.0	7.6
12347	82.0	17.6	0.4	74.5	23.5	2.0

The occurrence of mica is somewhat of a problem. Mica is abundant in rivers from the west African semihumid area and, in general, rare in wind-derived sediments. However, it can also be locally abundant off the arid Sahara. In core 39 (Fig. 5f) mica is abundant in the humid intervals and nearly disappears in the eolian silt-rich arid interval in the upper stage 2. The reduction in mica in the 350-400 cm interval is not well understood in terms of climatic variations.

Clay minerals contain indicators as to humid or arid climate and are latitude-dependent (Fig. 6). Variations in clay mineral composition in sediment cores might be interpreted in terms of climatic

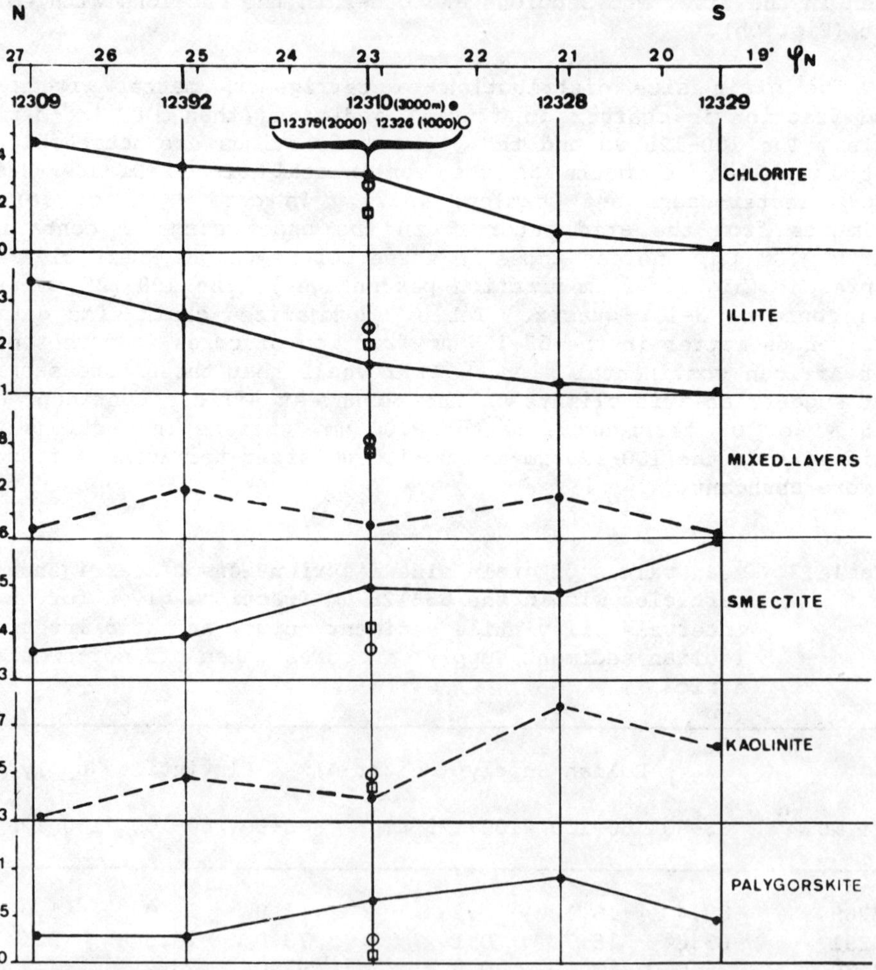

Fig. 6. Average composition of clay minerals off northwest Africa plotted against latitude of samples (for position see Fig. 2; from Chamley et al., 1977). By permission from 'Meteor'-Forsch.-Ergeb., Copyright (c) 1977, Gebr. Bornträger.

changes (Chamley et al., 1977). (Clay mineralogical investigations have been carried out up until now only in cores off the Sahara and will be extended in the future to the actually humid area off northwestern Africa [Lange, pers. comm.].)

The most distinctive indicators of fluvial sediment supply (completely absent in arid areas) are fluvial diatoms (Kiper, 1977) and wood fibers which have been found in the 40-125 μm fractions. Fig. 5g shows that in the arid upper stage 2 interval of core 39 there are no wood fibers at all, whereas in the river-influenced periods they form up to 1.5% of the 40-63 μm and up to 0.5% of the >63 μm fractions.

These parameters, especially the <2 μm content, the grain size distribution of terrigenous matter in the 63-125 μm fraction, and the wood fiber content, provide ways to distinguish arid from humid climate and, thus, eolian from fluvial sediment supply to the northwest African continental slope sediments, which appear to be influenced by increased fertility. As a result, increased fertility can be attributed to either upwelling or river discharge. Before proceeding with a climatic reconstruction, however, the effects of preservation on the productivity record in these two environments must be considered.

PRESERVATION OF BIOGENIC COMPONENTS OFF NORTHWEST AFRICA IN THE ARID AND HUMID CLIMATIC BELT

The components which indicate increased fertility have to pass several "filters" before being incorporated in the sediment. The mid-water filter is probably less important for our slope sediments than the "bottom water and sediment filter" (Seibold, 1982), where considerable changes occur in the particles. Preservation conditions for organic carbon, biogenous opal and $CaCO_3$ at the water-sediment interface and in the sediment are quite different in the rapidly deposited fine-grained river muds than in the more carbonate-rich sediments off the arid area off northwest Africa.

The preservation of organic matter is enhanced with increasing accumulation rate of the total sediments: a tenfold increase in accumulation rate leads to about a doubling of sedimentary organic carbon contents, if biological production at the sea-surface remains constant (Müller and Suess, 1979). Furthermore, the abundance and composition of clay minerals which are abundant off rivers, play an important role in organic matter preservation. The greater their specific surface areas, the more organic carbon may be preserved by sorption (Suess, 1973; Müller and Suess, 1977). Thus, it is clear that off arid areas on the northwest African continental margin preservation conditions of organic matter are less favorable than in front of river mouths. In water depths below 2000 m off the arid zone off northwest Africa, organic carbon contents are <1% (Müller,

1975) whereas in the sediments influenced by river supply they are 2-3% (Diester-Haass and Müller, 1979).

The preservation of biogenous opal is also favored by the high bulk accumulation rates (Johnson, 1974) and clay-sized sediments off rivers. Silica, normally diffusing out of the sediment into the ocean water (Berger, 1976), is protected by the fine-grained sediments which seal it from the opal-undersaturated bottom water (Diester-Haass and Müller, 1979). Thus, silica is protected aginst dissolution.

Calcium carbonate, on the other hand, is better preserved in areas off arid climates compared to river influenced areas. In the fine-grained river muds the high organic carbon/carbonate carbon ratios lead to strong carbonate dissolution (Diester-Haass and Müller, 1979; Emerson and Bender, 1981), while the oxic upper sediment layers in the northwest African upwelling area are less corrosive to carbonate constituents.

While the increased opal and organic carbon contents in river influenced sediments may appear to simply reflect better preservation and not any increased fertility, the following observations indicate that fluvial input actually does lead to enhanced productivity:

• Coastal waters in front of river mouths commonly support higher concentrations of plankton, benthos, microorganisms and fish than offshore waters (Lafond and Lafond, 1971).

• In front of the Amazon River estuary, which carries 40% of the dissolved silica brought into the Atlantic (Milliman and Boyle, 1975) abundant diatoms in the surface water (up to 2 mg/liter) reflect the direct fertilization and associated production from the dissolved river load.

• Sediment cores taken north of the Senegal River mouth contain no biogenous opal but do have abundant river mud (Chamley and Diester-Haass, in press) which is transported to these sites north of the river delta by the poleward flowing undercurrent (Mittelstaedt, 1976). Cores south of the Senegal mouth, however, contain both abundant biogenous opal and Senegal mud (Diester-Haass, 1975). This may be explained by southward transport of nutrient-rich Senegal River water in the surface flow of the Canary Current, thus effectively separating the dissolved load from the suspended load in a two-layered flow regime. This observation shows that preservation is not the only factor, but that the net production of opal is actually increased in areas influenced by favorable nutrient and light conditions associated with river discharge in the surface waters. Preservation alone is not responsible for the elevated opal and organic carbon contents.

As a result, production and preservation of biogenous opal and organic carbon is enhanced in river-influenced sediments where calcium carbonate is strongly dissolved. In marine continental margin sediments influenced by eolian supply, however, carbonate preservation is better and opal and organic matter are less well preserved.

NORTHWEST AFRICAN HISTORY OF UPWELLING AND CLIMATE DURING THE LATE QUATERNARY

The previous chapters showed that increased fertility can be detected in continental slope sediments off northwest Africa by means of increased accumulation rates of planktonic foraminifers, biogenous opal, benthic organisms and organic carbon, as well as by increased radiolarian/planktonic foraminifer ratios and benthos/plankton ratios of foraminifers.

Until now, attribution of the increased fertility either to upwelling influence or to river discharge has only been possible indirectly by finding criteria in favor of an arid (upwelling) or a humid climate (river discharge). Sediments containing abundant clay, plant debris and terrigenous particles coarser than 100 µm are river influenced. Sediments with small clay amounts, no wood fibers and no terrigenous particles coarser than 100 µm are probably deposited in areas without river influence. An additional indicator in favor of upwelling as the cause of increased fertility might be the presence of cold water species of planktonic foraminifers and radiolarians.

Application of these parameters for the recognition of increased fertility and for the distinction of eolian from fluvial sediment input to sediment cores from the northwest African continental slope makes possible a reconstruction of the Late Quaternary history of fertility and climate (Fig. 7) (Diester-Haass et al., 1973; Diester-Haass, 1975; 1976a and 1976b; 1977; 1978; 1980; 1981a and 1981b; in press; Chamley et al., 1977).

The area investigated between 12° and 27°N may be subdivided into four regions, each of which shows a different evolution of Late Quaternary climate and fertility. The delineation of the boundaries between the four regions (20-27°N, 19°, 13-18° and 12°N) (Fig. 2) is rather problematic for the following reasons:

- Cores presently available are widely spaced, thus more samples will probably lead to a change of the boundaries.

- The northeast trade wind blows eolian material to the southwest, the Harmattan blows eolian matter to the northwest (Sarnthein et al., 1981), which leads to a different position of climatic boundaries in the marine sediments compared to those on the continent.

- Marine currents redistribute terrigenous material, thereby obscuring the boundaries. The poleward flowing undercurrent trans-

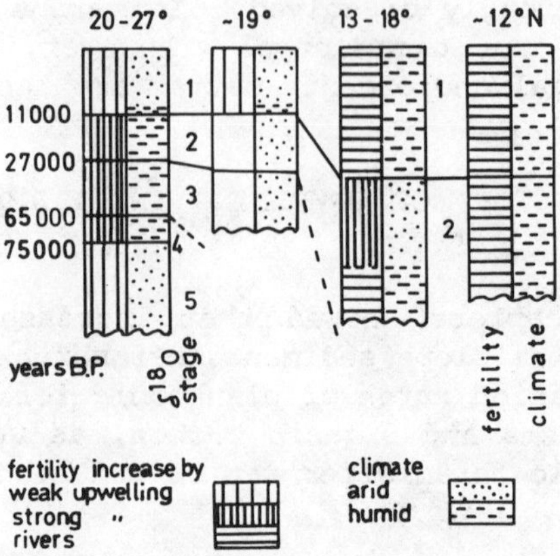

Fig. 7. Schematic graph showing history of climate and fertility in four areas off northwest Africa during the Late Quaternary (for position of cores and areas see Fig. 2). The basis for fertility and climate distinctions is given in the text.

ports Senegal River detritus at least 200 km to the north (Lange, 1975; Diester-Haass, in press). Thus a humid climate with river discharge is reflected in marine sediments at 18°N latitude, whereas the adjacent continent has an arid climate.

For the area between 20-27°N the Late Holocene has arid climatic characteristics, the Early Holocene and the last glacial period (oxygen isotope stages 2, 3 and 4) have a humid climate, and the penultimate interglacial period (oxygen isotope stage 5) arid climatic indicators. Results by Sarnthein and Koopmann (1980) and Sarnthein et al. (1981) point to an arid climate during the maximum of the last glaciation (about 18,000 yrs. B.P.) which cannot clearly be distinguished in the cores from 20-27°N, instead they reveal a rather uniform humid climate during the entire duration of stages 2, 3 and 4 (Chamley et al., 1977).

No fertility increase of the ocean waters can be observed in Holocene sediments off the northern Sahara, although oceanographic measurements point to upwelling today. Only off Cape Blanc do increased opal and organic matter contents indicate Holocene upwelling (Diester-Haass, 1978). This discrepancy between Cape Blanc and the northern areas around 25°N can be explained by a shallow pycnocline (50 m) off Cape Blanc, which allows high nutrient water to be tapped by upwelling and a deep pycnocline (300-400 m) which keeps high nutrient waters below the depth of origin of upwelled water (Schemainda et al., 1975).

During Glacial periods (oxygen isotope stages 2, 3 and 4) all sedimentological parameters point to strongly increased fertility. A humid climate then might have led to a fertility increase from river discharge. But other observations favor increased glacial upwelling intensity (Müller and Suess, 1979; Labracherie, 1980; Müller, Erlenkeuser and Schneider, this volume):

• The paleo-oceanographic reconstruction by the CLIMAP group (1976) reveals a 6°C negative surface water temperature anomaly for the northwest African margin during the last glacial maximum. This indicates increased upwelling.

• The $\delta^{13}C$ values are lighter in the Late Glacial sediments and thus might also point to increased upwelling (Berger et al., 1978).

• Indirect evidence that the fertility increase is caused by upwelling and not exclusively by river discharge is seen in the sediment texture. Sediments from stage 2, 3 and 4 do not contain the typical high percentages of <2 μm fraction associated with humid environments, as off the Senegal (Lange, 1975; Chamley et al., 1977), and they do not contain fluvial diatoms (Diester-Haass et al., 1973). This allows the conclusion that the rivers flowing from the northern Sahara to the sea during stages 2, 3 and 4 were perhaps of a Mediterranean type, draining areas with little chemical weathering and thus low in dissolved nutrients. They were probably flowing only during part of the year (differences in <6 μm fraction content in recent sediments are 70% off southern Morocco and 80-90% off Senegal/Gambia, Sarnthein and Koopmann, 1980).

For the area around 19°N only one core allows reconstruction of Late Quaternary fertility and climate. This core (13288) in about 1000 m water depth, reaches an age at depth of 36,000 yrs. B.P. (Diester-Haass, 1981b; C^{14} ages by H. Erlenkeuser, Kiel) and contains arid climatic indicators during all of this period except for a short humid event in Early Holocene times.

Increased fertility is not evident in this core, although oceanographic mesurements reveal strong upwelling at the present time (Weichart, 1974; Schemainda et al., 1975; Shaffer, 1976). A few radiolarians in the Holocene sediments may be indicative of upwelling, but the stage 2 and 3 sediments contain no opal in the sand fraction. This might be explained by a shift to the west of the maximum upwelling effects on the sediments during glacial periods when sea level was lower (Diester-Haass, 1978; Labracherie, 1980). It therefore is possible that Glacial sediments on the lower slope show stronger upwelling influence at ~19°N latitude, but these were not sampled by our cores.

This scarceness or absence of upwelling indicators in Holocene continental slope sediments off the Sahara desert is difficult to understand. One would expect higher opal and organic carbon contents than those found in the sediments because of high production in surface waters and because of a possible additional supply of organic particles due to the flow mechanism described by Smith (this volume); so organic matter and light diatoms could be transported to the continental slope. But strong onshore currents on the shelf (up to 20 cm/sec in about 60 m water depth; Smith, this volume) act in an opposite direction. The observations of Milliman (1977) that continental shelf sediments in this area do not reflect oceanographically measured upwelling conditions, can be explained by winnowing processes during this cross-shelf flow (Smith, this volume). Richert (1975) attributes the scarceness of diatoms in the sediments to the absence of downwelling. Miro de Orell (1973), however, found up to 16% biogenous sand-sized opal in sediments from 1000 m and 2% in sediments from 600 m water depth sampled a few kilometers south of our core. The difference may be explained by the observation of Shaffer (1976), that upwelling is concentrated near canyon heads and thus leads to local sedimentological differences.

For the area between 13 and 18°N latitude (Fig. 2) several cores yield identical results. The climatic parameters point to humid conditions during stage 1, whereas arid conditions characterize the upper portions of stage 2 and humid conditions the lower stage 2. During the upper stage 2, the Senegal River disappeared entirely and only eolian matter reached the core sites. Cores at 18°N also show this climatic evolution, although they actually were from off the southern Sahara. The Holocene indicators of a humid climate (abundant clay, plant debris, etc.) are brought in by the northward-flowing undercurrent (Mittelstaedt, 1976) which redistributes the Senegal River load in this direction.

Increased fertility was found in the sediments influenced by river load during stage 1 as well as during arid stage 2. It is possible that the southward migration of the arid belt and intensification of the Canary Current during stage 2 (CLIMAP, 1976) increased both upwelling intensity and its annual duration in this area off Senegal and Gambia. However, upwelling lasts only 5-6 months per year at the present time (Fig. 1, Schemainda et al., 1975).

For the area around 12°N, again only one core is available with autochthonous sedimentation (core 13218). Fluvial sedimentation is obvious during the cored section of stage 1 and 2. During stage 2 however, four intervals containing iron-stained coarse quartz, interlayered with fluvial mud, justify the interpretation of a climate with strong contrasts between arid and humid years. Perhaps during stage 2 conditions were similar to those in the Sahel zone of today, i.e., the transitional zone between the Sahara desert and the savannah zone in west Africa where, during arid years, dunes migrate to the south but rivers continue to flow. The lowered glacial sea level made possible the reworking of dunes formed on the shelf and distribution of dune sand to the continental slope. The formation of coastal dunes during stage 2 has also been observed by Michel (1960) in the Gambia-Casamance area.

Fertility indicators reveal increased fertility in stage 1 and 2 of this core at 12°N which can be attributed to river discharge. Further investigations have to consider the possibility of an additional upwelling influence.

The Late Quaternary climatic fluctuations in northwestern Africa can be summarized as follows: a south migration of the Sahara from its actual extension between about 18 and 27°N to about 13-19°N correlated to a narrowing of the arid belt during the last Glacial period. Fertility was strongly increased by upwelling during oxygen isotope stages 2-4 in the 20-27°N area and in stage 2 in the 13-18° area.

CONCLUSIONS

1. Some parameters that indicate increased ocean fertility in marine sediments have been discussed. These include increased accumulation rates of planktonic foraminifers, benthic foraminifers, biogenous opal (diatoms and radiolarians) and organic matter, and increased radiolarian/planktonic foraminiferal ratios as well as benthic/planktonic foraminiferal ratios.

2. Since upwelling and river discharge in subtropical areas have similar effects on ocean fertility, parameters have been discussed which directly allow distinguishing between fertility increases due to coastal upwelling from those due to increased river discharge. Cool water assemblages of planktonic foraminifers and radiolarians point here to upwelling, in addition to generally increased fertility.

3. Indirect ways to attribute increased fertility to either upwelling or river discharge are indicators for fluvial and eolian supply, where eolian indicators identify the wind system forcing coastal upwelling.

4. $\delta^{13}C$ signatures are not unambiguous as an indicator of upwelling. Further investigations need to be done on stable carbon isotopes as well as on the organic matter in order to find molecular markers which are typical for river-derived or river-influenced organic substances.

5. Preservation conditions for biogenous opal and organic matter as well as biogenous carbonate, which are quite different in upwelling and in river-influenced sediments, have to be evaluated for final criteria to differentiate the causes of increased fertility in coastal waters.

6. The application of fertility and climate parameters to northwest African continental slope sediments allows reconstruction of the Late Quaternary fluctuations in upwelling intensity and fertilization by rivers (Fig. 7).

ACKNOWLEDGEMENTS

I thank E. Seibold (Kiel, Bonn) for having made available core material from "Meteor" and "Valdivia" cruises, and him and my colleagues from the Geological Institute at Kiel for discussions. H. Lange made available unpublished results on clay mineralogy and sediment texture and H. Erlenkeuser ^{14}C dates. Thanks are also due to P. Rothe (Mannheim) and K. Heine (Saarbrücken) for providing working facilities. Sincere thanks are due to L. Krissek and E. Suess (Oregon State University) for reviewing the manuscript and giving very helpful suggestions for improvement. The financial support of the Deutsche Forschungsgemeinschaft is gratefully acknowledged.

REFERENCES

Berger, W.H., 1976, Biogenous deep-sea sediments: production, preservation and interpretation, in: "Chemical Oceanography," J.P. Riley and R. Chester, eds., vol. 5, 2nd edition, Academic Press, London, 265-388.

Berger, W.H., Diester-Haass, L. and Killingley, J.S., 1978, Upwelling off North-West Africa: the Holocene decrease as seen in carbon isotopes and sedimentological indicators, Oceanologica Acta, 1:3-7.

Chamley, H., Diester-Haass, L. and Lange, H., 1977, Terrigenous material in East Atlantic sediment cores as an indicator of NW African climates, "Meteor"Forschungs-Ergebnisse, C26:44-59.

Chamley, H. and Diester-Haass, L., in press, Effects du déplacement de l'embouchure du fleuve Sénégal, au Quaternaire supérieur, sur la sédimentation de la marge ouest-africaine, Comptes Rendus de l'Académie des Sciences, Paris.

CLIMAP Project Members, 1976, The surface of the ice age earth, Science, 191:1131-1144.

Deutsches Hydrographisches Institut, 1973, "Handbuch der Westküste Afrikas, I, Teil Nr. 2062," Hamburg, 512 pp.

Diester-Haass, L., 1975, Sedimentation and climate in the Late Quaternary between Senegal and the Cape Verde Islands, "Meteor" Forschungs-Ergebnisse, C20:1-32.

Diester-Haass, L., 1976a, Late Quaternary climatic variations in North-West Africa deduced from East Atlantic sediment cores, Quaternary Research, 6:299-314.

Diester-Haass, L., 1976b, Quaternary accumulation rates of biogenous and terrigenous components on the East Atlantic continental slope off NW Africa, Marine Geology, 21:1-24.

Diester-Haass, L., 1977, Radiolarian/planktonic foraminiferal ratios in a coastal upwelling region, Journal of Foraminiferal Research, 7(1):26-33.

Diester-Haass, L., 1978, Sediments as an indicator of upwelling, in: "Upwelling Ecosystems," R. Boje and M. Tomczak, eds., Springer, Berlin, 261-281.

Diester-Haass, L., 1980, Upwelling and climate off Northwest Africa during the Late Quaternary, in: "Paleoecology of Africa," vol. 12, E.M. van Zinderen-Bakker and J.A. Coetzee (eds.), Balkema, Rotterdam, 229-238.

Diester-Haass, L., 1981a, Les sédiments marins comme indicator des climats continentaux:liaison avec la limite sud du Sahara au Quaternaire supérieur, abstract volume, Réunion speciálisée de la Société Géologique de France, "Océans-Paléocéans", Lille.

Diester-Haass, L., 1981b, Factors contributing to Late Glacial and Holocene sedimentation on the continental shelf and slope off NW Africa, Banc d'Arguin, 19°N, "Meteor"Forschungs-Ergebnisse, C35:1-22.

Diester-Haass, L., in press, Late Quaternary sedimentation processes on the West-African continental margin and climatic history of West Africa (12-19°N), "Meteor"Forschungs-Ergebnisse, C.

Diester-Haass, L. and Müller, P.J., 1979, Processes influencing sand fraction composition and organic matter content in surface sediments off West Africa (12-19°N), "Meteor"Forschungs-Ergebnisse, C31:21-47.

Diester-Haass, L. and Chamley, H., 1982, Oligocene and post-Oligocene history of sedimentation and climate off Northwest Africa (DSDP Site 369), in: "Geology of the Northwest African Continental Margin," U. von Rad, K. Hinz, M. Sarnthein and E. Seibold, eds., Springer, 529-544.

Diester-Haass, L., Schrader, H.J. and Thiede, J., 1973, Sedimentological and paleoclimatological investigations of two pelagic ooze cores off Cape Barbas, North-West Africa, "Meteor" Forschungs-Ergebnisse, C16:19-66.

Emerson, S.E. and Bender, M., 1981, Carbon fluxes at the sediment-water interface of the deep-sea: calcium carbonate preservation, Journal of Marine Research, 31:139-162.

Johnson, T.C., 1974, The dissolution of siliceous microfossils in surface sediments of the eastern tropical Pacific, Deep-Sea Research, 21:851-864.

Killingley, J.S. and Berger, W.H., 1979, Stable isotopes in a mollusc shell: detection of upwelling events, Science, 205-186-188.

Kiper, M., 1977, "Sedimente und ihre Umwelt im Senegaldelta," Diplom Arbeit, Universität Kiel, 59 pp.

Labracherie, M., 1980, Les radiolaires témoins de l'évolution hydrologique depuis le dernier maximum glaciaire au large du Cap Blanc (Afrique du Nord-Ouest), Palaeogeography, Palaeoclimatology, Palaeoecology, 32:163-184.

Lafond, E.C. and Lafond, K.G., 1971, Oceanography and its relation to marine organic production, in: "Fertility of the Sea," J.D. Costlow, ed., Gordon and Breach Science Publications, New York, 241-265.

Lange, H., 1975, Herkunft und Verteilung von Oberflächensedimenten des westafrikanischen Schelfs und Kontinentalhanges, "Meteor" Forschungs-Ergebnisse, C22:61-84.

Michel, P., 1960, Recherches géomorphologiques en Casamance et en Gambie méridionale Bureau de la Recherche Géologique et Minière, Dakar, Etude entreprise par la Fédération du Mali, 67 pp.

Milliman, J.D. and Boyle, E., 1975, Biological uptake of dissolved silica in the Amazon river estuary, Science, 189:995-997.

Milliman, J.D., 1977, Effects of arid climate and upwelling upon the sedimentary regime off southern Spanish Sahara, Deep-Sea Research, 24:95-103.

Miro Orell, M. de, 1973, Sedimentos recientes del margen continental de Mauritania (expedicion Sahara II), Resultados Expediciones Cientificas del Buque oceanografico "Cornide de Saavedra", 2:1-12.

Mittelstaedt, E., 1976, On the currents along the northwest African coast south of 22°N, Deutsche Hydrographische Zeitschrift, 29(3):97-117.

Müller, P.J., 1975, Diagenese stickstoffhaltiger organischer Substanzen in oxischen und anoxischen marinen Sedimenten, "Meteor" Forschungs-Ergebnisse, C22:1-60.

Müller, P.J. and Suess, E., 1977, Interaction of organic compounds with calcium carbonate. III. Amino acid composition of sorbed layers, Geochimica et Cosmochimica Acta, 41:941-949.

Müller, P.J. and Suess, E., 1979, Productivity, sedimentation rate and sedimentary organic carbon content in the oceans, Deep-Sea Research, 26A:1347-1362.

Prell, W.L. and Curry, W.B., 1981, Faunal and isotopic indices of monsoonal upwelling: Western Arabian Sea, Oceanologica Acta, 4(1):91-98.

Richert, P., 1975, "Die räumliche Verteilung und zeitliche Entwicklung des Phytoplanktons, mit besonderer Berücksichtigung der Diatomeen, im nordwestafrikanischen Auftriebswassergebiet," Ph.D. Thesis, Universität Kiel, 260 pp.

Sarnthein, M., 1971, Oberflächensedimente im Persischen Golf und Golf von Oman. II. Quantitative Komponentenanalyse der Grobfraktion, "Meteor"Forschungs-Ergebnisse, C5:1-113.

Sarnthein, M. and Koopmann, B., 1980, Late Quaternary deep-sea record on northwest African dust supply and wind circulation, in: "Paleoecology of Africa," Vol. 12, E.M. van Zinderen-Bakker and J.A. Coetzee, eds., A.A. Balkema, Rotterdam, 239-253.

Sarnthein, M., Tetzlaff, G., Koopmann, B., Wolter, K., and Pflaumann, U., 1981, Glacial and interglacial wind regimes over the eastern subtropical Atlantic and North-West Africa, Nature, 293:193-196.

Schemainda, R., Nehring, D. and Schulz, S., 1975, Ozeanologische Untersuchungen zum Produktionspotential der nordwestafrikanischen Wasserauftriebsregion 1970-1973, Geodätische und Geophysikalische Veröffentlichungen, Reihe IV, Heft 16, 85 pp.

Seibold, E., 1972, Cruise 25/1971 of RV "Meteor": Continental margin of West Africa. General report and preliminary results, "Meteor"Forschungs-Ergebnisse, C10:17-38.

Seibold, E., 1982, Sediments in upwelling areas, particularly off Northwest Africa, Rapports Conseil International Exploration de Mer, 180:315-322.

Seibold, E. and Hinz, K., 1976, German cruises to the continental margin of North West Africa in 1975: General report and preliminary results from "Valdivia" 10 and "Meteor" 39, "Meteor" Forschungs-Ergebnisse, C25:47-80.

Shackleton, N.J., 1977, Carbon-13 in *Uvigerina*: tropical rainforest history and the equatorial Pacific carbonate dissolution cycles, in: "The Fate of Fossil Fuel CO_2," N.R. Andersen and A. Malahoff, eds., Plenum Press, New York, 401-427.

Shaffer, G., 1976, A mesoscale scale of coastal upwelling variability off NW Africa, "Meteor"Forschungs-Ergebnisse, A17:21-72.

Suess, E., 1973, Interaction of organic compounds with calcium carbonate. II. Organo-carbonate association in recent sediments, Geochimica et Cosmochimica Acta, 37:2435-2447.

Suess, E., 1981, Phosphate regeneration from sediments of the Peru continental margin by dissolution of fish debris, Geochimica et Cosmochimica Acta, 45:577-588.

Thiede, J., 1975, Shell- and skeleton-producing plankton and nekton in the eastern North Atlantic Ocean, "Meteor"Forschungs-Ergebnisse, C20:33-79.

Weichart, G., 1974, Meereschemische Untersuchungen im nordwestafrikanischen Auftriebsgebiet 1968, "Meteor"Forschungs-Ergebnisse, A14:33-70.

DISTRIBUTION OF ORGANIC CARBON IN THE GULF OF ALASKA NEOGENE AND HOLOCENE SEDIMENTARY RECORD

John M. Armentrout

Mobil Exploration and Producing Services, Inc.
P.O. Box 900
Dallas, Texas 75221, U.S.A.

ABSTRACT

Distribution of organic-carbon concentrations and organic-carbon types in Gulf of Alaska Holocene sediment reflects predominant terrestrial input. The dominance of terrestrial organic matter occurs despite the fact that the Gulf of Alaska is an area of upwelling and high marine productivity. The observed patterns are interpreted as resulting from removal of marine organic carbon within the water column and at the sediment/water interface, and the introduction of terrestrially derived organic carbon moving offshore.

The organic-carbon distribution patterns in the Holocene sediments and in the Neogene Yakataga Formation are similar, reflecting the late Miocene to Recent glaciomarine environment of the Gulf of Alaska. A Middle to early Late Miocene organic-rich claystone reflects a pre-glacial transgressive sequence. It is postulated that during the transgressive interval, an upwelling system moved shoreward resulting in the deposition of organic-rich sediment where the oxygen minimum zone impinged on the slope and shelf, and possibly extended over a shelf basin with restricted water circulation.

INTRODUCTION

This report compares the distributional pattern of organic-carbon concentrations and organic-carbon types between Holocene sediments and Neogene rocks from the northern Gulf of Alaska. The amounts and types of sedimentary organic carbon and their distributional patterns are interpretable as indicators of local oceanographic and terrestrial conditions. The Gulf of Alaska is a region of high primary marine productivity (Koblentz-Mishke, Volkovinsky and Kabanova, 1970; Degens and Mopper, 1976). The high productivity levels are attributed to nutrient availability due to deep-water upwelling (Dow, 1978). Sea floor surface sediment organic-carbon concentrations range from less than 0.5 wt-% to as high as 1.50 wt-% (Garshanovich, 1965; Armentrout, 1980b; Premuzic, 1980). The pattern of organic matter distribution is useful in interpreting the location of high productivity systems, such as upwelling, the relative supply and accumulation rates of marine versus terrestrial components, the degree of oxygenation of the water column, and the extent of benthic faunal activity (Rhoads and Morse, 1971; Welte, Cornford and Rullkötter, 1979; Demaison and Moore, 1980; Summerhayes, 1981; this volume). By comparing the Holocene distributional patterns of organic carbon with those of the Neogene rock record, inferences are made about the paleoceanographic conditions controlling Neogene deposition in the northern Gulf of Alaska.

The northern Gulf of Alaska has received glacially derived sediment since Miocene time (Bandy, Butler and Wright, 1969; Allison, 1978; Armentrout, Echols and Nash, 1978; Molnia and Sangray, 1979; Lattanzi, 1981). The sedimentary rocks of the Miocene to Pleistocene Yakataga Formation, cropping out in the coastal mountains of the Yakataga-Yakutat area, record this long history of tide-water glaciation (Plafker and Addicott, 1976; Rau, Plafker and Winkler, 1977; Armentrout, Rosenmeier and Rogers, 1979; Armentrout, 1980b). The several glacial facies of the Yakataga Formation occur in a stratigraphic sequence approaching 5000 meters in thickness.

Interpretation of the Gulf of Alaska late Neogene glacial record is best approached from a model based on the sediments and shelled organisms of similar modern environments. In the modern North Pacific, this model is best developed locally in the northern Gulf of Alaska. Although the intensity of glaciation is much less now than during the late Neogene and Quaternary glacial maxima, glaciers still reach tide water and are an important agent of erosion and of sediment supply in coastal areas of the Yakataga-Yakutat area. To provide a basis for interpretation of environments of deposition of the Yakataga Formation, an integrated study of Holocene sediments, organic matter and benthic faunas has been carried out (Armentrout, 1980a; Echols and Armentrout, 1980; Hickman and Nesbitt, 1980). This report describes and interprets the patterns of organic-carbon distribution.

PREVIOUS WORK

Studies of organic-carbon distribution in the Gulf of Alaska are limited. Maps of organic-carbon distribution are presented in Bezrukov et al. (1961), Garshanovich (1965), Romankevich (1968), and Armentrout (1980a). Bathymetric distribution of organic carbon is presented by Dow (1978). Sedimentology studies are reviewed by Armentrout (1980a) and Powell (1981) for the Holocene, and by Plafker and Addicott (1976) and Armentrout et al. (1979) for the Neogene and Quaternary.

Examination of foraminiferal faunas is reviewed by Echols and Armentrout (1980) for the Holocene, and by Rau, Plafker and Winkler (1977) and Lagoe (1978; in press) for the Neogene and Quaternary. Molluscan faunal analysis is discussed by Hickman and Nesbitt (1980) for the Holocene and by Plafker and Addicott (1976), Allison (1978) and Ariey (1978) for the Neogene. This study on Gulf of Alaska organic-carbon distribution draws heavily on data in Armentrout (1980a) and new data released by Mobil Oil Corporation.

DESCRIPTION OF STUDY AREA

The area examined in this study is along the continental shelf in the northern Gulf of Alaska, seaward of the Cape Yakataga-Malaspina Glacier shoreline, and between Pamplona Searidge on the west and Yakutat Seavalley on the east (Fig. 1). Sea floor sample traverses cross the shelf and upper slope of this area as well as Yakutat Seavalley and Yakutat Bay. The Gulf of Alaska region has been an area of alpine, piedmont and tidewater glaciation throughout the late Neogene and Quaternary. As a consequence, the subaerial and submarine topography of the present reflects the probable topography of the late Neogene. The following discussion of the modern and late Cenozoic topography is presented as a template for understanding physiographic constraints on organic matter distribution both in the Holocene and Neogene.

Holocene Sample Area

The continental shelf of the Yakataga-Yakutat area is an extension of the coastal plain: the width between the north bounding Chugach-St. Elias faults and the continental shelf/slope break is 60-80 km throughout the area (Plafker, 1967; Wright, 1968). Important geomorphic features include Pamplona Searidge and Yakutat Seavalley (Holtedahl, 1958; Jordan, 1958). Todd and Low (1967) have described the foraminifers and surficial sediments of Pamplona Searidge. Yakutat Seavalley is a broad northeastward-curving re-entrant into the Gulf of Alaska continental shelf. Molnia and Carlson (1978) suggest that Yakutat Seavalley is at least in part of glacial origin. The northeast end of the sea valley appears to be faulted (Wright, 1968, Fig. 10).

Yakutat Bay is a silled fjard; a fjard is a broad, relatively shallow embayment occupying a deglaciated terrain (Embleton and King, 1968). The sill is a bay mouth bar with a depth of 50 to 75 m which was formed as the terminal moraine of Hubbard Glacier when that gla-

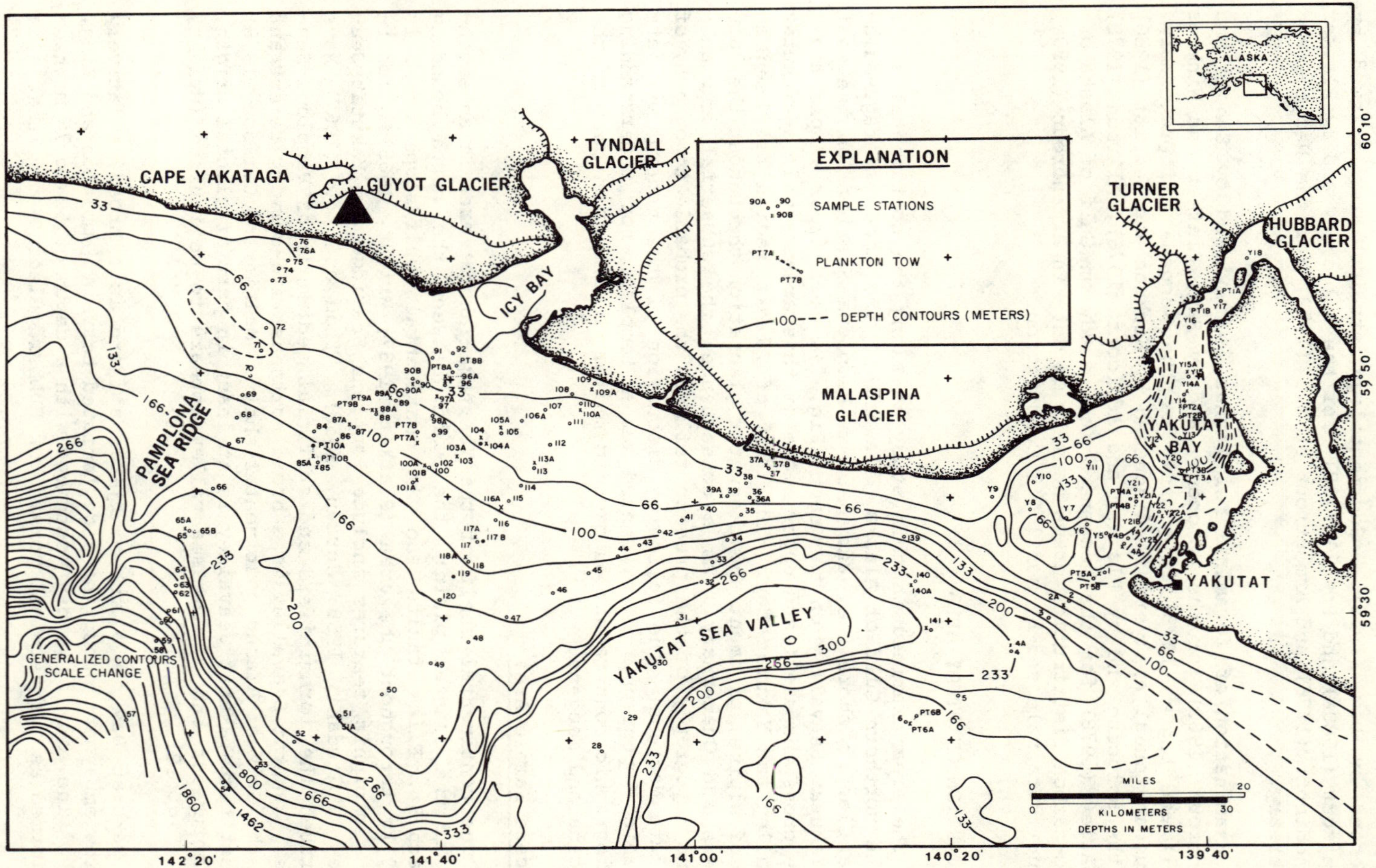

Fig. 1. Location of study area (insert) and sample stations. Sea floor sample data tabulated in Table 1. Surface section studied (▲) is located between Cape Yakataga and Icy Bay. By permission from Pacific Coast Paleogeogr. 4, Copyright (c) 1980 SEPM, Pacific Section, Los Angeles.

cier extended through Yakutat Bay onto the continental shelf between 700 and 1000 years ago (Plafker and Miller, 1958; Molnia and Sangray, 1979). The upper end of Yakutat Bay narrows into a fjord called Disenchantment Bay.

Malaspina Glacier, the world's largest non-polar piedmont glacier, is just northwest of Yakutat Bay. Meltwaters from the Malaspina Glacier drain across an outwash plain into the northwest side of Yakutat Bay as well as directly into the Gulf of Alaska. Turner and Hubbard glaciers, both tidewater alpine glaciers at the head of Yakutat Bay, calve icebergs into Yakutat Bay, and some of them drift out to the open sea.

The coastal plain relief surrounding Yakutat Bay, and the submarine topography within the bay, reflect recent glacial advances and retreats. High resolution seismic profiles within the bay show irregular, acoustically opaque rises with patches of stratified fill in the deeps (Wright, 1969). Wright interpreted the acoustically opaque rises as either bedrock or morainal material. Molnia (1979) identified three moraines within Yakutat Bay and these correspond to the acoustically opaque areas of Wright (1969). The stratified basins have been interpreted as areas of active, water laid deposition (Wright, 1969; Molnia, 1979). Molnia and Sangray (1979) have calculated rates of sedimentation for each of these basins.

Neogene Sample Area

The surface sections examined in this study occur on the coastal terraces and adjacent mountainous areas. Data are presented for a composite single section from the creek valleys and cirque headwalls, and on ridges between Poul Creek and the upper slopes of Munday Peak along Sullivan Anticline (Figs. 1 and 2). This section is representative of the typical lithofacies sequence of the upper Poul Creek and Yakataga Formations of the Robinson Mountains (Plafker and Addicott, 1976). The outcrop area of Sullivan Anticline is more than 80% denuded by glaciation, which affords a unique opportunity for recognizing the geometry of major stratigraphic units. The lower part of the section, consisting of the upper Poul Creek and lower Yakataga Formations, is composed of tabular siltstone, sandstone and claystone. These marine units were deposited in upper slope and shelf environments (Lagoe, in press) with periodic but subordinate glacial influence (Armentrout, 1980b), a depositional environment analogous to the present upper slope and outer shelf of the Gulf of Alaska (Armentrout, 1980a).

The upper Yakataga Formation consists of siltstone and sandstone units of both tabular and lenticular geometry. Facies analysis documents considerable glacial influence during neritic deposition of the upper Yakataga Formation (Plafker and Addicott, 1976; Armentrout et al., 1979; Armentrout, 1980b; Lagoe, in press). Major channel se-

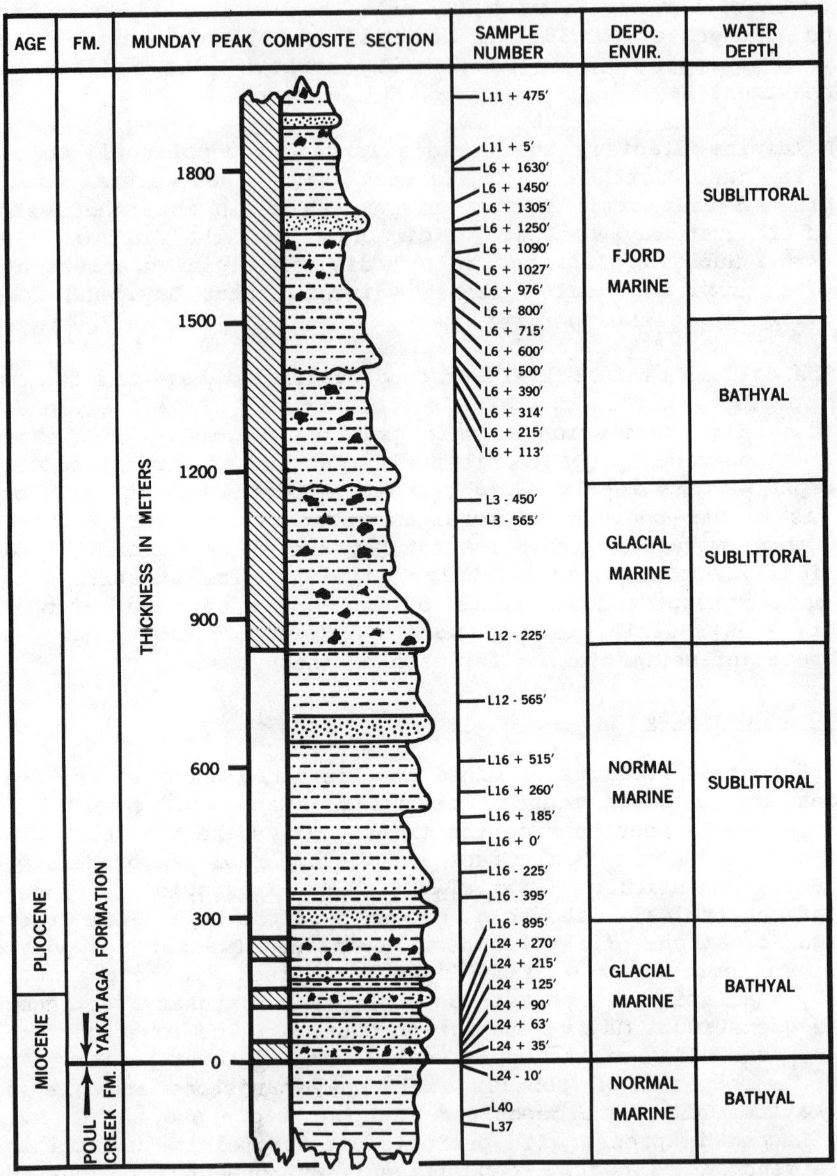

Fig. 2. Measured section at Munday Peak. Rock type is shown generalized from detailed measured sections. Dots represent sandstone; dots and dashes represent siltstone; dashes represent claystone. Glacial dropstones are shown as angular black blocks. Depositional sequences of glacial origin are noted by the oblique-line pattern to the left of the stratigraphic column. Sample numbers represent measured sections with sampled interval in feet above (+) or below (-) a zero datum for each measured interval.

quences (>90 m in depth) cut into and are filled with both fluvial and/or marine rocks of glacial origin. The channels are elongate and steep walled, and appear to be confined within or parallel to paleotopographic low areas. Detailed outcrop studies of five channels show maximum dimensions of 430 m in depth and about 3 km in width. Channel length was not determinable. The channels have been interpreted as fjords (Armentrout et al., 1979), and the modern fjords of Icy Bay and Yakutat Bay serve as a modern depositional analogue.

METHODS

Sampling

Sampling of the sea floor sediments was attempted at 155 stations (Fig. 1). Sampling methods and sample analyses are discussed in Armentrout (1980a). Data for 37 samples are presented in Table 1. Outcrop samples were collected within a detailed measured section shown schematically in Fig. 2. Data are presented in Table 2. All grain-size assignments in Tables 1 and 2 are based on outcrop or shipboard visual examination.

Table 1. Selected data set of Holocene sea floor sediment samples (from Armentrout, 1980a) (Gr) = gravel present in sediment sample.

SAMPLE NO.	SEDIMENT TYPE	TOC (wt.%)	DEPOSITIONAL ENVIRONMENT
Y18A	Clayey Silt (Gr)	0.24	Fjord - Sublittoral
Y17	Clayey Silt (Gr)	0.25	Fjord - Sublittoral
Y15	Clayey Silt (Gr)	0.40	Fjord - Bathyal
Y13	Silty Clay (Gr)	0.43	Fjord - Sublittoral
Y20	Silty Clay	0.49	Fjard - Sublittoral
Y22A	Clayey Silt	0.36	Fjard - Sublittoral
Y5	Clayey Silt	0.30	Fjard - Sublittoral
1	Silty Sand	0.12	Fjard - Sublittoral
2	Clayey Silt	0.25	Shelf - Sublittoral
3	Clayey Silt	0.29	Shelf - Sublittoral
4	Clayey Silt (Gr)	0.43	Shelf - Bathyal
141A	Clayey Silt	0.50	Canyon - Bathyal
32	Clayey Silt	0.41	Canyon - Bathyal
31	Clayey Silt	0.37	Canyon - Bathyal
30	Silty Sand	0.35	Canyon - Bathyal
29	Sandy Silt	0.30	Canyon - Bathyal
28	Silty Sand	0.26	Canyon - Bathyal
37	Sand	0.15	Shelf - Sublittoral
38	Sand	0.11	Shelf - Sublittoral
107	Silty Sand	0.13	Shelf - Sublittoral
39	Sand	0.18	Shelf - Sublittoral
41	Clayey Silt	0.27	Shelf - Sublittoral
43	Clayey Silt	0.29	Shelf - Sublittoral
101A	Clayey Silt	0.30	Shelf - Sublittoral
45	Clayey Silt	0.39	Shelf - Sublittoral
46	Clayey Silt	0.37	Shelf - Sublittoral
120	Clayey Silt	0.39	Shelf - Sublittoral
48	Clayey Silt	0.37	Shelf - Sublittoral
49	Clayey Silt	0.45	Shelf - Sublittoral
50	Clayey Silt	0.40	Shelf - Sublittoral
51	Silty Sand (Gr)	0.30	Shelf - Sublittoral
66	Clayey Silt (Gr)	0.49	Slope - Bathyal
65A	Clayey Silt	0.54	Slope - Bathyal
64	Sandy Silt	0.40	Slope - Bathyal
63	Sandy Silt	0.36	Slope - Bathyal
61	Clayey Silt	0.73	Slope - Bathyal
60	Clayey Silt	0.73	Slope - Bathyal

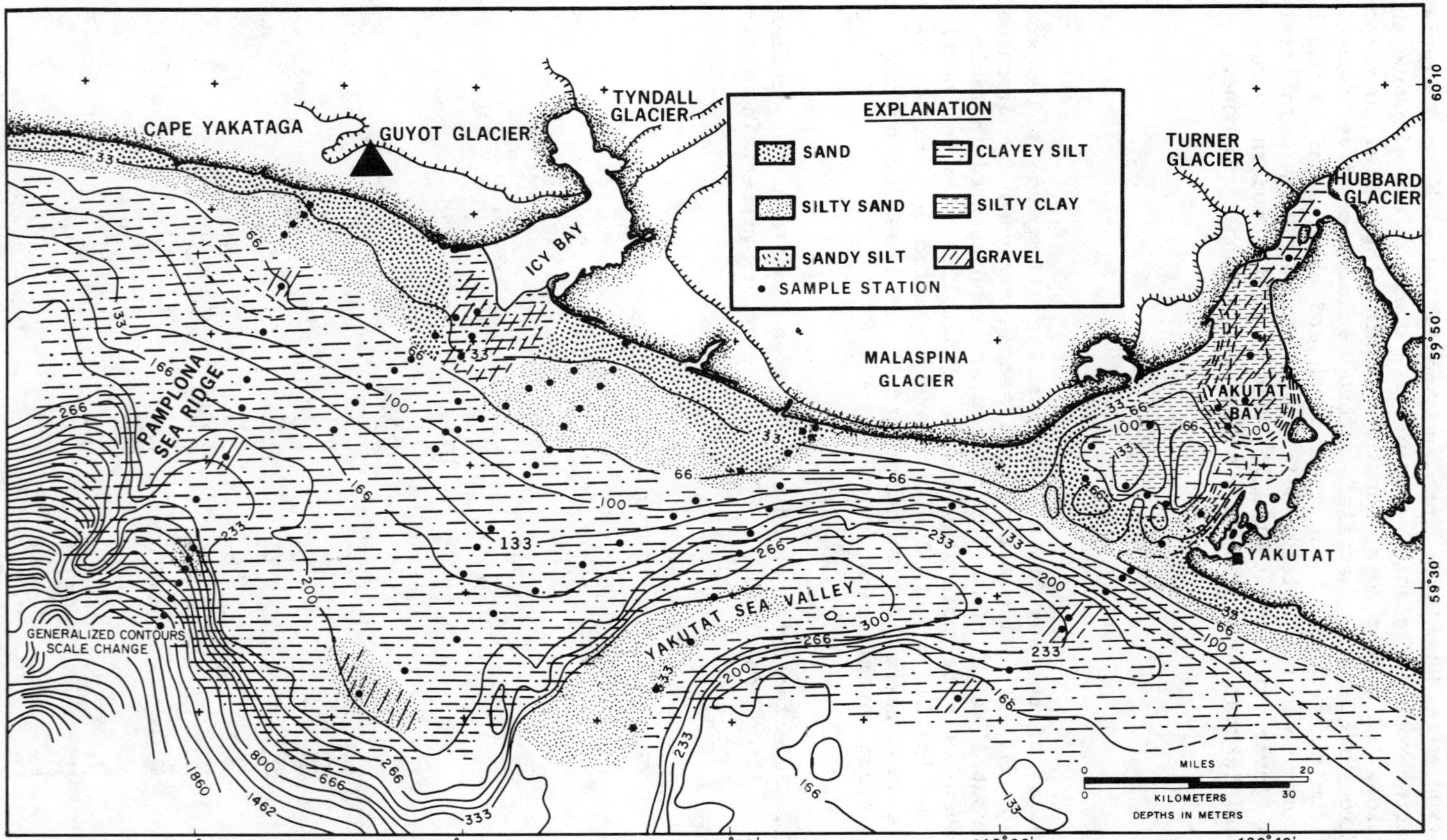

Fig. 3. Distribution of sediment types based on grain-size analysis (from Armentrout, 1980a). By permission from Pacific Coast Paleogeogr. 4, Copyright (c) 1980 SEPM, Pacific Section, Los Angeles.

Table 2. Sample stations and sediment analysis for Neogene outcrop sequence. Depositional environment is based on interpretation of lithofacies and benthic foraminiferal biofacies; normal refers to normal marine.

AGE	FORMATION	SAMPLE NO.	ROCK TYPE	TOC (Wt.%)	DEPOSITIONAL ENVIRONMENT
Pliocene	Yakataga	L11 + 475'	Siltstone	0.32	Fjord - Sublittoral
Pliocene	Yakataga	L11 + 5'	Siltstone	0.45	Fjord - Sublittoral
Pliocene	Yakataga	L 6 + 1630'	Siltstone	0.29	Fjord - Sublittoral
Pliocene	Yakataga	L 6 + 1450'	Siltstone	0.42	Fjord - Sublittoral
Pliocene	Yakataga	L 6 + 1305'	Siltstone	0.42	Fjord - Sublittoral
Pliocene	Yakataga	L 6 + 1250'	Siltstone	0.26	Fjord - Sublittoral
Pliocene	Yakataga	L 6 + 1090	Siltstone	0.39	Fjord - Sublittoral
Pliocene	Yakataga	L 6 + 1027'	Siltstone	0.32	Fjord - Sublittoral
Pliocene	Yakataga	L 6 + 976'	Siltstone	0.37	Fjord - Sublittoral
Pliocene	Yakataga	L 6 + 800'	Siltstone	0.38	Fjord - Sublittoral
Pliocene	Yakataga	L 6 + 715'	Siltstone	0.27	Fjord - Sublittoral
Pliocene	Yakataga	L 6 + 600'	Siltstone	0.36	Fjord - Bathyal
Pliocene	Yakataga	L 6 + 500'	Siltstone	0.33	Fjord - Bathyal
Pliocene	Yakataga	L 6 + 390'	Siltstone	0.33	Fjord - Bathyal
Pliocene	Yakataga	L 6 + 314'	Siltstone	0.33	Fjord - Bathyal
Pliocene	Yakataga	L 6 + 215'	Siltstone	0.40	Fjord - Bathyal
Pliocene	Yakataga	L 6 + 113'	Siltstone	0.48	Fjord - Bathyal
Pliocene	Yakataga	L 3 - 450'	Siltstone	0.33	Glacial - Sublittoral
Pliocene	Yakataga	L 3 - 575'	Siltstone	0.48	Glacial - Sublittoral
Pliocene	Yakataga	L12 - 225'	Siltstone	0.39	Glacial - Sublittoral
Pliocene	Yakataga	L12 - 575'	Siltstone	0.37	Canyon-Normal - Sublittoral
Pliocene	Yakataga	L16 + 515'	Siltstone	0.40	Normal - Sublittoral
Pliocene	Yakataga	L16 + 260'	Siltstone	0.39	Normal - Sublittoral
Pliocene	Yakataga	L16 + 185'	Siltstone	0.32	Normal - Sublittoral
Pliocene	Yakataga	L16 + 0'	Siltstone	0.46	Normal - Sublittoral
Pliocene	Yakataga	L16 - 225'	Siltstone	0.37	Normal - Sublittoral
Pliocene	Yakataga	L16 - 395'	Siltstone	0.40	Normal - Sublittoral
Pliocene	Yakataga	L16 - 895'	Siltstone	0.40	Glacial - Sublittoral
Late Miocene	Yakataga	L24 + 270'	Siltstone	0.40	Glacial - Sublittoral
Late Miocene	Yakataga	L24 + 215'	Siltstone	0.34	Glacial - Sublittoral
Late Miocene	Yakataga	L24 + 125'	Siltstone	0.53	Glacial - Bathyal
Late Miocene	Yakataga	L24 + 90'	Siltstone	0.54	Glacial - Bathyal
Late Miocene	Yakataga	L24 + 63'	Siltstone	0.47	Glacial - Bathyal
Late Miocene	Yakataga	L24 + 35'	Siltstone	0.44	Glacial - Bathyal
Late Miocene	Poul Creek	L24 - 10'	Siltstone	0.52	Normal - Bathyal
Late Miocene	Poul Creek	L40	Claystone	2.39	Normal - Bathyal
Late Miocene	Poul Creek	L37	Claystone	1.64	Normal - Bathyal

Laboratory Analysis

Dried sea floor sediment samples were analyzed for grain size by standard Rotap and pipette methods (Krumbein and Pettijohn, 1938). The results are tabled and mapped by Armentrout (1980a). The map of sediment types is reproduced here as Fig. 3. Outcrop samples were identified by handlens examination in the field. Concentration of organic carbon in both sea floor and outcrop samples was measured with a LECO induction furnace (Curl, 1963; Gross et al., 1972). Data on sea floor samples are tabled and mapped by Armentrout (1980a). The map of sea floor organic-carbon distribution is reproduced here as Fig. 4. Data for 37 sea floor samples are presented in Table 1. Outcrop data are presented in Table 2. Comparison of Holocene silt and Neogene siltstone organic-carbon concentrations of comparable depositional environments is presented in Table 3.

Table 3. Comparison of Holocene silt and Neogene siltstone organic-carbon concentrations of comparable depositional environments:

$\frac{\text{Minimum TOC (wt.\%) - Maximum TOC (wt.\%)}}{\text{Number of Samples}}$: Average TOC (wt.%)

DEPOSITIONAL ENVIRONMENTS		HOLOCENE	NEOGENE
GLACIATED TROUGH FJORD:	SUBLITTORAL	$\frac{0.24 - 0.49}{7}$: 0.35	$\frac{0.26 - 0.45}{11}$: 0.35
	BATHYAL	$\frac{0.40}{1}$: 0.40	$\frac{0.33 - 0.48}{6}$: 0.37
SUBSEA CANYON:	BATHYAL	$\frac{0.29 - 0.43}{9}$: 0.36	$\frac{0.37}{1}$: 0.37
GLACIAL MARINE:	SUBLITTORAL	$\frac{0.19 - 0.75}{7}$: 0.38	$\frac{0.33 - 0.48}{6}$: 0.39
	BATHYAL	$\frac{0.49}{1}$: 0.49	$\frac{0.44 - 0.54}{4}$: 0.50
NORMAL MARINE:	SUBLITTORAL	$\frac{0.19 - 0.46}{34}$: 0.32	$\frac{0.32 - 0.46}{6}$: 0.39
	BATHYAL	$\frac{0.36 - 0.75}{8}$: 0.60	$\frac{0.52}{1}$: 0.52

Organic matter for visual examination was recovered following standard palynological techniques (Gray, 1965). Organic-matter type was determined by standard optical microscopic methods for palynology (Combaz, 1964), fluorescence (Tissot and Welte, 1978; van Gizel, 1979; Durand, 1980) and reflectance (Stach, 1975; Dow, 1977). Data for selected samples are presented in Table 4 and Fig. 5.

RESULTS

Organic-Carbon Concentrations

The organic-carbon concentration in the Holocene sediments ranges from 0.06 wt-% in sand to 0.75 wt-% in clayey silt (Table 1, and Armentrout, 1980b). Organic-carbon concentrations in silts range from 0.19 to 0.75 wt-% (Table 3). The distributional pattern shows a seaward increase which correlates closely with decreasing sediment grain size (Figs. 3 and 4). Low organic-carbon concentrations extend further seaward offshore from the Malaspina Glacier foreland and Icy Bay than elsewhere along the study area. This pattern follows the distribution of the coarser grain-size sediments. Organic-carbon concentrations increase seaward with depth except for the floor of Yakutat Seavalley where slightly coarser sediment carpets the sea

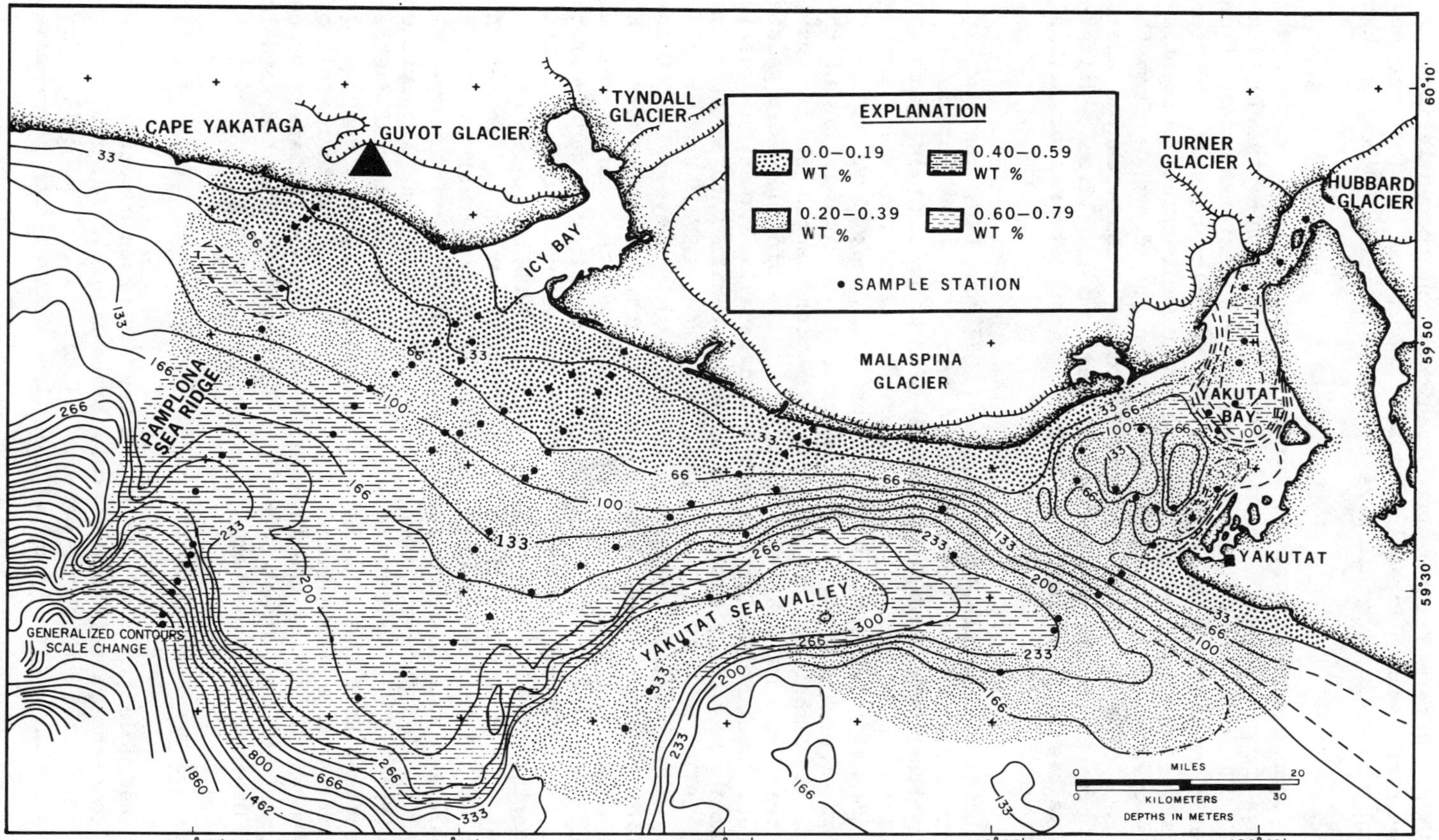

Fig. 4. Distribution of organic-carbon concentrations (weight percent) in surface sediment samples (from Armentrout, 1980a). By permission from Pacific Coast Paleogeogr. 4, Copyright (c) 1980 SEPM, Pacific Section, Los Angeles.

valley floor. Within Yakutat Bay organic-carbon concentrations are relatively high, with decreasing values toward the head of the bay even though the sediment grain size decreases, a trend opposite that on the open shelf.

The organic-carbon concentrations in the Neogene sedimentary rocks range from 0.26 wt-% in siltstone to 2.39 wt-% in claystone (Table 2). Organic-carbon concentrations in siltstones range from 0.26 to 0.54 wt-% (Table 3). Glacial marine Holocene silt and Neogene siltstone samples show an increase in organic-carbon concentration with greater depositional water depth and distance from the paleo-shore. Silt-size samples from fjord and subsea canyon areas show low organic-carbon concentrations with no significant differences in values between samples deposited in sublittoral or bathyal environments.

The Late Miocene claystone samples, L37 and L40, contain high concentrations of organic carbon; 1.64 and 2.39 wt-% respectively. These claystones are from an interval of the Poul Creek Formation often referred to as the claystone interval. Lagoe (in press) interprets the depositional environment of the Poul Creek claystone interval as bathyal. The Poul Creek claystone is a correlative of the Organic Shale Member of the Katalla Formation mapped in the more western Kayak Island and Katalla districts. The Middle to early Late Miocene Poul Creek Formation claystone and Katalla Formation Organic Shale Member are widely recognized as organic-rich bathyal deposits (Plafker, 1971; 1974; Plafker, Bruns and Page, 1975; Rau et al., 1977; Lagoe, in press). Samples L37 and L40 are typical of these late Middle to early Late Miocene organic-carbon-rich rocks. No analogous Holocene sediments were recovered from the Yakataga-Yakutat sea floor. The failure to find analogous seafloor organic-carbon concentrations may be due to the few samples recovered from mid-bathyal depths of 1000-2000 m (Armentrout, 1980b). Dow (1978) reports that in the North Pacific Ocean, organic-carbon concentrations greater than 1.0 weight percent were found in a few lower bathyal to abyssal samples from lower slope, rise and trench environments, all of which are significantly deeper than the environment interpreted for deposition of the Poul Creek Formation claystone interval. These results suggest that the organic-carbon concentrations of samples L37 and L40 represent a depositional environment in the northern Gulf of Alaska recognized only in the late Middle through early Late Miocene deposition of the Poul Creek Formation claystone interval and Katalla Formation Organic Shale Member.

Organic Carbon Type

The insoluble fraction of the sedimentary organic matter, called kerogen, consists principally of cellulosic-ligninic material, palynomorphs and lipid amorphous material (Table 4). The cellulosic-ligninic and palynomorph kerogen grains are derived from terrestrial

Table 4. Visual kerogen analysis. Kerogen types are reported as percent ranges; T = trace amounts. Data provided by R.J. Enrico, Mobil Exploration and Producing Services, Inc.

ORGANIC MATTER / SAMPLE	LIPID-WAXY			CELLULOSIC-LIGNINIC		
	AMORPHOUS	CUTICLE	PALYNOMORPHS	AMORPHOUS	BARK & WOODY DEBRIS	COALY FRAGMENTS
Y17				11-20		71-80
Y15				21-30		61-70
Y5			T	31-40	T	51-60
141A			T	41-50	21-30	11-20
107			11-20		21-30	41-50
106			1-10		21-30	61-70
101			T	51-60		21-30
120			1-10	21-30	11-20	31-40
49			T	61-70	T	21-30
60			T	41-50	11-20	21-30

ORGANIC MATTER / SAMPLE	LIPID-WAXY			CELLULOSIC-LIGNINIC		
	AMORPHOUS	CUTICLE	PALYNOMORPHS	AMORPHOUS	BARK & WOODY DEBRIS	COALY FRAGMENTS
L11+425	T		T	51-60	11-20	11-20
L6+976	T		T	51-60	11-20	11-20
L6+500			T	11-20	51-60	11-20
L6+215			T	61-70	21-30	T
L12-225			T	41-50	41-50	T
L16-395			T	0-10	51-60	21-30
L24+270	T		T	51-60	11-20	11-20
L24+90	T		T	11-20	51-60	11-20
L40	11-20		T	11-20	51-60	T
L37	61-70		T	21-30	T	

woody-plant matter; the lipid amorphous material is derived from algal matter (Tissot and Welte, 1978).

The predominance of cellulosic-ligninic kerogens in all but samples L37 and L40 indicates the dominance of terrestrially derived organic matter in both Holocene and most Neogene Gulf of Alaska samples. Samples Y17 and Y15, from within the fjord Disenchantment Bay, are mostly coaly fragments with associated non-fluorescing amorphous material of woody origin (van Gizel, 1979; Durand, 1980). Sample Y5 from the fjard, Yakutat Bay, is very similar, with the addition of pollen grains and bark and woody debris, a component characteristic of the more open marine bay, shelf and slope Holocene kerogen assemblages.

Kerogen assemblages of the Neogene samples are similar to those of the open marine Holocene samples, with a dominance of non-fluorescing amorphous material, bark and woody debris, coaly fragments and a trace of pollen grains. The rocks interpreted as being deposited in fjord environments (samples L6 through L11 intervals-Fig. 2) have a trace of pollen grains in the assemblage. Samples L11 + 425' and L6 + 976' also include fluorescing amorphous material interpreted as being of algal origin (van Gizel, 1979; Durand, 1980). Sample L37, from the late Miocene Poul Creek Formation claystone interval is organic-rich. Its kerogen assemblage is dominated by fluorescing amorphous material of algal origin, some non-fluorescing amorphous material of cellulosic-ligninic origin, and traces of pollen and woody debris. The upsection sequence of sample L37 through sample L16 + 395' shows a gradational sequence of kerogen assemblages from the uniquely algal-rich L37 assemblage to the cellulosic-ligninic assemblages analogous to the kerogen assemblages of open marine Holocene sediment.

Recycled Organic Matter

The Gulf of Alaska kerogen assemblages consist of a mixture of primary and recycled grains, a factor that must be carefully considered in all interpretations of organic-carbon data. The mixture of primary and recycled grains is suggested both by visual examination of the macroscopic organic matter in the modern sediment samples and microscopic examination of palynomorph and vitrinite grain assemblages.

Shipboard examination of recovered organic matter identified unaltered root, wood and leaf material, and carbonized woody material. The carbonized material is most probably from trees burned during forest fires, a not infrequent occurrence along the Gulf of Alaska coastal plain. Microscopic examination of Holocene sample kerogen assemblages identified pollen grains exhibiting colors indicative of four stages of alteration: (1) yellow = unaltered; (2) yellow-brown = slightly altered (diagenesis); (3) brown = extensively altered (catagenesis); and black = highly altered (metagenesis) (Tissot and Welte, 1978). The modern pollen can be recognized by its yellowish color, excellent preservation including cytoplasm within the exine,

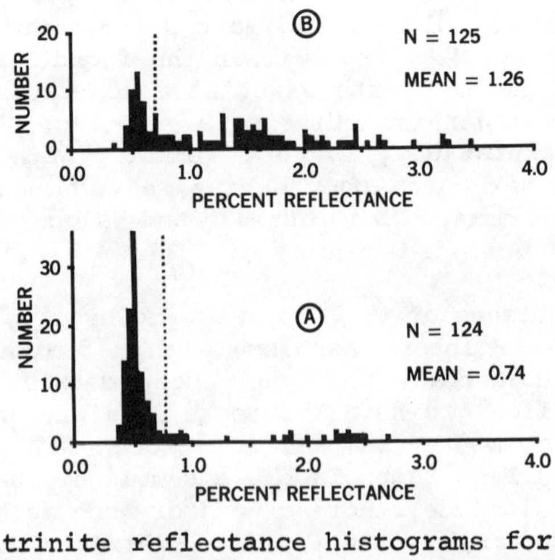

Fig. 5. Vitrinite reflectance histograms for two surface samples from the lower Yakataga Formation, Cape Yakataga, Alaska. The *in situ* grains are interpreted to be those to the left of the dashed line; all other grains are considered to be recycled.

and known nearby source plants, such as *Picea* and *Tsuga* (Viereck and Little, 1972; J.F. Stone, pers. comm.). Darker pollen grains suggest greater alteration, which may be due to inclusion of grains reworked from older rocks. Reworking is supported by the presence of *Fagus*-like pollen grains in the Holocene palynomorph assemblages. Beech (*Fagus*) trees are not reported as presently growing in Alaska (Vierek and Little, 1972) indicating that the *Fagus* pollen grains are probably reworked from older rocks (J.F. Stone, written comm., 1976).

Reworking of organic matter from older rocks is characteristic of both the Holocene and Neogene kerogen assemblages from the Gulf of Alaska. Histograms of vitrinite reflectance values for samples from the Neogene Yakataga Formation are typically broad based and polymodal (Fig. 5). Vitrinite is a product of the anaerobic diagenesis of woody material from terrestrial plants. The reflectivity of vitrinite increases with increasing alteration. The broad based polymodal histograms represent the mixing of several populations of vitrinitic grains, some of which can be correlated to the *in situ* vitrinite population of nearby Paleogene rocks from which they have been recycled. The vitrinite reflectance histograms of Fig. 5 are both from lower Yakataga Formation rock samples. In each histogram, the low reflectance mode is interpreted as the *in situ* vitrinite and grains of higher reflectance as reworked material (Dow, 1977). On that basis, reworked grains constitute 22% of sample 6A and 86% of sample 6B. These two samples approximate end members of a large data set.

DISCUSSION

The dominance of terrestrially derived organic carbon in Gulf of Alaska marine sediments seems incongruous with the recognized high levels of marine planktonic productivity of the north Pacific. Coastal Domain waters, extending over the continental shelf and slope (Dodimead, Favorite and Hirano, 1963), are estimated to have levels of annual primary productivity approaching 300 gC/m^2 (Cooney, 1972). This productivity is supported by high levels of nutrient availability supplied by deep-water upwelling (Romankevich, 1968; Dow, 1978). Local variations are certain to occur due to hydrographic constraints on upwelling systems (Smith, this volume), but the organic carbon produced should be uniformly high in the well-mixed surface waters of the Gulf of Alaska Coastal Domain. This high level of organic carbon should be reflected in the distribution of marine-algal-derived organic carbon in the underlying sea floor sediments.

Mapped organic-carbon values for Gulf of Alaska Holocene sea floor surface sediment are available from Garshanovich (1965), Premuzic (1980) and Armentrout (1980b). Values range from 0.5 to 1.50 TOC (wt-%). Levels of organic-carbon concentration between 1.0 and 1.50 TOC (wt-%) are restricted to areas along the continental slope and rise immediately west of and further southeast of the Yakataga-Yakutat study area (Garshanovich, 1965). Assuming that the

Coastal Domain surface water level of organic-carbon concentration is uniformly high, the problem then focuses on why the organic-carbon accumulations of the Yakataga-Yakutat outer shelf and upper slope are generally low.

Dow (1978) lists three prerequisites for the accumulation of organic-carbon-enriched sediments:

1. abundant nutrients to support high levels of organic productivity;
2. minimal exposure of the organic matter to degradation by biochemical oxidation; and
3. a sedimentation rate of mineral matter adequate for burial of the organic matter without excessive dilution.

Nutrient availability and high primary productivity have already been established for Coastal Domain waters in the Gulf of Alaska (Romankevich, 1968; Cooney, 1972). Paleogeographic modeling (Parrish, 1982; this volume) suggests that this high productivity system has persisted at least throughout the Cenozoic (Parrish, pers. comm., 1981). Thus, the factors controlling the distribution of organic carbon in waters of the Gulf of Alaska Coastal Domain must be degradation of organic carbon and differences in sedimentation patterns.

I interpret the available data to suggest that two principle factors control levels and types of organic carbon concentrations in the Gulf of Alaska Holocene sediments. First, the high concentrations of phytoplanktonic organic matter are greatly reduced both in the water column and at the sea floor by biochemical and physical degradation. Secondly, as a consequence of the removal of marine-algal-derived organic matter, terrestrially derived organic matter dominates the preserved assemblage. The concentrations of terrestrially derived organic matter are governed by availability of terrestrial organic matter versus mineral matter within the sediment source terrain and sediment size distribution patterns. Both of these observations have been made by previous workers (Dow, 1978; Demaison and Moore, 1980).

I further interpret the Yakataga-Yakutat organic-carbon distribution patterns to suggest that the Yakataga-Yakutat shelf Holocene sedimentation system is an appropriate analogue for latest Miocene through Pleistocene but that the marine-algal organic enrichment of late Middle and early Late Miocene required a markedly different oceanographic-sedimentation system related to intensified upwelling and eustatic sea level rise.

Marine Productivity and Preservation

Dead organic matter serves as a source of energy and nutrients for living organisms. Utilization by organisms and consequent degradation proceeds rapidly in aerobic (oxygen-rich) environments and is continued by anaerobic bacteria as the oxygen supply becomes exhausted (Demaison and Moore, 1980). Anaerobic degradation is thermodynamically less efficient than aerobic decomposition (Claypool and Kaplan, 1974) and results in more lipoid-rich and more reduced (hydrogen-rich) organic residue than does aerobic degradation (Demaison and Moore, 1980). The consequence of these degradation processes is the preferential enrichment of lipoid-rich organic carbon in sediments deposited under anerobic conditions.

The availability of oxygen for degradation of organic carbon is critical both within the water column and at the sediment-water interface. The longer the organic carbon resides in aerobic environments, the more thoroughly it becomes degraded. If the organic matter produced by phytoplankton sinks rapidly into anaerobic water column or sea floor environments, the higher will be the probability of its being less degraded.

The oxygen content of Gulf of Alaska surface oceanic waters ranges between 6.0 and 7.5 mℓ/ℓ of water, decreases at the halocline between depths of 100 to 250 m, and reaches an oxygen minimum zone of 0.4 to 0.5 mℓ/ℓ of water at depths of 700 to 800 m (Royer, 1972; Echols and Armentrout, 1980). Demaison and Moore (1980) define as "anoxic" any water containing less than 0.5 mℓ/ℓ which is the threshold below which the metazoan benthic biomass and bioturbation by deposit feeders becomes significantly depressed. If the oxygen minimum layer impinges on the sediment-water interface, those sediments deposited within the oxygen minimum should have potentially good quantitative preservation of organic matter (Dow, 1977; Demaison and Moore, 1980). Sea floor oxygen measurements are not available for the study area. Faunal studies of the Yakataga and Yakutat area suggest that all sampled areas are well oxygenated (Bergen and O'Neil, 1979; Echols and Armentrout, 1980; Hickman and Nesbitt, 1980; Quinterno, Carlson and Molnia, 1980).

Two other lines of evidence also suggest the presence of well-oxygenated bottom waters: sediment texture and organic-carbon concentrations. Anoxic conditions at the sediment-water interface often result in laminated sedimentation (Calvert, 1964; Demaison and Moore, 1980) although lamination of oxygen minimum zone sediments is not pervasive (Demaison and Moore, 1980). The eight sediment samples from depths equivalent to the Gulf of Alaska oxygen minimum zone of Royer (1972) were not laminated (Armentrout, 1980b).

Off Washington and Oregon, Gross et al. (1972) found a positive correlation between organic-carbon content of bottom sediments and oxygen depletion of bottom water. Where the dissolved oxygen concentrations fall below 1 mℓ O_2/ℓ, organic-carbon concentrations range between 1 and 3 wt-%. Between 1 and 2 mℓ O_2/ℓ of water, organic-carbon contents of bottom sediments fall around 1 wt-%. The maximum organic carbon samples from the data set in Table 1 approach 0.75 wt-% suggesting that they are not from an oxygen minimum environment analogous to that studied by Gross et al. (1972). If, as the above information suggests, the bottom water at the 700-800 m depth sediment-water interface is not at an oxygen minimum below 0.5 mℓ O_2/ℓ then the oxygen minimum zone identified by Royer (1972) for the Gulf of Alaska oceanic water is depressed below the Yakataga-Yakutat sample set, or is simply not developed along the sea floor beneath the Gulf of Alaska coastal domain.

Demaison and Moore (1980) present a discussion of deep ocean circulation as it relates to the development of an oxygen minimum zone. Two essential requirements must be met: (1) depletion of oxygen by biochemical oxidation of organic matter produced by high planktonic productivity; and (2) non-replenishment of the oxygen supply by deep water circulation. The Gulf of Alaska is an area of high productivity and therefore should have a high biochemical demand for oxygen. However, the Gulf of Alaska is also an area of strong bottom-current activity, a feature prevalent in cold high-latitude regions and on the western side of basins and these are the areas of most turbid bottom water (Hollister, Flood and McCae, 1978).

The presence of strong bottom circulation may provide oxygen replenishment at a rate exceeding oxygen depletion by biochemical oxidation and physical reworking of organic matter. If so, this circulation pattern would preclude development of an anoxic oxygen-minimum-zone within Gulf of Alaska coastal domain waters (Fig. 6). The absence of an anoxic water mass results in the biochemical and physical degradation of most of the lipoid-waxy kerogen component and consequently the absence of marine-algal-derived organic carbon in the kerogen assemblage of the Yakataga-Yakutat Holocene samples. The

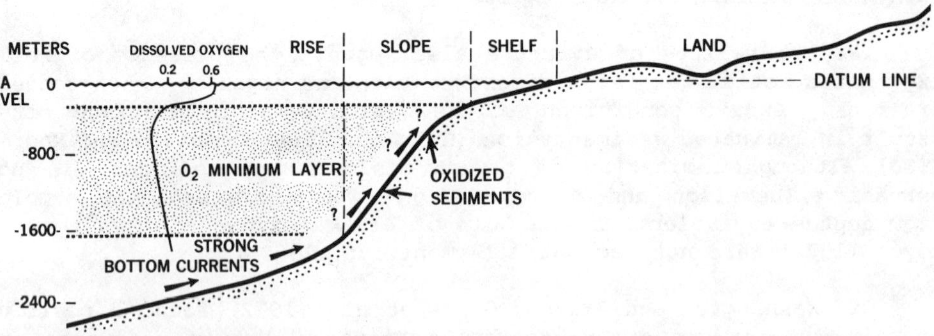

Fig. 6. Present day hydrographic profile of the Yakataga-Yakutat district. Data discussed in the text suggests that strong bottom currents re-oxygenate bottom waters keeping the oceanic oxygen minimum zone from impinging on the sea floor. Dissolved oxygen profile from Royer (1972).

Late Neogene rock kerogen assemblages also lack a significant marine-algal kerogen assemblage, suggesting deposition under oceanographic conditions analogous to the Holocene.

Sedimentation

Organic-matter composition and concentration in late Neogene and Holocene kerogen assemblages is a consequence both of organic matter availability and sediment transport and accumulation processes. The relative absence of marine-derived organic matter has already been discussed. Non-marine organic matter in the Gulf of Alaska comes from two sources: the plants living at the time of sedimentation and organic matter recycled from older sedimentary rocks.

The present flora of the Gulf of Alaska coastal plain is dominated by a coniferous forest of spruce (*Picea*) and hemlock (*Tsuga*); higher ice-free areas support an alpine tundra (Viereck and Little, 1972). Deciduous trees, such as alder (*Alnus*) and birch (*Betula*) are restricted to coastal and river valley areas. Palynomorph assemblages reflect this distribution, with few if any occurring in fjord samples adjacent to glaciated terrains, some in bay samples adjacent to coastal plains, and numerous palynomorph grains of the above genera occurring in offshore marine samples. Palynomorph assemblages from rocks of Oligocene to Pleistocene age are sufficiently similar to the Holocene assemblages to suggest that vegetation patterns have been rather similar for about 35 million years (J.F. Stone, pers. comm.). The seaward increase in palynomorph abundance (Table 4A) reflects their hydrodynamic properties because palynomorphs are more likely to accumulate in areas of low-energy, fine-grained sedimentation.

The other dominant organic matter component in the Gulf of Alaska samples is recycled kerogen (Fig. 5). The Gulf of Alaska Coastal Ranges have formed from numerous pulses of uplift since Middle Miocene time (Plafker, Bruns and Page, 1975). Erosion of these mountains, consisting dominantly of Mesozoic and Paleogene sedimentary rocks, has recycled organic matter along with the mineral matter. Geographic patterns of Neogene vegetation and glaciation suggest that much, if not most, of the Yakataga-Yakutat district Coastal Ranges terrain has been actively eroded during the Neogene to Holocene. As a consequence, recycled organic matter dominates the kerogen assemblages of most late Cenozoic rocks.

Sedimentation processes play a significant role in the concentration of organic matter. Table 3 shows the average concentration of organic matter by depositional environment for both Holocene and Neogene Gulf of Alaska samples. Two principal factors influence these organic matter sedimentation patterns: 1) grain size distributions; and 2) rate of sediment accumulation.

Sedimentary organic matter is usually fine-grained and thus accumulates with fine-grained mineral matter. This pattern is supported by comparing the distributions of sediment grain-size (Fig. 3) and organic-matter concentration (Fig. 4). The relatively low organic-matter concentrations of the inner sublittoral area and the bottom of Yakataga Seavalley reflect the relatively coarse-grained nature of the accumulated sediment. Most of the organic matter in the Holocene samples from these two areas consisted of macroscopic pieces of wood, root and bark.

Sediment accumulation rates for the Yakataga District Holocene have been estimated by Molnia and Sangrey (1979). Fjords may have sediment accumulation rates as high as 2000 m/1000 yrs.; fjards between 200 and 1400 m/1000 yrs.; and glacially influenced shelf environments from 0 to 16 m/1000 yrs. The rapid rate of fjord and fjard sediment accumulation would be expected to significantly dilute the organic matter content. The fact that dilution is not more significant is attributed to two factors: 1) most of the organic matter in the sediment is recycled from older rocks; and 2) much of the sediment accumulated in the fjords is glacial flour which is very fine grained.

The only organic-matter distribution anomaly in the data set is that of the upper Poul Creek Formation (Samples L37 and L40) claystones which represent the finest grained rocks recognized from the Yakataga-Yakutat district Neogene. In addition, using the chronostratigraphy of Armentrout et al. (1978) and Lagoe (in press) the rate of sediment accumulation for the organic-rich interval of the upper Poul Creek Formation in the Sullivan Anticline area is calculated at about 0.007 m/1000 yrs. (approximately 40 m deposited between 14 and 8 million years B.P.). This very slow rate of sediment accumulation is attributed to the relative absence of continentally derived mineral matter. The relatively abundant amorphous lipid-waxy kerogen in these samples (Table 4B) reflects the dominantly marine pelagic-hemipelagic depositional environment.

<u>Organic Enrichment of Upper Miocene Claystones</u>

The lower Upper Miocene organic-rich claystone samples L37 and L40 suggest deposition under anoxic conditions and thus, an oceanographic system markedly different from that of the glacially influenced late Late Miocene through Holocene. Samples L37 and L40 are from the Poul Creek Formation claystone interval. Within the Sullivan Anticline area of the Yakataga district, the claystone ranges from massive to finely laminated, contains a middle bathyal foraminiferal fauna, a few molluscan fossils, and is often phosphatic (Lagoe, in press). Age correlative rock to the west on Kayak Island is more organic-rich, more finely laminated, lacks macrofossils, has rare microfossils and is siliceous-rich. These lithofacies are similar to shale lithofacies of the Monterey Formation of California although the Poul Creek Formation rocks lack the varved diatomites so typical of many Monterey lithofacies (Pisciotto, 1978; Pisciotto and Garrison, 1981).

The Monterey Formation shales are interpreted as being deposited in response to climatically induced oceanographic changes (Ingle, 1973; 1981; Summerhayes, 1981). The middle Miocene intensification of Antarctic ice buildup (Woodruff, Savin and Douglas, 1980) resulted in oceanic cooling and a corresponding increase in the equator-to-pole thermal gradient (Summerhayes, 1981). The increased thermal gradient would have resulted in increased wind stress and a consequent increase in upwelling. The increase in upwelling led to high levels of primary productivity analogous to the Peru-Chile, southwest Africa and northern California coastal upwelling systems (Demaison and Moore, 1980; Summerhayes, 1981; Parrish, 1982). Recycling of the dead organic matter in the water column created a very high oxygen demand which brought about reduced oxygen and possibly anoxic conditions beneath the upwelling area. Impingement of the oxygen minimum conditions on the sediment-water interface may result in the deposition of organic and phosphatic-rich, laminated sediments. Few if any macrobenthic organisms can live in this low-oxygen to anoxic environment. Fluctuations in the oxygen content of the water may allow establishment of a benthic community which may bioturbate an interval of sediment before being killed by reestablishment of anoxic conditions (Pisciotto, 1978; Pisciotto and Garrison, 1981).

It is suggested that the Upper Middle to lower Upper Miocene Poul Creek Formation claystone interval and the Katalla Formation Organic Shale Member were deposited as a consequence of an oceanographic system similar to that controlling deposition of the Monterey Formation. Ingle (1981) has inventoried lithofacies equivalent rocks of Monterey age around the north Pacific margin and attributes their occurrence to a similar sequence of ocean-forced depositional events modified by local tectonics. Ingle's (1981) three major events are: (1) initial margin subsidence with deposition of gradational sequences of Oligocene non-marine or paralic lithofacies to lower Miocene bathyal lithofacies, (2) rapid subsidence to mid-bathyal depths with deposition of middle to upper Miocene siliceous sediments under anaerobic conditions, commonly with a basal phosphatic facies, and (3) introduction of rapidly deposited coarse terrigenous clastics during Pliocene-Pleistocene times, ultimately filling the basins.

A similar sequence of lithofacies occurs in the Gulf of Alaska Tertiary sequence of the Yakataga-Yakutat area (Plafker, 1974; Rau et al., 1977; Lagoe, in press): (1) Oligocene to early Miocene foraminiferal faunas record gradual basin deepening; (2) maximum depths of mid-bathyal range were reached during deposition of the organic-rich siliceous and/or phosphasic, laminated basinal shales of the upper Poul Creek Formation claystone interval and Katalla Formation Organic Shale Member; (3) Lower Yakataga Formation bathyal lithofacies are overlain by prograding sublittoral terrigeneous clastics including abundant glacially derived sediment. Tectonically, the Gulf of Alaska Neogene basin may have had a shelf-edge structural high (Bruns, 1977) which could have restricted shelf basin bottom water circulation (Fig. 7). Combined with intensified upwelling and consequent increased organic productivity and oxygen depletion of underlying waters, the slightly silled basin could have developed a depositional environment analogous to those of the Monterey Formation of California.

Coincident with the onset of mid-Miocene intensified upwelling (Summerhayes, 1981) was a major interval of sea level rise (Vail, Mitchum and Thompson, 1977). Besides deepening the water column over ocean basins, a rapid rise of sea level results in the entrapment of terrigenous sediments in estuaries and other nearshore environments, thus starving the continental shelf and ocean basins (Loutit and Kennett, 1981). This "starving" of the shelves and ocean basins results in a reduction of the mineral matter dilution of sea floor sediments and a net increase in the relative percentage of pelagic biogenic sediment, including organic carbon. Sediment accumulation rates of clay-size mineral matter is still adequate for burial and preservation of the organic matter (Ibach, 1982), especially if the oxygen content of the water column is depleted and anaerobic conditions develop. If all of the above physiographic factors acted in concert, the result would be deposition of an organic-rich laminated basinal shale. Lateral facies would show gradations toward more oxic environments or proximity to terrigenous clastic input. I suggest that the Poul Creek Formation claystone interval and correlative Katalla Formation Organic Shale Member were deposited in a restricted

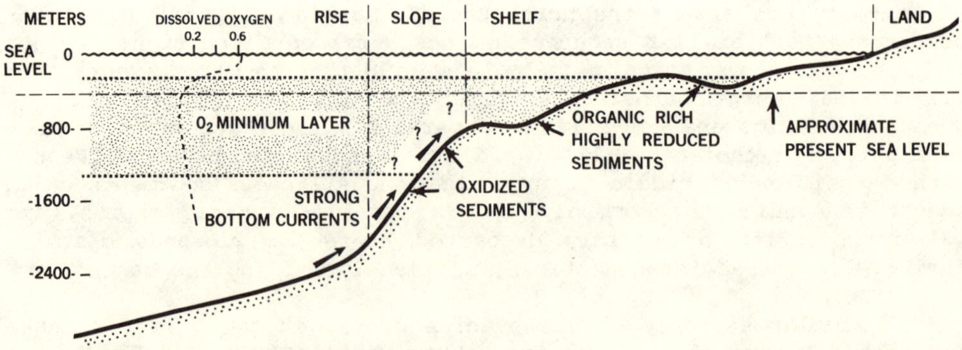

Fig. 7. Middle Miocene interpretative hydrographic profile for the Yakataga-Yakutat district. The organic-rich sediments of the upper Poul Creek Formation-Organic Shale Member are interpreted to have been deposited during a period of transgression with upwelling sufficiently intense to depress the oxygen minimum zone to the sea floor, possibly within a shelf margin basin.

shelf-margin basin in response to climatically induced intensified upwelling during a Middle Miocene to early Late Miocene rise in sea level (Fig. 7 - Miocene Transgression).

The depositional regime changed very quickly after deposition of the organic-rich facies. Late Miocene coast range tectonism (Ingle, 1973; Plafker et al., 1975), relative sea level drop (Vail et al., 1977) and the onset of glaciation (Armentrout et al., 1978; Armentrout, 1980b) resulted in basin shallowing and progradation of terrigenous clastics. This resulted in dilution of the organic-rich pelagic sediments by mineral matter, the infilling of any basin deep with restricted circulation thus eliminating silled basin anoxic environments, and seaward displacement of the upwelling system (Fig. 6 - Present Regression). Organic matter deposited in this depositional setting is dominantly terrestrial from plants and recycled kerogens, the marine-algal component being removed through biochemical and physical oxidative processes within the water column and at the sea floor.

SUMMARY

Two markedly different oceanographic-sedimentation systems are recorded in the Late Cenozoic stratigraphic record of the Yakataga-Yakutat district of southern Alaska. Latest Miocene through Holocene sedimentation patterns reflect well-oxygenated water and strong glacial influence. Sedimentary rocks and sediments are organically lean, well-oxidized, fossiliferous clastics. Organic carbon concentrations increase with decreasing grain size except toward the landward end of fjords where the high rate of sedimentation dilutes the organic carbon concentration. Dilution occurs despite the very fine-grained glacial flour sediments in the fjord and the dominance of recycled kerogen.

The generally low organic carbon concentrations and dominance of terrestrially derived organic matter are only superficially incompatible with the recognized high levels of upwelling-related marine planktonic productivity in the North Pacific. The absence of marine-derived organic carbon in the Holocene sea floor samples suggests that the organic matter produced in the photic zone does not reach the sea floor. Most probably, the marine-derived organic matter is consumed by the organisms living in the well-oxygenated water underlying the upwelling area. What marine-derived organic carbon reaches the sea floor is most probably consumed by benthic organisms which selectively feed on the more nutrient-rich marine organic matter rather than the oxidized terrestrial and recycled organic matter transported along with the clastic sediments.

The organic-rich, laminated, siliceous claystone of the lower Upper Miocene upper Poul Creek Formation suggests deposition under oceanographic conditions markedly different from those of the glacially influenced late Miocene through Holocene. The abundance of lipid-rich organic carbon in the claystone indicates preservation of algal-derived organic matter. The laminated sediments and general absence of fossils suggest low oxygen conditions. It is probable that the late Middle to early Late Miocene northern Gulf of Alaska oceanographic system consisted of: 1) high marine productivity due to upwelling nutrients; 2) development of an intensified oxygen-minimum zone due to oxidation of the abundant organic matter; 3) impingement of the oxygen-minimum zone on the sea floor; 4) relative absence of terrigenous clastic sediments due to global sea level transgression; and 5) accumulation of lipid-rich, laminated, poorly fossiliferous, siliceous claystones. This depositional regime is analogous to that controlling the Monterey Formation of California and other age equivalent North Pacific Miocene siliceous claystones.

Following the early Late Miocene anoxic depositional phase, the Yakataga-Yakutat district oceanographic/sedimentation system changed. Late Miocene tectonism and global sea-level lowering resulted in progradation of glacially influenced, organically lean, well-oxidized fossiliferous clastics deposited under sublittoral conditions. These conditions have persisted to the present.

ACKNOWLEDGEMENTS

This study was funded and released by Mobil Exploration and Producing Services, Inc. Discussion with J.T. Parrish of the U.S. Geological Survey, G.J. Demaison of Chevron International, and R.J. Echols of Mobil Oil Corporation are gratefully acknowledged. R.J. Echols, W.D. Waples, and E.H. Schot of Mobil reviewed the manuscript.

REFERENCES

Allison, R.C., 1978, Late Oligocene through Pleistocene molluscan faunas in the Gulf of Alaska region, The Veliger, 21:171-188.

Ariey, C., 1978, Molluscan biostratigraphy of the upper Poul Creek and lower Yakataga Formations, Yakataga District, Gulf of Alaska: Correlation of tropical through high latitude marine Neogene deposits of the Pacific Basin, Abstracts and Program, International Geological Correlation Program, Project 114, Stanford University Publications, Geological Sciences, 14:1-2.

Armentrout, J.M., 1980a, Surface sediments and associated faunas, Yakataga-Yakutat area, northern Gulf of Alaska, in: "Quaternary Depositional Environments of the Pacific Coast; Pacific Coast Paleogeography Symposium 4," M.E. Field, A.H. Bouma, I.P. Colburn, R.G. Douglas, and J.C. Ingle, eds., Pacific Section, Society of Economic Paleontologists and Mineralogists, April, 1980, Los Angeles, 241-255.

Armentrout, J.M., 1980b, Late Neogene depositional and climatic cycles in the Yakataga Formation, Gulf of Alaska (abstract), American Association of Petroleum Geologists, Bulletin, 64:671.

Armentrout, J.M., Echols, R.J. and Nash, K.W., 1978, Late Neogene climatic cycles of the Yakataga Formation, Robinson Mountains, Gulf of Alaska area. Correlation of tropical through high latitude marine Neogene deposits of the Pacific Basin, Abstracts and Program, International Geological Correlation Program, Project 114, Stanford University Publications, Geological Sciences, 14: 3-4.

Armentrout, J.M., Rosenmeier, F. and Rogers, J., 1979, Glacial origin of the mega-channels of the upper Yakataga Formation (Plio-Pleistocene), Robinson Mountains, Gulf of Alaska (abstract), American Association of Petroleum Geologists, Bulletin, 63:411.

Bandy, O.L., Butler, E.A. and Wright, R.C., 1969, Alaskan Upper Miocene marine glacial deposits and the *Tuborotalia pachyderma* datum plane, Science, 166:607-609.

Bergen, F.W. and O'Neil, P., 1979, Distribution of Holocene foraminifera in the Gulf of Alaska, Journal of Paleontology, 53: 1267-1292.

Bezrukov, P.L., Lisitzin, A.P., Romankevich, E.A., and Skorniakova, N.S., 1961, Recent sediments in the western part of the Pacific Ocean, Recent Marine and Oceanic Sediment, Izvestiya Academii Nauk S.S.S.R., Moscow, 98-123.

Bruns, T.R., 1977, Late Cenozoic structures of the continental margin, northern Gulf of Alaska, in: "The Relationship of Plate Tectonics to Alaskan Geology and Resources," A. Sisson, ed., Proceedings of the Sixth Alaska Geological Society Symposium, Anchorage, April 1977, I1:130.

Calvert, S.E., 1964, Factors affecting distribution of laminated diatomaceous sediments in Gulf of California, in: "Marine Geology of the Gulf of California," T.J. van Andel and G.G. Shor, eds., American Association of Petroleum Geologists, Memoir 3, 311-330.

Claypool, G.E. and Kaplan, I.R., 1974, The origin and distribution of methane in marine sediments, in: "Natural Gases in Marine Sediments," J.R. Kaplan, ed., Plenum Press, New York, 99-140.

Combaz, A., 1964, Les palynofacies, Revue de Micropaleontologie, 7(3):205-218.

Cooney, R.T., 1972, A review of the biological oceanography of the northeast Pacific Ocean, in: "A Review of the Oceanography and Renewable Resources of the Northern Gulf of Alaska," D.H. Rosenberg, ed., University of Alaska Institute of Marine Sciences Report, R72-23, 57-74.

Curl, H.C., Jr., 1963, Analyses of carbon in marine plankton organisms, Journal of Marine Research, 20:181-188.

Degens, E.T. and Mopper, K., 1976, Factors controlling the distribution and early diagenesis of organic material in marine sediments, in: "Chemical Oceanography," J.P. Riley and R. Chester, eds., vol. 6, Academic Press, New York, 59-113.

Demaison, G.J. and Moore, G.T., 1980, Anoxic environments and oil source bed genesis, American Association of Petroleum Geologists, Bulletin, 64(5):1179-1209.

Dodimead, A.J., Favorite, F. and Hirano, T., 1963, Salmon of the north Pacific Ocean; Part II: Review of the oceanography of the subarctic Pacific region, International North Pacific Fisheries Commission, Bulletin, 13:1-195.

Dow, W.G., 1977, Kerogen studies and geological interpretations, Journal of Geochemical Exploration, 7:79-99.

Dow, W.G., 1978, Petroleum source beds on continental slopes and rises, American Association of Petroleum Geologists, Bulletin, 62(9):1584-1606.

Durand, B., 1980, "Kerogen: Insoluble Organic Matter from Sedimentary Rocks," Editions Technip, Paris, 519 pp.

Echols, R.J. and Armentrout, J.M., 1980, Holocene foraminiferal distribution patterns on the shelf and slope, Yakataga-Yakutat area, northern Gulf of Alaska, in: "Quaternary Depositional Environments of the Pacific Coast; Pacific Coast Paleogeography Symposium 4," M.E. Field, A.H. Bouma, I.P. Colburn, R.G. Douglas, and J.C. Ingle, eds., Pacific Section, Society of Economic Paleontologists and Mineralogists, April 1980, Los Angeles, 281-303.

Embleton, C. and King, C.A.M., 1968, "Glacial and Periglacial Geomorphology," St. Martin's Press, New York, 608 pp.

Garshanovich, D. Ye., 1965, New data on the accumulation of organic matter in recent sediments in the extreme north of the Pacific, Oceanology, 5:85-89.

Gray, J., 1965, Extraction techniques -- Techniques in palynology, in: "Handbook of Paleontological Techniques," B. Kummel and D. Raup, eds., Freeman and Company, 530-587.

Gross, M.G., Carey, A.G., Jr., Fowler, G.A., and Kulm, L.D., 1972, Distribution of organic carbon in surface sediment, northeast Pacific Ocean, in: "The Columbia River Estuary and Adjacent Ocean Waters," A.T. Pruter and D.L. Alverson, eds., Bioenvironmental Studies, University of Washington Press, Seattle, 254-264.

Hickman, C.S. and Nesbitt, E.A., 1980, Holocene mollusc distribution patterns in the northern Gulf of Alaska, in: "Quaternary Depositional Environments of the Pacific Coast; Pacific Coast Paleogeography Symposium 4," M.E. Field, A. Bouma, I. Colburn, R.G. Douglas and J.C. Ingle, eds., Pacific Section, Society of Economic Paleontologists and Mineralogists, April 1980, Los Angeles, 305-312.

Hollister, C.D.R., Flood, R. and McCave, I.N., 1978, Geological aspects of the benthic boundary layer, Oceanus, 21:5-13.

Holtedahl, H., 1958, Some remarks on geomorphology of continental shelves off Norway, Labrador and southeastern Alaska, Journal of Geology, 66:461-471.

Ibach, L.E.J., 1982, Relationship between sedimentation rate and total organic content in ancient marine sediments, American Association of Petroleum Geologists, Bulletin, 66(2):170-188.

Ingle, J.C., Jr., 1973, Summary comments on Neogene biostratigraphy, physical stratigraphy, and paleo-oceanography in the marginal northeastern Pacific Ocean, in: "Initial Reports DSDP," 18, L.D. Kulm, R. von Huene et al., U.S. Government Printing Office, Washington, 949-959.

Ingle, J.C., Jr., 1981, Origin of Neogene diatomites around the Pacific Rim, in: "The Monterey Formation and Related Siliceous Rocks of California," R.E. Garrison and R.G. Douglas, eds., Pacific Section, Society of Economic Paleontologists and Mineralogists, May 1981, Los Angeles, 159-179.

Jordan, G.F., 1958, Pamplona Searidge, 1779-1957, International Hydrography Review, 35:3-13.

Koblentz-Mishke, D.I., Volkovinsky, V.V. and Kabanova, J.G., 1970, Plankton primary production of the world ocean, in: "Symposium of Scientific Exploration of the Southern Pacific," W.S. Wooster, ed., National Academy of Science, Washington, 183-193.

Krumbein, W.C. and Pettijohn, J., 1938, Manual of Sedimentary Petrography, D. Appleton-Century Company, New York, 91-181.

Lagoe, M.B., 1978, Foraminifera from the uppermost Poul Creek and lowermost Yakataga Formations, Yakataga District, Alaska; Correlation of tropical through high latitude marine Neogene deposits of the Pacific basin, Abstracts and Program, International Geological Correlation Program, Project 114, Stanford University Publications, Geological Sciences, 14:34-35.

Lagoe, M.B., in press, Oligocene through Pliocene foraminifera from the Yakataga Reef section, Gulf of Alaska Tertiary province, Alaska, Micropaleontology.

Lattanzi, R.D., 1981, Planktonic foraminiferal biostratigraphy of Exxon Company, U.S.A. wells drilled in the Gulf of Alaska during 1977 and 1978, Journal of Alaska Geological Society, 1:48-59.

Loutit, T.S. and Kennett, J.P., 1981, New Zealand and Australian Cenozoic sedimentary cycles and global sea-level changes, American Association of Petroleum Geologists, Bulletin, 65(9): 1586-1601.

Molnia, B.F., 1979, Sedimentation in coastal embayments, northeastern Gulf of Alaska, Proceedings of the 1979 Offshore Technology Conference, 6:665-676.

Molnia, B.F. and Carlson, P.R., 1978, Surface sedimentary units of northern Gulf of Alaska continental shelf, American Association of Petroleum Geologists, Bulletin, 62(4):633-643.

Molnia, B.F. and Sangray, D.A., 1979, Glacially derived sediments in the northern Gulf of Alaska - geology and engineering characteristics, Proceedings of the 1979 Offshore Technology Conference, 6:647-656.

Parrish, J.T., 1982, Upwelling and petroleum source beds, with reference to Paleozoic, American Association of Petroleum Geologists, Bulletin, 66:750-774.

Pisciotto, K.A., 1978, "Basinal Sedimentary Facies and Diagenetic Aspects of the Monterey Shale, California," Ph.D. Thesis, University of California, Santa Cruz, 450 pp.

Pisciotto, K.A. and Garrison, R.E., 1981, Lithofacies and depositional environment of the Monterey Formation, California, in: "The Monterey Formation and Related Siliceous Rocks of California," R.E. Garrison and R.G. Douglas, eds., Pacific Section, Society of Economic Paleontologists and Mineralogists, May 1981, Los Angeles, 97-122.

Plafker, G., 1967, Geologic map of the Gulf of Alaska Tertiary Province, Alaska, U.S. Geological Survey Miscellaneous Geological Investigations Map I-484.

Plafker, G., 1971, Possible future petroleum resources of Pacific-margin Tertiary basin, Alaska, in: "Future Petroleum Provinces of North America," I.H. Cram, ed., American Association of Petroleum Geologists, Memoir 15, 120-135.

Plafker, G., 1974, Preliminary geologic map of Kayak and Wingham Islands, Alaska, U.S. Geological Survey Open-File Map 74-82.

Plafker, G. and Miller, D.J., 1958, Glacial features and surficial deposits of the Malaspina district, Alaska, U.S. Geological Survey Miscellaneous Geologic Investigations Map No. I-271.

Plafker, G. and Addicott, W.O., 1976, Glaciomarine deposits of Miocene through Holocene age in the Yakataga Formation along the Gulf of Alaska margin, Alaska, in: "Recent and Ancient Sedimentary Environments in Alaska," T.P. Miller, ed., Proceedings of the Alaska Geological Society, Anchorage, Q1-Q23.

Plafker, G., Bruns, T.R. and Page, R.A., 1975, Interim report on the petroleum resource potential and geologic hazards in the outer continental shelf of the Gulf of Alaska Tertiary province, U.S. Geological Survey Open-File Report 75-592, 74 pp.

Powell, R.D., 1981, A model for sedimentation by tidewater glaciers, Annals of Glaciology, 2:129-134.

Premuzic, E.T., 1980, Organic carbon and nitrogen in the surface sediments of world oceans and seas; distribution and relationships to bottom topography, Brookhaven National Laboratory, Upton, New York, June 1980, (BNL 51084) 113 pp.

Quinterno, P., Carlson, P.R. and Molnia, B.F., 1980, Benthic foraminifers from the eastern Gulf of Alaska, in: "Quaternary Depositional Environments of the Pacific Coast: Pacific Coast Paleogeography Symposium 4," M.E. Field, A. Bouma, I. Colburn, R.G. Douglas and J.C. Ingle, eds., Pacific Section, Society of Economic Paleontologists and Mineralogists, April 1980, Los Angeles, 13-21.

Rau, W.W., Plafker, G. and Winkler, G.R., 1977, Preliminary foraminiferal biostratigraphy and correlation of selected stratigraphic sections and wells in the Gulf of Alaska Tertiary Province, U.S. Geological Survey Open-File Report 77-747.

Rhoads, D.C. and Morse, J.W., 1971, Evolutionary and ecologic significance of oxygen-deficient marine basins, Lethaia, 4:413-428.

Romankevich, Y.A., 1968, Organic carbon and nitrogen deposits in Recent and Quaternary sediments of the Pacific Ocean, Oceanology, 8(5):658-673.

Royer, T.C., 1972, Physical oceanography of the northern Gulf of Alaska, in: "A Review of the Oceanography and Renewable Resources of the Northern Gulf of Alaska," D.H. Rosenberg, ed., University of Alaska Institute of Marine Science Report R72-23, 5-22.

Stach, E., 1975, "Stach's Textbook of Coal Petrology," Gebrüder Borntraeger, Berlin, 428 pp.

Summerhayes, C.P., 1981, Oceanographic controls on organic matter in the Miocene Monterey Formation, offshore California, in: "The Monterey Formation and Related Siliceous Rocks of California," R.E. Garrison and R.G. Douglas, eds., Pacific Section, Society of Economic Paleontologists and Mineralogists, May 1981, Los Angeles, 213-219.

Tissot, B.P. and Welte, D.H., 1978, "Petroleum Formation and Occurrence," Springer-Verlag, New York, 521 pp.

Todd, R. and Low, D., 1967, Recent foraminifera from the Gulf of Alaska and southeastern Alaska, U.S. Geological Survey Professional Paper 573-A, 46 pp.

Vail, P.R., Mitchum, R.M. and Thompson, S., 1977, Seismic stratigraphy and global changes of sea level, part 4: global cycles of relative changes of sea level, American Association of Petroleum Geologists, Memoir 26, 83-98.

van Gizel, P., 1979, "Manual of Techniques and Some Geological Applications of Fluorescence Microscopy," American Association of Stratigraphic Palynologists, 1979 Annual Meeting, Core Laboratories, Dallas, 55 pp.

Viereck, L.A. and Little, E.L., 1972, Alaska Trees and Shrubs, U.S. Department of Agriculture - Forest Service, Washington, Agriculture Handbook No. 410, 265 pp.

Welte, D.H., Cornford, C. and Rullkötter, J., 1979, Hydrocarbon source rocks in deep sea sediments, Proceedings of the Offshore Technology Conference II, Houston, May 1977, 457-461.

Woodruff, F., Savin, S.M. and Douglas, R.G., 1980, Miocene stable isotope record: A detailed deep Pacific Ocean study and its paleoclimatic implications, Marine Micropaleontology, 5(1):3-11.

Wright, F.F., 1968, Sedimentation and heavy mineral distribution, northeastern Gulf of Alaska continental shelf, U.S. Geological Survey Progress Report, Office of Marine Geology and Hydrology, Contract No. 14-08-011-10885, 22 pp.

Wright, F.F., 1969, Sedimentation and gold distribution, Yakutat Bay, Alaska, U.S. Geological Survey Progress Report, Office of Marine Geology and Hydrology, Contract No. 14-08-011-10885, Report No. R69-9, 12 pp.

PRE-PLEISTOCENE
PHANEROZOIC TIME SCALES

ORGANIC GEOCHEMISTRY OF SEDIMENTS RECOVERED BY DSDP/IPOD LEG 75 FROM UNDER THE BENGUELA CURRENT

Philip A. Meyers

Department of Atmospheric and Oceanic Science
The University of Michigan
Ann Arbor, Michigan 48109, U.S.A.

Simon C. Brassell, University of Bristol
Bristol, England, United Kingdom
Alain Y. Huc, Institut Francais du Petrol
Rueil Malmaison, France
Eric J. Barron, National Center for Atmospheric Research
Boulder, Colorado, U.S.A.
Robert E. Boyce, Deep Sea Drilling Project
Scripps Institution of Oceanography
La Jolla, California, U.S.A.
Walter E. Dean, U.S. Geological Survey
Denver, Colorado, U.S.A.
William W. Hay, Joint Oceanographic Institutions Inc.
Washington, DC, U.S.A.
Barbara H. Keating, University of Hawaii
Honolulu, Hawaii, U.S.A.
Charles L. McNulty, The University of Texas
Arlington, Texas, U.S.A.
Mosato Nohara, Geological Survey of Japan
Ibaraki, Japan
Roger E. Schallreuter, University of Hamburg,
Hamburg, Federal Republic of Germany
Jean-Claude Sibuet, Centre National Exploration Oceanique
(CNEXO), Bretagne, France
John C. Steinmetz, University of South Florida
St. Petersburg, Florida, U.S.A.
Dorrik Stow, University of Edinburgh
Edinburgh, Scotland, United Kingdom
Herbert Stradner, Geological Survey of Austria
Vienna, Austria

ABSTRACT

During DSDP/IPOD Leg 75 in the southeastern Atlantic, biogenic sediments rich in organic carbon were recovered from two sites near the Benguela upwelling system. Sampling with the newly developed hydraulic piston corer provided a relatively undisturbed record of upwelling history from late Miocene times to Holocene times. Concentrations of organic carbon range up to 7%, increasing from late Miocene to late Pliocene deposits and then decreasing somewhat in younger sediments. This pattern reflects the onset and progressive intensification of upwelling to a maximum in late Pliocene time and subsequent weakening to present-day conditions. Organic carbon content fluctuates on 30 to 50 thousand year intervals which apparently reflect periodic variations in upwelling intensity. These data from DSDP/IPOD Sites 530 and 532, combined with earlier information from DSDP Site 362, provide a unique record of the development and history of a major upwelling system.

INTRODUCTION

The organic content of oceanic sediments provides a record of the history of origin and deposition of biogenic sediment over geologic time. The amounts and types of organic components reflect the two processes, production and preservation of organic matter, which determine accumulation of these components in sediments.

Production of marine organic matter depends principally upon the availability of dissolved nutrients needed for the growth of phytoplankton. Those areas of the oceans in which water motion carries nutrients upward from deeper waters into the photic zone have the potential for sustained formation of primary biomass. After the death of the phytoplankton, this material sinks to the ocean bottom and becomes part of the sediment record. Changes in productivity in the overlying water may be recorded as changes in the organic content of sediments deposited over time. However, most of the original mass of marine organic matter is remineralized while sinking through the upper few hundred meters of the oceans (Suess, 1980; Suess and Müller, 1980; Knauer and Martin, 1981), and so the material reaching the sediment surface is but a residual fraction of the original material produced by phytoplankton.

Preservation of organic matter is determined by processes occurring within the water column and within the ocean bottom. Rapid sinking of organic debris, by incorporation into fecal pellets for example, provides less time for remineralization to occur and enhances preservation (cf. Honjo, 1976). Production and subsequent sinking of large masses of organic carbon can depress dissolved oxygen concentrations within the water column by causing oxygen utilization rates to exceed replenishment rates. This process, too, will enhance organic matter preservation. Rapid rates of sediment accumulation also favor preservation by decreasing the exposure time of organic matter to the intense microbial degradation which occurs in surficial sediments. Sinking of biological debris into permanently anoxic waters, such as those of the Black Sea or the Cariaco Trench, is yet another route to enhanced preservation of organic matter, although one not typical of the open oceans.

Superimposed upon the patterns of marine organic matter production and preservation is the record of land-derived input to oceanic sediments. Rivers transport terrigenous plant debris to coastal areas, and winds carry dust particles and their associated organic materials to all parts of the seas.

Upwelling areas are good locations to study the conditions important to organic matter accumulation in marine sediments. In such regions, conditions are maintained that encourage sustained growth of abundant marine plants of the animals that feed upon them. The sinking of the resulting large amounts of biological debris leads to an expanded oxygen minimum zone and contributes to an increased sedimentation rate. As a consequence, bottom sediments beneath upwelling areas contain sufficiently high concentrations of organic matter deposited in rapidly accumulating sediments to allow detailed analyses of temporal fluctuations in input rates and character.

The Benguela Current off Namibia in the South Atlantic Ocean is an example of a major, permanent upwelling system. The Namib Desert forms the coastal lands bordering this part of the eastern South Atlantic and supplies negligible amounts of organic matter from land runoff (Gagosian and Farrington, 1978). Eolian inputs remain likely, but must come from distant sources and provide only minor contributions of organic debris. The Benguela Current is a coldwater current flowing northward along the coast of South Africa and Namibia. The establishment of this current is believed to be 10 m.y. B.P. in late Miocene time (Siesser, 1980), perhaps as a result of the growth of the Antarctic ice cap beginning in the mid-Miocene and reaching its maximum extent in the late Miocene (Schnitker, 1980). The maximum development of the Antarctic ice volume corresponds to the Messinian Salinity Crisis (6.2 to 5.2 m.y. B.P., Arthur, 1979) and may have caused the drying of the Mediterranean Sea by lowering sea level below the Gibraltar sill depth (Schnitker, 1980). Siesser (1980) and Diester-Haass and Schrader (1979) note decreases in biogenic sediment accumulation under the Benguela Current at this time. Between Miocene and Pliocene time, colder waters around Antarctica reduced seawater evaporation and caused a decrease in the polar ice volume and a northward expansion of Antarctic waters, possibly strengthening the Benguela Current. Northern hemisphere glaciation began about 3.2 m.y. B.P. in the late Pliocene and resulted in intensified global thermal gradients. Cycling of global climates increased in magnitude from this time up to the present (Schnitker, 1980). These events potentially influenced upwelling conditions in the southeastern Atlantic.

Several locations under the Benguela Current have been sampled as part of the Deep Sea Drilling Project/International Phase of Ocean Drilling (DSDP/IPOD). We report here the results of organic geochemical analyses performed aboard DV "Glomar Challenger" during DSDP/IPOD Leg 75. These shipboard measurements reveal changes in organic matter deposition with sediment depth which can be interpreted in terms of the history of the Benguela upwelling system and which provide a useful foundation for future investigations.

SAMPLING AND ANALYSIS

The drilling vessel "Glomar Challenger" occupied two sites under the western edge of the Benguela Current in August 1980. Sediments were sampled by continuous hydraulic piston coring (HPC) to a subbottom depth of 180 m (Pliocene) at Site 530 and to 291 m (late Miocene) at Site 532 (Fig. 1). Water depths at these locations are 4639 m and 1341 m, respectively. This newly developed coring procedure causes minimal sediment disturbance and allows detection and study of closely spaced variations in sediment character. Standard rotary drilling procedures were used to obtain samples deeper than 180 m at Site 530. A description of the Pleistocene to Miocene sediments from these two sites is given by Hay et al. (in press). Samples of these sediments were selected from representative sequences and also from sedimentary features believed to be of special interest to organic geochemistry.

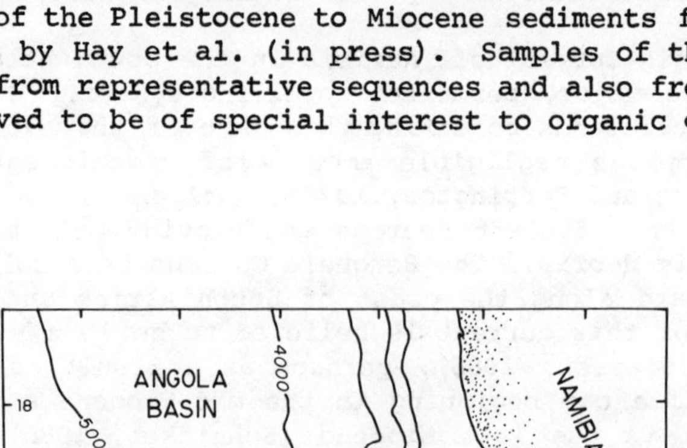

Fig. 1. Locations of DSDP Leg 75 Sites 530 and 532 and DSDP Leg 40 Site 362 in the southeastern Atlantic Ocean.

Analyses of organic carbon were done on board ship using a Hewlett-Packard 185-B CHN Analyzer. Portions of samples selected for carbonate measurements were treated with dilute HCl to remove carbonate, washed with deionized water, and dried at 110°C. A Cahn Electrobalance was used to weigh 20 mg samples of sediment for CHN analysis. Samples were combusted at 1050°C in the presence of an oxidant, and the volumes of the evolved gases determined as measures of the C, H and N contents of sediment organic matter. Concentrations of total organic carbon (TOC) and atomic C/N ratios were calculated using response factors determined from standards and were corrected for the small blank of the complete procedure.

Numerous samples were analyzed with the Girdel Rock-Eval instrument which uses the I.F.P.-Fina Process (Espitalie, Madec and Tissot, 1977) to measure both the free hdyrocarbons and the hydrocarbon generating potential of rock samples. This instrument was used routinely to monitor the nature and maturity of sediment organic matter. Small samples of bulk sediment (~0.5 g) were allowed to dry at room temperature and then were coarsely ground. One hundred mg portions of dry sediment were placed in the instrument and heated at 25°C/min in a helium stream from 250° to 500°C. Free hydrocarbons contained in the sediment are expressed as an S_1 peak; those released by the thermal breakdown of kerogen appear as an S_2 peak. Finally, an S_3 peak representing CO_2 produced from the kerogen appears and is a reflection of the oxygen content of the organic matter. Samples of two types of rocks were used to standardize the Rock-Eval. The first was Standard 27251, a lower Toarcian shale from the Paris Basin which has been widely used as a Rock-Eval standard. The second was USGS SDO-1, a Devonian black shale from Ohio which was standardized using 27251. Both rocks give similar Rock-Eval patterns, although SDO-1 contains about four times the organic carbon content of 27251.

RESULTS

Different types of sedimentary sequences were found at the two HPC sites. At Site 530 in the Angola Basin, mid-Miocene to Pleistocene sediments consist of turbidites and debris flow deposits that are believed to be redeposited biogenic oozes originally laid down on the Walvis Ridge. The late Miocene to Pleistocene sediments at Site 532 on the Walvis Ridge are made up of alternating sequences of light and dark pelagic oozes and are heavily bioturbated. The light-dark cycles represent an estimated time span of 30,000-50,000 years (Hay et al., 1982). The dark layers are richer in organic carbon, clay and pyrite. A possible explanation for this cyclicity is episodic increases in plankton productivity and organic matter preservation in the darker sediment layers. There is little evidence that the light-dark cycles were caused by carbonate dissolution within the sediment.

Although lithologically different, the sediments at Sites 530 and 532 share similar organic matter patterns that also resemble that

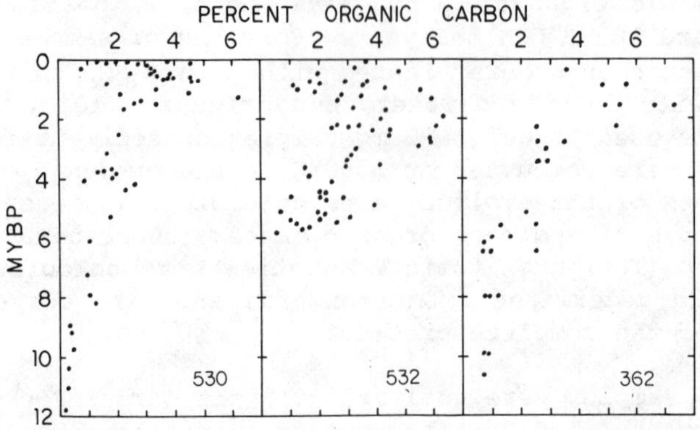

Fig. 2. Organic carbon concentrations of sediments of increasing age from DSDP Site 530 in the Angola Basin and DSDP Sites 532 and 362 on the Walvis Ridge. Data for Site 362 are from Erdman and Schorno (1978).

of Site 362 (Fig. 1). Fig. 2 shows values for organic carbon concentrations in individual samples. The percentages of organic carbon are relatively high, particularly in view of the amount of bioturbation evident in sediments from both Leg 75 locations. Maximum TOC values at 5 to 7% are found near the Plio-Pleistocene boundary at 1.7 m.y. B.P. at all three sites. Below this boundary, values tend to decrease with depth, but are still *ca.* 2% in late Miocene sediments. There is considerable variability in TOC of individual samples, with light-colored oozes possessing lower organic carbon content than the darker, olive-colored samples. The overall trends of organic carbon contents are clearly shown in Fig. 3, in which averaged values for one-million year intervals are plotted against sediment age. For sites 530 and 532, the increase in organic carbon values from upper Miocene-lower Pliocene (4 to 8 m.y. B.P.) sediments to maxima in upper Pliocene to lower Pleistocene (*ca.* 2 m.y. B.P.) oozes and subsequent decrease to upper Pleistocene strata matches the general trend of data from Site 362 reported by Erdman and Schorno (1978) and Kendrick, Hood and Castaño (1978).

In contrast to the variability and depth changes in the concentrations of organic carbon of Site 530 and 532 sediments shown in Fig. 2, the atomic C/N ratios change little from a value of 15 throughout the sediment sequence. This monotonous depth trend shows that the character of the organic matter has not changed significantly since the mid-Miocene. A C/N ratio of *ca.* 15, as found here, suggests a preferential loss of proteinaceous material from phytoplankton organic detritus (average C/N = 6, Goodell, 1972). Since C/N values of upper water column particulate matter average 7.3 in this area of the Atlantic Ocean (Bishop, Ketten and Edmond, 1978) a loss of nitrogen may occur after particles settle below the photic zone,

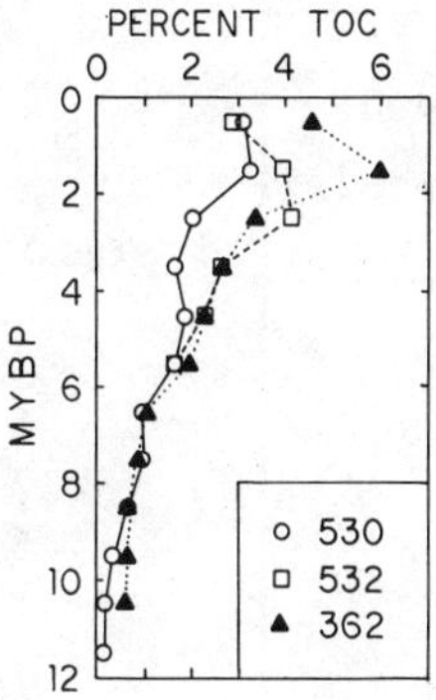

Fig. 3. Million-year averages of sediment organic carbon concentrations from DSDP Site 530 in the Angola Basin and DSDP Sites 562 and 362 on the Walvis Ridge. Data for Site 362 are calculated from data of Erdman and Schorno (1978).

but prior to the incorporation into the bottom sediments, as suggested by Suess and Müller (1980).

In Leg 75 samples, the lighter colored oozes generally possess lower concentrations of organic carbon and give smaller S_2 responses (thermogenic hydrocarbons) than the darker, olive-colored samples. These variations suggest that there are additional factors superimposed upon the overall trend of upwelling and productivity shown in Fig. 3, which have influenced the accumulation of organic matter in the underlying sediments. In many parts of the sedimentary section the color appears to change from dark to light in rhythmic cycles.

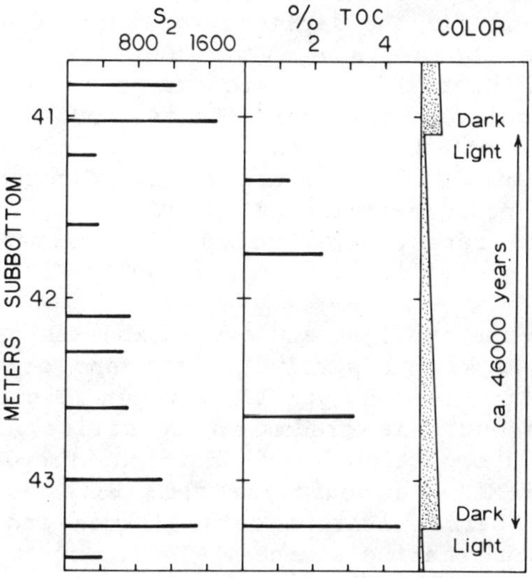

Fig. 4. Relationship between Rock Eval S_2 value, total organic carbon content (TOC), and sediment color in a representative light-dark cycle in Core 10 from DSDP Site 532 on the Walvis Ridge. S_2 value is expressed in mg pyrolytic hydrocarbons per 100 g dry sediment.

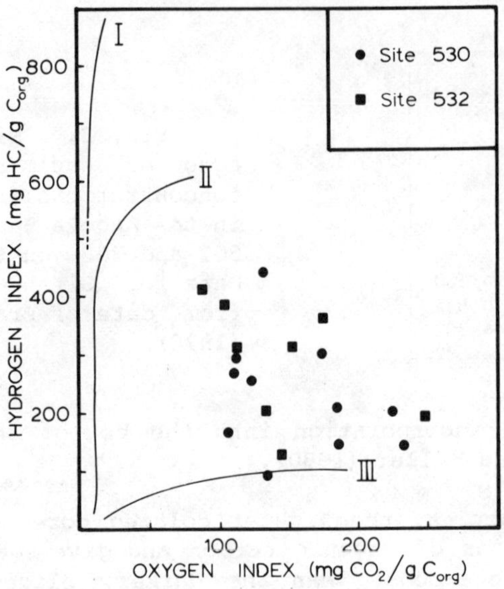

Fig. 5. Van Krevelen-type plot of Rock Eval Hydrogen and Oxygen Indices for Neogene sediments from DSDP Sites 530 and 532.

One such cycle in Core 10, Site 532, was studied by Rock-Eval pyrolysis to assess the relationship between sediment color and organic matter content. As Fig. 4 shows, the S_2 response is high (*ca.* 1500 mg HC/100 g sediment) for the darker colored (5Y 4/4) sediment horizons and decreases to low values (*ca.* 300) for the lighter colored (10Y 7/2) sediment intervals. Organic carbon follows a similar pattern. Hence, the color of the sediments is, indeed, related to their organic matter content, and the variation in this property may reflect episodes of midwater or bottom anoxia, fluctuations in upwelling strength or in pelagic biological populations, or changes in sediment accumulation rates. The period of time represented by this particular example of cyclicity is *ca.* 46,000 years.

The values of the hydrogen and oxygen indexes calculated for Leg 75 sediments from Rock-Eval pyrolysis data and organic carbon contents are between Type II and Type III kerogen (Fig. 5). Sedimentary organic matter evidently is predominantly marine and was deposited under slightly oxic conditions. In addition, the lipid contents in Pleistocene to mid-Miocene sediments from Site 362 (Boon et al., 1978), as well as carbon isotope data (Erdman and Schorno, 1978), indicate predominantly marine sources. Thus, our organic geochemical studies support the earlier conclusions that aquatic productivity is the primary origin of most of the organic matter in the sediments of this part of the southeastern Atlantic Ocean.

DISCUSSION

The averaged values of organic carbon in Fig. 3 are substantially higher than the average value of 0.2% for marine pelagic sediment (Degens and Mopper, 1976). This more typical, low concentration is not found in our cores until middle Miocene (*ca.* 10 m.y. B.P.) sediments are encountered. Even in the highly variable late Miocene to Pleistocene sequences, minimum TOC levels are around 1%; hence, exceptional amounts of organic matter are incorporated in these sediments. Throughout these organic-carbon-rich layers, the character of the organic matter does not change despite the variations in its concentration. Atomic C/N ratios of samples both rich and lean in organic carbon are about 15, and the kerogen type is consistently between Types II and III (Fig. 5). Because changes in the degree of organic matter preservation would probably cause alterations of organic matter character, the lack of such variability indicates that the fluctuations in organic carbon content are due not to different extents of preservation, but more likely to changes in rate of input. These patterns of changing organic input provide information about the history of biological productivity off southwest Africa. However, it is important to remember that no DSDP sites have been located under the present upwelling areas, which are generally closer to shore than the Benguela Current. Nonetheless, it appears the productivity of the offshore areas is influenced by nutrient availability in coastal waters, and so the records from Sites 362, 530 and 532 are related to upwelling history.

Other downcore changes in biogenic components of the sediments occur in addition to that of organic matter. The changes in contributions of opal, calcium carbonate, and organic matter to sediments at Sites 530 and 532 are summarized in Fig. 6. The combined increases in opal and organic content in these sediments record an overall increase in biological productivity from the late Miocene to the late Pliocene. Subsequent decreases record a lessening of productivity to the Holocene, but it remains at levels which exceed those of the Miocene. Sedimentation rates of these biogenic sediments reach as high as 65 m/m.y. in the late Pliocene and early Pleistocene and remain above 20 m/m.y. throughout these sediments (Hay et al., in press), reflecting the high productivity of surface waters.

Siesser (1980) presents micropaleontologic evidence from DSDP Site 362, adjacent to Site 532 on the Walvis Ridge, that South Atlantic plankton changed from warm-water to cold-water types in middle to late Miocene times. He interprets this as indicating the establishment of the modern Benguela Current system. At the same time that the sea surface cooled, surface productivity increased, evidently in response to nutrient upwellings associated with the Benguela system.

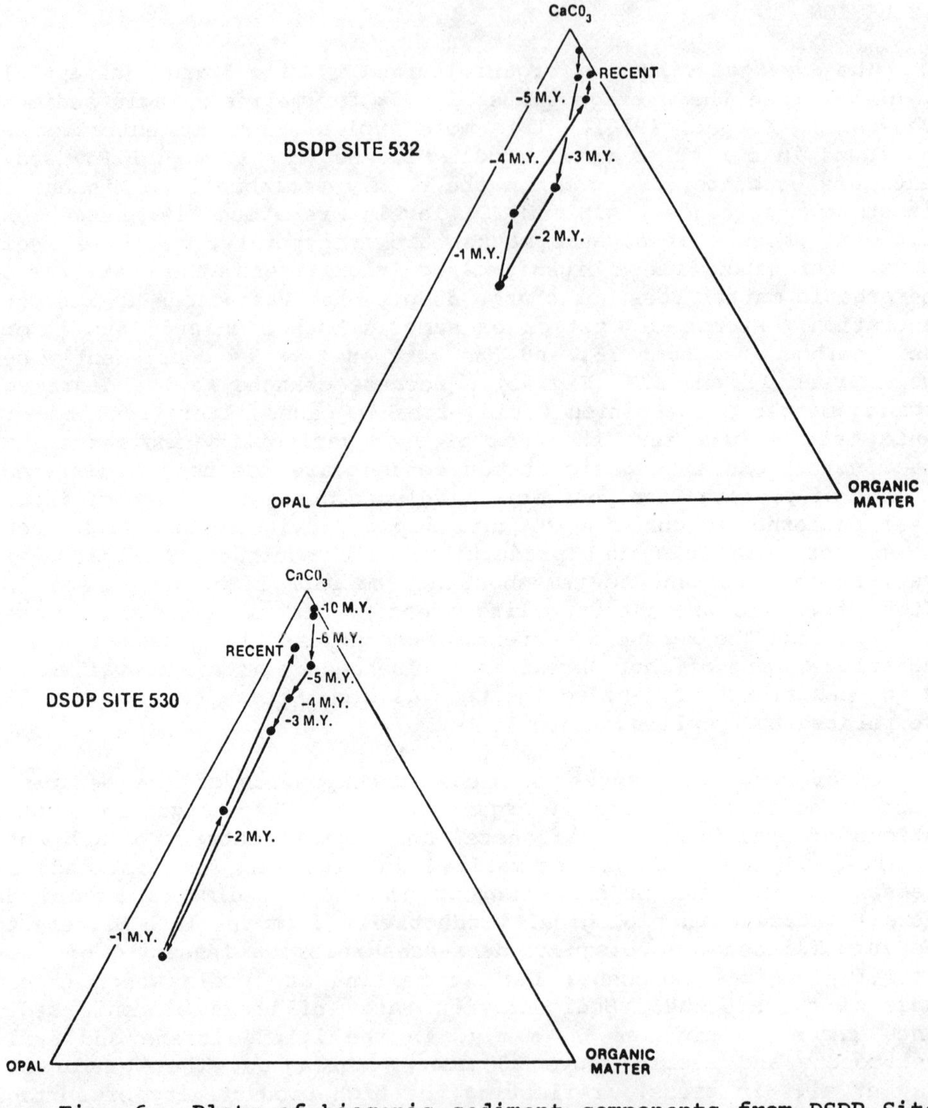

Fig. 6. Plots of biogenic sediment components from DSDP Sites 530 and 532. Organic matter is calculated from organic carbon measurements. Sediment age at which each increment terminates is indicated in millions of years before present. Both plots show maximum opal and organic matter in sediments deposited 2 million years ago.

Our data from Site 532 supports the conclusions that the onset of the Benguela upwelling system took place in the late Miocene (Diester-Haass and Schrader, 1979; Siesser, 1980). Because sampling at this site was done by hydraulic piston corer, much of the sediment disturbance which accompanies rotary drilling was avoided, and a more detailed sedimentary record of the history of this current system is available. An important new observation from this improved record is that the late Pliocene was a time of maximum biological productivity and that more recently there has been a decline. An explanation for the maximum and ensuing decrease in productivity remains elusive, yet changes in atmospheric circulation, hence sea surface currents, and sea level may have been important.

A significant modification in atmospheric conditions occurred in the middle Pliocene and may have led to the observed weakening in upwelling near the Plio-Pleistocene boundary. Northern hemisphere continental glaciation became established (Schnitker, 1980) and led to steeper thermal gradients between high and low latitudes. By the Pleistocene, higher wind velocities existed in both the northern hemisphere (Rea, 1982) and in the southern hemisphere (Diester-Haass and Schrader, 1979). The intensification of circulation may have been accompanied by shifting of wind belts, with the result that upwelling over the Walvis Ridge has not been as well developed since the end of the Pliocene as it was earlier.

Changes in sea level have been suggested by Summerhayes (1981) to explain fluctuations in organic content of sediments from under the California Current upwelling. Lowered sea level during the late Miocene/early Pliocene could have enhanced sediment organic matter preservation by lowering the oxygen minimum zone to the sea bottom beneath upwelling areas. Although the organic carbon variations in sediments on the Walvis Ridge probably reflect changes in supply of biogenic components rather than in their preservation, it is possible that sea level fluctuations contributed by shifting the upwelling zone closer to Sites 532 and 362 during low stages and farther away during subsequent higher stages.

The strong similarity in patterns of biogenic sediment components between Sites 530 and 532 (Fig. 6) gives support to our interpretation of a late Miocene onset of upwelling and late Pliocene upwelling maximum off southwest Africa. At Site 530 at the base of the Walvis Ridge, turbidites and debris-flow deposits rich in organic matter and in diatoms began to accumulate about 8 m.y. B.P. in the late Miocene and became more extensive through the Pliocene. These deposits are poor in carbonate components because of depth-related carbonate dissolution, yet otherwise closely resemble the sediments found on the Walvis Ridge. The fact that the deposits did not begin to accumulate in the Angola Basin until late Miocene times indicates that abundant amounts of biogenic sediments subject to slumping were not being formed before then. Because of the steep northern slope of the Walvis Ridge (see Fig. 1), transport to the Angola Basin and reburial must have been rapid; otherwise, organic matter would not be so well preserved.

Sediments from beneath the Benguela Current contain considerable variability in organic carbon concentrations as shown in Figs. 2 and 4. The 46,000-year cycle in sediment color and in thermogenic hydrocarbon content (S_2 value) in Core 10, Site 532, is typical. The range in cycle periodicities is from 30,000 to 50,000 years, based upon shipboard estimates of sedimentation rates. The dark layers may be the result of periodic increases in surface productivity induced by upwelling fluctuations. These layers are richer in siliceous components and organic matter and contain less carbonate material than the lighter colored layers.

Other workers have reported cyclic patterns in marine sediments. Dean, Gardner and Cepek (1981) report cycles of 44,000-year periodicity in Oligocene-Miocene biogenic sediments from the Sierra Leone Rise off northwest Africa. Rea (1982) has observed variations in terrigenous contents of Pleistocene sediments and infers 42,000-year cycles in tradewind intensity over the eastern equatorial Pacific. This periodicity corresponds to the tilt cycle of the earth, which determines the global distribution of incident solar radiation. Wind intensities respond to the resulting latitudinal temperature gradient. Stronger winds potentially cause increases in upwelling, so a similar periodicity in marine productivity can be expected (Rea, 1982). The cyclicity in biogenic sediment composition found at Site 532 may also be related to the tilt cycle of the earth. The dark, organic-rich layers may be sediments deposited during episodes of stronger winds and enhanced upwelling. Greater diatom production would create more siliceous material and dilute the carbonate component within these layers. At times of weaker winds, biological production would diminish, and less organic carbon and silica would accumulate in the sediments.

SUMMARY

Organic-carbon-rich sediments, which began to accumulate under the Benguela Current during the late Miocene, record the onset of upwelling off southwest Africa. Productivity increased to a maximum in late Pliocene time and has declined somewhat since then. Climatological factors are the likely causes of these changes. A worldwide mid-Miocene cooling trend accompanying growth and seaward expansion of the Antarctic ice cap may have led to development of the present-day Benguela Current and the availability of nutrient-rich water to upwelling systems off Namibia. Subsequent strengthening or shifting of wind systems resulted in progressively better-developed upwelling from late Miocene to late Pliocene times. Biological productivity has moderated after peaking in the late Pliocene, but still remains high. The decline from maximum productivity closely follows full development of northern hemisphere continental glaciation and concomitant intensification of winds, which evidently caused a departure from optimal upwelling conditions over the Walvis Ridge. Throughout the organic-rich sediments, cycles apparently reflecting lower and higher sea surface productivity appear. These occur at intervals of 30,000 to 50,000 years and may reflect the variations in global wind intensities related to the tilt cycle of the earth.

ACKNOWLEDGEMENTS

We thank D.K. Rea for critically reviewing this paper and for providing information about tilt cycles of the earth. K.A. Kvenvolden suggested a number of improvements to our manuscript. We are grateful to DSDP/IPOD for the opportunity to participate as shipboard scientists aboard DV "Glomar Challenger" during Leg 75.

REFERENCES

Arthur, M.A., 1979, Paleoceanographic events-recognition, resolution, and reconsideration, Reviews of Geophysics and Space Physics, 17:1474-1494.

Bishop, J.K.B., Ketten, D.R. and Edmond, J.M., 1978, The chemistry, biology and vertical flux of particulate matter from the upper 400 m of the Cape Basin in the southeast Atlantic Ocean, Deep-Sea Research, 25:1121-1161.

Boon, J.J., van der Meer, F.W., Schuyl, P.J.W., de Leeuw, J.W., Schenck, P.A., and Burlingame, A.L., 1978, Organic geochemical analyses of core samples from Site 362, Walvis Ridge, Deep Sea Drilling Project Leg 40, in: "Initial Reports DSDP," 40 (Supplement), H.M. Bolli and W.G.F. Ryan, U.S. Government Printing Office, Washington, 627-637.

Dean, W.E., Gardner, J.V. and Cepek, P., 1981, Tertiary carbonate-dissolution cycles on the Sierra Leone Rise, eastern equatorial Atlantic Ocean, Marine Geology, 39:81-101.

Degens, E.T. and Mopper, K., 1976, Factors controlling the distribution and early diagenesis of organic material in marine sediments, in: "Chemical Oceanography, Volume 6," J.P. Riley and R. Chester, eds., Academic Press, New York, 59-113.

Diester-Haass, L. and Schrader, H., 1979, Neogene coastal upwelling history off northwest and southwest Africa, Marine Geology, 29: 39-53.

Erdman, J.G. and Schorno, K.S., 1978, Geochemistry of carbon: Deep Sea Drilling Project Leg 40, in: "Initial Reports DSDP," 40 (Supplement), H.M. Bolli and W.G.F. Ryan, U.S. Government Printing Office, Washington, 651-658.

Espitalie, J., Madec, M. and Tissot, B., 1977, Source rock characterization method for petroleum exploration, Offshore Technology Conference, 399-404.

Gagosian, R.B. and Farrington, J.W., 1978, Sterenes in surface sediments from the southwest African shelf and slope, Geochimica et Cosmochimica Acta, 42:1091-1101.

Goodell, H.G., 1972, Carbon/nitrogen ratio, in: "Encyclopedia of Geochemistry and Environmental Science," R.W. Fairbridge, ed., Van Nostrand Reinhold, New York, 136-142.

Hay, W.W., Sibuet, J.-C. and Leg 75 Shipboard Party, 1982, Sedimentation and accumulation of organic carbon in the Angola Basin and on Walvis Ridge: Preliminary results of the Deep Sea Drilling Project Leg 75, Geological Society of America, Bulletin, 93: 1038-1055.

Honjo, S., 1976, Coccoliths: Production, transportation and sedimentation, Marine Micropaleontology, 1:65-79.

Kendricks, J.W., Hood, A. and Castaño, J.R., 1978, Petroleum-generating potential of sediments from Leg 40, Deep Sea Drilling Project, in: "Initial Reports DSDP," 40, H.M. Bolli and W.B.F. Ryan, U.S. Government Printing Office, Washington, 671-676.

Knauer, G.A. and Martin, J.H., 1981, Primary production and carbon-nitrogen fluxes in the upper 1,500 m of the northeast Pacific, Limnology and Oceanography, 26:181-186.

Rea, D.K., 1982, Fluctuation in eolian sedimentation during the past five glacial-interglacial cycles: A preliminary examination of data from Deep Sea Drilling Project Hole 503B, eastern equatorial Pacific, in: "Initial Reports DSDP," 68, W.L. Prell and J.V. Gardner et al., U.S. Government Printing Office, Washington, 409-415.

Schnitker, D., 1980, Global paleoceanography and its deep water linkage to the Antarctic glaciation, Earth-Science Reviews, 16:1-20.

Siesser, W.G., 1980, Late Miocene origin of the Benguela upwelling system off northern Namibia, Science, 208:283-285.

Suess, E., 1980, Particulate organic carbon flux in the oceans--surface productivity and oxygen utilization, Nature, 288:260-263.

Suess, E. and Müller, P.J., 1980, Productivity, sedimentation rate and sedimentary organic matter in the oceans II - Elemental fractionation, Biogeochimie de la Matiere Organique a l'Interface Eau-Sediment Marin, Colloques Internationaux du Centre National de la Recherche Scientifique, No. 293:17-26.

Summerhayes, C.P., 1981, Oceanographic controls on organic matter in the Miocene Monterey Formation, offshore California, The Monterey Formation and Related Siliceous Rocks of California, Society of Economic Paleontologists and Mineralogists, Tulsa, 213-219.

POTENTIAL DEEP-SEA PETROLEUM SOURCE BEDS RELATED TO COASTAL UPWELLING

Jürgen Rullkötter

Kernforschungsanlage Jülich GmbH
D-5170 Jülich, Federal Republic of Germany

Vassil Vuchev, Kansas Geological Survey
Lawrence, Kansas, U.S.A
Karl Hinz, Bundesanstalt für Geowissenschaften und Rohstoffe, Hannover, Federal Republic of Germany
Edward L. Winterer and Peter O. Baumgartner
Scripps Institution of Oceanography
La Jolla, California, U.S.A.
Martin J. Bradshaw, University of Aston
Birmingham, United Kingdom
James E.T. Channell, Lamont-Doherty Geological Observatory, Palisades, New York, U.S.A.
Michel Jaffrezo, Université Pierre et Marie Curie
Paris, France
Lubomir F. Jansa, Geological Survey of Canada
Dartmouth, Nova Scotia, Canada
Robert M. Leckie, University of Colorado
Boulder, Colorado, U.S.A.
Johnnie M. Moore, University of Montana
Missoula, Montana, U.S.A.
Carl Schaftenaar, Texas A&M University
College Station, Texas, U.S.A.
Torsten H. Steiger, Universität München
München, Federal Republic of Germany
George E. Wiegand, Florida State University
Tallahassee, Florida, U.S.A.

ABSTRACT

Settling of organic matter through oxic deep ocean waters destroys most of the hydrogen-rich material and thus is not favorable for the accumulation of hydrocarbon source rocks. Such rocks can be expected only where mass flows down the continental slope from an oxygen minimum zone at the shelf edge, in areas of high bioproductivity associated with upwelling, produce organic-carbon-rich sediments or where oxygen-depleted waters reach extremely deep. Deep-sea drilling off the northwest African continental margin and in the eastern North Pacific has encountered such sediments of Tertiary age with organic carbon contents up to five percent and kerogens rich in algal matter with only minor to moderate admixtures of terrigenous material. Similar, but slightly less organic - carbon-rich sediments in the mid-Cretaceous may have a similar primary origin, however they are more strongly influenced by terrigenous organic matter and are only gas- and not oil-prone.

INTRODUCTION

Coastal upwelling regions are concentrated in the trade wind belts along the eastern boundaries of the world's oceans. Upwelling circulation, driven by steady, predominantly equatorward winds according to a conceptual model supported by observations made on different continental margins, consists of an offshore Ekman flow in the surface water layer, an onshore subsurface Ekman flow and a geostrophic alongshore current in the direction of the wind, all together leading to three-dimensional upwelling 'centers' (Smith, this volume). The actual circulation pattern depends on the morphology of the continental margins; for example, upwelling occurs closer to the coastline on the steep Peruvian margin than on the relatively wide and shallow northwest African shelf (Smith, this volume).

Upwelling fed by the onshore subsurface Ekman flow brings up cold waters enriched in oxygen and mineral nutrients. This has a strong effect on the fertility of the ecosystem in the surface water layers (Ryther, 1969; Walsh, 1981 and many others). The enhanced bioproductivity in areas of upwelling increases the chance for significant amounts of organic matter to be preserved in the underlying sediments. Buried to an appropriate depth, sediments rich in marine organic matter are commonly considered potential source rocks for petroleum generation (Tissot and Welte, 1978; Hunt, 1979).

The intensity and spacial pattern of coastal upwelling is sensitive to relatively short-term influences, i.e., daily to seasonal weather changes, whereas the sedimentary record preserves only the average long-term influence. Although the effect of upwelling on organic matter content in oceanic sediments probably is more pronounced in the nearshore continental shelf areas, an influence can also be expected to a certain extent further offshore. This may be derived, for example, from the large offshore extension of oxygen-depleted water layers in the northeast Pacific (Summerhayes, 1981a). Furthermore, the presence of well-preserved diatom and radiolarian assemblages and of phosphorite grains in Neogene sediments on the continental slope off northwest Africa has been ascribed to the influence of strong coastal upwelling in the geologic past (Diester-Haass and Schrader, 1979). In this paper, we would like to emphasize downslope transport of organic-matter-rich sediment masses as an important mechanism for transfer of the effect of coastal upwelling into the deep-water environment. Consequently, the chance of finding potential deep-sea hydrocarbon source rocks appears to be especially great in those parts of the ocean where coastal upwelling has occurred in the past.

Deep sea drilling by the current Deep Sea Drilling Project on western continental margins in latitudes affected by coastal upwelling has provided the opportunity to trace possible imprints of upwelling on the sediment. Here we discuss data for Cretaceous and Tertiary sediments from the eastern North Atlantic off northwest Africa, DSDP Legs 47a (south of the Canary Islands), 50 and 79 (off Morocco), the latter two possibly being at the northern boundary of the present upwelling area in the eastern North Atlantic, and from the eastern North Pacific, DSDP Leg 63 (off southern California and Baja California).

ORGANIC MATTER DEPOSITION ON CONTINENTAL MARGINS

A schematic model for the sedimentation of organic matter on continental margins has been developed by Cornford (1979) for the northwest African continental margin, and a slightly modified version is shown in Fig. 1. Eolian dust and fluviatile transport provide land-derived organic matter in amounts depending on onshore vegetation, river flow and wind directions. In the same way, recycled organic matter from erosion of uplifted sedimentary rocks may be transported into the ocean. Both types are mixed with autochthonous marine organic matter and subsequent incorporation into the sediment occurs depending on the environmental conditions at the site of deposition. Sedimentation and preservation of organic matter is not usually favored under deep-water conditions. Settling through oxygen-rich water preferentially destroys the hydrogen-rich organic lipid material by consumption within the food chain, including microbial action, and oxidation. The residue reaching the ocean floor consists mostly of refractory inert organic matter, normally dominated by more resistant recycled and terrigenous material without any significant potential for hydrocarbon generation (Welte, Cornford and Rullkötter, 1979). Pelagic and hemi-pelagic sediments in these cases contain only low amounts of total organic carbon, typically represented by the mean value of 0.3% C-org calculated by McIver (1975) from a large number of deep-sea sediment analyses. The processes of destruction and preservation of organic matter are modulated by the sedimentation rate, i.e., oxidation is more effective at slow sedimentation rates, whereas organic matter preservation is favored at high sedimentation rates (Aizenshtat, Baedecker and Kaplan, 1973; Müller and Suess, 1979; Welte et al., 1979).

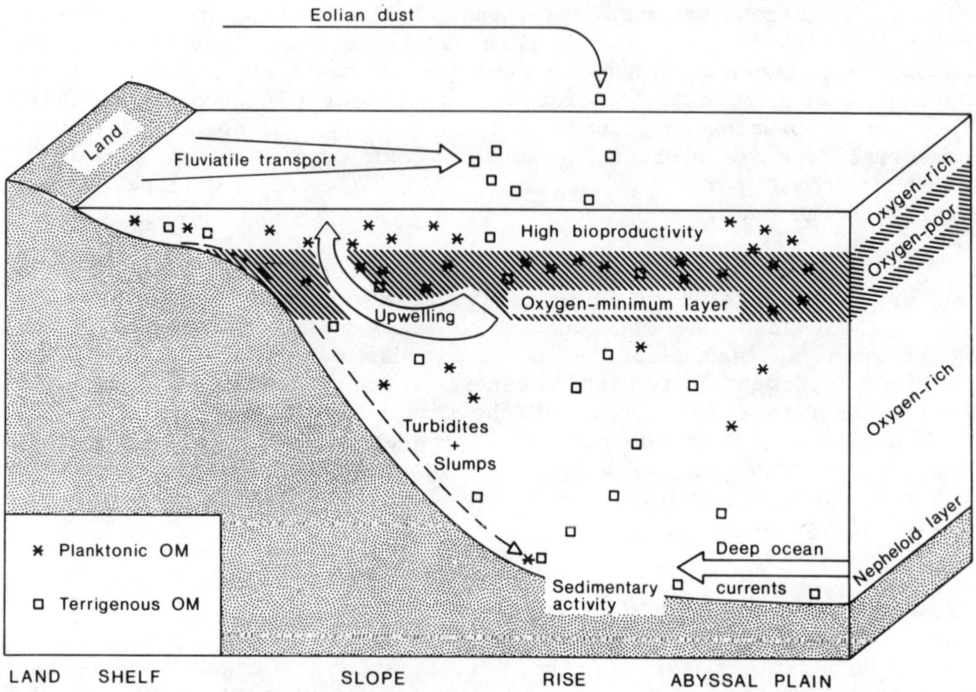

Fig. 1. Schematic diagram showing the factors which affect the organic matter content of marine continental shelf, slope and rise sediments (modified after Cornford, 1979).

Exceptional situations exist, however, which allow organic matter-rich sediments to be found in the deep-water area. Among these are oceanic basins with restricted water circulation, like the Cariaco Trench off Venezuela, which have developed anoxic bottom waters. Another scenario occurs on steep continental margins, as described by Cornford (1979) for the northwest African continental margin, where sediment masses are transported down the slope as slumps or turbidites and deposited at the continental rise or in the abyssal plain. If primary deposition of these sediments occurred on the outer shelf or upper slope within an oxygen-depleted subsurface water layer related to local high bioproductivity (Fig. 1), the re-deposited sediments may be rich in organic matter. Rapid burial at the final site of deposition allows preservation of most of this organic matter in the deep sea even if the bottom water is oxic.

Bioproductivity generally is higher on the continental margins than in the open ocean, but it is especially high in coastal upwelling areas. This does not necessarily mean that organic carbon contents are highest in the sediments deposited on the shelf and in the deep-water regions of these areas because organic matter accumulation

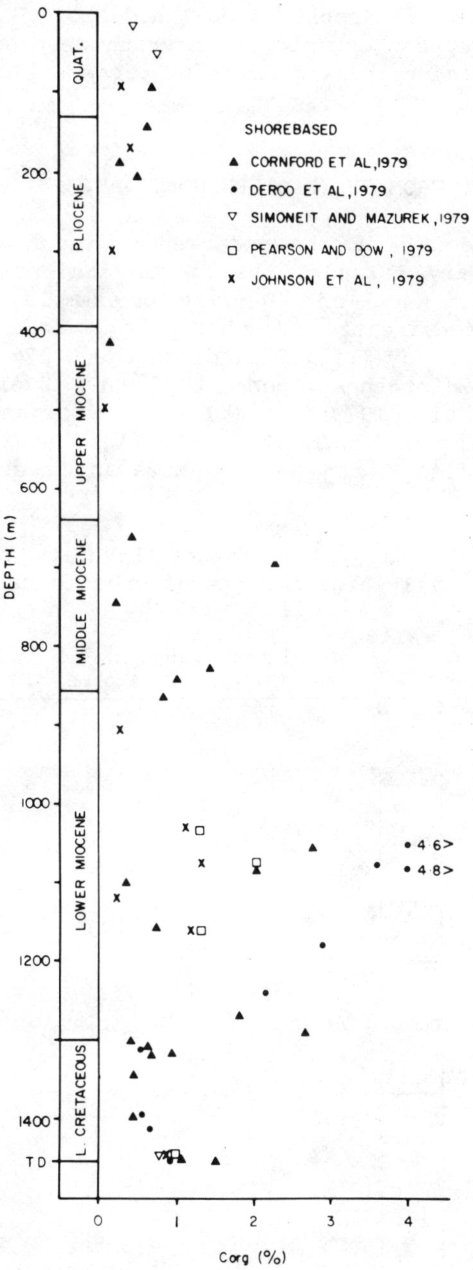

Fig. 2. Total organic carbon versus depth profile for sediments from DSDP Site 397, south of the Canary Islands (after Rullkötter, Cornford and Welte, 1982). By permission from Geology of NW African Continental Margin, Copyright (c) 1982, Springer Verlag.

depends on a number of factors as outlined above. There is, however, a great probability for organic-matter-rich deep-sea sediments to be deposited in coastal upwelling areas by processes schematically shown in Fig. 1.

DEEP-SEA SEDIMENTS FROM THE EASTERN NORTH ATLANTIC

At DSDP Site 397 on the northwest African continental margin south of the Canary Islands, the Miocene and post-Miocene sections down to a subbottom depth of 690 m are an example of continuous deposition of upper continental rise hemipelagic sediments. The organic carbon contents are low (Fig. 2) and average 0.17% for the Miocene to Pliocene and 0.45% for the Pliocene to Recent intervals, respectively (von Rad, Ryan et al., 1979). Similarly low organic carbon contents were determined for the bulk of the Tertiary sediments at DSDP Site 547 (Leg 79) further north on the Mazagan Escarpment off Morocco (Fig. 3).

In contrast to this, some other Tertiary sediments at these sites contain unusually high amounts of organic matter. At Site 397,

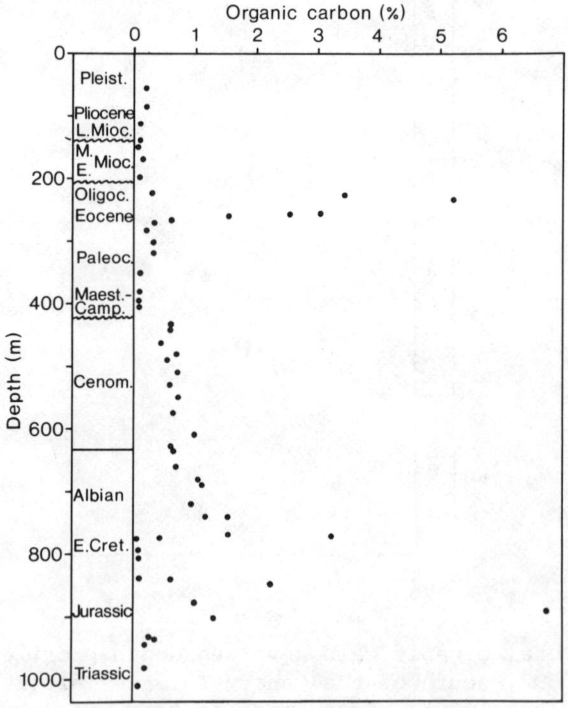

Fig. 3. Total organic carbon versus depth profile for sediments from DSDP Site 547 on the Mazagan Escarpment.

organic carbon values up to nearly 5% were measured mainly in the Lower Miocene sections (Fig. 2). The characteristics of the sediments show that they were deposited as debris flows. The type of microfauna present indicates that original deposition occurred in the outer shelf or upper slope environment and that slumping and turbidite transport resulted in final deposition of the organic carbon-rich Miocene mudstones in the oxidizing waters of the upper continental rise (Cornford, 1979; Cornford, Rullkötter and Welte, 1979). Contrasting modes of deposition are indicated by the light-colored organic-carbon-lean hemipelagic sediments interbedded with dark grey to olive organic-carbon-rich allochthonous units. Influence of upwelling in this area during the Early and Middle Miocene was detected at DSDP Site 369, a short distance upslope from DSDP Site 397, by Diester-Haass and Schrader (1979). They found well-preserved diatom and radiolarian assemblages, low values for the ratio of planktonic to benthic foraminifers, and increased amounts of fish debris and phosphorite grains, all thought to be indicative of upwelling.

Mud-supported debris flows were found in the Eocene at DSDP Site 547 on the Mazagan Escarpment. Organic carbon values are high in some of the allochthonous clasts and occasionally exceed 5% (Fig. 3). Some of the clasts are at least several meters in diameter. The possible primary origin of these sediments still has to be studied in detail. Eocene sediments enriched in organic carbon were also found at DSDP Site 367 and 368 further south in the Cape Verde Basin and on the Cape Verde Rise, respectively, as rhythmic thin beds of black clay interbedded with organic-carbon-lean greenish gray clay. At Site 368, this cyclic sedimentation was ascribed to turbidite flows whereas the cycles at Site 367 were interpreted as reflecting fluctuations in the input of terrigenous organic matter from the African continent (Dean et al., 1978). Unfortunately, no detailed organic geochemical investigations of these sediments have yet been reported. However, rhythmic sedimentation of laminated black shales and bioturbated organic-carbon-lean green (and red) claystones in the Albian to Cenomanian of the Angola Basin, very similar to the sequence at Site 367, were recently shown to be turbidite deposits (Hay, Sibuet et al., in press). Thus, there is the possibility that all the organic-carbon-rich Eocene deep-sea sediments found along the northwest African continental margin result from downslope transport events. Eocene strata in this area which are rich in phosphorite were described by Dillon and Sougy (1974) and give supporting evidence that coastal upwelling has occurred in the Eocene off northwest Africa and that the model in Fig. 1 is principally valid for the Eocene organic-carbon-rich deep-sea sediments in this area as well.

Most of the Cretaceous deep-sea sediments off northwest Africa are characterized by organic carbon contents of between about 0.5 and 1.5%. The values are at the lower end of this range of organic carbon concentrations in the Hauterivian sediments from DSDP Site 397 (Fig. 2) and in the Cenomanian sediments from DSDP Site 547 (Fig. 3);

higher values were found in the Aptian to Albian section at Site 547 including a small slump clast near the base of this section with an organic carbon content of about 3% (Fig. 3).

An attempt has been made to classify such deep-sea sediment in terms of different organo-facies according to their total organic carbon content and the type of organic matter present (Tissot, Deroo and Herbin, 1979; Tissot et al., 1980; Cornford, Rullkötter and Welte, 1980; Summerhayes, 1981b; Rullkötter, Cornford and Welte, 1982). A variety of organic geochemical and organic petrographic analyses have revealed that the hemipelagic Tertiary sediments at DSDP Site 397 contain a dominance of detrital terrigenous organic matter with abundant particles severely altered during long-distance transport (Cornford et al., 1979). Occasional well-preserved algal bodies indicate only chance preservation of planktonic components.

In contrast to this, the organic-carbon-rich Miocene slumps and turbidite flows at Site 397 contain a highly variable kerogen type (Cornford, 1979) with significant amounts of algal and amorphous liptinites of marine origin intermixed with terrestrial higher plant debris (spores, pollen, vitrinite, inertinite) (Cornford et al., 1979). The richly heterogeneous nature of the kerogen was confirmed by pyrolysis results and H/C ratios indicating a mixture of type III (vitrinite) and type II (spores/pollen/algal) kerogen (Deroo et al., 1979). The unconformably underlying Hauterivian sediments at Site 397 contain a purely terrigenous vitrinitic kerogen with subsidiary spores and pollen (Cornford et al., 1979), whereas Aptian-Albian sediments at Site 369 further upslope contain purely marine organic matter (Tissot et al., 1979; 1980).

For the sediments from DSDP Site 547 only data from Rock-Eval pyrolysis are available at present to characterize the type of organic matter. During Rock-Eval pyrolysis (Espitalié et al., 1977), a rock sample is gradually heated up in an oven under a stream of helium, and the amount of hydrocarbon-type compounds released is monitored with a flame ionization detector and yields a "hydrogen index" (IH; mg hydrocarbons per g organic carbon). Part of the pyrolysis products are split off to monitor the carbon dioxide evolved, which is thought to represent the oxygen content of the kerogen and is expressed as the "oxygen index" (IO; mg carbon dioxide per g organic carbon).

Among the DSDP Site 547 sediments three different organo-facies types can be differentiated in the hydrogen index versus oxygen index diagram (Fig. 4), in addition to the extremely degraded (oxidized) kerogens outside the encircled areas which comprise mainly the autochthonous pelagic and hemipelagic Tertiary sediments at this site. Within the Tertiary, the slump samples at DSDP Site 547 are like those from DSDP Site 397. Most of them fall into the area close to the kerogen type II trend line (Fig. 4) together with a Jurassic

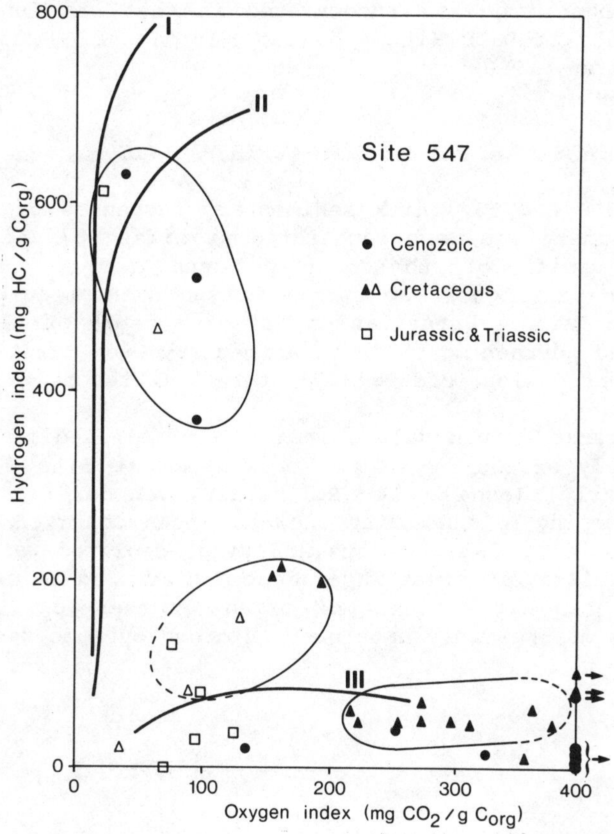

Fig. 4. Results of Rock-Eval pyrolysis displayed as hydrogen index (IH) versus oxygen index (IO) diagram for sediments from DSDP Holes 547A (closed symbols) and 547B (open symbols). Encircled areas mark different organofacies types.

"black shale" found in the lower section of Hole 547B. The high hydrogen and low oxygen index values indicate that the kerogens are predominantly composed of hydrogen-rich marine lipid material and are well preserved, i.e., not affected by oxidation. The Aptian/Albian sediments from Site 547 plot in the encircled area above the kerogen type III trend line, indicating that they contain a mixture of terrigenous and marine organic matter with the former predominating. The relatively low oxygen contents (IO between 100 and 200 mg CO_2/g C-org) indicate environmental conditions favorable for the preservation of organic matter, possibly also related to downslope transport. Finally, most of the Cenomanian sediments contain a hydrogen-deficient and oxygen-rich kerogen. They plot just below the kerogen type III trend line (Fig. 4), and it may be concluded that the organic matter is of terrigenous origin and partly oxidized during transport.

A similar kerogen type was encountered in the Late Cenomanian and Albian sediments at DSDP Site 415 (Leg 50) in the Agadir Canyon off Morocco (Boutefeu, 1980).

DEEP-SEA SEDIMENTS FROM THE EASTERN NORTH PACIFIC

During DSDP Leg 63, thick sedimentary sequences of Middle Miocene to Quaternary age were penetrated at sites 467 and 471 in the eastern North Pacific off the coast of North America. Hole 467 was drilled in the San Miguel Gap of the Patton Escarpment on the outer California Continental Borderland at a water depth of 2146 m. Hole 471 was drilled further south in a sedimentary wedge west of the foot of the continental slope off Baja California (water depth 3115 m).

Total organic carbon values in the Site 467 sediments are high and occasionally exceed 5% (Fig. 5), whereas at Site 471 they are fairly constantly close to 1% (Rullkötter, von der Dick and Welte, 1981). No slumping or turbidite flow has been observed at Site 467 (Yeats, Haq et al., 1981a), thus an oxygen-depleted water body extending to a relatively great water depth is assumed to explain preservation of these high amounts of organic matter (cf. Summerhayes, 1981a). This may possibly have been enforced by some restriction of

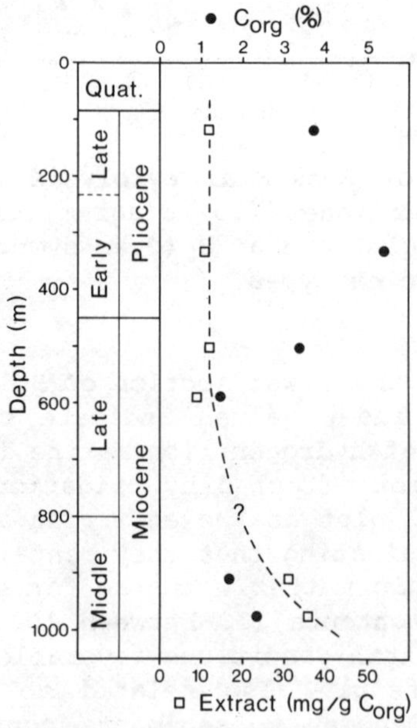

Fig. 5. Total organic carbon (●) and extract (□) versus depth profiles for sediments from DSDP Site 467 off southern California (after Rullkötter et al., 1981).

water circulation in the San Miguel Gap. The sediments at DSDP Site 471 were deposited as turbidites (Yeats, Haq et al., 1981b), which may have led to dilution of organic matter by clastic mineral matter on the one hand and may have supported organic matter preservation by rapid burial on the other.

Rock-Eval pyrolysis shows higher hydrogen indices for the sediments from the San Miguel Gap indicating a higher contribution of marine organic matter, compared with Site 471 (Fig. 6) where there is a stronger influence of terrigenous organic matter. An exception at Site 467 is the sample from Core 467-63 which yielded a low hydrogen and a high oxygen index (Fig. 6) and which contains humic (coaly) organic matter according to organic petrographic investigation (Rullkötter et al., 1981).

The organic-carbon-rich Miocene sediments found in DSDP Hole 467 correlate well with the Miocene Monterey Shales on the California Continental Borderland which contain sections extremely enriched in organic matter and which have been shown by organic geochemical

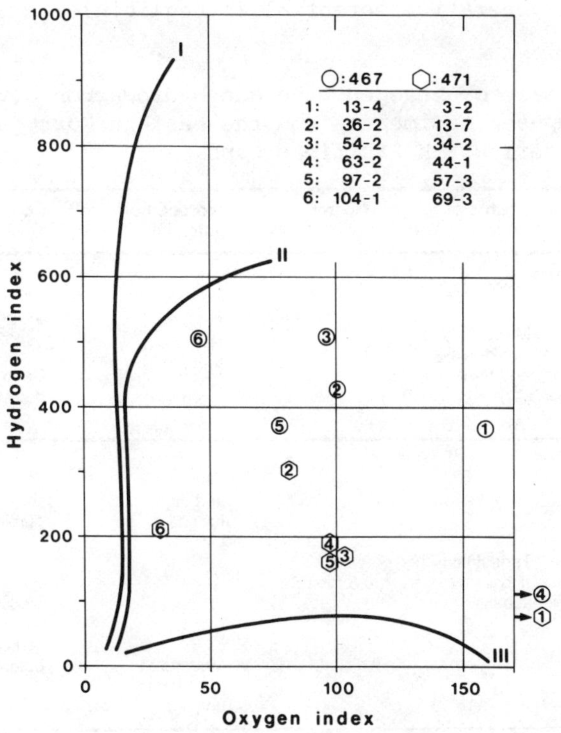

Fig. 6. Results of Rock-Eval pyrolysis displayed as hydrogen index (IH) versus oxygen index (IO) diagram for sediments from DSDP Holes 467 and 471 (from Rullkötter et al., 1981).

analyses to be the source rocks of at least part of the Californian crude oils (Seifert and Moldowan, 1978). Deposition of the Monterey Shales was suggested to have occurred as the result of intense coastal upwelling (Ingle, 1973). The high organic carbon contents in these shales are caused both by the high bioproductivity related to coastal upwelling and by deposition in an extended oxygen-minimum zone which developed as a result of the high bioproductivity (Summerhayes, 1981a).

HYDROCARBON POTENTIAL

There is no *a priori* reason why hydrocarbon source beds should not exist in continental slope and rise sediments as they do in sediments on continental shelves, although commercial drilling has not advanced into the deep-water area and much of the geology has to be inferred (Dow, 1978). A high total organic carbon content alone, however, is not always an indication of a high hydrocarbon potential. Incorporation of major quantities of recycled or oxidized organic matter, which is dominantly resistant and inert, may lead to sediments rich in organic carbon, but because of the low hydrogen content the hydrocarbon generation potential is negligible.

Table 1. Summary of kerogen type and hydrocarbon potential for deep-sea sediments from the eastern North Atlantic and eastern North Pacific oceans

DSDP Site (Interval)	Source of organic matter	Mean C_{org}	Hydrocarbon potential	Comments
397, 547 (allochthonous Miocene and Eocene slumps)	Mixed marine planktonic and terrestrial	1 %–5 %	High for oil and gas	From outer shelf/ upper slope oxygen minimum
467 (Miocene and post-Miocene)				Deep oxygen depletion?
547 (Aptian/Albian)	Terrestrial with subsidiary aquatic	0.5 %–1.5 %	Low for oil and high for gas	
471 (Miocene and post-Miocene)				Turbidites
397, 415, 416 (Cretaceous)			Moderate to low for gas only	Oxidation during transport reduces hydrocarbon potential
547 (Cenomanian)				
397, 415, 547 (autochthonous Tertiary)	Residual terrestrial	<0.5 %	Nil	

Table 1 summarizes the organofacies classification of the deep-sea sediments from the eastern North Atlantic and eastern North Pacific oceans discussed in this paper, ascribing a different hydrocarbon potential to each group. A high potential for liquid and gaseous hydrocarbon generation may be inferred for the organic matter in the Miocene and Eocene slump sequences at DSDP Sites 397 and 547, respectively, as well as in most of the Miocene and post-Miocene sediments recovered at Site 467. In each case the amount of lipid-rich organic matter preserved is relatively high.

A number of sediments containing terrigenous with subsidiary aquatic organic matter have a significantly lower hydrocarbon potential. Due to good preservation of the organic matter, there is a low potential for oil but high potential for gas generation in the Aptian/Albian sediments from Site 547 in the Mazagan Escarpment and the Miocene and post-Miocene turbidites from Site 471 off Baja California. Oxidation before final burial has reduced the hydrocarbon potential in the Cretaceous sediments at DSDP Sites 397, 415, and 416 as well as in the Cenomanian sediments at Site 547. Finally, the autochthonous hemipelagic Tertiary sediments at Sites 397, 415, and 547 have no hydrocarbon potential at all since the organic carbon contents are low and most of the organic matter consists of residual, inert terrigenous material.

Not included in Table 1 are the so-called black shales with extreme organic carbon contents sometimes in excess of 20%. Black shales were not encountered in the sediments studied by us except a few thin layers in the Jurassic at DSDP Site 547. Within the upwelling zones of the Atlantic Ocean, Cretaceous black shales were found in deep-sea sediments off Senegal and in the Angola and Cape basins. Their organofacies characteristics and hydrocarbon potential has been described in detail by Tissot et al. (1980) and Summerhayes (1981b). The environmental conditions leading to the deposition of the black shales, however, are not yet clear. Schlanger and Jenkyns (1976) have suggested worldwide Cretaceous anoxic events as the cause of the formation of black shales, whereas Tissot et al. (1980) have assumed restricted circulation in the early Atlantic Ocean to be the cause. Recent preliminary investigations on Cretaceous black shales from the Angola Basin have shown, however, that they were deposited from turbidite flows and not under stagnant water conditions (Hay et al., in press).

The zone of thermal hydrocarbon generation has not been reached in any of the holes drilled on the northwest African and northwest American continental margins because of safety precautions. Vitrinite reflectance measurements indicated values of 0.4 to 0.5 R_m at bottom-hole depth in each case (Cornford et al., 1979; Cornford, 1980; Rullkötter et al., 1981), i.e., values just below the onset of hydrocarbon generation from marine organic matter (Tissot and Welte, 1978). A comparative study of northeast Atlantic and northeast Pacific sediments revealed, however, that due to the higher geothermal gradient off California and Baja California the maturation of the organic matter in sediments of comparable age (Miocene) has proceeded further in the northeast Pacific (Rullkötter and Welte, in press). An indication of this can be gained from the increase in the amount of extractable organic matter with depth in the sediments from DSDP Site 467 (Fig. 5), an effect which is not observed in the northwest African continental margin sediments.

CONCLUSIONS

The importance of coastal upwelling for the deposition of organic-matter-rich sediments on the continental shelves at low latitudes on the eastern margins of the world's oceans and the significance of these sediments for the genesis of potential petroleum source beds have been pointed out by Demaison and Moore (1980). In contrast, typical deep-sea sediments contain very little organic matter because most of it is destroyed during settling through oxygen-rich oceanic water. In this paper it has been shown, however, that exceptional situations exist where the effect of ancient coastal upwelling could be traced in the organic matter content and composition of deep-sea sediments. High bioproductivity in upwelling areas results in a mid-water oxygen-minimum layer. Sedimentation of organic matter within this zone on the outer shelf or upper slope leads to high preservation rates. Subsequent downslope transport, as has been shown for two different sites on the northwest African continental margin, provides a mechanism for transfering major amounts of lipid-rich marine organic matter into the deep-sea environment. Alternatively, particularly strong upwelling during the Miocene off Southern California seems to have extended oxygen depletion into the deep-water area of the San Miguel Gap.

Where the content of marine organic matter in deep-sea sediments is high, a high potential for oil and gas generation can be ascribed to the related kerogens. More terrigenous influence and less efficient preservation of marine organic matter, depending on the extent to which this occurs, reduces the hydrocarbon potential.

ACKNOWLEDGEMENTS

The authors thank the National Science Foundation, Washington, D.C., for the opportunity to participate in Deep Sea Drilling Project Leg 79 on the DV "Glomar Challenger". Financial support by the Deutsche Forschungsgemeinschaft, Bonn, grant No. We 346/25, is gratefully acknowledged.

REFERENCES

Aizenshtat, A., Baedecker, M.T. and Kaplan, I.R., 1978, Distribution and diagenesis of organic compounds in JOIDES sediment from Gulf of Mexico and western Atlantic, Geochimica et Cosmochimica Acta, 37:1881-1898.

Boutefeu, A., 1980, Pyrolysis study of organic matter from Deep Sea Drilling Project Sites 370 (Leg 41), 415, and 416 (Leg 50), in: "Initial Reports DSDP," 50, Y. Lancelot, E.L. Winterer et al., U.S. Government Printing Office, Washington, 555-566.

Cornford, C., 1979, Organic deposition at a continental rise: organic geochemical interpretation and synthesis at DSDP Site 397, eastern North Atlantic, in: "Initial Reports DSDP," 47, Part 1, U. von Rad, W.B.F. Ryan et al., U.S. Government Printing Office, Washington, 503-510.

Cornford, C., 1980, Petrology of organic matter, Deep Sea Drilling Project Sites 415 and 416, Morrocan Basin, eastern North Atlantic, in: "Initial Reports DSDP," 50, Y. Lancelot, E.L. Winterer et al., U.S. Government Printing Office, Washington, 609-614.

Cornford, C., Rullkötter, J. and Welte, D.H., 1979, Organic geochemistry of DSDP Leg 47a, Site 397 eastern North Atlantic: organic petrography and extractable hydrocarbons, in: "Initial Reports DSDP," 47, Part 1, U. von Rad, W.B.F. Ryan et al., U.S. Government Printing Office, Washington, 511-522.

Cornford, C., Rullkötter, J. and Welte, D.H., 1980, A synthesis of organic petrographic and geochemical results fron DSDP sites in the eastern central North Atlantic, in: "Advances in Organic Geochemistry - 1979," Progress in Physics and Chemistry of the Earth Series, A.G. Douglas and J.R. Maxwell, eds., Pergamon Press, Oxford, 445-452.

Dean, W.E., Gardner, J.V., Jansa, L.F., Cepek, P., and Seibold, E., 1978, Cyclic sedimentation along the continental margin of Northwest Africa, in: "Initial Reports DSDP," 41, Y. Lancelot, E. Seibold et al., U.S. Government Printing Office, Washington, 965-989.

Demaison, G.J. and Moore, G.T., 1980, Anoxic environments and oil source bed genesis, American Association of Petroleum Geologists, Bulletin, 64:1179-1209.

Deroo, G., Herbin, J.P., Roucaché, J., and Tissot, B., 1979, Organic geochemistry of some organic-rich shales from DSDP Site 397, Leg 47a, eastern North Atlantic, in: "Initial Reports DSDP," 47, Part 1, U. von Rad, W.B.F. Ryan et al., U.S. Government Printing Office, Washington, 523-529.

Diester-Haass, L. and Schrader, H.J., 1979, Neogene coastal upwelling history off Northwest and Southwest Africa, Marine Geology, 29: 39-53.

Dillon, W.P. and Sougy, J.M.A., 1974, Geology of West Africa and Canary and Cape Verde Islands, in: "The Ocean Basins and Margins, 2. The North Atlantic," E.M. Nairn and F.C. Stehli, eds., Plenum Press, New York, 315-390.

Dow, W.G., 1978, Petroleum source beds on continental slopes and rises, in: "Geological and Geophysical Investigations of Continental Margins," J.S. Watkins, L. Montadert and P.W. Dickerson, eds., American Association of Petroleum Geologists, Memoir 29:423-442.

Espitalié, J., Laporte, J.L., Madec, M., Marquis, F., Leplat, P., Paulet, F., and Boutefeu, A., 1977, Méthode rapide de caractérisation des roches-mères, de leur potentiel pétrolier and de leur degré d'évolution, Revue de l'Institut Français du Pétrole, 32:23-42.

Hay, W.W., Sibuet, J.-C. et al., in press, Sedimentation and accumulation of organic carbon in the Angola Basin and on Walvis Ridge: Preliminary results of Deep Sea Drilling Project Leg 75, Geological Society of America, Bulletin.

Ingle, J.C., 1973, Summary comments on Neogene biostratigraphy, physical stratigraphy, and paleo-oceanography in the marginal northeastern Pacific Ocean, in: "Initial Reports DSDP," 18, L.D. Kulm, R. von Huene et al., U.S. Government Printing Office, Washington, 949-959.

Hunt, J.M., 1979, "Petroleum Geochemistry and Geology," Freeman and Company, San Francisco, 617 pp.

Johnson, D.L., McIver, R.D. and Rogers, M.A., 1979, Insoluble organic matter bitumens in Leg 47 samples, in: "Initital Reports DSDP," 47, part 2, J.C. Sibuet, W.B.F. Ryan et al., U.S. Government Printing Office, Washington, 543-546.

McIver, R., 1975, Hydrocarbon occurrences from JOIDES Deep Sea Drilling Project, in: "Proceedings of the 9th World Petroleum Congress (Tokyo)," Vol. 2, Applied Science Publishers, Barking, U.K., 269-280.

Müller, P.J. and Suess, E., 1979, Productivity, sedimentation rate and sedimentary organic matter in the oceans. I. Organic carbon preservation, Deep-Sea Research, 26A:1347-1362.

Pearson, B. and Dow, W.G., 1979, Geochemical analysis of the samples from Sites 397 and 398, in: "Initial Reports DSDP," 47, part 2, J.C. Sibuet, W.B.F. Ryan et al., U.S. Government Printing Office, Washington, 533-541.

Rullkötter, J. and Welte, D.H., in press, Maturation of organic matter in areas of high heat flow: a study of sediment from DSDP Leg 63, offshore California and Leg 64, Gulf of California, in: "Advances in Organic Geochemistry-1981," M. Bjorøy et al., eds., John Wiley and Sons, Chichester, U.K.

Rullkötter, J., von der Dick, H. and Welte, D.H., 1981, Organic petrography and extractable hydrocarbons of sediments from the eastern North Pacific Ocean, Deep Sea Drilling Project Leg 63, in: "Initial Reports DSDP," 63, R.S. Yeats, B.U. Haq et al., U.S. Government Printing Office, Washington, 819-836.

Rullkötter, J., Cornford, C. and Welte, D.H., 1982, Geochemistry and petrography of organic matter in Northwest African continental margin sediments: quantity, provenance, depositional environment, and temperature history, in: "Geology of the Northwest

African Continental Margin," U. von Rad, K. Hinz, M. Sarnthein, and E. Seibold, eds., Springer-Verlag, Berlin, 686-703.

Ryther, J.H., 1969, Photosynthesis and fish production in the sea, Science, 166:72-76.

Schlanger, S.O. and Jenkyns, H.C., 1976, Cretaceous anoxic events - causes and consequences, Geologie en Mijnbouw, 55:179-184.

Seifert, W.K. and Moldowan, J.M., 1978, Application of steranes, terpanes and monoaromatics to the maturation, migration and source of crude oils, Geochimica et Cosmochimica Acta, 42:77-95.

Simoneit, B.R.T. and Mazurek, M.A., 1979, Search for eolian lipids in the Pleistocene off Cape Bojador and lipid geochemistry of a Cretaceous mudstone, DSDP/IPOD Leg 47a, in: "Initial Reports DSDP," 47, part 1, U. von Rad, W.B.F. Ryan et al., U.S. Government Printing Office, Washington, 541-545.

Summerhayes, C.P., 1981a, Oceanographic controls on organic matter in the Miocene Monterey formation, offshore California, in: "The Monterey Formation and Related Siliceous Rocks of California," Society of Economic Paleontologists and Mineralogists, Tulsa, 213-219.

Summerhayes, C.P., 1981b, Organic facies of middle Cretaceous black shales in deep North Atlantic, American Association of Petroleum Geologists, Bulletin, 65:2364-2380.

Tissot, B.P. and Welte, D.H., 1978, "Petroleum Formation and Occurrence. A New Approach to Oil and Gas Exploration," Springer-Verlag, Berlin, 538 pp.

Tissot, B.P., Deroo, G. and Herbin, J.P., 1979, Organic matter in Cretaceous sediments of the North Atlantic: contribution to sedimentology and paleogeography, in: "Deep Drilling Results in the Atlantic Ocean: Continental Margins and Paleoenvironment," Maurice Ewing Series 3, M. Talwani, W. Hay and W.B.F. Ryan, eds., American Geophysical Union, Washington, 362-374.

Tissot, B.P., Demaison, G., Masson, P., Delteil, J.R., and Combaz, A., 1980, Paleoenvironment and petroleum potential of middle Cretaceous black shales in Atlantic basins, Amercian Association of Petroleum Geologists, Bulletin, 64:2051-2063.

von Rad, U., Ryan, W.B.F. et al., 1979, Site 397, in: "Initial Reports DSDP," 47, part 1, U. von Rad, W.B.F. Ryan et al., U.S. Government Printing Office, Washington, 17-217.

Walsh, J.J., 1981, A carbon budget for overfishing off Peru, Nature, 290:300-304.

Welte, D.H., Cornford, C. and Rullkötter, J., 1979, Hydrocarbon source rocks in deep sea sediments, Proceedings of the 11th Annual Offshore Technology Conference (Houston), 1:457-464.

Yeats, R.S., Haq, B.U. et al., 1981a, Site 467, in: "Initial Reports DSDP," 63, R.S. Yeats, B.U. Haq et al., U.S. Government Printing Office, Washington, 23-112.

Yeats, R.S., Haq, B.U. et al., 1981b, Site 471, in: "Initial Reports DSDP," 63, R.S. Yeats, B.U. Haq et al., U.S. Government Printing Office, Washington, 269-349.

CRETACEOUS UPWELLING OFF NORTHWEST AFRICA: A SUMMARY

Gerhard Einsele and Jost Wiedmann

Geologisch-Paläontologisches Institut der Universität
Sigwartstrasse 10
D-7400 Tübingen, Federal Republic of Germany

ABSTRACT

As early as Late Cretaceous times, but mainly during the Turonian, bituminous marls deposited in marginal basins off Morocco indicate the onset of coastal upwelling in the eastern North Atlantic. Principal evidence for this is the occurrence of a northern temperate, partly endemic faunas, organic matter of marine origin, and chert.

INTRODUCTION

One of the oldest and, as we believe, most convincing examples for coastal upwelling is that of the Cretaceous Tarfaya-Aaiun Basin, southern Morocco. Since the features of the Tarfaya mid-Cretaceous deposits (Cenomanian to Coniacian) have been discussed elsewhere (Einsele and Wiedmann, 1975; 1982; Wiedmann, Butt and Einsele, 1978; Wiedmann, Einsele and Immel, 1978), this paper presents only a brief review of the available data and their interpretation.

OBSERVATIONS

Most sections of the Tarfaya Cretaceous exhibit continental, nearshore, or inner shelf depositional environments. The sedimentary sequence starts with a thick series of continental to deltaic conglomerates and sandstones ("weald" facies) of Early Cretaceous age. The subsequent marine transgression is of Late Middle and Upper Albian age during which time a series of greenish marls and fossiliferous shell beds were deposited (Figs. 1A, 2D). A similar type of

Fig. 1. The marine Cretaceous of Tarfaya (from Einsele and Wiedmann, 1982).

A. Composite simplified section of the complete marine Cretaceous sequence.

B. Paleobathymetric interpretation of sequence.

C. Turonian to Lower Coniacian section of Oued Ouaâr/Tarfaya in which laminated bituminous marls alternate with micritic limestones or nodules of limestone and chert.

D. Selected short sections showing different bedding features.

Lithology: 1 sandstone; 2 sandy marl with gypsum (y); 3 silty clay and marl; 4 bituminous marl, well-laminated; 5 marl and marly limestone; 6 limestone; 7 nodules of limestone and chert; 8 bioturbation; 9 shell-bearing limestones and marls; 10 shell beds; 11 oyster reefs.

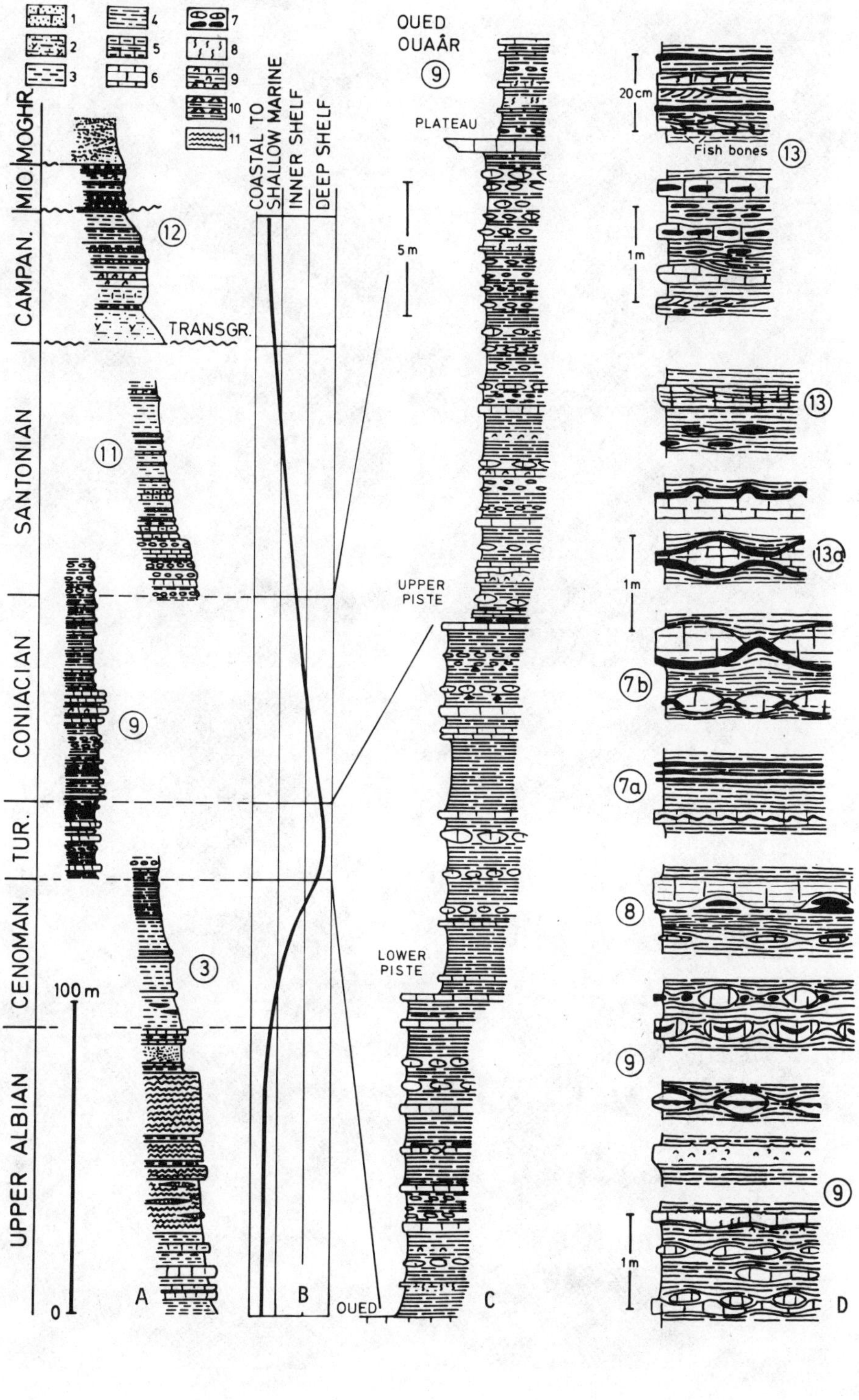
1
2
3
4
5
6
7
8
9
10
11
MIO. MOGHR.
CAMPAN.
SANTONIAN
CONIACIAN
TUR.
CENOMAN.
UPPER ALBIAN
100 m
0
TRANSGR.
A
COASTAL TO SHALLOW MARINE
INNER SHELF
DEEP SHELF
B
OUED OUAÂR
PLATEAU
5 m
UPPER PISTE
LOWER PISTE
OUED
C
20 cm
Fish bones
1 m
D

B
D
A
C

Fig. 2. Lithologies of Moroccan Mid-Cretaceous deposits.

A. Black shales and well-bedded limestones of Upper Cenomanian age, Sabkha Tazra/Tarfaya.

B. Fine laminations in bituminous marls with chert nodules containing accumulations of smaller benthic molluscs (*Astarte*, *Rostellaria*).

C. For comparison: thick oyster shell beds of the Upper Cenomanian of High Atlas Mountains near Agadir.

D. Middle Albian shell bed (with *Ptychomya*, *Trigonia*) south of Tantan Plage/Tarfaya.

B
D
A
C

Fig. 3. Microfaunas of the marine Cretaceous deposits of Tarfaya.

A. Upper Albian microfauna of inner shelf environment with planktonics (*Hedbergella washitensis*, *Ticinella raynaudi*, *Globigerinelloides breggiensis*) and benthics (*Trochammina*, *Lenticulina*, *Nodosaria*, *Bairdia*, *Cytherella*). Tassagdelt, magnification 40/1.

B. Fish remains in the Upper Turonian black shales from Houiselgua E, with planktonic foraminifers (*Globotruncana schneegansi*, *Hedbergella planispira*, *Heterohelix globulosa*) and reduced number of benthics (*Veenia* ?), magnification 10/1.

C. Late Cenomanian microfauna of black shale facies and deep shelf environment with planktonic foraminifera (*Rotalipora cushmani*, *Hedbergella delrioensis*, *H. paradubia*, *Heterohelix globulosa*) and fish remains. Sabkha Houiselgua S, magnification 20/1.

D. Lower Santonian inner shelf environment of Sabkha Tah, with planktonic foraminifera (*Globotruncana concavata*, *Gl. angusticarinata*) and benthics (*Palmula cushmani*, *Veenia* ?), magnification 20/1.

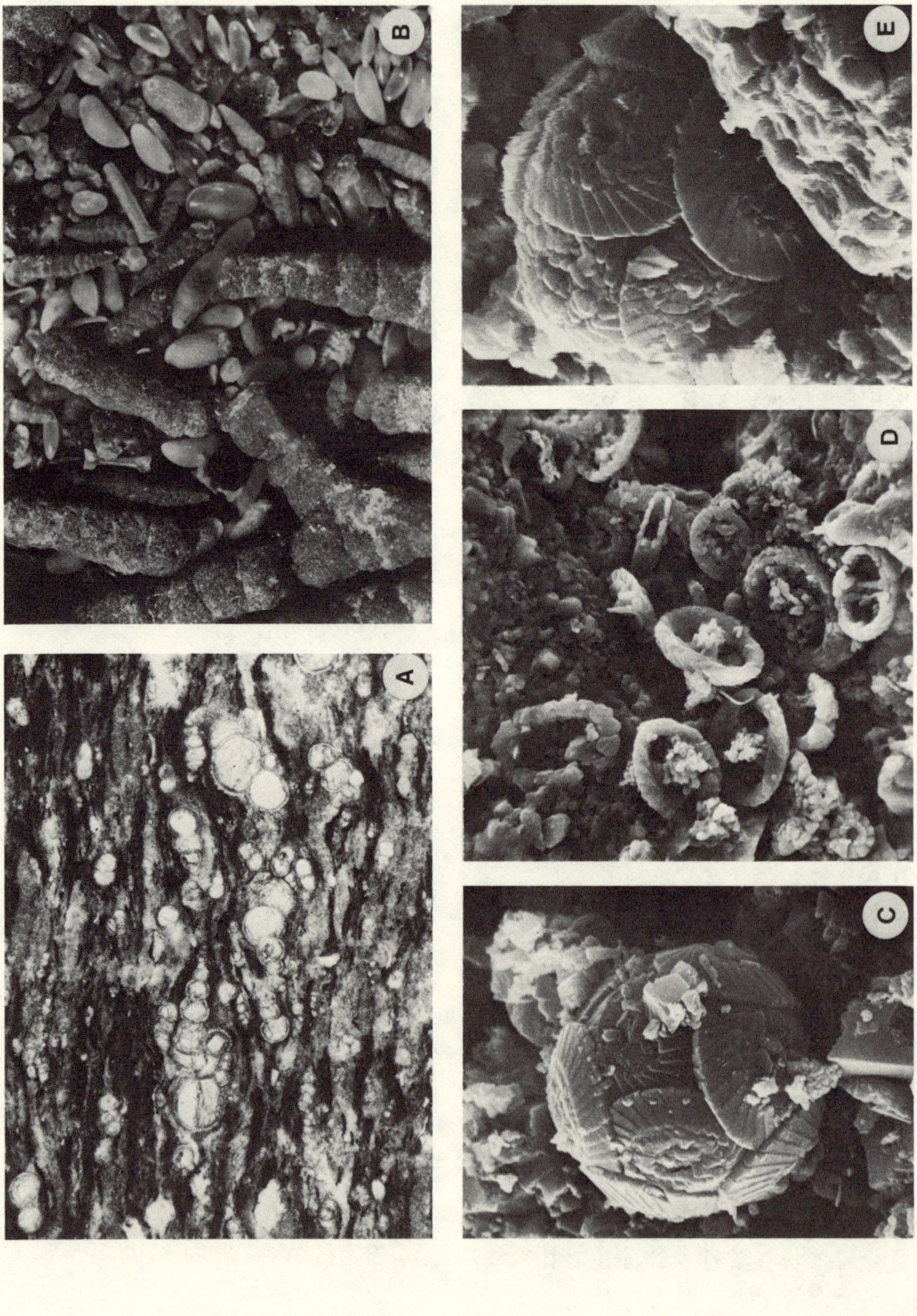
B
E
D
A
C

Fig. 4. Microfaunas and nannofloras from Tarfaya and High Atlas Mountains.

A. Thin section of Upper Cenomanian black shales with typical micro-flaser bedding and intercalated layers of foraminifers (*Heterohelix*). Sabkha Houiselgua S, magnification 90/1.

B. Microfauna of inner shelf Upper Cenomanian, north of Agadir, High Atlas Mountains. Predominance of benthics (*Cribratina, Spiroplectammina, Cytherella, Brachycythere*) and rare planktonics (*Clavihedbergella simplex*), magnification 20/1.

C. Coccosphere *Watznaueria barnesae* (Black), Lower Turonian black shales of Oued Ouaâr/Tarfaya, magnification 4000/1. GPIT type collection G483-77150.

D. *Zygodiscus diplogrammus* (Deflandre) from Lower Turonian black shales of locality Es-Zeiba/Tarfaya, magnification 4500/1. GPIT type collection 3640-63851.

E. Coccosphere *Watznaueria biporta* Bukry, Middle Turonian black shales of locality Es-Zeiba/Tarfaya, magnification 5500/1. GPIT type collection 3650-64057.

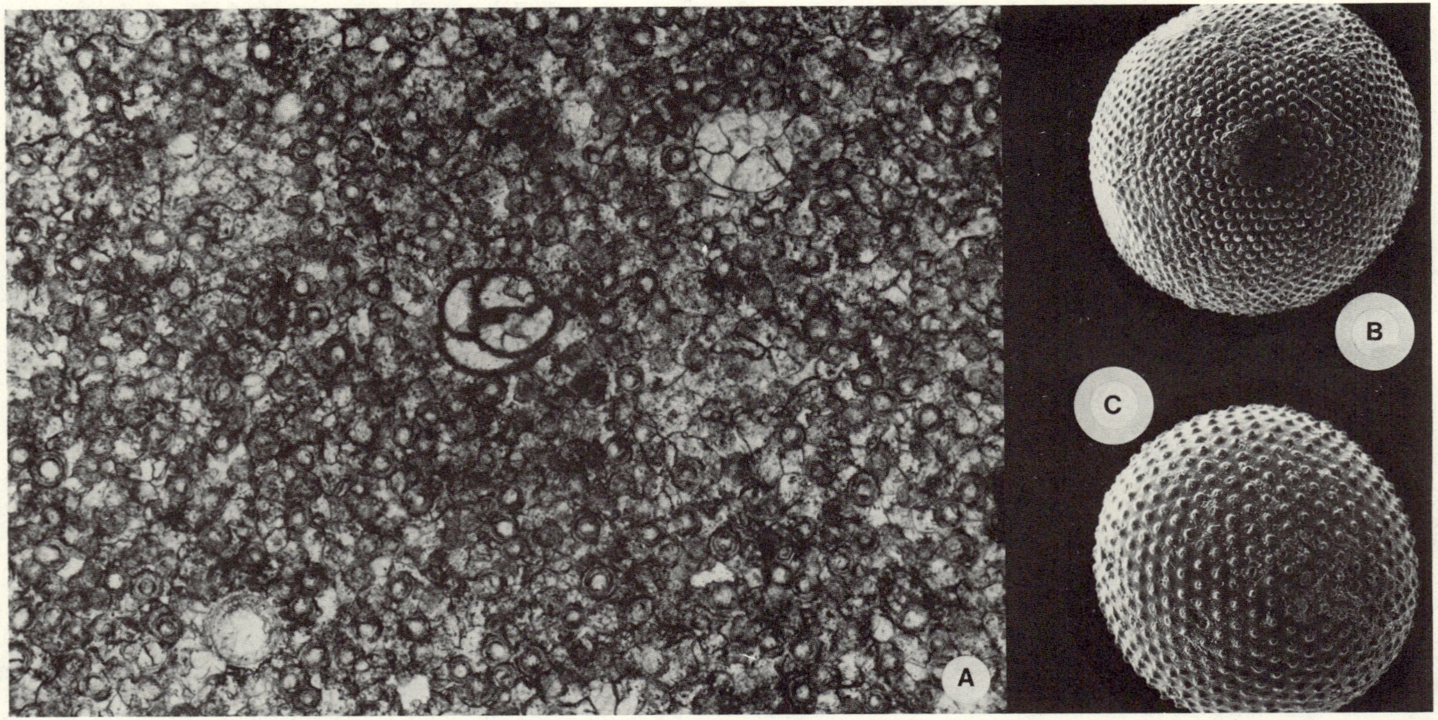

Fig. 5. Calcispheres and siliceous planktonics of Tarfaya black shales.
A. Thin section of Lower Turonian, Sabkha Houiselgua W, composed by calcispheres, magnification 150/1. Steinkerns of siliceous planktonics (radiolarians?) from Upper Cenomanian of Sabkha Houiselgua S, magnification 200/1 (B). GPIT type collection 3638-63386 and Lower Turonian of Es-Zeiba, magnification 200/1 (C). GPIT type collection 3638-63390.

Fig. 6. Lower Turonian index ammonites from southern and northern Morocco.

A. *Selwynoceras reymenti* Collignon, index species of Tarfaya Lower Turonian black shale facies, lowermost Turonian of Oued Quaâr section/Tarfaya, magnification 1/1. GPIT type collection $C_1$231271.

B. *Nigericeras lamberti* Schneegans, index species of Tethyan Lower Turonian, Sidi Hajaj near Settat, Moroccan Meseta, magnification 1/1. Collection Service Géologique Rabat SH10.

C. Lower Turonian limestone bedding plane with ammonite (*Benuites*, *Watinoceras*) and gastropod (*Rostellaria*) juvenile mass mortality. Sabkha Houiselgua W, 3/4 natural size.

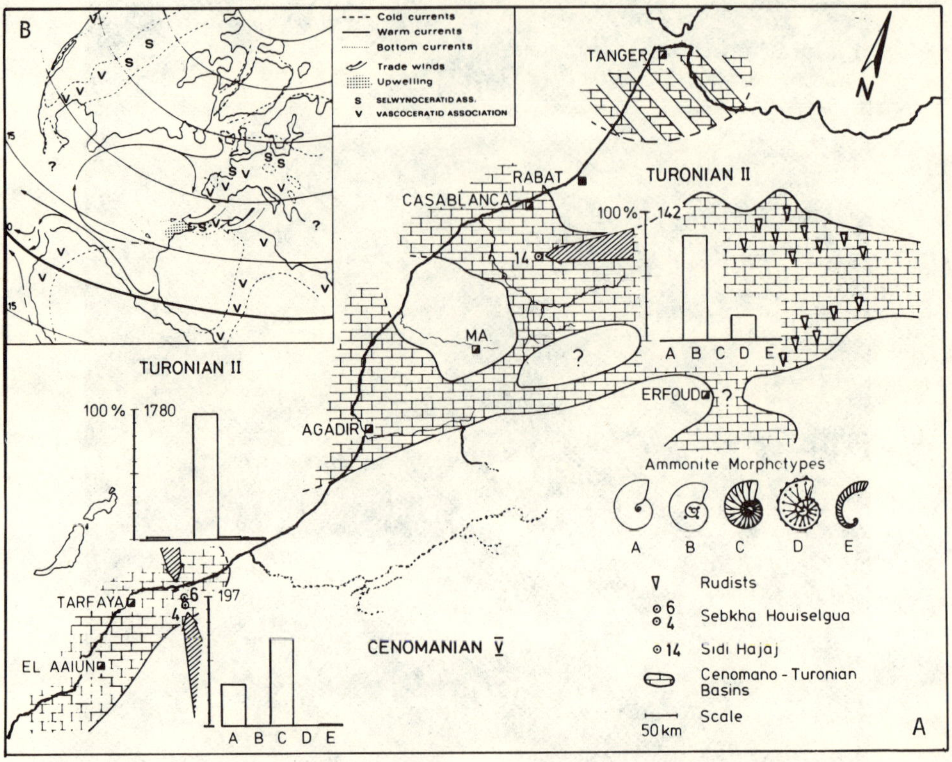

Fig. 7. Cenomanian-Turonian coastal basins of northwest Africa and the divergence of ammonites in the Tarfaya Basin from those of the Moroccan Meseta, demonstrated by quantitative spectra of main ammonite morphotypes (A) and paleogeographic implication (B) (from Einsele and Wiedmann, 1982). By permission from Geology of NW African Continental Margin, Copyright (c) 1982, Springer Verlag.

facies and environment characterize the Santonian-Campanian Epochs. During the Cenomanian times, however, quite a different facies type becomes established which dominates throughout the Turonian and continues into the Coniacian. In particular, this facies consists of thin-bedded black marls and shales, which alternate with well-bedded limestones or concretionary, often silicified layers (Figs. 1C, 1D, 2A, 2B). The largely planktonic microfaunas of these black marls (Figs. 3B, 3C, 4, 5) point to an open, deep-shelf environment. Even more important is the appearance and high abundance of north temperate molluscs (Figs. 6A, 6C, 7) and the complete absence of Tethyan faunas (Fig. 6B), the latter typical for northern Morocco and the entire area of the Mediterranean Basin.

INTERPRETATION

In light of our present knowledge about recent coastal upwelling and its imprint on the underlying sediment record, we believe that such an oceanographic process was responsible for the sedimentological and faunistic characteristics of the Mid-Cretaceous deposits of Morocco, illustrated above.

Sedimentological Evidence

- Finely laminated sediments (Fig. 2A, 2B),
- Bioturbation limited to a few layers,
- Absence of shell beds which otherwise are typical for the underlying and overlying strata as well as for the contemporaneous deposits in the neighboring Agadir (Atlas) Basin (Fig. 2C, 2D),
- High concentration of organic matter of marine origin (up to 9.8% C-org) compared to approximately 0.1% C-org in the Albian marls (analyses by J. Rullkötter, Aachen),
- High silica content, in the form of cherts, which is undoubtedly derived from opaline skeletons of diatoms and radiolarians,
- High concentrations of phosphates ranging from 250-4600 ppm phosphorous,
- Enrichment of heavy metals, Zn, Ni, Cr.

The high calcium carbonate content, varying between 35 and 99%, indicates an environment with low terrestrial input but situated well above the calcium carbonate compensation depth. A similar situation is observed with sediments of recent coastal upwelling. In contrast, the Albian facies contain only between 1-50% calcium carbonate.

Faunistic Evidence

The most surprising feature of the Tarfaya Mid-Cretaceous deposits is the appearance and very high abundance of northern temperate molluscs. This is clearly evident in the Late Cenomanian-Early Turonian ammonites (Figs. 6A, 6C, 7) which are accompanied by northern inoceramids. On the other hand, the absence of Tethyan ammonites (Fig. 6B) or other molluscs, such as oysters and rudists, is striking. As mentioned above, planktonic forms prevail among the microfauna, such as Rotaliids and Heterohelicids within the foraminifers (Figs. 3B, 3C, 4A) often accompanied by rock-forming coccoliths (Fig. 4C-4E) or calcispheres (Fig. 5A). Benthic micro- and macrofaunas are very scarce and limited to some apparently more oxygenated layers. Within the planktonic foraminifers, globigerinoid species of the genera *Hedbergella*, *Globigerinoides*, *Rugoglobigerina* dominate over the keeled species of *Globotruncana*; furthermore, cool water sculptures are developed in foraminifers and ostracodes.

Bioturbation is rare, which correlates with the scarcity of benthics. The black marl samples contain high percentages of fish debris (Fig. 3B). Within the black marl faunas an increase of endemic forms can be observed; the faunal density is extremely high (Figs. 3-6), while species diversity is low (Figs. 3C, 6C). Finally, the small size of molluscs (Fig. 6C) as well as microfaunas (Fig. 3C) may indicate low oxygen environment. Fig. 6C shows an example of mass mortality of ammonites. Palaeobiogeographically, the faunistic relationships of foraminifers and ammonites point to North America and northern Europe provinces (Fig. 7B), while relationships to the Algerian and Tunesian Atlas ranges are very limited.

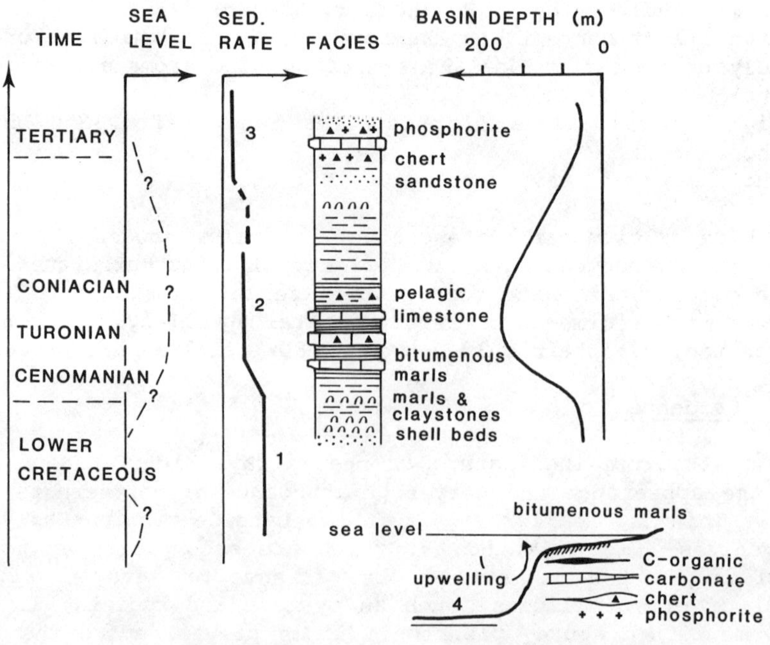

Fig. 8. Relationship between sea level, sedimentation rate, paleo-water depth and facies in Moroccan coastal basins prior to, during, and after the Turonian-Coniacian black shale deposition (very much simplified). Inserted cross-section of the continental margin shows general trends in relative proportions of different sediment components across the coastal basin; thick bars signify high percentages. 1, Sedimentation rate approximately balances subsidence; 2, low sedimentation rate; 3, very low sedimentation rate, condensation or hiatuses; 4, oxygenated deep-sea sediments or hiatus.

SUMMARY AND CONCLUSIONS

The Upper Cretaceous, chiefly Turonian, bituminous marls of the Moroccan coastal basins take an exceptional position compared with the black shales of the deep Atlantic. These marls were deposited in a water depth of 200 to 300 m during a period when deep Atlantic water circulation in an already wide ocean led to oxygenated sea floor conditions and incomplete sediment sequences (Fig. 8). The black shales of the Moroccan shelf are characterized by carbonaceous laminated marls with intercalated pelagic limestones and cherts. Their occurrence in marginal basins coincides with the peak of the Cretaceous marine transgression and with decreasing terrigenous sediment supply from the continent (Fig. 8). Their sedimentation rate is on the order of 5-20 m per million years and the fauna contains elements of the northern temperate zones. The faunas are interpreted as an indicator of the onset of coastal upwelling along the northwest African continental margin, although long distance cold bottom currents probably did not yet exist during that time period (Fig. 7B). The lack of a middle and Upper Cretaceous black shale facies in the conjugate North American shelf basins, however, may be explained by the already existing influence of a trade wind system which generated coastal upwelling only in the eastern North Atlantic.

REFERENCES

Einsele, G. and Wiedmann, J., 1975, Faunal and sedimentological evidence for upwelling in the Upper Cretaceous coastal basin of Tarfaya, Morocco, 9th Congress, International Association of Sedimentologists, Publication, Nice, Theme 1:67-74.

Einsele, G. and Wiedmann, J., 1982, Turonian black shales in the Moroccan coastal basins: first upwelling in the Atlantic Ocean? in: "Geology of the Northwest African Continental Margin," U. von Rad, K. Hinz, M. Sarnthein, and E. Seibold, eds., Springer, Berlin, 396-414.

Wiedmann, J., Butt, A. and Einsele, G., 1978, Vergleich von marokkanischen Kreide-Küstenaufschlüssen und Tiefseebohrungen (DSDP), Stratigraphie, Paläoenvironment und Subsidenz an einem passiven Kontinentalrand, Geologische Rundschau, 67:454-508.

Wiedmann, J., Einsele, G. and Immel, H., 1978, Evidence faunistique et sédimentologique pour un upwelling dans le Bassin Côtier de Tarfaya/Maroc dans le Crétacé supérieur, Actes VI-éme Colloque Africain de Micropaléontologie, Tunis 1974, Annales des Mines et de la Géologie, Tunis, 28, II:415-441.

FACIES PATTERNS OF A CRETACEOUS/TERTIARY SUBTROPICAL UPWELLING SYSTEM (GREAT SYRIAN DESERT) AND AN APTIAN/ALBIAN BOREAL UPWELLING SYSTEM (NW GERMANY)

Edwin Kemper

Bundesanstalt für Geowissenschaften und Rohstoffe
Stilleweg 2
D-3000 Hannover 51, Federal Republic of Germany

Winfried Zimmerle

Laboratorium für Erdölgewinnung, Deutsche Texaco A.G.
D-3109 Wietze, Federal Republic of Germany

ABSTRACT

Facies patterns of two ancient upwelling systems in different climatic belts are compared, one in the Late Cretaceous/Early Tertiary of the Great Syrian Desert (Syria, northeast Jordan, northwest Iraq) and the other in the Late Early Cretaceous of northwest Germany. The first case history is an example of an upwelling regime in a warm, subtropical ocean under arid climate. The second case history is an example of a rather cool boreal sea under humid climatic conditions. In both cases three depositional belts were distinguished displaying a remarkable facies asymmetry. The typical features are: (1) abundance of siliceous sediments (radiolarians, diatoms) in a nearshore belt, (2) presence of phosphorite accumulations, and (3) a rather high supply of organic matter causing oxygen deficiency in the environment of the second belt. The sedimentary rocks of the Great Syrian Desert are a light-colored chalky limestone/chert association, those of northwest Germany a dark-colored glauconitic claystone association. Sedimentary rock associations as well as faunal and floral assemblages observed are considered to be diagnostic of upwelling.

According to our observations, coastal upwelling was important in Phanerozoic times, even at unexpected localities, such as in small offshoots of cool epicontinental seas at higher latitudes. In such a case the typical features, however, are less distinct. This is explained by less intensive upwelling, by the limited regional extent of the epicontinental sea, and by considerable admixture of terrigenous debris in a humid climate. While Recent belts of coastal upwelling have a north-south orientation, the two examples described are part of west-east extending belts.

INTRODUCTION

Complicated facies patterns, which cannot be explained by normal depositional conditions and whose origin until recently remained obscure, repeatedly appear in the sedimentary record. Their most important feature is a pronounced asymmetry, i.e., certain rocks are confined over one part of the basin only. Correlations may be impossible, even on short distances. As it will be demonstrated, these special facies patterns are frequently the result of coastal upwelling. The purposes of this study are (1) to analyze facies patterns for which normal bathymetrical conceptions do not work, (2) to compare two different types of upwelling systems from two climatic belts: the Tethyal Upper Cretaceous/Lower Tertiary of the Great Syrian Desert in Syria, Jordan, and Iraq, and the boreal Aptian/Albian of NW Germany, and (3) to stress the importance of upwelling in earth history.

METHOD OF INVESTIGATION

This comparative study is based on previous paleogeographic and biostratigraphic studies of the Lower Cretaceous of northwest Germany and on recent work in the Great Syrian Desert by the first author. The second author contributed lithostratigraphic, petrographic, and mineralogical data in order to supplement this study. The analytical work included megascopic inspection of sedimentary rocks in the field, micropaleontological analysis, thin section examination, scanning electron microscopy (SEM), x-ray fluorescence analysis (XRF), and x-ray diffractometry (XRD). The results of the XRD and XRF analyses of chert and phosphorite samples from the Great Syrian Desert are tabulated in Tables 1-3.

UPPER CRETACEOUS/LOWER TERTIARY FACIES PATTERNS IN THE GREAT SYRIAN DESERT

Analytical Data

In addition to numerous micropaleontological mounts and thin sections, eighteen samples from the Great Syrian Desert comprising Maastrichtian, Paleocene, and Eocene phosphorites and Maastrichtian cherts were analyzed for major and trace elements and ten samples for mineral composition. In the following, these analyses of major and trace elements (Tables 1 and 2) are briefly compared with coeval phosphorites from the "Mediterranean Phosphorite Belt." They reveal the several geochemical trends.

Table 1. Major element concentrations in weight-% by x-ray fluorescence analysis

Rock	Sample	SiO_2	TiO_2	Al_2O_3	Fe_2O_3	MnO	MgO	CaO	Na_2O	K_2O	P_2O_5	SO_3	LOI	Sum
Locality: Syria-Al Bardeh														
Age: Maastrichtian														
Phosphorites	B 19	21.36	0.02	0.75	0.55	0.01	0.19	41.68	0.34	0.07	23.18	0.63	10.60	99.38
	B 20	33.59	0.05	1.33	0.68	0.00	0.11	34.67	0.14	0.09	24.32	0.40	4.00	99.39
	B 21	26.93	0.05	1.67	0.54	0.01	0.43	36.62	0.62	0.09	22.80	0.65	9.10	99.50
Opoka	B 22a	83.24	0.16	4.20	1.32	0.00	0.20	4.15	0.11	0.27	0.89	0.08	5.20	99.81
	B 22b	75.31	0.13	3.21	1.11	0.01	0.35	8.23	0.23	0.22	0.43	0.34	9.40	98.96
	B 22c	79.69	0.38	8.09	2.84	0.00	0.51	0.96	0.24	0.50	0.55	0.05	6.00	99.82
Locality: Syria-Knefis														
Age: Maastrichtian														
Phosphorites	A 1	1.18	0.00	0.12	0.19	0.00	0.22	54.07	0.16	0.01	22.87	0.83	19.80	99.47
	A 2	1.28	0.00	0.14	0.18	0.00	0.22	53.80	0.16	0.00	22.51	1.15	20.40	99.84
	A 3	32.32	0.00	0.08	0.10	0.00	0.09	36.51	0.33	0.01	24.60	0.65	4.60	99.29
Opoka	A 4	70.03	0.05	1.19	0.46	0.01	0.32	14.10	0.18	0.12	0.79	0.09	12.80	100.13
	A 5	83.04	0.06	1.43	0.51	0.00	0.29	6.45	0.17	0.13	0.64	0.07	7.20	99.99
Locality: NW Iraq-near H3														
Age: Paleocene														
Phosphorites	I 4	3.55	0.02	0.49	0.21	0.01	3.57	49.50	0.06	0.00	9.87	0.77	32.00	100.03
	I 5	4.73	0.04	0.83	0.44	0.01	1.03	48.86	0.24	0.02	16.18	3.15	23.40	98.93
	I 6	1.53	0.00	0.14	0.17	0.00	0.38	52.91	0.28	0.00	21.18	1.51	21.30	99.41
Locality: NE Jordan														
Age: Eocene														
Phosphorites	U 1	63.53	0.00	0.13	0.20	0.00	0.10	17.64	0.34	0.03	11.19	0.72	4.80	98.68
	Kd 1	0.58	0.00	0.07	0.12	0.01	0.23	54.22	0.27	0.01	19.39	1.23	23.30	99.42
	K 1-2	47.17	0.00	0.10	0.29	0.00	0.13	27.35	0.24	0.02	10.73	1.32	11.70	99.05
	U 2	70.32	0.00	0.09	0.11	0.01	0.08	14.96	0.30	0.04	9.48	0.86	3.70	99.95

Table 2. Minor and trace element concentrations in ppm by x-ray fluorescence analysis; the contents of Bi, Nb, Pb, Sc, Ta, and W are <10 ppm, the content of Sn was <20 ppm

Rock	Sample	Ba	Ce	Co	Cr	Cu	La	Mo	Ni	Rb	Sr	Th	U	V	Y	Zn	Zr
Locality: Syria-Al Bardeh																	
Age: Maastrichtian																	
Phosphorites	B 19	286	25	3	91	24	135	7	48	3	2455	8	65	59	112	95	31
	B 20	299	0	0	58	15	64	8	31	4	1960	<5	58	60	40	114	37
	B 21	244	0	12	62	11	84	6	39	4	2032	9	47	55	30	26	37
Opoka	B 22a	26	0	4	76	21	118	10	48	13	206	5	6	124	4	185	26
	B 22b	18	0	1	51	16	63	12	46	10	392	<5	5	100	4	152	21
	B 22c	41	16	6	146	47	70	23	69	23	133	<5	8	184	8	295	51
Locality: Syria-Knefis																	
Age: Maastrichtian																	
Phosphorites	A 1	114	0	0	64	6	94	<3	42	<3	1445	8	75	152	82	158	14
	A 2	141	0	1	52	7	88	7	26	<3	1524	9	116	147	73	140	13
	A 3	55	0	9	119	<5	115	<3	23	<3	1227	7	74	114	86	126	19
Opoka	A 4	3	0	0	42	11	78	3	31	7	231	8	3	73	4	111	10
	A 5	4	0	3	43	6	53	4	32	9	161	8	3	94	<3	110	12
Locality: NW Iraq-near H3																	
Age: Paleocene																	
Phosphorites	I 4	0	0	6	118	<5	116	<3	33	<3	743	6	26	52	94	71	34
	I 5	0	0	3	161	11	142	4	48	<3	902	<5	45	111	170	108	33
	I 6	0	0	11	107	<5	157	4	37	<3	1083	7	41	68	232	53	31
Locality: NE Jordan																	
Age: Eocene																	
Phosphorites	U 1	2048	0	4	47	12	66	<3	23	<3	983	<5	39	138	28	31	9
	Kd 1	1931	0	9	12	<5	80	5	38	<3	1099	10	44	46	52	43	15
	K 1-2	92	0	8	35	30	94	<3	44	<3	638	9	56	236	29	142	10
	U 2	99	0	3	1	23	116	4	20	<3	687	6	13	78	12	140	10

The P_2O_5 values of the rather pure phosphorites from Al Bardeh and Knefis range from 22 to 25%, the silicified phosphorites show low P_2O_5 values between 9 and 20%. The phosphorites rich in calcite matrix (I 4-I 6) show intermediate P_2O_5 values. The new phosphorite deposit in NW Iraq (I 4-5) is characterized by fluctuating P_2O_5 values at high carbonate and LOI values. Similar P_2O_5 and other major element values in phosphorites were reported from Jordan (Bashir, 1977), from Israel (Shirav-[Schwartz] and Ginzburg, 1978), from southern Israel (Nathan et al., 1979) and from southeast Turkey (Lucas et al., 1979; 1980). The TiO_2, Al_2O_3, Fe_2O_3, and K_2O values are very low indicating only minor admixtures of clay minerals. The sulfur contents in the phosphorites fluctuate.

The chert samples are characterized by high SiO_2 values and relatively high values of TiO_2, Al_2O_3, and Fe_2O_3 (clay minerals). The P_2O_5 values of chert, however, are below 1% which conforms with the thin section analyses. Moreover, the chert samples (B 22a,b,c) are characterized by relatively high contents of Cu, Mo, Ni, Rb, Zn, and V and low contents in Ba, U, and Y, which demonstrated, in spite of the light color, an affinity in trace element content towards "black shales".

Marked differences in trace element content such as Ba, Sr, U and Y allow grouping for the same lithologies of the different locations as well as for different lithologies. Apatite is capable of concentrating considerable amounts of various trace elements (Altschuler, 1980; Bentor, 1980). Thus, Sr is concentrated in phosphorites (compare Bashir, 1977; Prévôt and Lucas, 1979).

Minor or moderate differences between phosphorites and cherts analyzed are shown by the following trace elements: Cu, Mo, Ni, Rb, Zn, Zr, Ce, Cr, La, and V. Mn values are low to nil in all rock types analyzed; they are much lower than those of the phosphorites from the Lower Saxony Basin.

The x-ray diffractometer analyses are compiled in Table 3. Apatite is uniform in composition. Normal hydroxyl-fluor-apatites (± carbonate) prevail as common in the Cretaceous/Tertiary phosphorites of the Mediterranean area (compare Prévôt and Lucas, 1979). Quartz predominates or is present in seven samples. Sample A 5 is characterized by opal-CT (lussatite), a poorly ordered crystalline modification of opal. This silica modification presumably points to a high-diagenetic alteration. Two phosphorite samples (Kd 1, I 4) lack silica. Calcite is the only carbonate present. Traces of kaolinite occur in the Maastrichtian samples (B 21, B 22a, B 22b) and of muscovite/illite in the Paleocene sample (I 4).

Paleogeographic Remarks

Observations are based on geological mapping within the Great Syrian Desert in the frontier area between south Syria, east Jordan

Table 3. Mineral composition determined by x-ray diffractometry

Rock	Sample	Main Composition	Minor Composition	Traces
Locality: Syria-Al Bardeh				
Age: Maastrichtian				
Phosphorite	B 20	Apatite, Quartz		
	B 21	Apatite	Quartz	Calcite, Kaolinite
Opoka	B 22a	Quartz		Calcite, Apatite, Kaolinite
	B 22b	Quartz	Calcite	Apatite, Kaolinite
Locality: Syria-Knefis				
Age: Maastrichtian				
Phosphorite	A 1	Apatite, Calcite		Quartz
Opoka	A 5	Opal-CT	Quartz, Calcite	Apatite
Locality: NW Iraq-near H3				
Age: Paleocene				
Phosphorite	I 4	Calcite	Dolomite, Apatite	Muscovite-Illite
Locality: NE Jordan				
Age: Eocene				
Phosphorite	U 2	Quartz	Apatite	
	Kd 1	Calcite, Apatite		
	K 1-2		Quartz Calcite Apatite	

and northwest Iraq (Fig. 1). They are supplemented by previous studies in Israel (Flexer, 1968; 1971; Nathan et al., 1979; Soudry and Nathan, 1980), in Jordan (Beerbaum, 1977), and in Syria (Atfeh and Faradjev, 1963). However, most of these publications refer to the lithology only; the origin of facies is discussed either not at all or incompletely.

During Late Cretaceous and Early Tertiary the Afro-Arabian Shield was rimmed by a broad carbonate shelf trending east-west (Winnock, 1980). The southern shelf of the Tethys was subdivided by shoals into small basins. Because of the complicated submarine relief, it is impossible to reconstruct the normal bathymetrical zonation of the depositional environment. Thus, a more descriptive classification of the facies types into outer, middle and inner shelf deposits is preferred. The typical association of sedimentary rocks in this shelf are (1) carbonate rocks of various types, mainly light-colored limestones, (2) chert beds and limestone layers with flint nodules, and (3) phosphorite accumulations. The phosphorites are part of the "Mediterranean Phosphorite Belt" that extends from Morocco to Turkey and forms the biggest concentration of phosphorite known in earth history (McKelvey, 1978; Bodelle, 1980; Notholt, 1980; Sheldon, 1980). The southern shelf of the Tethys is known best in northwest Africa from Morocco to Tunisia (Bodelle, 1980).

Reliable observations about the paleogeography of the Lower Maastrichtian in Syria were made only in the mountainous areas. Facies conditions in the Great Syrian Desert are unknown, except for

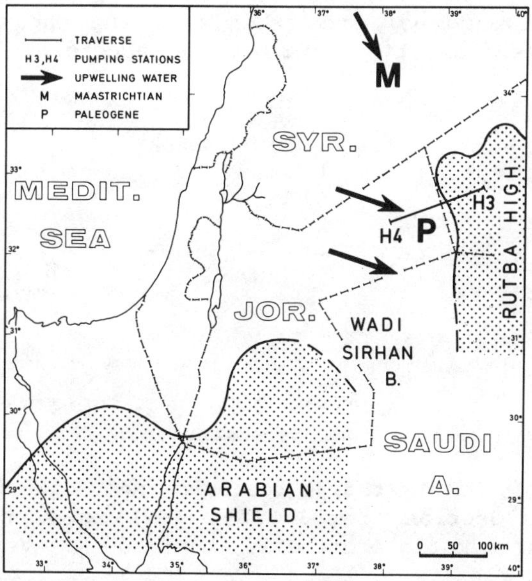

Fig. 1. Location map and general geological setting of the North Arabian shelf.

special features such as lagoonal sedimentation, etc., in northwest Iraq (compare Arambourg et al., 1959). Consequently, a facies reconstruction is not yet possible for this area. However, we assume that the facies resembles the one described by Flexer (1971). Better known is the lateral facies trend during the Eocene which was studied in east Jordan (E. Kemper, unpublished report, 1980). In the area of investigation the Wadi Sirhan Basin separates the Rutba High from the Arabian Shield proper (Fig. 1). The paleogeographic setting of the Upper Cretaceous/Lower Tertiary interval is exemplified best by two stages, the Maastrichtian and the Paleogene.

Maastrichtian Facies Patterns

The facies distribution of the Early Maastrichtian in the area under consideration is complex; shoals and basins show repeated lateral facies changes. Flexer (1971, Fig. 8) designed a model of the complicated shelf configuration for the Israel portion (Fig. 2). The main rock types are chalk and marly chalk; cherts occur mainly on shoals and in a nearshore belt. Cherts were called opoka or porcellanite in previous publications on the Maastrichtian of this area.

Of economic importance are the Lower Maastrichtian phosphorite accumulations (in Israel called Phosphorite Unit, see Atfeh and Faradjev, 1963; Nathan et al., 1979; Soudry and Nathan, 1980). The rock association of the northern Negev Desert with phosphorites, cherts, and chalks (classified best as a middle shelf deposit), resembles well the rock association observed in the Great Syrian Desert and in the Palmyrids, Syria.

The samples examined were taken from the phosphorite deposit Knefis, now in exploitation, (Khneifiss according to Atfeh and

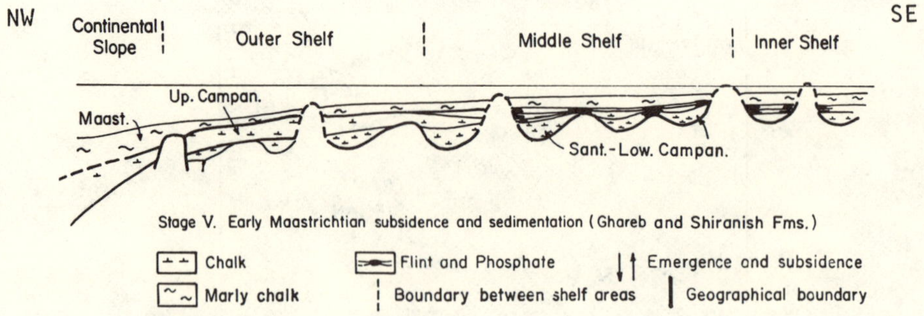

Fig. 2. Early Maastrichtian tectonic and sedimentary setting as shown in a cross section from Lebanon (NW) to southeastern Syria (SE). Length of schematic profile is about 500 km; vertical exaggeration 200X (after Flexer, 1971). By permission from Paleogeogr. Palaeochim. Palaeoecol., Copyright (c) 1971, Elsevier Publ. Co.

Faradjev, 1963) in Syria and from outcrops in the Palmyrids near Al Bardeh. Atfeh and Faradjev (1963) list the stratigraphic sequence of these phosphorite deposits. Typical of both locations is the alternation between chert and phosphorite horizons. These phosphorite deposits accumulated in a high-energy environment and consist of reworked sediments comparable with placer deposits. They are consolidated, fairly or poorly sorted, fine to coarse-grained, rarely granular to conglomeratic, locally also bimodal sandstones cemented by calcite or more rarely by silica. Their main components are detrital grains of reworked rounded phosphate crusts, phosphate nodules, phosphate pellets and ooids, fragments with phosphatized radiolarians, and bone phosphate (fish remnants). Quartz and other detrital minerals of sand size, however, are absent. Calcite dominates as cement; silica cement occurs also. The silica-cemented phosphorites show a somewhat higher P_2O_5 content than the calcite-cemented phosphorites (Table 1). Phosphate does not occur as transparent cement. The site of primordial formation or sedimentation of the detrital phosphorite components such as phosphate crusts, phosphate concretions and nodules, phosphate pellets, and bone phosphate is uncertain, but is probably nearby. Reworking and trapping (e.g., in small basins or other structural positions) are the necessary mechanisms for upgrading some of the phosphorite deposits. Such reworked deposits were formed on top and on the flanks of shoals analogous to the Miocene phosphorite sediments of Florida (Riggs, 1980). The thickness of the phosphorite horizons fluctuates markedly from a few centimeters to several meters. An autochthonous benthic fauna is rare. The amount of vertebrate remnants is rather high in the Maastrichtian. Particles of phosphatized radiolarians and foraminifers are absent at Knefis. The close association of these sand-grain sized phosphorites with laminated chert horizons indicates that high-energy conditions existed only for a short time and on a local scale.

Chert is one of the main sedimentary rocks of this time interval and characteristic of the middle shelf area. It is light-colored, light weight, porous, and laminated (mm-dimension), resembling a specific variety of chert called gaize in Europe, and is thoroughly silicified (60-80% SiO_2). Some cherts are composed of quartz, others of opal-CT and quartz. Bioturbation is absent. Siliceous microorganisms like radiolarians and diatoms, often poorly preserved, are the main constituents of the chert. In that respect the chert of the Syrian Maastrichtian may be compared with the radiolarian cherts of the Lower Carboniferous of Central Europe. On bedding planes phosphatic fish remains are common. Phosphatic micronodules are sparse and of microscopic size only. Silicification of biogenic limestone is inferred by Atfeh and Faradjev (1963).

The microfauna consists of very few species and small ones only. Specimens of benthic Buliminacea are common. Only a few planktonic foraminifers (*Globigerinelloides*) occur. Conspicuous is the absence of groups elsewhere predominant such as *Globotruncana*.

PLATE 1

Varietial features of Maastrichtian chert, Great Syrian Desert

1 Cryptocrystalline chert with biogenic phosphate particles (white), dispersed pyrite, and foraminifers. Bar length: 200 µm. Photomicrograph, parallel nicols, TS 32401. Maastrichtian, Al Bardeh (B-22a).

2 Cryptocrystalline chert with limonitic conic radiolarian and biogenic phosphate particles (black arrows). Bar length: 100 µm. Photomicrograph, parallel nicols, TS 32401. Maastrichtian, Al Bardeh (B-22a).

3 Cryptocrystalline chert with a lobate contact between a dark-colored (opal CT) and a light-colored (quartz + clay) layer. The lobate front resemble a "hardground" as common in carbonate rocks. Bar length: 200 µm. a = parallel nicols; b = crossed nicols. Photomicrograph, TS 32408. Maastrichtian, Knefis (A-5).

4 Siliceous matrix of chert displaying a characteristic "cauliflower texture". Bar length: 1 µm. Scanning electron micrograph, magnification about x 20000. Maastrichtian, Al Bardeh (B-22a).

5 Chert with minute network of submicroscopic chain-like fibers, less than 0.1 µm in diameter, composed of silica. A biogenic origin of this silica precipitation seems to be likely. Bar length: 0.5 µm. Scanning electron micrograph, magnification about x 50000. Maastrichtian, Knefis (A-5).

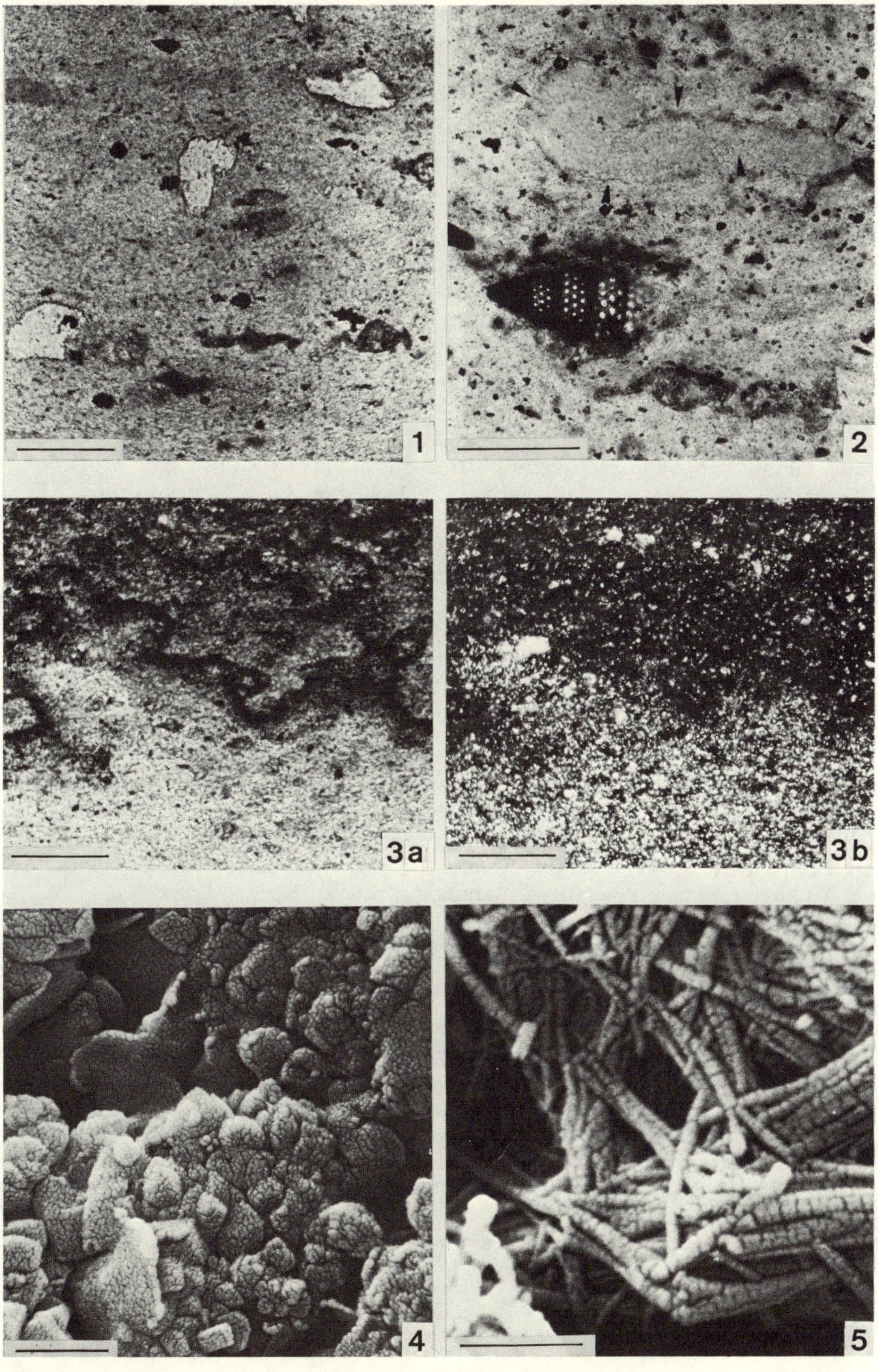
1
2
3a
3b
4
5

Varietal features of the Maastrichtian chert as seen in thin sections and under the scanning electron microscope are depicted in Plate 1. Lithology and faunal content indicate that the chert was deposited in a low-energy and low-oxygen environment. Presumably, this chert is coeval with the oil shale observed on the North Arabian shelf (compare Shirav-[Schwartz] and Ginzburg, 1978). The genetic relationship between oil shales and phosphorite accumulations, as stressed in the "Geologie comparée des gisements de phosphates et de pétrole" (Bodelle, 1980), is well documented in this stratigraphic interval and area.

Paleogene Facies Patterns

Paleogene strata were studied in the region between the Rutba High (NW Iraq) and the Hauran basalts (NE Jordan). The stratigraphic sequence ranges from Paleocene near pump station H3 to Late Middle Eocene near pump station H4 (Fig. 1). The sediments of this area were deposited in an epicontinental basin, the so-called Wadi Sirhan Basin. Based on mapping, subsurface data from wells, and micropaleontological studies, three facies belts can be distinguished (Fig. 3).

Belt 1, corresponding to the neritic to littoral zone, is composed of light-colored dolomitic limestones and coarse crystalline limestones, nonfossiliferous or with rich benthic communities (nummulites and bivalves).

Belts 2 and 3 contain sediments of the deeper neritic zone. Belt 2 is mainly composed of cherts and dolomites with a fossil association indicating slight oxygen deficiency (i.e., Buliminacea assemblage with no other benthic fossils). Siliceous microplankton (radiolarians and diatoms) and pteropods occur in belt 2 as well as in belt 3. Belt 3 comprises radiolarian marls and chalky limestones with flint nodules and marls. The benthos is more diversified. In belt 3 phosphorites are not as abundant as in belts 1 and 2.

The Paleocene phosphorite deposit in belt 1 north of pump station H3 (NW Iraq), unknown up to now, is several meters thick and composed of friable to slightly consolidated, fairly to poorly sorted, fine to coarse-grained, slightly bimodal phosphate sand cemented by calcite. The detritus consists mainly of phosphate grains comprising reworked rounded phosphate crusts, phosphate nodules, oval phosphate pellets and ooids, and rounded bone phosphate. Phosphatized foraminifers and radiolarians are absent. Solitary dolomite rhombs and detrital quartz grains are dispersed in the calcite matrix. Silica cement is absent. The P_2O_5 content is only moderate due to the high calcite content; it fluctuates between 10 and 21% (Table 1). Carbonate-rich layers alternate with phosphate-rich layers.

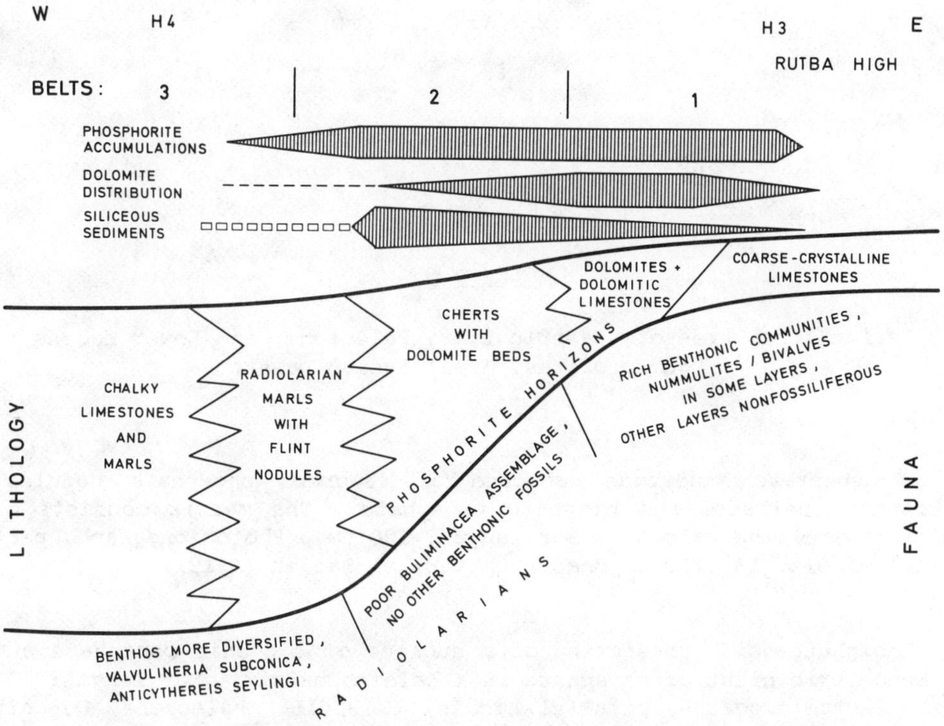

Fig. 3. Paleogene facies pattern showing three facies belts in a west-east cross section, Great Syrian Desert.

In contrast with the above phosphorite deposit, the Lower Eocene phosphorite deposits of belt 2 (5 miles NW from At Tanf, Syria and Wadi Ruweished, Risha region, NE Jordan) are thinner and much harder. These phosphorites are fairly to poorly sorted, fine to coarse-grained, locally conglomeratic sandstones with primary calcite cement. The detrital components comprise mainly well-rounded, detrital bone phosphate (fish remains), phosphatic micronodules, and phosphate pellets with pteropods, foraminifers, or radiolarians. The original calcite matrix subsequently has been more or less silicified. The P_2O_5 content fluctuates between 10 and 20%. The silica content reaches up to 70% (Table 1).

Varietal features of the Maastrichtian and Paleogene phosphorites as seen in thin sections, are depicted in Plate 2.

Interpretation as Upwelling in the Southern Tethys (Arid Zone)

The Upper Cretaceous/Lower Tertiary sedimentary sequence of the Great Syrian Desert was deposited at a paleolatitude of about 25° north (after Owen, 1976). Accordingly, a subtropical climate is pos-

PLATE 2

Varietal features of Maastrichtian, Paleocene, and Lower Eocene phosphorites, Great Syrian Desert

1 Phosphorite sandstone composed of reworked phosphate nodules, phosphate pellets, and biogenic phosphate. The matrix consists of microcrystalline calcite. Bar length: 200 µm. Photomicrograph, parallel nicols, TS 32398. Maastrichtian, Al Bardeh (B-19).

2 Phosphate ooids consisting of a nucleus of biogenic phosphate and a concentric crust of phosphate in a calcite matrix. Bar length: 100 µm. Photomicrograph, parallel nicols, TS 32415. Paleocene, Deposit of H3 (NW Iraq).

3 Calcitic foraminifers and radiolarians in a calcite matrix. The chamber filling consists of crypto-crystalline phosphate (black, under crossed nicols). This petrographic assemblage resembles the "phosphate peloids" depicted by Soudry and Nathan (1980). Bar length: 200 µm. Photomicrograph, TS S843, a = parallel nicols; b = crossed nicols. Lower Eocene, Wadi Ruweishid (Kd-1).

4 Close-up view of detrital phosphorite particles cemented by microcrystalline silica. Note detrital quartz grain (center) surrounded by secondary quartz overgrowth in optical continuity. Bar length: 100 µm. Photomicrograph, TS 32406, a = parallel nicols; b = crossed nicols. Maastrichtian, Knefis (A-3).

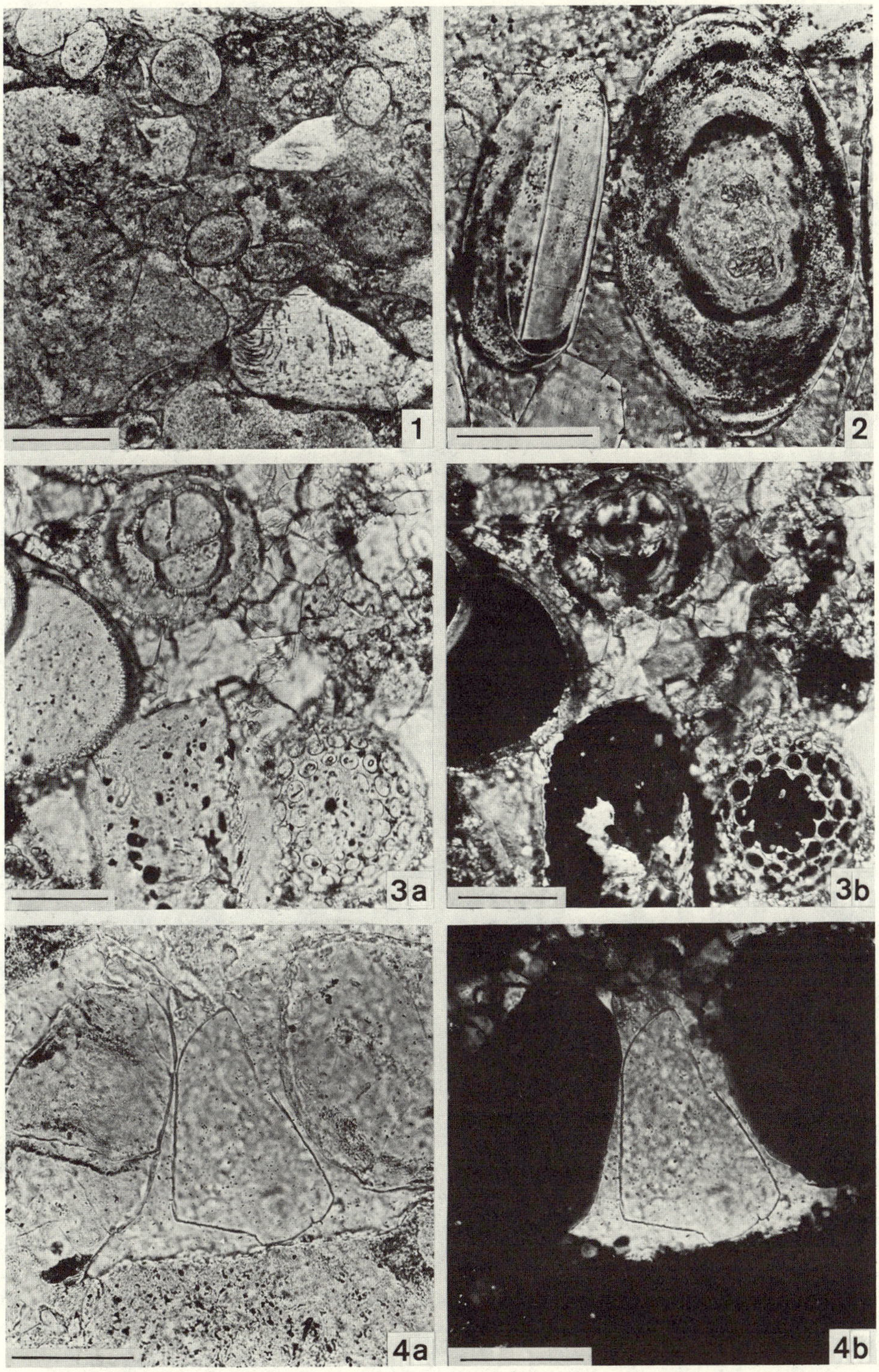
1
2
3a
3b
4a
4b

tulated. Sediment transport from the ancient continent was minor and sedimentation was chiefly restricted to chemical precipitation and accumulation of marine organisms.

It is emphasized that in spite of the low paleolatitude, benthic warm water faunas are absent. This refers especially to coral reefs and large foraminifers with complex test structures such as *Alveolina*, *Coskinolina*, *Dictyoconus*, *Somalina*, *Lockhartia*, etc., which are characteristic of warm water deposits. They are abundant on the eastern shelf of the Arabian Shield.

The benthic faunas are little diversified. In belt 1 communities of *Nummulites* and *Operculina* or associations of bivalves occur. Belt 2 is characterized by the predominance of Buliminacea and the absence of arenaceous foraminifers. The benthic faunas of belt 3 are richer and more diversified than those of belts 1 and 2.

Among planktonic microorganisms radiolarians are frequent, however their skeletons are commonly corroded or completely altered. Diatoms, presumably abundant at deposition, were also altered. Relics of frustules are still observed. The silica of these skeletons is concentrated in flint nodules or chert beds. Even carbonate rocks became silicified as a consequence of this silica mobilization. The Buliminacea associations, poor in individuals, are indicative of a poorly oxygenated environment of the deeper water, especially in belt 2. In the Great Syrian Desert area the Buliminacea associations are among the few indications of an oxygen deficiency. In other regions of the North Arabian shelf during Upper Cretaceous/Lower Tertiary, not only were black shales deposited, but also oil shales rich in organic carbon (Jordan, Israel, Tunisia).

The benthic faunas of belts 2 and 3, which are rather poor in individuals and species, contrast markedly with the enrichment of relics of large nekton, especially fishes. Such a concentration can be explained only by mass extermination or low sedimentation rate of any other material.

Glauconite, however, is absent or sparse in all belts; it thus contrasts with the Aptian/Albian of the boreal realm in NW Germany. Another feature typical of this lithofacies is the phosphorite deposits which consist of reworked bone phosphates (fish remains) and phosphate pellets, ooides, nodules, and crusts. The high-grade phosphorites, mostly fine to very coarse-grained or conglomeratic phosphatic sandstones, are of detrital origin and mechanically upgraded. Late diagenetic silicification of the phosphorites downgrades them.

According to Bentor (1980) "low values of Ce in apatite from phosphorites reflect the influence of deep-ocean water." Thus, the absence or the low values of Ce (Table 2) provide additional support for the assumption that the phosphorites formed in areas of oceanic upwelling. The phosphorite-chert association is typical of the "Mediterranean Phosphorite Belt." It occurs in Morocco, Algeria, and Tunisia (Winnock, 1980) and markedly contrasts with the facies development in the northern part of the Tethys (facies asymmetry). However, the cyclic arrangement of phosphorite sequences alternating with high silica rocks is not yet well understood (Bentor, 1980).

A cross section of Maastrichtian, Paleocene, and Lower Eocene sediments from the Atlantic to the High Atlas was shown by Il'in (1978) with the emphasis on the origin of north African phosphorites. The facies patterns described indicate upwelling for the following reasons: rather cool water faunas at low paleolatitude, local oxygen deficiency, common siliceous microfossils (diatoms and radiolarians), abundance of phosphatic fish relics, ubiquity of upgraded phosphorites, the low Ce content of the phosphorites, and the large-scale facies asymmetry across the Tethys.

APTIAN/ALBIAN FACIES PATTERNS IN THE LOWER SAXONY BASIN, NW GERMANY

Analytical Data

A detailed facies analysis of Upper Aptian and Lower Albian sedimentary rocks from the Lower Saxony Basin in northwest Germany will

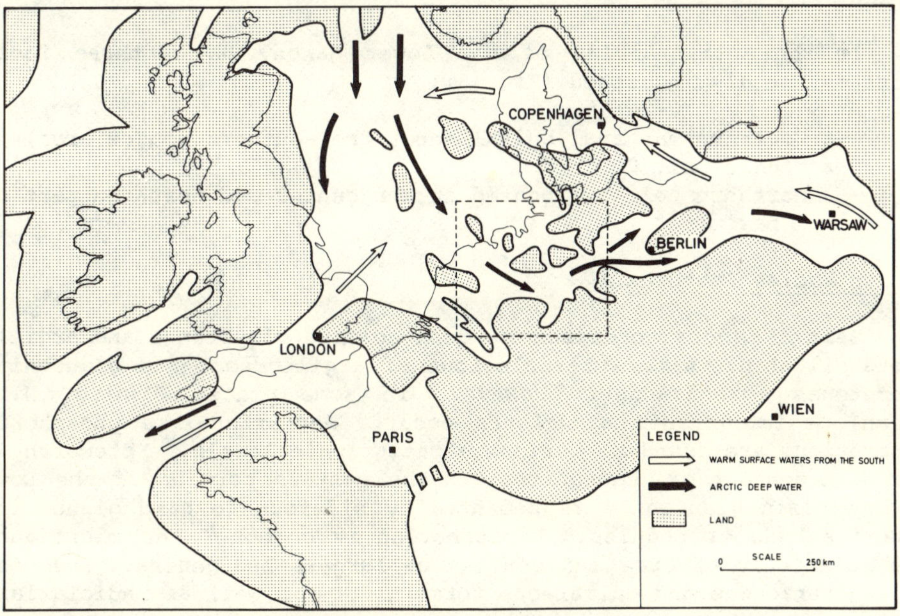

Fig. 4. Lower Cretaceous paleogeography of northwestern Europe with major currents (modified from Ziegler, 1981). Dashed rectangular insert shows contour of Fig. 5.

be published soon in a special volume (Kemper, in press). This volume will contain contributions on the micro- and macropaleontology, sedimentology, thin section and x-ray analysis, inorganic and organic geochemistry of the above interval. These unpublished data have been considered in the following evaluation; however, no additional data are included as in the Analytical Data in the prior section.

Paleogeographic Remarks

The paleogeographic setting of northwest Europe during Early Cretaceous is a triangular depositional area with a southern east-west trending coast of lobate configuration (Fig. 4). The area of investigation proper comprised mainly the quadrangle Hannover - Braunschweig - Goslar - Alfeld in the eastern portion of the Lower Saxony Basin (Fig. 5), northwest Germany. The Lower Saxony Basin was a marginal sea of 300 km length and 60-70 km width connected with the North Sea of that time. The basin was strongly subsiding from Late Jurassic to Late Early Cretaceous. During Early Cretaceous, sedimentation of pelitic sediments predominated; marlstones, siltstones, and sandstones are sparse. Sandstones were brought into the basin mainly from the south. The Pompeckj Swell to the north did not contribute large quantities of sediments. A schematic north-south cross section (Kemper, 1979) through the basin shows the asymmetric distribution of facies (Fig. 6). For such a small epicontinental sea, one that is smaller than the Baltic Sea, normal bathymetric zonation cannot be applied.

In the Aptian/Albian of the Lower Saxony Basin three facies belts may be distinguished (Fig. 5):

--a southern marginal belt 1 (near the southern source area)
--a middle belt 2
--a northern belt 3 (located in the center and northern part of the basin)

Aptian Facies Patterns

Belt 1 is characterized by dark-colored claystones and argillaceous siltstones with intercalations of argillaceous and glauconitic sandstones (greensands). However, in some places (Osning, Hils) clean, medium-grained sandstones occur. Septarian-like phosphorite concretions are found at various stratigraphic levels. Noteworthy is the high content of glauconite. In the western portion of the Lower Saxony Basin glauconite is cemented by siderite to hard glauconite-siderite beds or nodules. Bioturbation is common. The microfaunas are relatively diverse and consist of large-sized genera. They comprise arenaceous and calcareous foraminifers as well as radiolarians. In several layers the microfaunas are poor in species. Relics of siliceous sponges occur locally (Krüger, 1978). Giant ammonites with siderite steinkerns are characteristic megafossils.

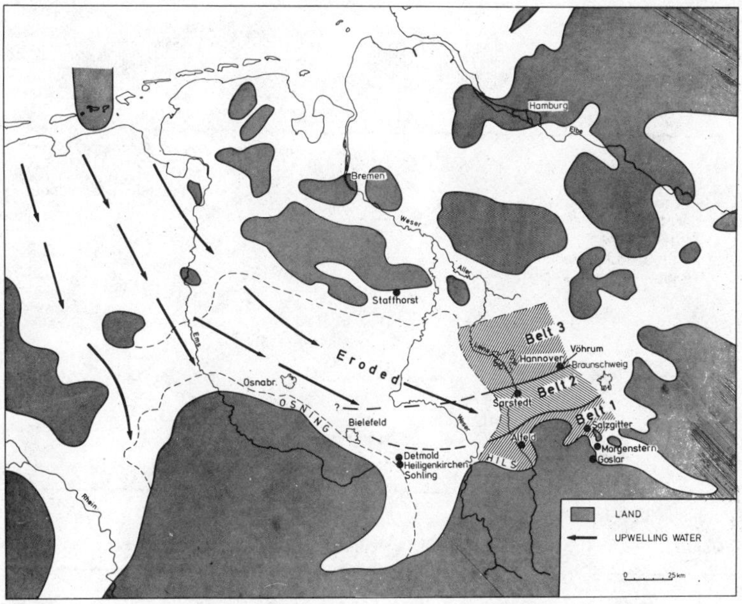

Fig. 5. Coast lines during Aptian/Albian in the Lower Saxony Basin, northwest Germany. The arrows show the general direction of the inferred upwelling. The three depositional belts are marked. Locations cited in the text.

Belt 2 is mainly characterized by dark-colored claystones. C-org values are slightly higher than in the adjacent belts. Glauconite is nearly absent. Conspicuous is the high content of smectite among the clay minerals (Gaida et al., 1981). The claystones are commonly composed of fecal pellets. Layers of actual laminated "black shales" do not occur. The microfauna is characterized by an association of arenaceous foraminifers of low diversity and siliceous microplankton (diatoms and radiolarians). Diatoms are mainly preserved as pyrite steinkerns as the tender frustules of the siliceous plankton tended to be dissolved during early diagenesis. Sponge spicules may occur. Calcareous foraminifers and macrofossils are absent or rare. The faunal and floral assemblage is indicative of a rather specialized oxygen-poor microenvironment. Accretionary phosphorite nodules, commonly associated with clay ironstone concretions, are less common than in the northern and southern belt. Bioturbation is common.

The lower Upper Aptian (Gargasian) section of belt 3 is mainly composed of light-colored, partly varicolored marlstones. The upper Upper Aptian (Clansayesian), however, consists of dark-colored claystones with a moderate content of organic carbon and with numerous horizons of earthy phosphorite or siderite-phosphorite concretions.

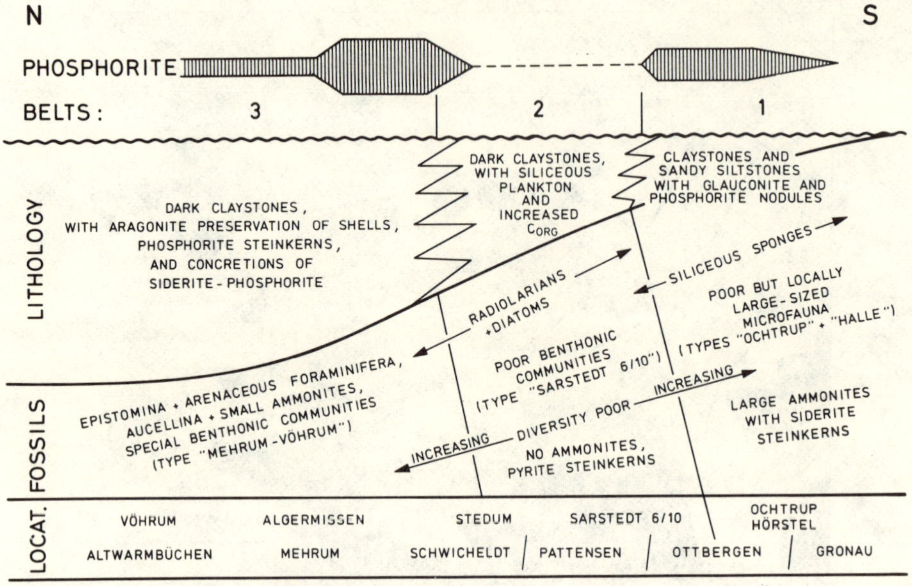

Fig. 6. Schematic north-south cross section showing the facies patterns in the three depositional belts of the boreal Aptian/Albian, northwest Germany.

Smectite is absent or rare. The macrofauna consists of rich associations of epibiontic suspension feeders, especially *Aucellina*, and numerous small ammonites, commonly with steinkerns of siderite-phosphorite. The microfaunas are rich and diverse, but arenaceous foraminifers still predominate. Bioturbation is widespread. Glauconite is rare; locally it shows spotty and lenticular distribution, especially in tuffaceous horizons (Gaida, Kemper and Zimmerle, 1978). It is less granular than in the southern belt.

Albian Facies Patterns

The facies asymmetry of the Albian resembles that of the Aptian. Belts 1 and 2 are narrow and occur on the southern third of the Lower Saxony Basin; belt 3 covers two-thirds of the basin, e.g., the central and northern parts. The asymmetry in the Albian is also documented by the different lithostratigraphic terms for the central/northern part and for the southern part of the Lower Saxony Basin (Fig. 7).

Belt 1 is characterized by greensands and glauconitic clays (Minimum Greensand, Hils Clay, Osning Greensand, Altenbeken Greensand, etc.) which contain siliceous sponge spicules and radiolarians.

Diatoms, presumably present originally, are not preserved. Spiculitic sandstones with chert nodules and low in glauconite (Hils Sandstone) were deposited locally. Phosphorite nodules are common. Smectite also occurs in varying concentrations (Kull, 1979; Gaida et al., 1981); however, coeval sediments in the northern portion of the basin (e.g., Staffhorst mine shaft) do not contain smectite.

Belt 2 is mainly composed of marlstones and the so-called Flammenmergel of Late Albian age which was deposited in the Osning, Hils, and Salzgitter area (Lehmann, 1954; Jordan, 1968; Jordan and Schmidt, 1968). The Flammenmergel is a biogenic siliceous marlstone that is pyritic, glauconitic, and thoroughly bioturbated. Phosphorite nodules are rare. The clay size fraction is characterized by a marked smectite content (Jordan, 1968). Noncalcareous silt or sand particles are sporadic.

Dominant biogenic constituents are siliceous sponge spicules, frequently associated with radiolarians and diatoms (Krüger, 1978). Planktonic foraminifers, the chambers of which are filled with pyrite, are common. Arenaceous foraminifers occur mainly in the argillaceous layers of the Flammenmergel. The benthic communities are characterized by low diversity. The siliceous matrix consists of chalcedony or quartz with a particle size of a few microns. Locally, silica is concentrated in flint nodules. Opaline silica of the sponge spicules is rarely preserved; mostly it is replaced by calcite or transformed into glauconite or chalcedony. Abundant siliceous sponge relics indicate the presence of suspension feeders. In addition, the strong bioturbation shows the existence of rich animal life on the sea floor. Thus, oxygen and nutrient supply were sufficient to support benthos life. Oxygen depletion was minor.

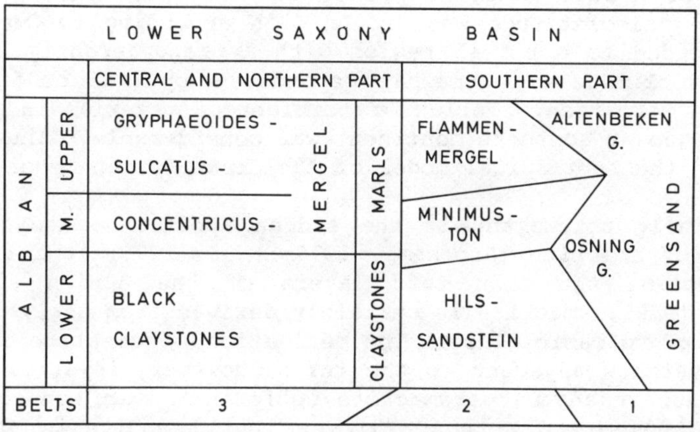

Fig. 7. Albian rock units in the Lower Saxony Basin, northwest Germany.

In the Anglo-Paris Basin, coeval sediments are known as malmstone or as gaize in Poland (Sokolowski, 1976) and in the USSR (Nalivkin, 1973) as opoka. This conspicuous sediment type is widespread in northern and central Europe within an east-west extending belt.

Belt 3 is mainly composed of thoroughly bioturbated, light-gray marlstones, varicolored near the base. Small phosphorite concretions are common. Smectite occurs in varying amounts. Epibiontic suspension feeders such as *Inoceramus* (in the middle and basal Upper Albian) and *Aucellina* (in the uppermost Upper Albian) predominate. Ammonites are frequent. They are commonly preserved as phosphorite steinkerns. The microfaunas are rich and highly diverse. The faunal assemblages point to optimal biotopes, sufficiently aerated and rich in nutrients.

Varietal features of some Aptian and Albian phosphorites and of the Flammenmergel from the Lower Saxony Basin are depicted in Plate 3. Note the striking similarity between the submicroscopic structures of the Flammenmergel (Plate 3, Fig. 5) and the chert (Plate 1, Fig. 4) regardless of the stratigraphic position.

Interpretation as Upwelling of Boreal Waters (Humid Zone)

The Aptian/Albian facies belts described are considered to be part of an upwelling system that is directed from north to south. According to our opinion, the system was essentially influenced by (1) the physiography of the marginal sea and (2) by the paleocirculation system nurtured by cool waters coming from the north. Its extension and intensity changed in the course of time responding to climatic changes which influenced circulation intensity. A strong drop of temperature (Kemper and Schmitz, 1981) which caused the influx of cool Arctic waters happened in the Late Aptian (Clansayesian). The paleolatitude was around 47°N according to Owen (1976). It corresponded to a boreal region with a preponderantly humid climate. This climate was somewhat beyond the optimal conditions for phosphorite formation. Influx of terrigenous material in suspension from the adjacent southern continent was considerable. Thus, it contrasts with the subtropical model of the Great Syrian Desert.

Especially noteworthy is the sudden appearance and widespread occurrence of smectite (Brockamp, 1976; Kull, 1979) at the beginning of the Aptian. In thin tuff layers of the Aptian and Albian (Zimmerle, 1979), smectite is certainly derived from early diagenetic alteration of volcanic glass. The derivation of smectite in the ordinary claystones adjacent to the tuffs, however, is still disputed. The first author advocates smectite neoformation in connection with upwelling; the second author believes that smectite is of detrital and diagenetic origin.

The organic carbon content (C-org) is low to moderate (0.5-1.5%) (Gaida et al., 1981). Glauconite is abundant on the southern margin of the basin. Marginal occurrence of glauconite is also known from the Aptian and Albian of Great Britain, France, and Poland.

Phosphorite concretions are common and widespread except for the dark-colored claystones of belt 2 with elevated C-org values. The phosphorites are characterized by a globular apatite matrix (Paproth and Zimmerle, 1980). Similar microstructures describe Burnett, Veeh and Soutar (1980). The association of phosphate, manganese carbonate, and siderite, a so-called remineralization assemblage, might well indicate, in accordance with observations in the Baltic Sea by Suess (1979), a terrestrially influenced marginal sea environment. But the environment of the Aptian/Albian is not as poorly oxygenated as that in the Landsort Deep of the central Baltic Sea. In structure and composition the Upper Aptian phosphorite nodules (Vinken, 1977; Paproth and Zimmerle, 1980) resemble Recent phosphorite nodules as described by Dietz, Emery and Shephard (1942), Birch (1980), Cullen (1980), and Burnett et al. (1980) except for the considerable amounts of Ca-Fe-Mn-carbonates in the Aptian/Albian phosphorites.

The synsedimentary to early diagenetic mineral assemblage manganese carbonate, siderite, phosphate, smectite, and glauconite together with the organic carbon and low calcite content reflect specific pH and Eh conditions. The particular faunal and floral assemblages, i.e. arenaceous foraminifers, siliceous microplankton, siliceous sponges in belts 1 and 2, and abundant suspension feeder communities in belt 3, are mainly controlled by nutrient supply, but presumably also by pH and Eh (compare Moorkens, 1976). Both patterns are indicative of upwelling (Diester-Haass, 1978).

Similar sedimentary rock and mineral associations (Sokolowski, 1976) as well as faunal and floral fossil assemblages, appeared in most European basins in Late Aptian and Albian times as a consequence of increasing current activity triggered by cooling (Kemper and Schmitz, 1981). These associations extend from Great Britain and France in the west to as far as the Caucasus, Caspian Sea, and even Iran in the east. Also, glauconite is widespread and abundant in the Aptian/Albian interval throughout Europe. Likely, the presence of smectite favored the neoformation of glauconite (Jeans et al., 1982). Presumably, intensity and distribution of this unique rock and mineral assemblage varies depending upon the local orography and the paleocurrent systems. Future geochemical studies should help to support these interpretations.

CONCLUSIONS

The facies patterns described above can be explained neither by bathymetrical zonation alone nor by unilateral supply from the conti-

PLATE 3

Varietal features of Aptian and Albian phosphorites and of the Flammenmergel from the Lower Saxony Basin in northwest Germany

1 Accretionary siderite-phosphorite nodule as seen in a polished cross section. Note the accretion layers of microcrystalline siderite with fluctuating apatite admixture. Light-colored crust shows highest concentration of P_2O_5. Lower Albian, Vöhrum Claypit.

2 Phosphorite nodule with crypto- to microcrystalline phosphate-clay-carbonate matrix, a detrital feldspar grain with secondary overgrowth, and a detrital pigmented volcanic rock fragment. Bar length: 100 µm. Photomicrograph, parallel nicols, TS 29153. Upper Aptian, Morgenstern Openpit Mine.

3 Thoroughly bioturbated Flammenmergel shown in polished handspecimen. Upper Albian, Morgenstern Openpit Mine.

4 Several round radiolarians (black arrows) embedded in dark-colored claystone completely replaced by blocky clinoptilolite (dark gray to black under crossed nicols). Bar length: 200 µm. Upper Aptian, Sarstedt Claypit (O. Gott). Photomicrograph, TS 29969, a = parallel nicols; b = crossed nicols.

5 Siliceous matrix of the Flammenmergel displaying a characteristic "cauliflower texture". Bar length: 1 µm. Scanning electron micrograph, magnification about x2000. Upper Albian, Heiligenkirchen-Schling.

6 Recrystallized concretion in the Flammenmergel displaying euhedral quartz crystals with rough crystal faces. Bar length: 2 µm. Scanning electron micrograph, magnification about x1000. Upper Albian, Heiligenkirchen-Schling.

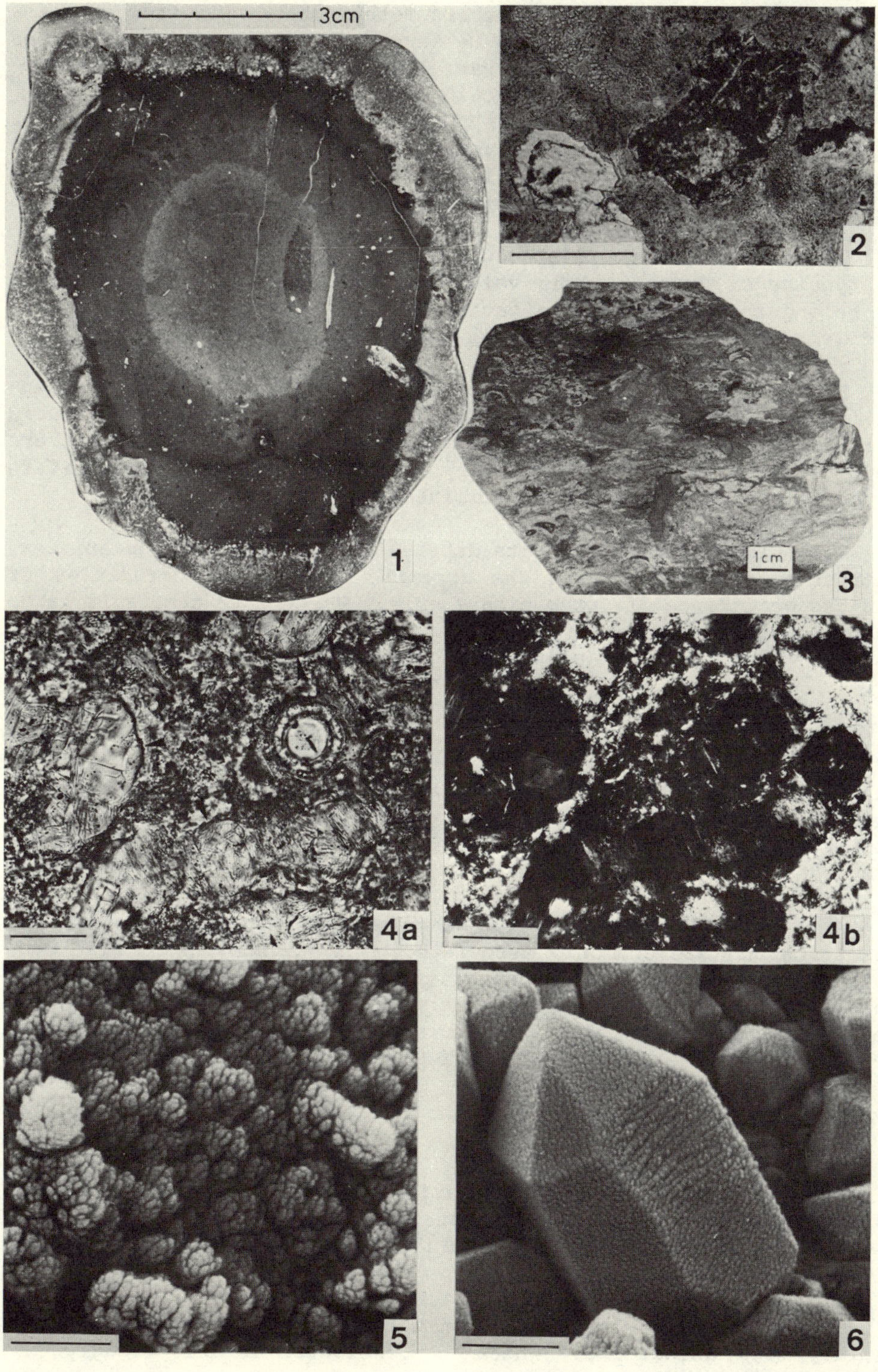
3cm
1
2
1cm
3
4a
4b
5
6

nent. For instance, in the southern Tethys the supply from the continent was negligible. Thus, we believe that both examples of facies associations and their lateral variations are caused by coastal upwelling as they exhibit the following features:

(1) Abundance of siliceous plankton (diatoms and radiolarians) in a nearshore belt; however, the siliceous fossils have in part been transformed into flint nodules or chert beds.

(2) Presence of phosphorite accumulations (in part of the accretionary type *sensu* Burnett, this volume). The phosphorite samples studied belong to the "phosphate facies" (Notholt, 1980), which is regarded to be diagnostic of coastal upwelling.

(3) Rather high supply of organic matter leading locally to the deposition of black shales, or even oil shales. Even chert beds devoid of benthic fossils or with poor assemblages of benthic foraminifers are indicative of oxygen deficiency (certain arenaceous foraminifers in the Lower Saxony Basin and Buliminaceae in the Tethys).

(4) Accumulations of vertebrate debris, locally in large quantities, contrasting remarkably with the absence of benthic communities or with the occurrence of foraminifers only. Vertebrate debris in large quantities indicates mass mortalities, which are known to happen, for instance, as a result of plankton blooms.

(5) Rich association of suspension feeders common in more seaward belts.

Each single criterion might be explained otherwise but all criteria together and the lateral succession of the different facies patterns point strongly to coastal upwelling. Moreover, our examples demonstrate some other important facts:

(1) Authigenic phosphorites and claystones with marked contents of organic carbon are commonly associated, but maximum phosphorite concentrations are not found in beds with maximum concentrations of organic carbon. Environments with very high C-org values even impede precipitation of authigenic phosphorite as shown by Diester-Haass (1978) and Slansky (1980). Also off Peru, peak phosphorite formation is confined to the upper and lower boundary of the oxygen minimum layer of the open ocean (Soutar and Burnett, 1979; Burnett et al., 1980).

(2) The phosphorite-chert association may be typical of some upwelling areas such as those postulated for the Phosphoria Formation, the "Mediterranean Phosphorite Belt", and the Chatam Rise (South Pacific).

(3) In the boreal realm, such as in the Lower Saxony Basin, the sedimentary rock association, which is considered to be strongly in-

fluenced by upwelling, is characterized by a marginal facies rich in glauconite. These greensands and glauconitic claystones contain phosphorite concretions ("phosphate-glauconite-association" *sensu* Burnett, 1980 and Notholt, 1980) and locally also sponge spicules. Upwelling may facilitate the formation of glauconite, even if some glauconite is derived from volcanogenic debris as postulated for the Lower Cretaceous glauconites of England (Jeans et al., 1982).

(4) Smectite is an important clay mineral in the upwelling systems described here. But more observations are necessary to explain the origin of smectite in upwelling systems.

Parameters for upwelling, classified according to basic factors and dependent features, as well as their distribution in the three facies belts of the Tethyal Arabian Shelf and of the Boreal Lower Saxony Basin were compiled in a comparative chart (Fig. 8). Basic factors controlling the depositional environments in the different facies belts of upwelling systems are (1) richness of nutrients (phosphorus, silica) associated with upwelling of cool water leading (2) to mass production of planktonic organisms. The dependent features are related to the fate of the biomass and to the availability of oxygen. In the area of maximum bioproduction (belt 2) oxygen consumption by decaying organic matter is so high that life on the bottom is either impossible or very restricted. The existence of well-developed oxygen-minimum layers is known from many sites in the open ocean. Dark rock colors prevail. Anoxia may lead to black shale deposition. Oxygen consumption by decaying organic matter was lower

		NORTH ARABIAN SHELF	LOWER SAXONY BASIN
BASIC FACTORS	PALEOLATITUDE / FAUNAL REALM	25° N / TETHYAL	47° N / BOREAL
	CLIMATE	SUBTROPICAL - ARID	TEMPERATE - HUMID
	TEMPERATURE OF SURFACE WATER	WARM	TEMPERATE - COOL
	CHARACTERISTIC GEOGRAPHICAL FEATURES	EXTENSIVE SHELF OF A MAJOR OCEAN SUBDIVIDED BY SILLS	SOUTHERN MARGIN OF AN EPICONTINENTAL SEA
DEPENDENT FEATURES	INTENSITY OF UPWELLING	HIGH	MODERATE
	DOMINANT ROCK TYPE	CARBONATES (CHALKY LIMESTONES)	CLAYSTONES
	ROCK COLOR	LIGHT	DARK
	SOURCE OF DETRITAL PARTICLES	LESS TERRIGENOUS	MORE TERRIGENOUS
	IMPORTANT MARGINAL FACIES	CHERT, FLINT	GLAUCONITE
	TYPE OF PHOSPHORITE	DETRITAL (BONES, PELLETS, MICRONODULES)	CONCRETIONARY (ASSOCIATED WITH SIDERITE)
	BENTHONIC FORAMINIFERA	EXCLUSIVELY CALCAREOUS	PREDOMINANTLY ARENACEOUS

Fig. 8. Comparison of the two ancient upwelling systems from warm and cool oceans.

in belts 1 and 3. Hence, living conditions were good and faunal diversity is high. In belt 1 the environments varied. Endobiontic substrate feeders are frequent in some places. Suspension feeders are scarce. If present, they are represented by siliceous sponges in the boreal realm. On the North Arabian shelf bivalve and *Nummulites-Operculina* communities may occur. Living conditions are especially good in belt 3 which is indicated by the abundance of suspension feeders in the boreal seas and the deposition of chalky limestones in the Tethys. Additionally, Fig. 8 shows the sites where phosphorite, glauconite, and silica accumulated.

The upwelling systems of the North Arabian Shelf, representing a fossil example of a warm shallow sea, is compared to the Lower Saxony Basin, representing a fossil example of a cool shallow sea (Fig. 9). The sedimentary rock association of the Tethyal Late Cretaceous/Early Tertiary is characterized by light-colored, non-glauconitic limestones and dolomites, cherts, and detrital phosphorite concentrations. The Tethyal upwelling system was of larger scale than that of the Late Lower Cretaceous of NW Germany. The boreal Aptian/Albian upwelling system described is characterized by dominant claystones, which are locally silty and very glauconitic, minor siliceous intercalations, and concretionary phosphorites. Most features known from upwelling systems at low latitude are also observed at higher latitudes, however they are less distinct. This is explained by the weaker intensity of upwelling at higher latitudes and by the marked admixture of terrigenous particles in the rainy-humid climate.

In concluding, it has to be stressed that the actualistic approach is limited. The major coastal upwelling sites of Recent oceans are oriented north-south; moreover, they are located on narrow shelves with rather steeply dipping continental slopes. Contrary to the best known Recent examples, the ancient examples of coastal upwelling described here occur in belts extending east-west. This is most obvious for the Upper Cretaceous/Lower Tertiary "Mediterranean Phosphorite Belt". Another contrast with Recent open oceans is the limited extent of ancient seas in which upwelling is thought to have occurred, and their strong subdivision by sills and platforms. Thus, the effects of upwelling, such as the formation of phosphorite, are much more pronounced in small epicontinental seas or highly subdivided oceans of limited extent (e.g., the western Tethys) than in Recent open oceans. Apparently, cold waters from greater depths cross sills or penetrate even through narrow seaways into marginal seas of the east-west trending major oceans, forming there reservoirs of cold bottom water.

According to this hypothesis, differences from the present day circulation pattern are postulated for Late Cretaceous/Early Tertiary times. Part of the Canary Current, which is today passing down the continental margins of the Iberian Peninsula/North Africa to the

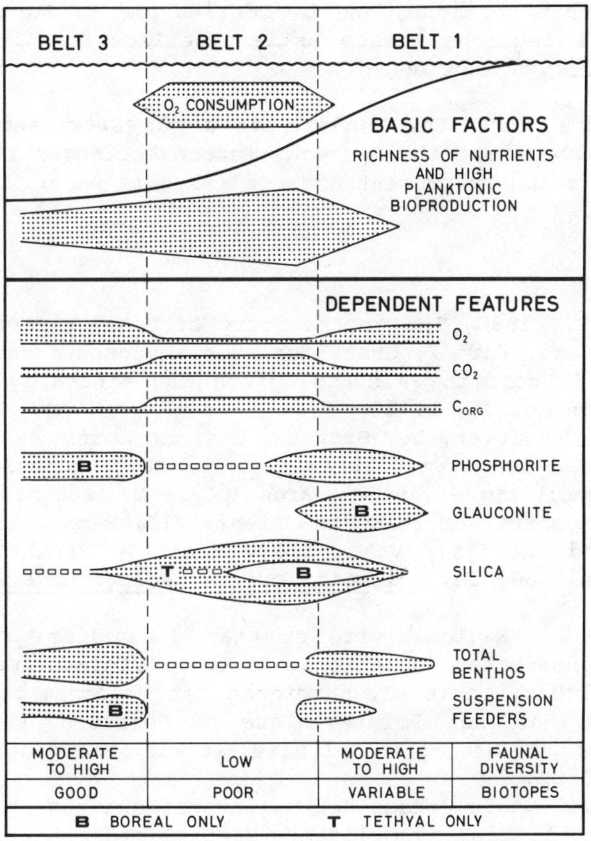

Fig. 9. Synopsis and comparison of major parameters indicating upwelling in the Tethyal Arabian Shelf (Upper Cretaceous/Lower Tertiary) and in the boreal Lower Saxony Basin (Aptian/Albian).

south, could have easily penetrated into western Tethys via the Vascogotic Trough or other depressions of the Pyrenean region and produced upwelling between Algeria and Turkey. Details on the subdivision of the southern part of the western Tethys are shown by Winnock (1980). The final conclusion is that coastal upwelling played an important role along continental margins and in marginal seas during Phanerozoic times, even under unexpected circumstances: in small offshoots of cool epicontinental seas at higher latitudes.

ACKNOWLEDGMENT

H. Raschka and H. Rösch, both Bundesanstalt für Geowissenschaften and Rohstoffe, Hannover, carried out the x-ray fluorescence and x-ray diffraction analyses. H. Hemme did the drafting, W. Zilian processed the thin sections, and L. Dörries and E. Rantze typed the manuscript (all Deutsche Texaco Aktiengesellschaft, Wietze). Their help is gratefully acknowledged.

The authors thank the Bundesanstalt für Geowissenschaften und Rohstoffe, Hannover, and the Deutsche Texaco Aktiengesellschaft, Hamburg, for permission to present and publish this paper.

REFERENCES

Altschuler, Z.S., 1980, The geochemistry of trace elements in marine phosphorites. Part I. Characteristic abundances and enrichment, Society of Economic Paleontologists and Mineralogists, Special Publication No. 29, 19-30.

Arambourg, C., Dubertret, L., Signeux, J., and Sornay, J., 1959, Contributions a la stratigraphie et a la paléontologie du Crétacé et du Nummulitique de la marge NW de la Péninsule Arabique, Notes et Memoirs sur le Moyen-Orient, VII, Paris, 193-251.

Atfeh, S.A. and Faradjev, V.A., 1963, Position stratigraphique des phosphates en Syrie, Bulletin Société geologique France, 7:1076-1084.

Bashir, S., 1977, "Radioactivité et état d'équilibre radioactif de quelques phosphates jordaniens. Leurs caractères pétrologiques, minéralogiques, cristallochimiques et géochimiques," Thèse, L'Institut National Polytechnique de Lorraine, École Nationale Supérieure de Géologie Apliquée et de Prospection Minière, Nancy, 193 pp.

Beerbaum, B., 1977, Die Genese der marin-sedimentären Phosphat-Lagerstätte von Al Hasa (westliches Zentraljordanien), Geologisches Jahrbuch, Reihe D, 24:3-55.

Bentor, Y.K., 1980, Phosphorites - The unsolved problems, Society of Economic Paleontologists and Mineralogists, Special Publication No. 29, 3-18.

Birch, G.F., 1980, A model of penecontemporaneous phosphatization by diagenetic and authigenic mechanisms from the western margin of southern Africa, Society of Economic Paleontologists and Mineralogists, Special Publication No. 29, 79-100.

Bodelle, J., 1980, Géologie comparée des gisements de phosphates et de pétrole, Colloque international, Orléans, 6-7 novembre 1979, Documents du Bureau de Recherches Géologiques et Minières, no. 24, 268 pp.

Brockamp, O., 1976, Nachweis von Vulkanismus in Sedimenten der Unter- und Oberkreide in Norddeutschland, Geologische Rundschau, 65: 162-174.

Burnett, W.C., 1980, Apatite-glauconite associations off Peru and Chile: palaeo-oceanographic implications, Journal of the Geological Society, London, 137:757-764.

Burnett, W.C., Veeh, H.H. and Soutar, A., 1980, U-series, oceanographic and sedimentary evidence in support of Recent formation of phosphate nodules off Peru, Society of Economic Paleontologists and Mineralogists, Special Publication No. 29, 61-71.

Cullen, D.J., 1980, Distribution, composition and age of submarine phosphorites on Chatam Rise, east of New Zealand, Society of Economic Paleontologists and Mineralogists, Special Publication No. 29, 139-148.

Diester-Haass, L., 1978, Sediments as indicators of upwelling, in: "Upwelling Ecosystems," R. Boje and M. Tomczak, eds., Springer, Berlin, 261-281.

Dietz, R.S., Emery, K.O. and Shepard, F.P., 1942, Phosphorite deposits on the sea floor off southern California, Geological Society of America, Bulletin, 53:815-847.

Flexer, A., 1968, Stratigraphy and facies development of Mount Scopus Group (Senonian-Paleocene) in Israel and adjacent countries, Israel Journal Earth-Science, 17:85-114.

Flexer, A., 1971, Late Cretaceous palaeogeography of northern Israel and its significance for the Levant geology, Palaeogeography, Palaeoclimatology, Palaeoecology, 10:293-316.

Gaida, K.-H., Kemper, E. and Zimmerle, W., 1978, Das Oberapt von Sarstedt und seine Tuffe, Geologisches Jahrbuch, Reihe A, 45: 43-123.

Gaida, K.-H., Gedenk, R., Kemper, E., Michaelis, W., Scheuch, R., Schmitz, H.H. and Zimmerle, W., 1981, Lithologische, mineralogische und organisch-geochemische Untersuchungen an Tonsteinen und Tonmergelsteinen der Unterkreide Nordwestdeutschlands (unter besonderer Berücksichtigung der Schwarzschiefer), Geologisches Jahrbuch, Reihe A, 58:15-47.

Il'In, A.V., 1978, Sea floor spreading in the Atlantic and accumulation of phosphate, Doklady Akademie Nauk, USSR, 240(6):1414-1417.

Jeans, C.V., Merriman, R.J., Mitchell, J.G., and Bland, D.J., 1982, Volcanic clays in the Cretaceous of southern England and northern Ireland, Clay Minerals, 17:105-156.

Jordan, H., 1968, Gliederung und Genese des Flammenmergels (Alb) in Hils- und Sackmulde (Süd-Hannover), Zeitschrift deutsche geologische Gesellschaft, 1965, 117:391-424.

Jordan, H. and Schmidt, F., 1968, Zur Altersstellung und Gliederung des Flammenmergels (Oberalb) im Sackwald, Geologisches Jahrbuch, 85:55-66.

Kemper, E., 1979, Die Unterkreide Nordwestdeutschlands. Ein Überblick, Aspekte der Kreide Europas, International Union Geological Sciences, Series A, 6:1-9.

Kemper, E., editor, in press, Das späte Apt und frühe Alb NW-Deutschlands, Versuch der vollständigen Analyse einer Schichtenfolge, Geologisches Jahrbuch, Reihe A, Hannover.

Kemper, E. and Schmitz, H.H., 1981, Glendonite - Indikatoren des polar-marinen Ablagerungsmilieus, Geologische Rundschau, 70:759-773.

Krüger, S., 1978, Zur Taxonomie und Systematik isolierter Schwammskleren mit Beispielen aus der Unter-Kreide Ostniedersachsens, Mitteilungen aus dem Geologischen Institut der Technischen Universität Hannover, 15:3-84.

Kull, H., 1979, "Sedimentpetrographische Untersuchungen von Nebengesteinsproben aus der Schachtanlage Konrad," Diplom-Arbeit, Technische Universität, Clausthal, 43 pp.

Lehmann, S., 1954, "Die Oberstufe des Oberalbs (Flammenmergel) im Subherzynen Becken und im westlichen Harzvorland," Inauguraldissertation Naturwissenschaftlich-Philosophische Fakultät, Technische Hochschule Braunschweig, 133 pp.

Lucas, J., Prévôt, L., Ataman, G., and Gündoğdu, N., 1979, Étude mineralogique et géochimique de la série phosphatée du Sud-Est de la Turquie (Mazidaği-Mardin), Science Géologique, Bulletin, 32:59-68.

Lucas, J., Prévôt, L., Ataman, G. and Gündoğdu, N., 1980, Mineralogical and geochemical studies of the phosphatic formations in southeastern Turkey (Mazidaği-Mardin), Society of Economic Paleontologists and Mineralogists, Special Publication No. 29:149-152.

McKelvey, V.E., 1978, Phosphate in sediments, in: "The Encyclopedia of Sedimentology, Encyclopedia of Earth Sciences," Vol. 6, R.W. Fairbridge and J. Bourgeois, eds., Dowden, Hutchinson & Ross, Stroudsburg, Pennsylvania, 574-579.

Moorkens, T.L., 1976, Palökologische Bedeutung einiger Vergesellschaftungen von sandschaligen Foraminiferen aus dem NW europäischen Alttertiär und ihre Beziehung zu Muttergesteinen, Compendium 75/76, Ergänzungsband Erdöl und Kohle, Leinfelden-Echterdingen, 77-95.

Nalivkin, D.V., 1973, "Geology of the U.S.S.R.", Oliver & Boyd, Edinburg, 855 pp.

Nathan, Y., Gal, I., Deutsch, Y., Shiloni, Y., and Roded, R., 1979, The geochemistry of the northern and central Negev phosphorites, Bulletin, Geological Survey Israel, 73:1-41.

Notholt, A.J.G., 1980, Economic phosphatic sediments: mode of occurrence and stratigraphical distribution, Journal of the Geological Society, London, 137:793-805.

Owen, H.G., 1976, Continental displacement and expansion of the earth during the Mesozoic and Cenozoic, Philosophical Transactions, Royal Society of London, A, Mathematical and Physical Sciences, 281(1303):223-291.

Paproth, E. and Zimmerle, W., 1980, Stratigraphic position, petrography, and depositional environment of phosphorites from the Federal Republic of Germany, Mededelingen Rijks Geologische Dienst, 32:81-95.

Prévôt, L. and Lucas, J., 1979, Comportement de quelques éléments traces dans les phosphates, Science Géologique, Bulletin, 32:91-105.

Riggs, S.R., 1980, Intraclast and pellet phosphorite sedimentation in the Miocene of Florida, Journal of the Geological Society, London, 137:741-748.

Sheldon, R.P., 1980, Episodicity of phosphate deposition and deep ocean circulation - a hypothesis, Society of Economic Paleontologists and Mineralogists, Special Publication No. 29, 239-247.

Shirav-(Schwartz), M. and Ginzburg, D., 1978, A guidebook to the oil shale deposits in Israel, Prepared for the International Symposium on oil shale chemistry and technology, Jerusalem, Oct. 30-Nov. 2, 1978, Geological Survey of Israel, Mineral Resource Division, Jerusalem, 20 pp.

Slansky, M., 1980, Géologie des phosphates sédimentair, Mémoire du Bureau de Recherches Géologiques et Minières, No. 114, 92 pp.

Sokołowski, St., 1976, "Geology of Poland, Vol 1: Stratigraphy, Part 2: Mesozoic," Wydawnictwa Geologiczne, Warszawa, 859 pp.

Soudry, D. and Nathan, Y., 1980, Phosphate peloids from the Negev phosphorites, Journal of the Geological Society, London, 137: 749-755.

Soutar, A. and Burnett, W., 1979, Distribution of phosphorite in relation to the oxygen minimum layer off the west coast of South America, in: "Report on the Marine Phosphatic Sediments Workshop, February 9-11, 1979, Honolulu, Hawaii," W.C. Burnett and R.P. Sheldon, eds., 24-25.

Suess, E., 1979, Mineral phases formed in anoxic sediments by microbial decomposition of organic matter, Geochimica et Cosmochimica Acta, 43:339-352.

Vinken, R., 1977, Geologische Karte Niedersachsen 1:25,000, Erläuterungen Blatt Hämelerwald Nr. 3626, 142 pp.

Winnock, E., 1980, Les dépôts de l'Eocene au Nord de l'Afrique: Apercu paléogéographique de l'ensemble, in: "Géologie comparée des gisements de phosphates et de pétrole, Colloque international, Orléans, 6-7 novembre 1979, Documents du Bureau de Recherches Géologiques et Minières, no. 24, 219-242.

Ziegler, P.A., 1981, Evolution of sedimentary basins in North-West Europe, Petroleum Geology of the Continental Shelf of North-West Europe, Institute of Petroleum, London, 3-39.

Zimmerle, W., 1979, Lower Cretaceous tuffs in northwest Germany and their geotectonic significance, in: "Aspekte der Kreide Europas," E. Kemper, ed., International Union of Geological Sciences, A6:385-402.

INDICATIONS OF UPWELLING IN THE LOWER ORDOVICIAN OF SCANDINAVIA

Maurits Lindström and Walter Vortisch

Institut für Geologie und Paläontologie
Lahnberge, D-3550 Marburg
Federal Republic of Germany

ABSTRACT

In the Early and Middle Ordovician the Precambrian shield area of northern Europe formed a deep shelf that sloped very gently toward the Caledonian mobile zone fringing the Iapetus Ocean. Shelf sediments are mainly represented by a facies of bedded limestone ("Orthoceratite limestone") which was deposited extremely slowly. It consists of varying proportions of arthropod and echinoderm fragments, and fine-grained matrix with calcite cement. In the Upper Arenigian to Lower Llanvirnian faunal differentiation, invasion of sessile plankton feeders, a greater rate of skeletal sediment production, and general coarsening of sediment are evidence of a major regression correlated with the American Whiterock Regression and hence regarded as eustatic. The greatest lowering of sea level in North America was at least 150 m, and the same is expected for northern Europe. During the regression an upwelling regime is inferred to have formed on the southeast flank of a rise situated in southern Sweden. Upwelling led to a strong increase of average phosphorus content in the sediments. The phosphorus occurs as fine-grained apatite.

INTRODUCTION

Starting with Kazakov's work (1937), a great deal of evidence has developed that upwelling and the formation of many submarine phosphate deposits are intimately connected (McKelvey, Swanson and Sheldon, 1953; Cook, 1976; Christie, 1978; Baturin and Bezrukov, 1979; Kress and Veeh, 1980; Sheldon, 1980; 1981; Parrish, 1982). It is well known that the Lower Paleozoic of Scandinavia contains phosphate deposits at different levels (Andersson, 1896; Hadding, 1927).

Based on our own work on the Lower Ordovician of Sweden we propose that upwelling was an important factor in the formation of some of these deposits. More specifically, we are dealing with phosphate deposition in connection with a major regression at the transition from Arenigian to Llanvirnian times.

MODELS OF MARINE PHOSPHATE DEPOSITION

The upwelling hypothesis states that a subsurface current impinging on the upper continental slope and shelf supplies nutrients, including phosphorus, which greatly enhance the organic productivity of the affected waters. On the bottom, more organogenic sediment will accumulate than can be destroyed by available oxygen (see for instance Burnett, Veeh and Soutar, 1980). If the organic matter becomes embedded before phosphorus is released by decay, the interstitial water of the sediment will transport the phosphorus after its ultimate release from the organic matter. Under suitable chemical conditions the phosphorus will accumulate as disseminated phosphate within the sediment. Reworking of the sediment may concentrate the phosphate particles thus forming larger aggregates (Baturin and Bezrukov, 1979). It has been demonstrated that this model works at the present in areas of eastern boundary current upwelling of South Africa and Peru (Baturin, Merkulova and Chalov, 1972; Burnett et al., 1980; Burnett, Roe and Piper; O'Brien and Veeh, all this volume). One of the most convincing older records of this process appears to be the Permian Phosphoria Formation of the western United States (McKelvey et al., 1953).

The Phosphoria was laid down in relatively deep water bordering a continent, whereas several other marine phosphate deposits are intracratonal and cannot easily be explained by oceanic upwelling (Bushinski, 1964; Youssef, 1965; Christie, 1978). The association of phosphate and riverborne clastics in many phosphate deposits led Bushinski to assume that phosphate is precipitated in shallow seas near the mouth of rivers, and the rivers are responsible for bringing the phosphate into the ocean. Contrary to this model, the upwelling hypothesis does not limit phosphate deposition to the vicinity of land. The wide acclaim of the upwelling model is indicated by the fact that even phosphorites of generally shallow-water sequences such as the intracratonal Upper Ordovician of North America (Dubuque Shale) are interpreted by some as due to upwelling of subsurface cold nutrient-rich waters (Brown, 1974).

In present oceans there is a lowermost depth of marine phosphate deposition usually at the lower boundary of the oxygen minimum zone. Among other things, this is due to the more extensive destruction of organic material and the release of phosphorus at greater depth in more oxygenated waters. On the other hand, organic matter would also be destroyed and the phosphorus contained in it set free in sediments exposed to well aerated water at very shallow depths. In view of these circumstances, Bushinski (1964) puts the depth of phosphorite formation at 30-200 m, whereas Bromley (1967) favors 30-300 m. Burnett et al. (1980) find phosphorite to be forming at 200-600 m off the Peru coast. In earlier periods, however, with more restricted vertical circulation and less oxygenated deep water, it is conceivable that organic sedimentation could have taken place at greater depths. Such conditions might have prevailed during part of the Early Paleozoic (Taylor and Cook, 1976).

GEOLOGICAL SETTING

During the Early Paleozoic the Baltic Shield was subject to very extensive marine transgressions that laid down a relatively thin sedimentary cover. To the west and south, the epicontinental sea covering the Baltic Shield was open towards deeper seas. To the west was the Iapetus Ocean, which separated the Baltic Shield from the Laurentian Shield; to the south, the northernmost part of the Proto-tethys Sea, covering what is now Central Europe. This area was characterized by argillaceous (graptolitic) to arenaceous sedimentation and evidently included marginal sea basins as well as sediment-yielding land areas.

In the Early Ordovician the Baltic Shield was the site of deposition of a relatively monotonous succession of bedded limestones. The upper Tremadocian to basal Llandeilian part is about 20 m thick (Bohlin, 1949; Hadding, 1958; Thorslund, 1960). For these limestones the name Orthoceratite limestone is frequently used--though not paleontologically adequate, this term is better than "*Orthoceras* limestone" since the cephalopods, colloquially orthoceratites, contained in it are mainly endoceroids and do not belong to the genus *Orthoceras*. The Orthoceratite limestones are mainly calcilutitic but also contain varying amounts of biocalcarenitic material, chiefly trilobite and echinoderm fragments. Glauconite is abundant at certain levels. Discontinuity surfaces, many of which are demonstrably hardgrounds, are very frequent (Jaanusson, 1961; Lindström, 1979). In the Arenigian dominant colors are either gray or red; red colors predominate in the Llarnvirnian part.

In the marginal parts of the Baltic Shield the Orthoceratite limestone is replaced by graptolite-carrying argillites as well as by other facies with a greater content of terrigenous material. For the 20 million years or more during which it was deposited, the Orthoceratite limestone appears to have been below wave base (Lindström, 1963; 1979). No unambiguously littoral sediments can be identified in association with this calcareous facies.

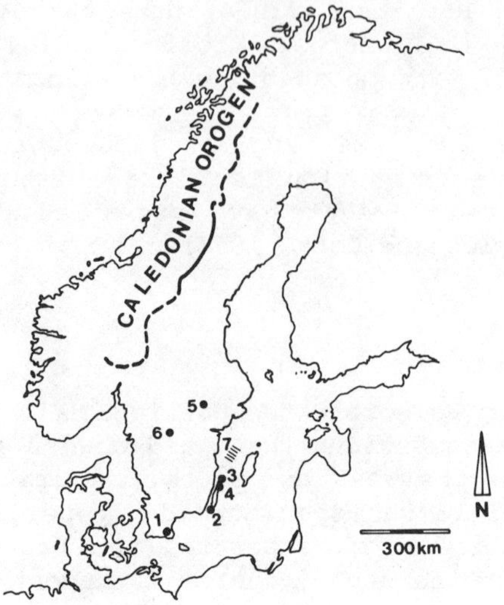

Fig. 1. Position of localities studied or referred to in text. 1: Gislövshammar. 2: Möckleby. 3: Horns Udde. 4: Gillberga. 5: Yxhult. 6: Skövde. 7(diagonal shading): *Strophomena jentschi* conglomerate.

PHOSPHATE OCCURRENCE

The two principal phosphate occurrences in the Lower Ordovician of the Baltic Shield are at the base of the Orthoceratite limestone (diachronous: upper Tremadocian to lowermost Arenigian), and in the uppermost Arenigian to basal Llanvirnian. The basal Ordovician occurrences consist mostly of angular to subrounded fragments of massive brownish phosphorite, abundantly scattered throughout a thin bed of glauconitic limestone. They were well described by Andersson (1896). In the phosphate fragments several Upper Cambrian trilobites have been found; the evidence suggests that they originated through the phosphatization of Upper Cambrian limestone during exposure of the Upper Cambrian at the sediment surface for a time interval in the Early Ordovician.

Upper Arenigian to basal Llanvirnian phosphorite deposits were likewise described by Andersson (1896). He occupied himself principally with a phosphate breccia ("*Strophomena jentschi* conglomerate") that evidently formed a lag deposit in areas of the present Baltic to the north of Öland (Fig. 1). Jaanusson (1955) gave phosphate analyses from a section through the Orthoceratite limestone on northern Öland. There are generally low phosphate concentrations in the lower and upper parts of this section (<0.3% P_2O_5) and several times higher concentrations (0.3-1.1% P_2O_5) in the "*Lepidurus* Limestone," the "*Expansus* Limestone," and the lower part of the "*Raniceps* Limestone,"

Table 1. Stratigraphy of the Arenig and part of the Llanvirn on the Baltic Shield

<table>
<tr><th>SERIES</th><th>BALTOSCANDIAN STAGES & SUBSTAGES</th><th colspan="2">CONODONT ZONES</th></tr>
<tr><td rowspan="3">LLAN-VIRN</td><td>Aseri</td><td rowspan="2">Eoplacognathus suecicus</td><td>U.</td></tr>
<tr><td rowspan="4">Kunda</td><td>L.</td></tr>
<tr><td rowspan="2">Amorphognathus variabilis</td><td>U.</td></tr>
<tr><td rowspan="8">ARENIG</td><td>L.</td></tr>
<tr><td colspan="2">Microzarkodina parva</td></tr>
<tr><td rowspan="3">Volkhov</td><td colspan="2">Paroistodus originalis</td></tr>
<tr><td colspan="2">Baltoniodus navis</td></tr>
<tr><td colspan="2">Baltoniodus triangularis</td></tr>
<tr><td rowspan="2">Billingen</td><td colspan="2">Oepikodus evae</td></tr>
<tr><td colspan="2">Prioniodus elegans</td></tr>
<tr><td>Hunneberg</td><td colspan="2">Paroistodus proteus</td></tr>
<tr><td colspan="4">TREMADOC</td></tr>
</table>

respectively, corresponding to the upper Arenigian conodont zone of *Microzarkodina parva*, the basal (uppermost Arenigian) part of the conodont zone of *Amorphognathus variabilis*, and to the lower ranges of the upper *A. variabilis* zone, or basal Llanvirnian (see Table 1).

We have investigated sections from Gislövshammar (Site 1, Fig. 1) at the southern boundary zone of the Baltic Shield; Möckleby (Site 2, Fig. 1) on southern Öland; Gillberga (Site 4, Fig. 1) and Horns Udde (Site 3, Fig. 1) on northern Öland; Skövde (Site 6, Fig. 1) in south-central Sweden; and Yxhult, 100 km to the northeast of Skövde (Site 5, Fig. 1). Together, our sections should be representative of an area of about 500 x 300 km^2 of the Baltic Shield. The lithologic features and correlations are shown in Fig. 2.

Phosphate was identified as apatite and semiquantitatively determined by x-ray diffractometry. In all sections the samples below the *parva* zone (excepting the ones at the very base of the Ordovician) are essentially barren of apatite at the level of identification. Since the Yxhult section does not reach the *parva* zone, this section yielded no significant apatite. In the three sections from Öland the first continuous apatite enrichment (reaching circa 1% P_2O_5 of the whole rock or circa 3% apatite) occurs in the *parva* zone. After a minimum near the *parva-variabilis* boundary, a somewhat greater enrichment occurs in the lower *variabilis* zone. This maximum of apatite enrichment can also be identified in Skövde and

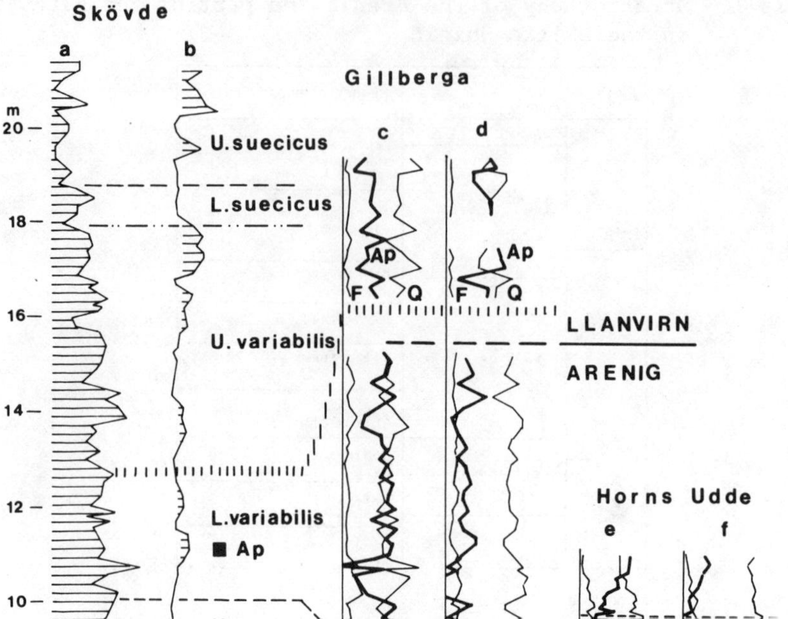

Fig. 2. Logs of facies and mineralogy features of Skövde, Gillberga, and Horns Udde sections. Conodont zones are indicated, and their boundaries marked by different signatures, where identifiable (cf. Table 1; ev = *Evae* Zone, tr = *Triangularis* Zone). a: running 3-sample average of percent identifiable biogenic components for Skövde sections. b: running 3-sample average of trilobite/echinoderm ostracomass ratio, Skövde section. c-f: percent quartz (Q), feldspar (F), and apatite (Ap) of (Q + F + Ap); semiquantitatively determined by x-ray diffractometry (peak intensity ratios) of non-carbonate

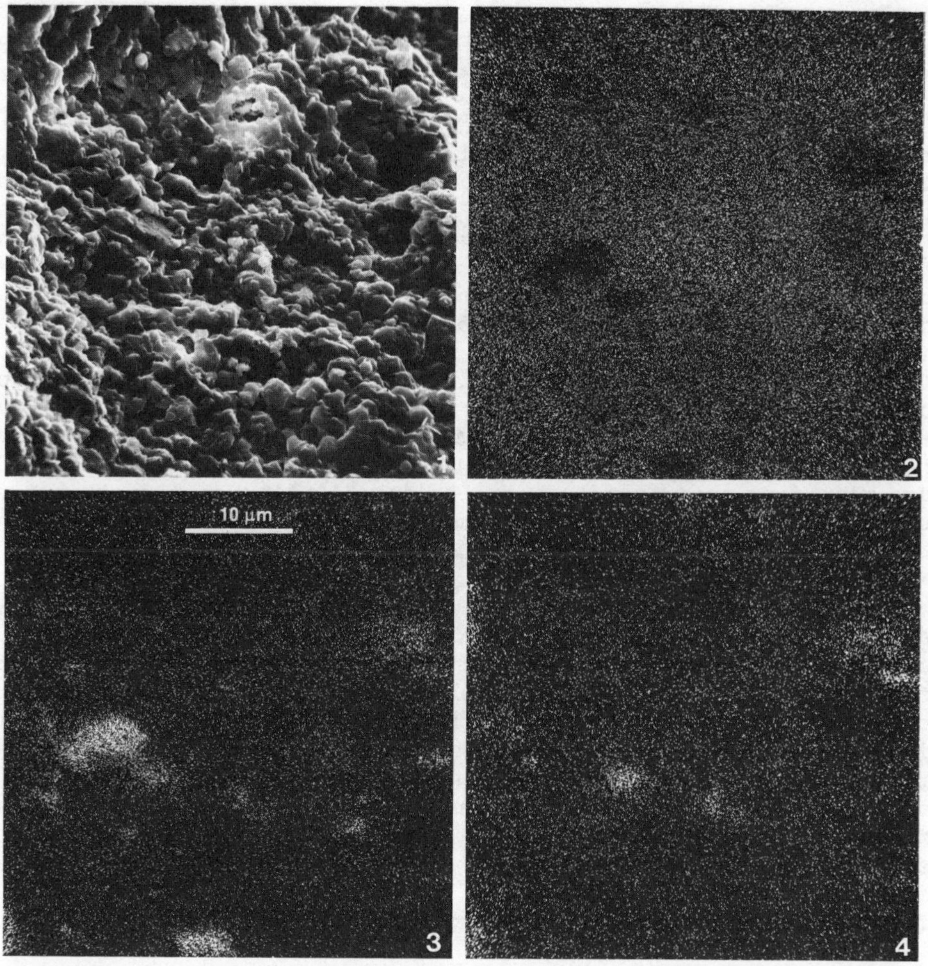

Fig. 3. Energy dispersive x-ray analysis of massive phosphate enrichment in pore space of grain-supported fabric, with foreign particles floating in the phosphate; sample Gillberga 8i. Same magnification in all cases (see scale bar). 1: SEM micrograph of area shown in 2-4; 2: phosphorus; 3: silicon; 4: aluminum.

←

Fig. 2 (cont.):
fractions. c: Gillberga section, <2 µm. d: Gillberga section, 2 - 6.3 µm. e: Horns Udde section, <2 µm. f: Horns Udde section, 2 - 6.3 µm.

Gislövshammar. The upper *variabilis* zone shows decreasing apatite enrichment. The apatite content in Horns Udde and Gillberga is appreciably higher than in the other sections.

Phosphate occurs in essentially three modes in our samples: (1) as a finely disseminated part of the non-carbonate fraction <6.5 μm; (2) as irregularly scattered microcrystalline parts of aggregates of non-carbonate matter, chiefly clay minerals; and (3) as massive phosphate aggregates. The first category can amount to over 10% of the total phosphate (Fig. 3). Phosphate of the second category may have poor morphological definition (Fig. 4: 1-4). After leaching of carbonates in weak acetic acid, phosphates appear as small patches mixed with clay minerals and silica into a spongy, coherent structure. The structure probably formed in an interstitial water solution of low mobility, which indicates that the source of the phosphorus may have been close to the point of precipitation. The mechanisms of phosphate formation, leading to structures similar to the ones described, have been discussed in detail by Sheldon (1981); for a further discussion of precipitation of phosphates from pore water solution, see Suess (1979).

Massive aggregates formed in place of matrix filling or blocky calcite cement in pores between skeletal particles in particle supported biocalcarenite (Fig. 5). It can contain isolated small clots of clay minerals and quartz particles (Fig. 3: 1-4), as well as non-corroded bioclastic carbonate particles. The structure is uniformly microcrystalline. The occurrence of this phosphorite suggests that it is neither replacement of carbonate matrix nor of carbonate in any other form (Cook, 1970). It appears to be the first formed mineral precipitate in the pores in which it occurs, in which case the source of phosphorus and other elements might be an organic colloid that filled the pores and perhaps also intermittently covered the surface of the sediment (cf. Baturin and Bezrukov, 1979).

Dissolution of phosphatic skeletal elements is a conceivable source of phosphate (Suess, 1981). Some of the investigated beds contain abundant phosphatic microfossils, such as conodonts and small, inarticulate brachiopods. However, because of their small size these skeletal particles form a negligible portion of the rocks, and they show hardly any evidence of dissolution.

A fourth mode of phosphate occurrence in coeval beds was described by Andersson (1896). The "*Strophomena jentschi* conglomerate" contains redeposited phosphate fragments some of which yielded Upper Cambrian trilobites. Evidently this residual deposit contains clasts of Upper Cambrian limestone which was exposed and phosphatized in adjacent parts of the Baltic basin. Phosphatization of limestone at the top of the Cambrian is known to have occurred during the hiatus between the Cambrian and the earliest Ordovician.

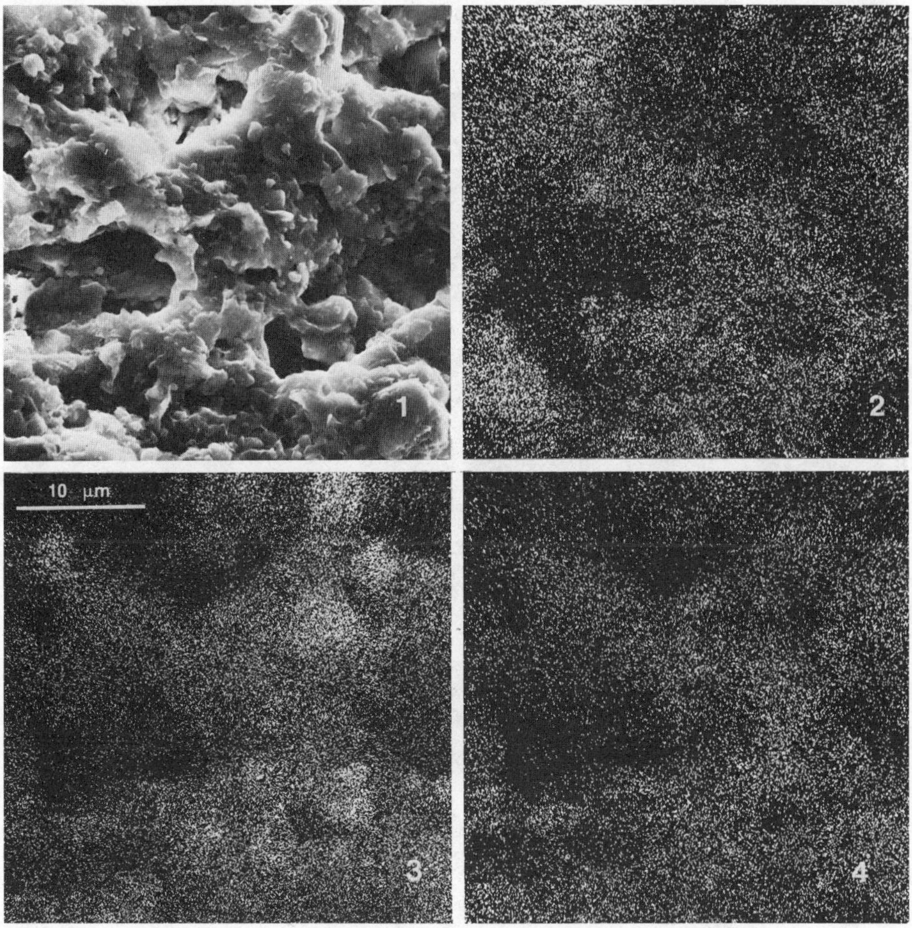

Fig. 4. Energy dispersive x-ray analysis of disseminated, fine-grained phosphate as part of matrix of silicates and quartz; sample Gillberga 8i. Same scale in all cases (see scale bar). 1: SEM micrograph of area shown in 2-4 (note spongy matrix structure surrounding hollow spaces left after dissolved calcite); 2: phosphorus; 3: silicon; 4: aluminum.

THE WHITEROCK REGRESSION

The phosphatized interval of the Orthoceratite limestone differs from overlying and underlying beds in several respects. Whereas lower and middle Arenigian and middle and upper Llanvirnian beds have a monotonous fauna dominated by asaphoid trilobites, the upper Arenigian to basal Llanvirnian interval had a rich fauna including many sessile benthic forms (Bohlin, 1949; 1955; Tjernvik, 1956). This is

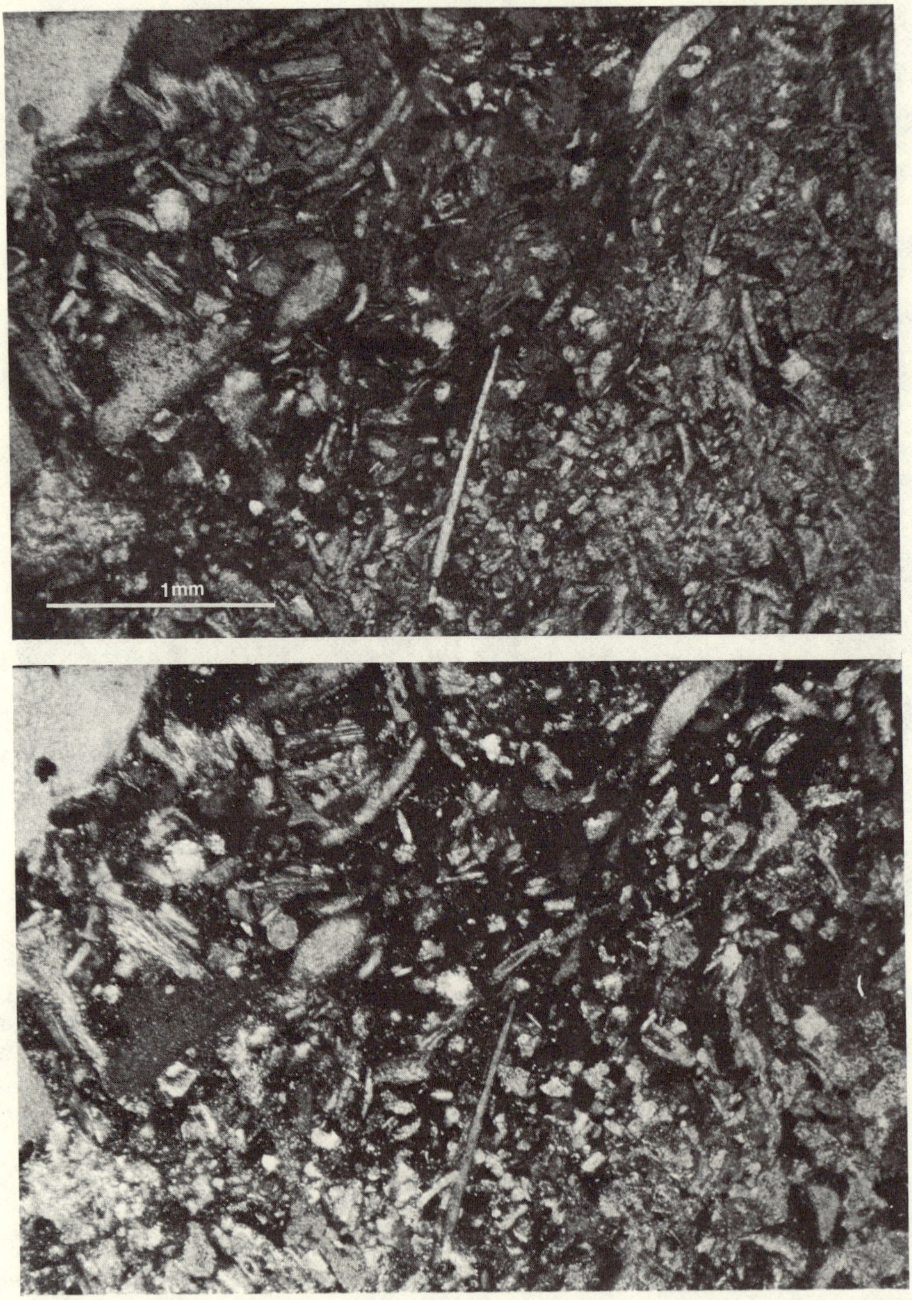

Fig. 5. Phosphorite (apatite) as cement in particle-supported biocalcarenite, sample Gillberga 8i. Top, plain polarized light, bottom, crossed nicols. Particles are mainly fragments of echinoderms and trilobites; scale 1 mm.

indicated by an increase of the bioclastic content of the sediments, dominance of echinoderms over arthropods as sediment producers, and an abundant occurrence of bryozoans and, in some sections, gastropods, major brachiopods, and *Receptaculites*. It should be stressed, however, that this biogenic debris appears to be transported an unknown distance. In spite of the existence of apparently suitable substrates, such as hardgrounds, sessile benthos is extremely rare. Bohlin (1949) suggested that the beds formed around the Arenigian to Llanvirnian transition time were laid down in appreciably shallower water than the older and younger beds. We concur with him in this conclusion.

It is interesting to note that a coeval regression occurred extensively in North America during much of the Whiterock Stage, which is late Arenigian (Bergström and Cooper, 1973) to Llanvirnian (Bergström, Ethington and Jaanusson, 1973; Ross, 1976). The following transgression did not take place until late in the Llanvirnian with the deposition of the St. Peter Sandstone (Shaw, 1974). In Illinois the Whiterock regression created a karst topography reaching 150 m deep into the underlying sediments (Buschbach, 1964). Since regressions have their most obvious expression on platform areas, it is worthwhile to note that the Siberian Platform has its most important Ordovician hiatus roughly corresponding to the uppermost Arenigian and Llanvirnian (Chugaeva, 1976). Late Arenigian to Llanvirnian regression, including karst formation, is indicated also for the shallow water sequence of the Canning Basin of Australia (McTavish and Legg, 1976). The coeval occurrence of a major regression on several continents indicates that this regression --for which we use the name Whiterock Regression-- was global and eustatic. The depth of karstification in the Upper Mississippi Valley suggests that the lowering of sea level was of the order of 150 meters. The upper Arenigian to lower Llanvirnian phosphates of Sweden formed during the Whiterock Regression.

A MODEL FOR THE BALTOSCANDIAN CASE

It is geophysically necessary that there be a limit below which a cratonic area like the Baltic Shield cannot be lowered relative to sea level. All possible mechanisms of local and eustatic sea level oscillation having been considered, it is improbable that areas belonging to the Baltic Shield were ever submerged to greater depths than about 500 m. There is no known evidence of great post-Ordovician tectonic movements within the Baltic Shield. If one assumes, for simplicity, that the topography of the Precambrian basement was very roughly the same in the Ordovician as it is now (although Mattsson, 1962, presents evidence of a certain amount of geomorphologic evolution in almost every detail that can be checked), there could have been areas that were as much as 300 m higher than the areas in which the Orthoceratite limestones was deposited. These areas could have been the main sites of organic productivity from which the bioclastic material was transported relatively short distances into the depositional areas (Lindström, 1979).

Assuming that the Whiterock Regression was eustatic, and that the sea level oscillation of 150 m or more recorded in the Upper Mississippi Valley (Buschbach, 1964) is valid for Scandinavia as well, the elevated production areas may have been somewhat deeper than 150 m before and after the regression; during the regression

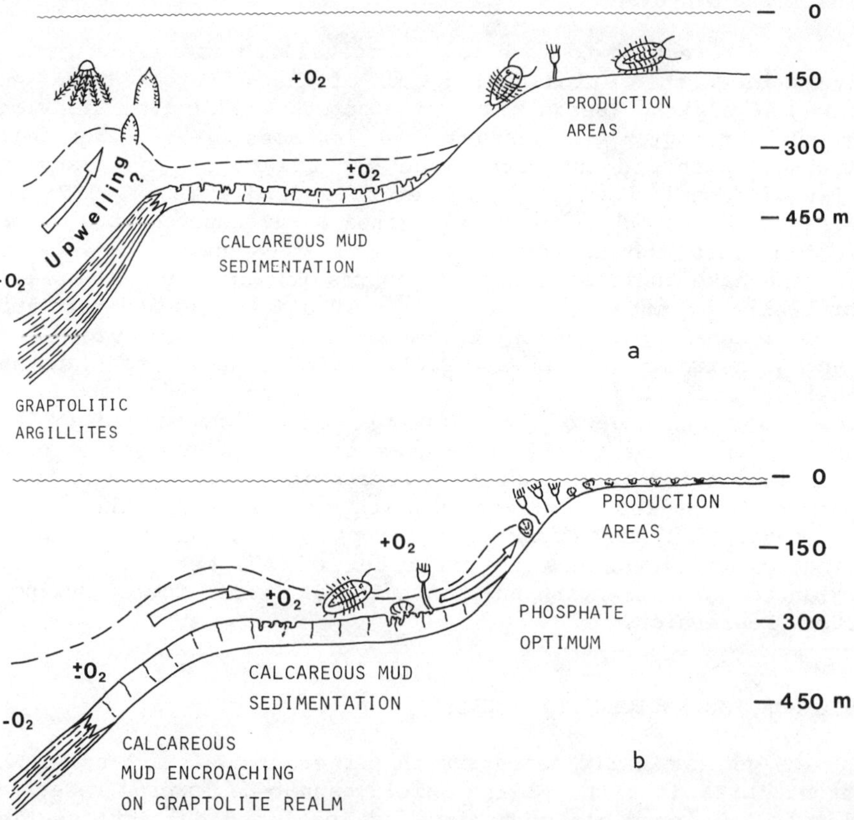

Fig. 6. Sketches of the paleoceanographic situation before and after the Whiterock Regression (a), and during the regression (b). Hypothetical depths of the sea are indicated to the right. a shows a relatively sparse fauna of trilobites and echinoderms on the higher ground which is postulated to have been the main site of biogenic production; graptolites are believed to have thrived over the outer slope where upwelling may have occurred. b shows richer fauna of echinoderms and other sessile benthos, as well as trilobites. Production is believed to have been greatest on the rises, but may have extended over deeper parts of the shelf as well. This is the situation in which apatite formed in the shelf areas.

they became quite shallow. The deeper phases were characterized by a specialized and relatively depauperate trilobite and echinoderm fauna, the skeletal remains of which were transported into the depositional basins in which they are preserved. During the regression the shallow rise biotopes had a rich fauna, characterized by abundant sessile benthos. The debris of these biota likewise spread into the basins. By simplistic reckoning, these basins were perhaps 300-500 m deep before and after the regression and 150-350 m deep during the regression.

According to Bergström and Noltimier (1982) the Baltic Shield in the Early Ordovician was about 60°S. Since the relative position of the South Pole was to the southeast of the present position of Baltoscandia, it follows that Baltoscandia's boundary towards the Iapetus Ocean faced northwards. In all probability, winds and the ocean currents, were generally westerly. Before and after the Whiterock Regression, surface currents are likely to have passed unimpeded over large parts of the Baltic Shield. This situation evidently did not cause upwelling on the shield; however, the northern and western marginal areas were the sites of deposition of a muddy graptolitic facies that might indicate upwelling, or conditions resembling upwelling. The profuse graptolite faunas are evidence of thriving planktonic life. As suggested by Berry (1977), the graptolite facies may be largely a continental slope facies. In this case, upwelling or conditions resembling upwelling, developed in some situations. Significantly, the depositional areas off the southern margin of the Baltic Shield not only contain abundantly fossiliferous graptolite shales; they also contain thin beds of high grade phosphorite, for instance in the lower Arenigian Skelbro limestone (Poulsen, 1966); in the "*bronni* beds" (= "*coscinorrhinus* beds") (Funkquist, 1919) of Llandeilian age (Bergström, 1973); at the base of the *Nemagraptus gracilis* beds (Hadding, 1913; Bergström and Nilsson, 1974). The Oslo area, a depositional area with rather special features situated on the western margin of the Baltic Shield, has generally high phosphate content in Lower Ordovician sediments, concentrated particularly at hiatuses (Björlykke, 1974).

During the Whiterock Regression the graptolitic facies was pushed toward the north and west in front of tongues of calcareous, skeletal mud. It is possible that the surface current system was more strongly influenced by the morphology of the Baltic Shield during the regression and that this caused widespread upwelling within the Shield area (Fig. 6).

SUMMARY

We conclude that the paleogeographic situation of southern Sweden in the early parts of the Ordovician was one in which deep nutrient-rich water flowed eastwards along the northern margin of the Baltic Shield craton. This water normally deposited black graptolite mud with occasional enrichments of phosphorite. On the Baltic Shield itself a fairly monotonous limestone facies was normal which was deposited slowly.

During the Whiterock Regression, areas characterized by Orthoceratite limestone became the site of phosphate precipitation, during an eustatic event of Late Arenigian to earlier Llanvirnian age. The phosphate occurs mainly in biocalcarenitic limestones and is either very finely disseminated in the matrix or forms void fillings in place of calcite cement or matrix. Breccia-like lag deposits of phosphorites have also been described. The phosphorus probably was derived from abundant organic matter which was either included in the sediment while this was being deposited or formed a slurry resting on the carbonate sediment. Probably both kinds of accumulation of organic matter could occur, although the first appears to be the more likely one. We believe that the excessive accumulation of organic matter was caused by upwelling, and that this upwelling was induced by submarine rises which developed into barriers to the bottom currents during the regression.

ACKNOWLEDGEMENTS

We thank E. Suess and A.J. Boucot (both Corvallis, Oregon) and M. Arthur (Columbia, South Carolina) for stimulating discussions and helpful criticism. The electron optic expertise of K. Fecher (Marburg) was important to the success of our study. We acknowledge support from Deutsche Forschungsgemeinschaft (Projekt Li 174/12).

REFERENCES

Andersson, J.G., 1896, Über cambrische und silurische, phosphoritführende Gesteine aus Schweden, Bulletin of the Geological Institution of the University of Upsala, 2(1895):133-238.

Baturin, G.N. and Bezrukov, P.L., 1979, Phosphorites on the sea floor and their origin, Marine Geology, 31:317-332.

Baturin, G.N., Merkulova, K.I. and Chalov, P.I., 1972, Radiometric evidence for recent formation of phosphatic nodules in marine shelf sediments, Marine Geology, 13:M37-M41.

Bergström, S.M., 1973, Correlation of the Lasnamägian Stage (Middle Ordovician) with the graptolite succession, Geologiska Föreningens i Stockholm Förhandlingar, 95:9-18.

Bergström, S.M. and Cooper, R.A., 1973, *Didymograptus bifidus* and the trans-Atlantic correlation of the Lower Ordovician, Lethaia, 6:313-340.

Bergström, S.M. and Nilsson, R., 1974, Age and correlation of the Middle Ordovician bentonites of Bornholm, Bulletin of the Geological Society of Denmark, 23:27-48.

Bergström, S.M. and Noltimier, H.C., 1982, Latitudinal positions of Ordovician continental plates based on paleomagnetic evidence, in: Abstracts for meetings 20, 21 and 23 August 1982, IV International Symposium on the Ordovician System, D.L. Bruton and S.H. Williams, eds., Paleontological Contributions of the University of Oslo, 280:8.

Bergström, S.M., Ethington, R.L. and Jaanusson, V., 1973, On the stage subdivision of the North American lower-Middle Ordovician: age of strata at the top of Whiterock reference sequences in Nevada, Geological Society of America, Annual Meeting, Abstract, 5:299.

Berry, W.B.N., 1977, Graptolite biostratigraphy: A wedding of classical principles and current concepts, in: "Concepts and Methods of Biostratigraphy," E.G. Kauffman and J.E. Hazel, eds., Doyden, Hutchinson and Ross, Stroudsburg, Pennsylvania, 321-338.

Björlykke, K., 1974, Depositional history and geochemical composition of the Lower Palaeozoic epicontinental sediments from the Oslo region, Norges Geologiske Undersøkelse, 305, 81 pp.

Bohlin, B., 1949, The *Asaphus* Limestone in northernmost Öland, Bulletin of the Geological Institution of the University of Upsala, 33:529-570.

Bohlin, B., 1955, The Lower Ordovician limestones between the *Ceratopyge* Shale and the *Platyurus* Limestone of Böda Hamn. With a description of the microlithology of the limestones by V. Jaanusson, Bulletin of the Geologial Institution of the University of Uppsala, 35:111-173.

Bromley, R.G., 1967, Marine phosphorites as depth indicators, Marine Geology, 5:503-509.

Brown, C.E., 1974, Phosphatic zone in the lower part of the Maquoketa Shale in northeastern Iowa, Journal of Research of the U.S. Geological Survey, 2:219-232.

Burnett, W.C., Veeh, H.H. and Soutar, A., 1980, U-series, oceanographic and sedimentary evidence in support of recent formation of phosphate nodules off Peru, Society of Economic Paleontologists and Mineralogists, Special Publication, 29:61-71.

Buschbach, T.C., 1964, Cambrian and Ordovician strata of northeastern Illinois, Illinois State Geological Survey, Report of Investigations 218, 90 pp.

Bushinski, G.I., 1964, On shallow water origin of phosphorite sediments, in: "Deltaic and Shallow Marine Sediments: Developments in Sedimentology," L.M.J.U. van Straaten, ed., Elsevier Publishing Company, Amsterdam, 62-70.

Christie, R.L., 1978, Sedimentary phosphate deposits: An interim review, Geological Survey of Canada, Paper 78-20, 9 pp.

Chugaeva, M.N., 1976, Ordovician in the northeastern U.S.S.R., in: "The Ordovician System: Proceedings of a Palaeontological Association Symposium," M.G. Bassett, ed., Birmingham, September 1974, 283-292.

Cook, P.J., 1970, Repeated diagenetic calcitization, phosphatization and silicification in the Phosphoria Formation, Geologial Society of America, Bulletin, 81:2107-2116.

Cook, P.J., 1976, Sedimentary phosphate deposits, in: "Handbook of Strata-bound and Stratiform Ore Deposits," 7, K.H. Wolf, ed., Elsevier Publishing Company, Amsterdam, 505-535.

Funkquist, H.D.A., 1919, Asaphusregionens omfattning i sydöstra Skåne och på Bornholm, Lunds Universitets Årsskrift, N.F., Avdelning 2, 16(1):55 pp.

Hadding, A.R., 1913, Undre dicellograptusskiffern i Skåne jämte några därmed ekvivalenta bildningar, Lunds Universitets Årsskrift, N.F., Avdelning 2, 9(15):90 pp.

Hadding, A.R., 1927, The pre-Quaternary sedimentary rocks of Sweden. 2. The Paleozoic and Mesozoic conglomerates of Sweden, Lunds Universitets Årsskrift, N.F., Avdelning 2, 23(5):42-171.

Hadding, A.R., 1958, The pre-Quaternary sedimentary rocks of Sweden. 7. Cambrian and Ordovician limestones, Lunds Universitets Årsskrift, N.F., Avdelning 2, 54(5):262 pp.

Jaanusson, V., 1955, Description of the microlithology of the Lower Ordovician limestones between the *Ceratopyge* shale and the *Platyurus* limestone, Bulletin of the Geological Institution of the University of Uppsala, 35:153-173.

Jaanusson, V., 1961, Discontinuity surfaces in limestones, Bulletin of the Geological Institution of the University of Uppsala, 40: 221-241.

Kazakov, A.A., 1937, The phosphorite facies and the genesis of phosphorites, Transactions of the Scientific Institute of Fertilizers Insecto-Fungicides, 142:95-115.

Kress, A.G. and Veeh, H.H., 1980, Geochemistry and radiometric ages of phosphatic nodules from the continental margin of northern New South Wales, Australia, Marine Geology, 36:143-157.

Lindström, M., 1963, Sedimentary folds and the development of limestones in an Early Ordovician sea, Sedimentology, 2:243-292.

Lindström, M., 1979, Diagenesis of Lower Ordovician hardgrounds in Sweden, Geologica et Palaeontologica, 13:9-30.

Mattsson, Å., 1962, Morphologische Studien in Südschweden und auf Bornholm über die nichtglaziale Formenwelt der Felsenskulptur, Meddelanden Lunds Universitets geografiska Institution, Avhandlingar, 39, 357 pp.

McKelvey, V.E., Swanson, R.W. and Sheldon, R.P., 1953, The Permian phosphorite deposits of western United States, Congres géologique international, Comptes Rendus 19 Session Alger 1952, Section 11, Origine des gisements de phosphates de chaux, 11:45-64.

McTavish, R.A. and Legg, D.P., 1976, The Ordovician of the Canning Basin, Western Australia, in: "The Ordovician System: Proceedings of a Palaeontological Association Symposium," M.G. Bassett, ed., Birmingham, September 1974, 447-478.

Parrish, J.T., 1982, Upwelling and petroleum source beds, with reference to the Palaeozoic, American Association of Petroleum Geologists, Bulletin, 66:750-774.

Poulsen, V., 1966, Cambro-Silurian stratigraphy of Bornholm, Meddelelser fra Dansk Geologisk Forening, 16:117-137.

Ross, R.J., Jr., 1976, Ordovician sedimentation in the western United States, in: "The Ordovician System: Proceedings of a Paleontological Association Symposium," M.G. Bassett, ed., Birmingham, September 1974, 73-105.

Shaw, F.C., 1974, Simpson Group (Middle Ordovician) trilobites of Oklahoma, Journal of Paleontology 48, Supplement, 54 pp.

Sheldon, R.P., 1980, Episodicity of phosphate deposition and deep ocean circulation -- a hypothesis, Society of Economic Paleontologists and Mineralogists, Special Publication 29:239-247.

Sheldon, R.P., 1981, Ancient marine phosphorites, Annual Reviews of Earth and Planetary Sciences, 9:251-284.

Suess, E., 1979, Mineral phases formed in anoxic sediments by microbial decomposition of organic matter, Geochimica et Cosmochimica Acta, 43:339-352.

Suess, E., 1981, Phosphate regeneration from sediments of the Peru continental margin by dissolution of fish debris, Geochimica et Cosmochimica Acta, 45:577-588.

Taylor, M.E. and Cook, H.E., 1976, Continental shelf and slope facies in the Upper Cambrian and lowest Ordovician of Nevada, Brigham Young University Geology Studies, 23:part 2, 181-214.

Thorslund, P., 1960, The Cambro-Silurian, in: "Description to accompany the map of the pre-Quaternary rocks of Sweden," N.H. Magnusson, P. Thorslund, P. Brotzen, B. Asklund, and O. Kulling, eds., Sveriges Geologiska Undersökning, Serie Ba, 16:69-110.

Tjernvik, T., 1956, On the Early Ordovician of Sweden, Bulletin of the Geologial Institutions of the University of Uppsala, 36: 107-284.

Youssef, M.I., 1965, Genesis of bedded phosphates, Economic Geology, 60:590-600.

UPWELLING IN THE PALEOZOIC ERA

Judith Totman Parrish* and Alfred M. Ziegler

Department of Geophysical Sciences, University of Chicago
Chicago, Illinois 60637, U.S.A.

Roger G. Humphreville

Amoco Production Company (Int'l), P.O. Box 4381
Houston, Texas 77210, U.S.A.

*Present address:
U.S. Geological Survey, Denver Federal Center
Denver, Colorado 80225, U.S.A.

ABSTRACT

During Paleozoic times the continents were in geographic positions substantially different from today, and models keyed to continental drift are required to understand sedimentary upwelling records along Paleozoic continental margins. This applies particularly to the Early Paleozoic when east-west trending continental margins predominated. After construction of paleogeographic maps for various Paleozoic time slices, it is important to model the impact of the atmospheric circulation which is controlled in its vigor by the temperature gradient between equatorial and polar regions. This temperature gradient and the rotation of the earth create a pattern of barometric pressure with high pressure at the poles and at 30° latitude, and low pressure at the equator and 60° latitude. The resulting surface winds are zonal and constitute the most important and stable features of atmospheric circulation. These wind patterns can then be applied to ancient paleogeographic situations, and upwelling can be expected along west-facing coasts at low latitudes and along east-west trending coastlines that were favorably oriented to the zonal surface wind regimes. The distribution of predicted upwelling zones has then been tested by comparison with the distributions of marine hydrocarbon source beds, phosphorites and cherts which are frequently generated in upwelling zones. These deposits often tracked the upwelling zones as the continents drifted in and out of the favorable atmosphere circulation settings.

INTRODUCTION

Upwelling is a complex process and it is difficult to scale upwelling models from the short-term phenomena that we can study directly to the longer-term phenomena we can observe in the rock record. Nevertheless, it is desirable from an economic point of view to be able to predict where upwelling occurred in the past because economic deposits, namely hydrocarbon source beds and phosphorites, are sometimes laid down in upwelling zones (Trask, 1932; Brongersma-Sanders, 1948; McKelvey, Swanson and Sheldon, 1952; and many subsequent workers). Chert derived from biogenic siliceous oozes can also indicate past upwelling. The lithologic association of organic-rich rocks, chert and phosphorite has been termed "bioproductite" because the association occurs today in areas of high biologic productivity (e.g., Calvert and Price; Bremner; both this volume). The association is also found in ancient deposits that are likely to have been deposited in areas of high biologic productivity, for example, the Monterey Formation of Middle-Late Miocene age, approximately 10-5 million years before present (Bramlette, 1946; Pisciotto and Garrison, 1981) and the Phosphoria Formation of Late Permian age, approximately 270 million years before present (McKelvey et al., 1952; 1967).

The three elements of bioproductite--organic-rich sediment, phosphorite, biogenic siliceous sediment--are not always intimately associated. For example, Burnett (1977) demonstrated that phosphate nodules in the Peru upwelling zone tend to be deposited in the sediment belt that is transitional between the wholly anoxic region under the upwelling locus and the oxygenated waters farther up and down the slope. This general pattern is confirmed by the distribution of characteristic sediments in the South African upwelling (Bremner, this volume). The sediment where the nodules are precipitated is less organic-rich than the sediment in the anoxic zone, so the phosphate nodules and the organic-rich sediments represent different coeval facies of the upwelling system. A sea level rise would permit the vertical juxtaposition of organic-rich sediments over the upslope phosphate facies and a vertical juxtaposition of normal marine sediments over the downslope phosphate facies. Thus, in any given locality, one might find only one bioproductite element rather than two or all three. Moreover, the original assemblage of sediments in an upwelling zone might lack one of the expected elements. For example, the upwelling sediments in the Gulf of California are rich in organic matter and siliceous sediment, but no phosphate nodules have been reported up to now. Indeed, variability seems to be the rule for upwelling sediments, as several workers have demonstrated (Fütterer; Diester-Haass; both this volume). Therefore, we feel justified in considering the distributions of the individual elements separately.

Contrary to Valentine and Moores (1971), who claimed that "[ocean] current reconstruction cannot be predicted [for the past] from first principles", we have found that it is possible to predict the distribution of ancient upwelling currents from models of atmospheric circulation. Several studies have demonstrated the utility of these models (Parrish, 1982; Parrish and Curtis, in press; Ziegler et al., 1977; 1981). In this paper, we will briefly summarize the findings of Parrish (1982) on the Paleozoic distribution of organic-rich rocks and present data on the distribution of phosphorites heretofore given only in talks and described in abstracts. In addition, we will compare the distribution of predicted upwelling zones to the distribution of chert deposits from information presented by Drewry, Ramsay and Smith (1974).

METHODS

For simplicity in Figs. 1-7, we have included only the upwelling zones predicted from the models of atmospheric circulation and upwelling presented by Parrish (1982). The method of constructing the circulation maps was described in detail in the original papers (Parrish, 1982; Parrish and Curtis, in press). The circulation maps were based on basic principles of atmospheric circulation as they are understood from present circulation. The circulation models were qualitative, that is, they did not predict nor take into account temperature distributions, wind speeds, and so on. The purpose of the models was to provide a general guide to wind direction patterns. We assumed that major climatic changes, especially decreases in the equator-to-poles temperature gradient, would not have significantly changed the wind patterns, although we concede that wind speeds, and therefore upwelling current speeds, might have been affected (e.g., Berger, Diester-Haass and Killingley, 1978; Diester-Haass and Schrader, 1979; Brass et al., in press). In a sense, we are trying to predict all possible upwelling zones; whether each individual predicted upwelling zone actually developed is a matter for further study.

Data for the organic-rich rocks were provided by Amoco Production Company of Tulsa and Amoco International Oil Company and were also culled from the literature. The data consisted of geochemical analyses that determined at least the total organic carbon (TOC) content of the rocks. Most of the data are of this type, and a minimum of 1.0% TOC was required for the data to be included. About a third of the data also included geochemical analyses designed to determine the type of organic matter (Philippi, 1974; Tissot et al., 1974; Tissot and Welte, 1978), and the TOC cut-off for these data was 0.4% (Momper, 1978). The method of petroleum source bed data collection and analysis is explained further in Parrish (1982).

Phosphorite data were taken from the literature. A tabulation of the phosphorite data used for this study is included in Appendix I. Unlike many previous workers, we have included non-economic deposits of phosphorite. The average bulk rock phosphate content ranges from 0.08% to 0.72% (Wedepohl, 1969, Table 8-4). The phosphate content was not reported for all of the deposits we used here, but the ones that had been analyzed had phosphate contents much higher than the "background" phosphate content. It is not possible to gauge the significance of economic deposits versus non-economic deposits in terms of their depositional environments. Economic deposits are by their nature anomalous; therefore, we feel that by considering only those deposits, we may be losing important information regarding phosphorite deposition. Much work needs to be done contrasting the environments of deposition of economic and non-economic phosphorite deposits.

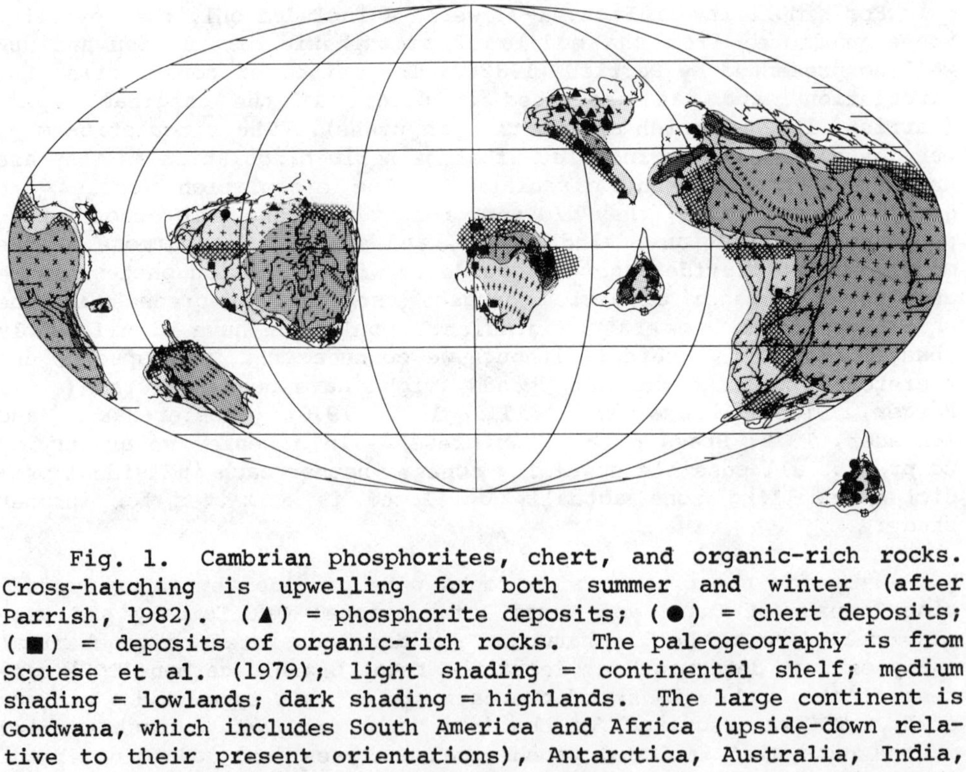

Fig. 1. Cambrian phosphorites, chert, and organic-rich rocks. Cross-hatching is upwelling for both summer and winter (after Parrish, 1982). (▲) = phosphorite deposits; (●) = chert deposits; (■) = deposits of organic-rich rocks. The paleogeography is from Scotese et al. (1979); light shading = continental shelf; medium shading = lowlands; dark shading = highlands. The large continent is Gondwana, which includes South America and Africa (upside-down relative to their present orientations), Antarctica, Australia, India, southern Europe, Turkey and Iran, and Tibet. The very small continental fragments off eastern Gondwana are Avalon (Nova Scotia and eastern Newfoundland) in the north and England in the south. From west to east and south to north, the remaining continents are Baltica (northern Europe west of the Urals), Laurentia (most of present-day North America), Siberia (upside-down relative to its present position), Kazakhstan (part of present-day USSR), and China. The inset shows the chert occurrences of Kazakhstan separately.

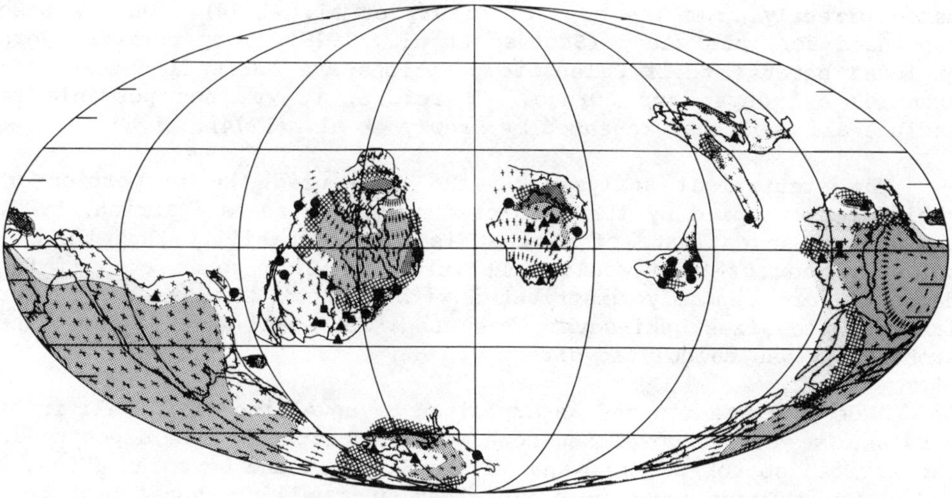

Fig. 2. Ordovician phosphorites, chert, and organic-rich rocks. Symbols, shading and continents as in Fig. 1.

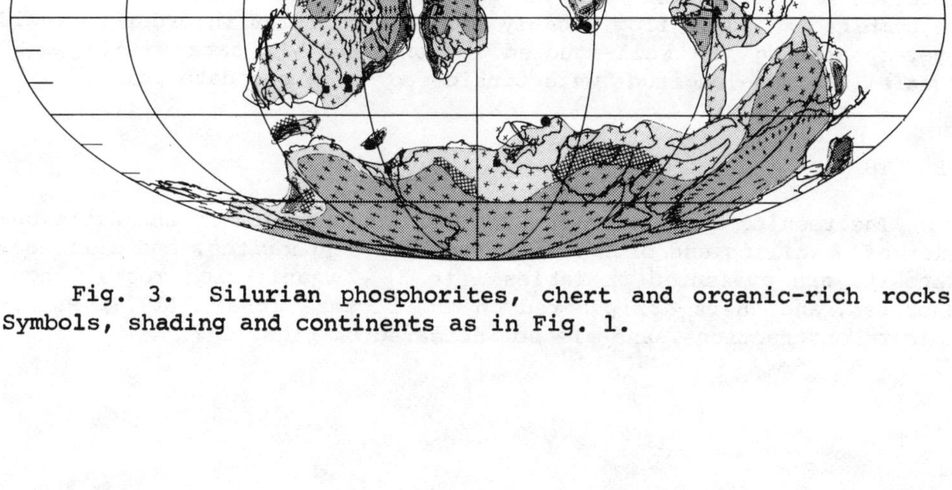

Fig. 3. Silurian phosphorites, chert and organic-rich rocks. Symbols, shading and continents as in Fig. 1.

Data on chert were partly recompiled from the references given by Drewry et al. (1974), with additional data from Vinogradov (1968; 1969). A few of the data points for cherts shown on Figs. 1-7 were taken directly from the maps of Drewry et al. (1974). On the base maps used for this study (Scotese et al., 1979), many terrains were excluded because their Paleozoic positions are not well understood. Examples are Japan and Borneo. Therefore, it was not possible to include all the data presented by Drewry et al. (1974).

The statistical analyses of the data used the proportion of shelf area occupied by the predicted upwelling zones (Parrish, 1982, Table 2) as an estimate of p, that is, the probability that the observed number of data would fall into upwelling zones even if the deposits were randomly distributed with respect to upwelling. For large sample sizes, chi-square was used; for small sample sizes, the probability was calculated directly.

A deposit was counted as explained by upwelling if it fell in or very near a modeled upwelling zone. The exceptions were deposits in the African sector of Gondwana in the Silurian and Devonian (Figs. 3 and 4), which were counted as explained by upwelling even though they were distant from the upwelling zones. This was done for the following reason. Bathymetry is extremely important in the control of upwelling location (Smith, this volume). Because there is as yet little bathymetric control for the Paleozoic continental shelves, Parrish (1982) modeled the upwelling zones at the shorelines. However, on very broad shelves with low depth gradients, the upwelling would actually have occurred some distance from the geographic shoreline. We have left the modeled upwelling zones where they were originally placed, but are certain the actual loci of upwelling were much farther offshore.

Organic-rich rock (Parrish, 1982), phosphorite, and chert localities were clustered for the purposes of the statistical analyses. Individual data points were not evenly distributed, some areas having received more attention from geologists than other areas. Therefore, we combined the data from closely spaced localities in order to avoid overemphasizing the well-studied regions and all data within a 5° square latitude-longitude were considered to be one data point.

RESULTS

The results of the statistical comparisons between the distribution of upwelling and organic-rich rocks, phosphorites, and siliceous deposits are presented in Tables 1 to 3. Organic-rich rocks, phosphorites, and chert are plotted on the Scotese et al. (1979) Paleozoic reconstructions, and are so indicated on Figs. 1-7.

Table 1. Results of statistical tests of correlation between modeled upwelling zones and petroleum source bed distribution (modified from Parrish, 1982). For all the Paleozoic data together, the statistic is chi-square; for the other analyses, the numbers represent calculated probabilities of the observed correspondence.

Time	Sources Explainable by Upwelling (Total) Marine Sources)	Statistical Results
Paleozoic	46 (77)	55.61*
Cambrian°	2 (6)	0.32
Late Ordovician°	3 (6)	0.43
Silurian°	1 (4)	0.70
Early Devonian°	3 (8)	0.11
Middle Devonian	5 (7)	0.001*
Late Devonian	13 (17)	3.3×10^{-8}*
Early Mississippian°	3 (5)	0.02*
Late Mississippian	4 (4)	0.0002*
Early Pennsylvanian	3 (3)	0.009*
Late Pennsylvanian°	2 (2)	0.04*
Early Permian	3 (4)	0.04*
Late Permian°	4 (11)	0.005*

* Values significant to at least $p < 0.05$. Data included on Figs. 1-7 are from time periods indicated by (°).

Organic-Rich Rocks

As reported in Parrish (1982), there is a significant correspondence between the distributions of the predicted upwelling zones and marine hydrocarbon source beds. This is true if all Paleozoic data are combined, and is also true for many individual geologic periods analyzed separately, although in the latter case, small sample sizes often inhibited interpretation of the statistical results (Table 1). The correspondence is particularly striking for the middle Devonian onwards. Before that time, there is no apparent correlation.

Phosphorites

For all the Paleozoic data together, there was a significant correspondence between the modeled upwelling zones and the distribution of phosphorites. For the individual geologic periods, significant correspondences were obtained for the Cambrian, Devonian, and

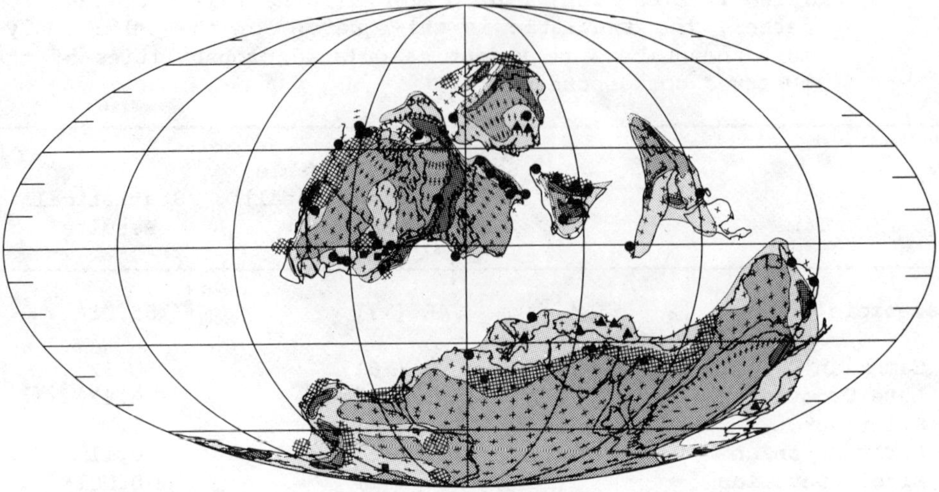

Fig. 4. Devonian phosphorites, chert, and organic-rich rocks. Symbols and shading as in Fig. 1. Baltica, Laurentia, Avalon, and England have collided to form Laurussia.

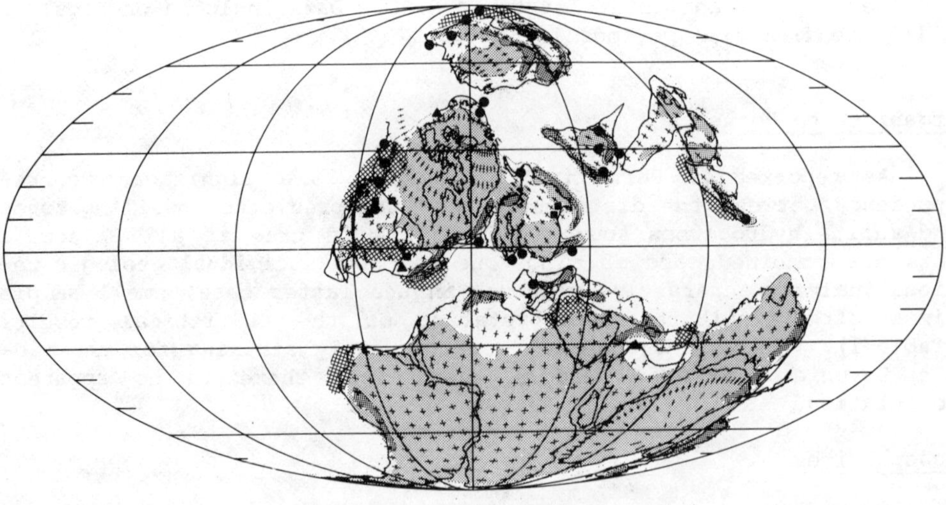

Fig. 5. Mississippian phosphorites, chert and organic-rich rocks. Symbols, shading and continents as in Fig. 1.

Permian Periods (Table 2). With the exception of the Ordovician Period, the remaining periods had rather small sample sizes. The Ordovician anomaly is not explained at this time, although the bathymetric argument invoked above to account for the African deposits might also apply to the Ordovician North American deposits. Unfortunately, it was not possible to analyze such narrow time slices as was done for the organic-rich rocks, and in this sense, the analysis of phosphorite distribution is much less reliable. However, the observed correspondence, particularly for the extremely phosphogenic Cambrian Period, is suggestive of a relationship between upwelling and phosphorite distribution. Most of the Cambrian phosphorites are of Early Cambrian age. Recently, we have constructed an Early Cambrian paleogeographic map and plotted the phosphorites (Parrish et al., in press). Despite the differences between the Early Cambrian paleogeography and that of the Late Cambrian reconstruction used here, the phosphorite deposits still mostly fell into the modeled upwelling zones.

Chert

The correspondence between modeled upwelling zones and the distribution of chert was particularly striking (Table 3). Extensive belts of siliceous rocks, such as in the North American Cordillera in the Devonian and Permian (Figs. 4 and 7), were precisely correlated with extensive upwelling zones.

Table 2. Results of statistical tests of correspondence between modelled upwelling zones and the distribution of phosphorites. The statistic for all the Paleozoic data together is chi-square; for the individual periods, the probabilities for the observed correspondences were calculated directly.

Time	Phosphorites Explainable by Upwelling (Total Phosphorites)	Statistical Results
Paleozoic	50 (103)	21.54*
Cambrian	19 (40)	0.009*
Ordovician	10 (22)	0.30
Silurian	1 (5)	0.76
Devonian	8 (15)	0.0007*
Mississippian	2 (4)	0.08
Pennsylvanian	2 (5)	0.28
Permian	8 (12)	0.002*

* Results significant to at least $p < 0.05$.

Table 3. Results of statistical tests of correspondence between modeled upwelling zones and the distribution of chert (data from Drewry et al., 1974). The statistics for the individual time periods are the probabilities calculated directly.

Time	Siliceous Deposits Explainable by Upwelling (Total) Deposits)	Statistical Results
Paleozoic	63 (120)	40.10*
Cambrian	10 (19)	0.02*
Ordovician	12 (20)	0.04*
Silurian	5 (9)	0.06
Devonian	14 (30)	0.00005*
Mississippian	6 (22)	0.04*
Pennsylvanian	4 (7)	0.04*
Permian	11 (12)	1.2×10^{-6}*

* Results significant to at least $P < 0.05$.

DISCUSSION

Organic-rich rocks, phosphorites and chert often occur together regionally. For example, the Permian of the North American Cordillera has extensive deposits of all three bioproductite constituents (Fig. 7). Chert, organic-rich rocks, and phosphorites are also coextensive in the mid-continent region of North Amnerica in the Devonian and the North American Cordillera on the Mississippian. Chert, phosphorites, and "black shales" co-occur in China in the Cambrian (Fig. 1 and Appendix I; the Chinese Cambrian black shales were not included in Parrish [1982] because their organic content could not be determined).

There are several regions in which two bioproductite elements co-occur. Although there are exceptions, in general, organic-rich rocks and cherts tend to occur with phosphorites rather than with each other when only two of the elements are present. Organic-rich rocks and phosphorites occur together, without chert, in the Cambrian, Ordovician, and Mississippian of eastern North America (Figs. 1, 2, 5), the Cambrian of northern Siberia (Fig. 1), the Ordovician and Permian of Australia (Figs. 2, 7), and the Permian of northwesternmost India (Fig. 7). Phosphorites and cherts occur together (without organic-rich rocks) in numerous localities, for example, the

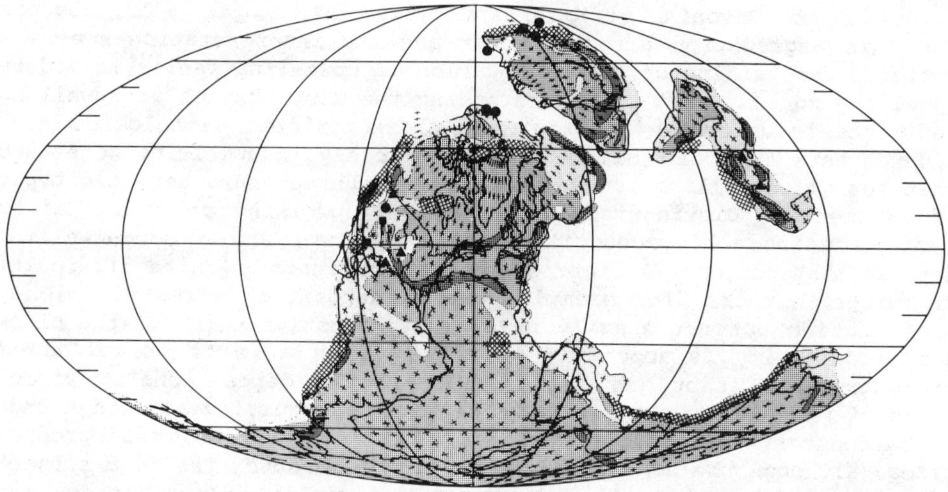

Fig. 6. Pennsylvanian phosphorites, chert and organic-rich rocks. Symbols and shading as in Fig. 1. Laurussia and Gondwana have collided and Kazakhstan and Siberia have collided, starting the assembly of Pangaea.

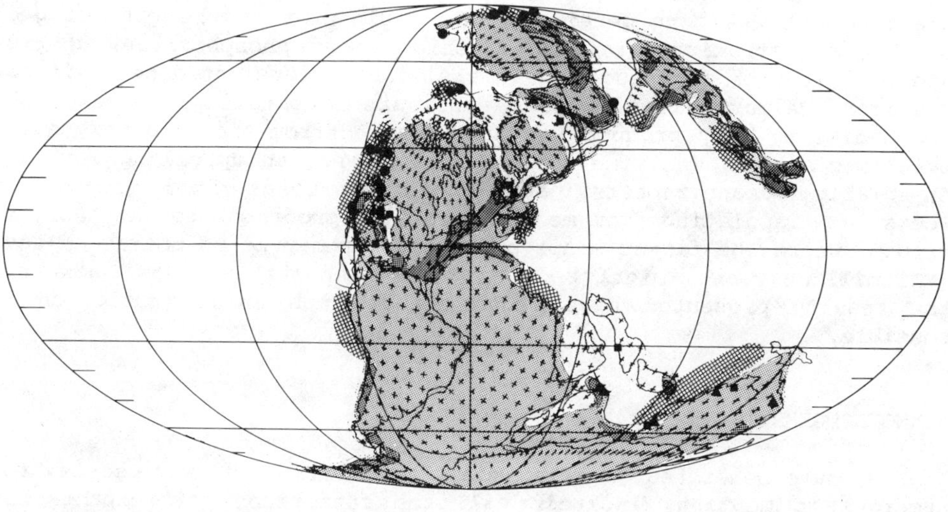

Fig. 7. Permian phosphorites, chert and organic-rich rocks. Symbols and shading as in Fig. 1. With the exception of China, Pangaea is assembled.

Cambrian of northeastern North America (Fig. 1) and the Ordovician, Silurian, and Devonian of Kazakhstan (Figs. 2, 3, 4). The reasons for this segregation are not clear and its interpretation awaits a better understanding of the segregation of upwelling facies in modern upwelling zones. Part of understanding the distribution of upwelling sediments is understanding that the characteristic lithologies considered here might not all be deposited solely in upwelling zones and that not all upwelling zones will have all three characteristic types of sediments. Distinguishing between an upwelling deposit that is merely missing a bioproductite element or two, and a non-upwelling deposit that contains a bioproductite element may be quite difficult. Organic-rich rocks, for example, can be deposited in restricted basins that do not necessarily have high productivity (e.g., the Black Sea productivity is about 100 $gC/m^2/yr$, comparable to normal shelf productivity [Tissot and Welte, 1978]). The depositional environments of phosphorites are also quite varied, occurring as hardgrounds in carbonates, for example (Jarvis, 1980), as well as in bioproductites, although the non-upwelling origin of phosphorites is a subject of debate. Biogenic siliceous rocks, especially diatomites and radiolarites, may be more closely related to upwelling than are the other two types of deposits. Modern marine diatom and radiolarian oozes occur under zones of high productivity in the oceans. Although there are exceptions, the data presented here indicate that the phosphorites and organic-rich rocks tend to be deposited at lower latitudes, whereas cherts occur at all latitudes. This tendency is best illustrated in the Carboniferous (Figs. 5 and 6).

The approach presented here is a first step to understanding the distribution of upwelling and its characteristic sediments in the distant past. A more detailed characterization of the precise geographic and stratigraphic relationships among phosphorites, cherts, and organic-rich rocks, in the global context presented here, is now required. Almost all of what we know about upwelling processes and the sediments they produce has been learned from the study of modern upwelling zones. The time scales to which modern upwelling workers, especially oceanographers, are accustomed range from minutes to weeks. Extrapolating from modern upwelling processes to the geologic record, where the finest scale of resolution may be as coarse as several million years, might seem an almost impossible task. However, the results presented here suggest that such an extrapolation is possible.

ACKNOWLEDGEMENTS

We have benefited greatly from discussions with many people, too numerous to mention, including all the conference participants and organizers. Funding for this project was supplied by Amoco Production Co. (International), Mobil Exploration and Producing Services, Shell Development Co., Chevron Inc., Exxon Production Research Co., and the National Science Foundation. We also acknowledge the inspiration of International Correlation Project 156 (Phosphorites). L. Lashmet aided in the collection of the data on phosphorites and drafted the figures. M. Westphall and C. Wenkam aided in the collection of data on chert.

REFERENCES

Adyshev, M.M., Kalmurzayev, K. Ye. and Shabalin, V.V., 1967, Distribution of phosphorus Cambrian-Ordovician sediments of the Sary Dzhaz River Basin (central Tien Shan), Doklady Akademia Nauk S.S.S.R., 173:190-191.

Ali, S.T., 1977, Phosphate deposits of Pakistan, in: "Phosphate Rock in the CENTO Region (Iran, Pakistan, and Turkey)," A.J.G. Notholt, ed., Working Group on Phosphates, Ankara, Turkey, 42-62.

Andersson, J.G., 1895, Über cambrische und silurische, phosphoritfürende Gesteine aus Schweden, Bulletin Geological Institution Uppsala, 2:133-238.

Baskakov, M.P., 1959, Phosphatic facies in the Paleozoic rocks of Kyzyl Kum, Doklady Akademia Nauk S.S.S.R., 124:177-178.

Berger, W.H., Diester-Haass, L. and Killingley, J.S., 1978, Upwelling of north-west Africa: The Holocene decrease as seen in carbon isotopes and sedimentological indicators, Oceanologica Acta, 1:3-7.

Bhatti, N.A., 1977, Phosphorite deposits of the Kakul-Mirpur area, Hazara District, N.W. Frontier Province, Pakistan, in: "Phosphate Rock in the CENTO Region (Iran, Pakistan and Turkey)," A.J.G. Notholt, ed., Working Group on Phosphates, Ankara, Turkey.

Blisset, A.H. and Callen, R.A., 1967, Myponga phosphorite deposits, Mineral Resources Review, 127:93-103.

Bramlette, M.N., 1946, The Monterey Formation of California and the origin of its siliceous rocks, United States Geological Survey Professional Paper 212, 57 pp.

Brasier, M.D., 1980, The Lower Cambrian transgression and glauconite-phosphate facies in western Europe, Journal of the Geological Society of London, 137:695-703.

Brass, G.W., Hay, W.W., Holser, W.T., Peterson, W.H., Saltzman, E., Sloan, J.L., II, and Southam, J.R., in press, Ocean circulation, plate tectonics, and climate, in: "Pre-Pleistocene Climates," W.H. Berger and J.C. Crowell, National Research Council Special Publication.

British Sulphur Corporation, 1971, "World Survey of Phosphate Deposits," 3rd edition, British Sulphur Corp., Ltd, London, 180 pp.

Brongersma-Sanders, M., 1948, The importance of upwelling water to vertebrate paleontology and oil geology, Verhandelingen der Koninklijke Nederlandsche Akademie van Wetenschappen, Afd. Naturkunde, 45:1-112.

Brown, D.A., Campbell, K.S.W. and Crook, K.A.W., 1968, "The Geological Evolution of Australia and New Zealand," Pergamon Press, New York, 409 pp.

Burnett, W.C., 1977, Geochemistry and origin of phosphorite deposits from off Peru and Chile, Geological Society of America, Bulletin, 88:813-823.

Bushinskii, G.I., 1969, "Old Phosphorites of Asia and Their Genesis," Akademia Nauk SSSR, Israel Program for Scientific Translation, Jerusalem, 266 pp.

Cathcart, J.B. and Schmidt, D.L., 1977, Middle Paleozoic sedimentary phosphate in the Pensacola Mountains, Antarctica, United States Geological Survey Professional Paper, 456E:E1-E18.

Chalyshev, V.I., 1968, The phosphorite assemblage of the Permian and Triassic deposits of the northern pre-Ural downwarp, Lithology Mineral Resources, 2:174-183.

Ch'ang Ta, 1959, "The Geology of China," Peking. Translated by the United States Department of Commerce, Office of Technical Services, Washington, 623 pp.

Chiang, N.-J., Wang, T.-C. and Chen, Y.-K., 1968, Cambrian stratigraphy of eastern Yunnan, International Geology Review, 10: 1153-1172.

Christie, R.L., 1980, Phosphate rock: selected references and deposits of interest, International Geological Correlation Project 156, Newsletter no. 6.

Cook, P.J., 1972, Petrology and geochemistry of the phosphate deposits of northwest Queensland, Australia, Economic Geology, 67:1193-1213.

Davies, D.C., 1875, The phosphorite deposits of North Wales, Quarterly Journal of the Geological Society of London, 31:357-367.

Diester-Haass, L. and Schrader, H., 1979, Neogene coastal upwelling history off northwest and southwest Africa, Marine Geology, 29: 39-53.

Donov, N.A., Edemskii, E.V., Elyanov, A.A., Ilyin, A.V., and Muzalevskii, M.M., 1969, Cambrian phosphorites in the Mongolian People's Republic, Soviet Geology, 3:55-60.

Douglas, R.J.W., 1970, Geology and economic minerals of Canada, Geological Survey of Canada Economic Geology Report No. 1, 838 pp.

Drewry, G.E., Ramsay, A.T.S. and Smith, A.G., 1974, Climatically controlled sediments, the geomagnetic field, and trade wind belts in Phanerozoic time, Journal of Geology, 82:531-553.

Dvortsova, K.I., 1958, Cambrian phosphorite deposits in the Kandyktas Moutains (south Kazakhstan), Doklady Academia Nauk S.S.S.R., 123: 961-962.

Fleming, P.J.G., 1974, Origin of some Cambrian bedded cherts, and other aspects of silicification in the Georgina Basin, Queensland, Geological Survey of Queensland Publication 358, 9 pp.

Geological Survey of Western Australia, 1975, The geology of Western Australia, Geological Survey of Western Australia Memoir 2, 541 pp.

Gimmel'farb, B.M. and Tushina, A.M., 1966, Principal phosphorite ore deposits of the Kara-Tau, Lithology Mineral Resources, 4:88-102.

Gimmel'farb, B.M. and Egorova, O.P., 1969, Geology of the Khubsugul phosphorite deposit, Mongolian People's Republic, Lithology Mineral Resources, 2:139-144.

Harrington, J.F., Ward, D.E. and McKelvey, V.E., 1966, Sources of fertilizer minerals in South America: a preliminary study, United States Geological Survey Bulletin 1240, 66 pp.

Heckel, P.H., 1977, Origin of phosphatic black shale facies in Pennsylvanian cyclothems of Mid-Continent North America, American Association of Petroleum Geologists, Bulletin, 61:1045-1068.

Hicks, H., 1875, On the occurrence of phosphates in the Cambrian rocks, Quarterly Journal of the Geological Society of London, 31(123):368-383.

Howard, P.F., 1972, Exploration for phosphorite in Australia--a case history, Economic Geology, 67:1180-1192.

Hsieh, C.Y. and Chao, C.H., 1948, Note on the phosphate deposits of China, Bulletin of the Geological Society of China, 28(1-2):71-74.

Hutchinson, R.D., 1962, Cambrian stratigraphy and trilobite faunas of southeastern Newfoundland, Geological Survey of Canada, Bulletin 88, 156 pp.

Isokangas, P., 1980, Finland, Mineral Deposits of Europe, 1:40-92.

Jarvis, I., 1980, Geochemistry of phosphatic chalks and hardgrounds from the Santonian to early Campanian (Cretaceous) of northern France, Journal of the Geological Society of London, 137:705-721.

Johns, R.K., 1961, South Australian rock phosphate deposits, Mining Review, Department of Mines South Australia, 114:22-30.

Kholodov, V.N., 1969, Secondary alterations of bedded phosphorites of Malyi Karatau in the hypergene zone, Lithology Mineral Resources, 3:176-288.

Kholodov, V.N. and Khoryakin, A.S., 1961, Origin of phosphatic conglomerate-breccia in Lesser Karatau, Doklady Akademia Nauk S.S.S.R., 135:1140-1142.

Knipper, A.L., 1958, Geosynclinal phosphorites in southern Ulu Tau, central Kazakhstan, Doklady Academia Nauk S.S.S.R., 115:609-611.

Latif, M.A., 1972, An occurrence of Palaeozoic phosphate rock in Hazara District, West Pakistan, Transactions of the Institute of Mining and Metallurgy, 81:B50-B53.

Martinsson, A., 1974, The Cambrian of Norden, in: "The Cambrian of the British Isles, Norden, and Spitsbergen," C.H. Holland, ed., Wiley and Sons, New York, 185-283.

McGugan, A. and Rapson, J.E., 1961, Stratigraphy of the Rocky Mountain Group (Permo-Carboniferous) Banff area, Alberta, Journal Alberta Society of Petroleum Geologists, 9:73-106.

McKelvey, V.E., Swanson, R.W. and Sheldon, R.P., 1952, The Permian phosphorite deposits of western United States, 19th International Geological Congress, Algiers, 11:45-64.

McKelvey, V.E., Williams, J.S., Sheldon, R.P., Cressman, E.R., Cheney, T.M., and Swanson, R.W., 1967, The Phosphoria, Park City, and Shedhorn Formations in western phosphate field, in:

"Anatomy of the Western Phosphate Field," L.A. Hale, ed., Intermountain Association of Geologists 15th Annual Field Conference, Salt Lake City, 15-34.

Meng, H.-H., 1959, The petrography of phosphorites of the Karatau Basin, Doklady Akademia Nauk S.S.S.R., 126:524-526.

Momper, J.A., 1978, Oil migration limitations suggested by geological and geochemical considerations, in: "Physical and Chemical Constraints on Petroleum Migration," W.H. Roberts, III and R.J. Cordell, eds., American Association of Petroleum Geologists Short Course, B1-B60.

Notholt, A.J.G., 1972, Paleozoic phosphate rocks in southwest Asia and the Near East, Transactions of the Institute of Mining and Metallurgy, Section B, Vol. 81, 789:B158-B163.

Notholt, A.J.G., 1979, Resources of Precambrian and Cambrian sedimentary phosphate rock, in: "Proterozoic-Cambrian Phosphorites," P.J. Cook and J.H. Shergold, eds., 1st International Field Workshop and Seminar, August, Australia, International Geological Correlation Project 156.

Parrish, J.T., 1982, Upwelling and petroleum source beds, with reference to the Paleozoic, American Association of Petroleum Geologists, Bulletin, 66:750-774.

Parrish, J.T. and Curtis, R.L., in press, Atmospheric circulation, upwelling and organic-rich rocks in the Mesozoic and Cenozoic, Palaeogeography, Palaeoclimatology, Palaeoecology.

Parrish, J.T., Ziegler, A.M., Scotese, C.R., and Kirschvink, J.L., in press, Early Cambrian paleogeography, paleoceanography, and phosphorites, in: "Proterozoic and Cambrian Phosphorites," J.H. Shergold and P.J. Cook, eds., Cambridge University Press, Cambridge and International Geological Correlation Project 156.

Patton, W.W., Jr. and Matzko, J.J., 1959, Phosphate deposits in northern Alaska, United States Geological Survey Professional Paper 302A, 17 pp.

Philippi, G.T., 1974, The influence of marine and terrestrial source material on the composition of petroleum, Geochimica et Cosmochimica Acta, 38:947-966.

Pisciotto, K.A. and Garrison, R.E., 1981, Lithofacies and depositional environments of the Monterey Formation, California, in: "The Monterey Formation and Related Siliceous Rocks of California," R.E. Garrison and R.G. Douglas, eds., Society of Economic Paleontologists and Mineralogists, Pacific Section, 97-122.

Polyanskii, N.V., 1977, Discovery of phosphate shows in lower Paleozoic deposits of the Chingiz-Tarbagatai meganticlinorium in eastern Kazakhstan, Lithology Mineral Resources, 12:108-110.

Poulsen, C., 1960, The Palaeozoic of Bornholm, 21st International Geological Congress Guide to Excursions A46 and C41, 15 pp.

Prian, J.-P., 1980, Caractérisation des paléoenvironments des phosphorites cambriennes du versant septentrional de la Montagne Noire (Sud du Massif central francais), Géologie Compareé des Gisements de Phosphates et de Pétrole, Bureau Recherches Géologique et Minères Document, 24:93-112.

Regnéll, G. and Hede, J.E., 1960, The lower Palaeozoic of Scania, the Silurian of Gotland, 21st International Geological Congress Guide to Excursions, A22, C17, 89 pp.

Rushton, A.W.A., 1974, The Cambrian of Wales and England, in: "The Cambrian of the British Isles, Norden and Spitsbergen," C.H. Holland, ed., Wiley and Sons, New York, 43-121.

Russell, R.T. and Trueman, N.A., 1971, The geology of the Duchess phosphate deposits, northwestern Queensland, Australia, Economic Geology, 66:1186-1214.

Samimi, M. and Ghasemipour, R., 1968, Phosphate deposits in Iran, in: "Phosphate Rock in the CENTO Region (Iran, Pakistan, and Turkey)," A.J.G. Notholt, ed., Working Group on Phosphates, Ankara, Turkey, 1-41.

Scotese, C.R., Bambach, R.K., Barton, C., Van der Voo, R., and Ziegler, A.M., 1979, Paleozoic base maps, Journal of Geology, 87:217-277.

Shcherbakova, M.N., 1968, Phosphorites from volcanic-sedimentary deposits of the Famennian Stage (north Balkhash area, central Kazakhstan), Lithology Mineral Resources, 6:765-767.

Sheldon, R.P., 1964, Paleolatitudinal and paleogeographic distribution of phosphorite, United States Geological Survey Professional Paper, 501C:C106-C113.

Shkol'nik, E.L., 1976, Phosphate presence in Silurian deposits of the Zeya-Selemdzha interfluve region, Geology and Geophysics, 17(7):93-96.

Shkol'nik, E.L. and Antipenko, V.G., 1976, On the primary phosphorites of the Uda-Shantar Basin, Geology and Geophysics, 17:114-117.

Smirnov, A.I., 1959, New data on elemental constitution of the phosphorites in the Karatau Basin, Doklady Akademia Nauk S.S.S.R., 125: 205-207.

Sokolova, Ye. A., 1961, The distribution of manganese and phosphorus in various types of rocks of the Usinsk Formation (Lower Cambrian of Kuznetsk Ala-Tau), Doklady Akademia Nauk S.S.S.R., 135: 1081-1083.

Southgate, P.N., 1980, Cambrian stromatolitic phosphorites from the Georgina Basin, Australia, Nature, 285:395-397.

Tissot, B. Durand, B., Espitalié, J., and Combaz, A., 1974, Influence of nature and diagenesis of organic matter in formation of petroleum, American Association of Petroleum Geologists, Bulletin, 58:499-506.

Tissot, B.P. and Welte, D.H., 1978, "Petroleum Formation and Occurrence," Springer-Verlag, New York, 538 pp.

Trask, P.D.,, 1932, "Origin and Environment of Source Sediments of Petroleum," Gulf Publishing Company, Houston, 323 pp.

Tuchkov, I.I., 1966, Phosphorites on the lower reaches of the Lena River, Lithology Mineral Resources, 4:103-118.

Tushina, A.M., 1968, Lithofacies characteristics of the formation of the phosphorite stratum of Karatau, Lithology Mineral Resources, 4:456-465.

Valentine, J.W. and Moores, E.M., 1971, Provinciality and diversity across the Permian-Triassic boundary, in: "The Permian and Triassic Systems and Their Mutual Boundary," Canadian Society of Petroleum Geologists Memoir, 2:759-766.

Vinogradov, A.P. (ed.), 1968, "Atlas of the Lithological-Paleogeographic Maps of the USSR," Vol. I. Precambrian, Cambrian, Ordovician, and Silurian, Akademia Nauk S.S.S.R., Moscow.

Vinogradov, A.P. (ed.), 1969, "Atlas of the Lithological-Paleogeographic Maps of the USSR," Vol. II. Devonian, Carboniferous, and Permian, Akademia Nauk S.S.S.R., Moscow.

Volkov, B.N., 1970, Some aspects of the formation of phosphorite deposits, Doklady Akademia Nauk S.S.S.R., 192:63-65.

Wang, C.C., 1942, The phosphate deposits of Talungten, Kunming, Yunnan (Summary), Geological Bulletin of the National Geological Survey of China, 35:39-40.

Wedepohl, K.H., 1969, Composition and abundance of common sedimentary rocks, in: "Handbook of Geochemistry," K.H. Wedepohl, ed., Vol. 1, Springer-Verlag, New York.

Zaitsev, N.S., 1970, The Khubsugul phosphate deposit, Doklady Akademia Nauk S.S.S.R., 192:391-394.

Zanin, Yu. N. and Krivoputskaya, L.M., 1977, Fine crystallographic structure of apatitic material in phosphorites and its geological interpretation, Lithology Mineral Resources, 12:312-323.

Ziegler, A.M., Hansen, K.S., Johnson, M.E., Kelley, M.A., Scotese, C.R., and Van der Voo, R., 1977, Silurian continental distributions, paleogeography, climatology, and biogeography, Tectonophysics, 40:13-51.

Ziegler, A.M., Bambach, R.K., Parrish, J.T., Barrett, S.F., Gierlowski, E.H., Parker, W.C., Raymond A., and Sepkoski, J.J., Jr., 1981, Paleozoic biogeography and climatology, in: "Paleobotany, Paleoecology, and Evolution," K.H. Niklas, ed., Praeger Publishers, New York, 231-266.

Appendix I. Data on Paleozoic phosphorites used in this study. Owing to the uneven distribution of data, individual localities all occurring with a 5° square latitude-longitude were combined into one data point for the maps. The term "composite" under the latitude-longitude indicates where data have been combined. "e"=Early, "m"=Middle, and "l"=Late; "undiff." indicates the age was not differentiated by the author(s). The locality numbers for Cambrian data taken from Bushinskii (1969) are also given. "B.S.C."=British Sulphur Corporation.

Plate	Lat.-Long.	Ages	Form in which the phosphorite occurs	Associated lithologies	References
CAMBRIAN					
China	$42^{\circ}N$ $82^{\circ}E$	e	nodules, beds, breccia	"coaly", siliceous shale; dolomite; bituminous marlstone	Bushinskii, 1969, loc. 10
China	$40^{\circ}N$ $92^{\circ}E$	e	nodules, beds	dark, argillaceous limestone	Bushinskii, 1969, loc. 11
China	$39^{\circ}N$ $107^{\circ}E$ (composite)	e	nodules, conglomerate, sand	shale, chert, cherty limestone, sandstone, marl, glauconite	Bushinskii, 1969, loc. 56-58; B.S.C., 1971, p. 160
China	$30^{\circ}N$ $105^{\circ}E$	e	beds, lenses, grains, flat-pebble conglomerate, nodules	"coaly", siliceous shale and marlstone; glauconite	Bushinskii, 1969, loc. 43; B.S.C., 1971, p. 161
China	$33^{\circ}N$ $112^{\circ}E$ (composite)	e	beds, nodules, lenses, cement	limestone, shale, marlstone, glauconite	Bushinskii, 1969, loc. 26, 27, 41, 42; B.S.C., 1971, p. 158
China	$32^{\circ}N$ $117^{\circ}E$	e	nodules	limestone, dolomite, shale, pyrite, glauconite	Bushinskii, 1969, loc. 28
China	$27^{\circ}N$ $102^{\circ}E$ (composite)	e	beds, nodules, lenses	sandstone; black shale; chert; marlstone; siliceous, dolomitic limestone; glauconite	Bushinskii, 1969, loc. 44, 45, 47; Wang, 1942; B.S.C. 1971, p. 161
China	$27^{\circ}N$ $107^{\circ}E$ (composite)	e	nodules, beds, oolite, pellets	organic-rich, siliceous shale; chert	Ch'ang Ta, 1959; Bushinskii, 1969, loc. 34-37
China	$28^{\circ}N$ $111^{\circ}E$ (composite)	e	beds, nodules, grains	"coaly", pyritic, black shale; chert	Bushinskii, 1969, loc. 38-40

Appendix I (cont.)

Plate	Lat.-Long.	Ages	Form in which the phosphorite occurs	Associated lithologies	References
China	29°N 116°E (composite)	e	beds, nodules	siliceous dolomite; siliceous and carbonaceous shale, chert	Ch'ang Ta, 1959; Bushinskii, 1969, loc. 29, 30
China	24°N 104°E (composite)	e	nodules, beds	dolomitic, pyritic, "coaly" shale; chert	Chiang et al., 1968; Bushinskii, 1969, loc. 46, 48; B.S.C., 1971, p. 160-161
China	23°N 113°E (composite)	e	nodules, pebbles	"coaly" shale, sandstone, shale, conglomerate	Bushinskii, 1969, loc. 32, 33
China	40°N 116°E	e,u	oolite, nodules, conglomerate	shale, conglomerate, pyrite, glauconite	Hsieh and Chao, 1948; B.S.C., 1971, p. 158-159
Kazakhstan	48°N 69°E (composite)	e,m	beds, nodules, veins, oolite, lenses	chert; carbonate; siliceous, pyritic shale; sandstone; siltstone	Bushinskii, 1969, loc. 1, 3; Notholt, 1972
Kazakhstan	43°N 69°E (composite)	e,m	beds, nodules, lenses, oolite, concretions, flat-pebble conglomerate	dolomite; radiolarian chert; siliceous, "coaly" shale	Bushinskii, 1969, loc. 6; Gimmel'farb and Tushina, 1966; Kholodov, 1969; Kholodov and Khoryakin, 1961; Smirnov, 1959; Tushina, 1968; Vinogradov, 1968;
Kazakhstan	43°N 73°E (composite)	m	concretions, oolite	dolomite; siliceous, "coaly" shale; black limestone; radiolarian chert; sandstone	Bushinskii, 1969, loc. 5, 7; Vinogradov, 1968
Kazakhstan	42°N 78°E (composite)	m	nodules	organic-rich, siliceous, pyritic, limestone and shale	Bushinskii, 1969, loc. 8, 9
Kazakhstan	52°N 73°E (composite)	e,m,l	oolite, lenses, beds, nodules	sandstone, limestone, siltstone, dolomite, conglomerate, siliceous shale	Bushinskii, 1969, loc. 2; Dvortsova, 1958; Knipper, 1958
Kazakhstan	49°N 79°E	m	not reported	siliceous (radiolarians), carbonaceous sediments	Polyanskii, 1977
Siberia	45°N 133°E (composite)	e	not reported	siliceous, "coaly" shale	Vinogradov, 1968; Bushinskii, 1969, loc. 21

Siberia	54°N 145°E	e	beds, lenses, breccia, phosphatic fossils	limestone, volcanics, chert, quartz	Vinogradov, 1968; Shkol'nik and Antipenko, 1976
Siberia	52°N 101°E (composite)	e	phosphatic sediments	sandstone, siliceous rocks, bituminous carbonates	Bushinskii, 1969, loc. 15, 16; Donov et al., 1969; Gimmel'farb and Egorova, 1969; Zaitsev and Ilyin, 1970
Siberia	72°N 127°E	undiff.	not reported	siliceous carbonate	Tuchkov, 1966
Siberia	54°N 89°E (composite)	e	nodules, breccia	carbonates; calcareous and siliceous, black shale	Bushinskii, 1969, loc. 12–14; Vinogradov, 1968; Sokolova, 1961
Laurentia	71°N 22°W	e,m	nodules	conglomerate, limestone, shale, sandstone, glauconite	Brasier, 1980
Laurentia	36°N 84°W	undiff.	pellets	not reported	B.S.C., 1971, p. 28
Laurentia	58°N 122°W	l	nodules	shale	Christie, 1980
Laurentia	68°N 137°W	undiff.	not reported	not reported	Christie, 1980
Baltica	61°N 11°E (composite)	e,m	conglomerate	sandstone, shale, glauconite	Brasier, 1980; Martinsson, 1974
Baltica	56°N 15°E (composite)	e,m,l	nodules, conglomerate, phosphatic sand	sandstone, conglomerate, black limestone, chert, pebbles, pyrite, glauconite	Brasier, 1980; Andersson, 1895; Regnell and Hede, 1960; Martinsson, 1974
Baltica	61°N 17°E	m	conglomerate	sandstone	Martinsson, 1974
Baltica	67°N 23°E	e	phosphatic sediments	conglomerate, sandstone	Martinsson, 1974
Gondwana	35°S 139°E (composite)	e	lenses, nodules, beds	limestone, shale, sandstone, laterite	Blisset and Callen, 1967; Callen, 1969; Johns, 1961
Gondwana	21°S 135°E (composite)	m	not reported	limestone, sandstone, shale, dolomite	B.S.C., 1971, p. 168
Gondwana	21°S 140°E (composite)	m	pellets, stromatolitic laminae	chert, limestone, siltstone, shale, dolomite	Cook, 1972; Howard, 1972; B.S.C. 1971, p. 168; Fleming, 1974, 1977; Russell and Trueman, 1971; Southgate, 1980

Appendix I (cont.)

Plate	Lat.-Long.	Ages	Form in which the phosphorite occurs	Associated lithologies	References
Gondwana	37°S 146°E	l	not reported	not reported	B.S.C., 1971, p 171
Gondwana	44°N 3°E (composite)	e,m	nodules, hardground	sandstone, limestone, black shale, pyrite, glauconite	Prian, 1980; Brasier, 1980
Avalon	47°N 54°W	m	nodules	siltstone, black slate, pyrite	Douglas, 1970; Hutchinson, 1962
Avalon	45°N 66°E	e,l	nodules	sandstone, black shale, limestone	Brasier, 1980; B.S.C., 1971, p. 3; Douglas, 1970
England	52°N 3°W (composite)	e	cement, phosphatic sediments	black shale, sandstone, argillaceous limestone, pyrite, glauconite	Hicks, 1875; Rushton, 1974; Brasier, 1980
ORDOVICIAN					
China	29°N 103°E	undiff.	phosphatic shells	shale, sandstone	Bushinskii, 1969
Kazakhstan	42°N 73°E	m	not reported	sandstone, shale, limestone, conglomerate	Vinogradov, 1968
Kazakhstan	42°N 78°E	e	not reported	not reported	Adyshev et al., 1967
Siberia	67°N 92°E	m	not reported	sandstone, shale, marl	Vinogradov, 1968
Siberia	62°N 97°E	m	not reported	shale, sandy limestone	Vinogradov, 1968
Siberia	58°N 107°E	m	not reported	sandstone, shale, siltstone	Vinogradov, 1968
Siberia	61°N 118°E	m	not reported	siltstone, shale	Vinogradov, 1968
Siberia	63°N 92°E	m	not reported	sandstone, shale, marl, limestone	Vinogradov, 1968
Siberia	57°N 102°E	m	not reported	sandstone, shale	Vinogradov, 1968
Siberia	74°N 90°E	m	not reported	sandy limestone, shale, dolomite	Vinogradov, 1968

Laurentia	35°N	98°W	undiff.	oolite, cement	not reported	B.S.C., 1971, p. 24
Laurentia	43°N	91°W	l	pellets, fossils, nodules	shale	B.S.C., 1971, p. 24-25
Laurentia	37°N	86°W	undiff.	not reported	limestone	B.S.C., 1971, p. 26
Laurentia	33°N	86°W	undiff.	shells, nodules	siliceous, glauconitic sandstone	B.S.C., 1971, p. 25
Baltica	56°N	14°E	m,l	nodules	shale, limestone, pyrite	Poulsen, 1960; Regnell and Hede, 1960
Baltica	59°N	29°E	e	not reported	sandstone	B.S.C., 1971, p. 141
Gondwana	18°N	67°W	undiff.	not reported	sandstone	Harrington et al., 1966; B.S.C., 1971, p. 65
Gondwana	23°S	130°E	m	pellets	sandstone, siltstone, glauconite	Brown et al., 1968
Gondwana	36°S	148°E	e,l	nodules, oolite, pellets, breccia	black shale, chert	Brown et al., 1968; Howard, 1972
Gondwana	15°S	130°E	e	not reported	glauconitic sandstone	Howard, 1972; Geol. Surv. W. Australia, 1975
Avalon	44°N	66°W	undiff.	nodules	limestone	B.S.C., 1971, p. 3
England	53°N	4°W	e	nodules	pyrite	Rushton, 1974
SILURIAN						
China	29°N	116°E	undiff.	not reported	shale	Ch'ang Ta, 1959
Kazakhstan	43°N	65°E	undiff.	not reported	limestone; siliceous, bituminous shale	Baskakov, 1959
Siberia	53°N	129°E	undiff.	*Lingula* shells	siltstone, shale	Shkol'nik, 1976

Appendix I (cont.)

Plate	Lat.-Long.	Ages	Form in which the phosphorite occurs	Associated lithologies	References
Baltica	54°N 58°E	m	not reported	sandstone, shale, chert	Volkov, 1970
England	53°N 3°W	undiff.	not reported	limestone, shale, chert	Davies, 1875
DEVONIAN					
Kazakhstan	48°N 75°E (composite)	l	nodules, shell replacement	sandstone	Notholt, 1972; Shcherbakova, 1968
Kazakhstan	42°N 76°E	undiff.	nodules	shale	Notholt, 1972
Siberia	68°N 89°E (composite)	e	not reported	siliceous and calcareous shale, dolomitic marl and limestone	Vinogradov, 1969
Siberia	67°N 100°E	e	not reported	calcareous shale, dolomitic marl	Vinogradov, 1969
Laurussia	54°N 58°E	m	not reported	shale, chert	Volkov, 1970
Laurussia	40°N 77°W	e,l	nodules	clay, sandstone, chert	B.S.C., 1971, p. 29
Laurussia	69°N 25°E	undiff.	not reported	dolomite, pyrite	Isokangas, 1980
Gondwana	34°N 73°E (composite)	e,l	pellets	dolomite, limestone, claystone, chert	Latif, 1972; Ali, 1975
Gondwana	32°N 56°E (composite)	l	nodules	sandstone, limestone, shale, bauxite	Notholt, 1972; Samini and Ghasemipour, 1968
Gondwana	28°N 52°E (composite)	l	thin beds	silty, black shale; sandstone	Samimi and Ghasemipour, 1968
Gondwana	35°N 52°E (composite)	l	nodules	sandstone, conglomerate, shale, limestone	Samimi and Ghasemipour, 1968
Gondwana	38°N 47°E (composite)	l	pellets	sandstone, shale	Samimi and Ghasemipour, 1968; B.S.C., 1971, p. 131

Gondwana	37°N 35°E (composite)	undiff.	nodules	clay, sandstone, chert	Notholt, 1972
Gondwana	43°N 2°W	undiff.	nodules	black shale	B.S.C., 1971, p. 136
Gondwana	84°S 56°W	undiff.	pebbles, cement	siltstone, sandstone, conglomerate	Cathcart and Schmidt, 1977
MISSISSIPPIAN					
Laurussia	42°N 112°W	l	not reported	shale, limestone	B.S.C., 1971, p. 8
Laurussia	36°N 88°W	undiff.	oolite	sandstone, shale, conglomerate	B.S.C., 1971, p. 26-27
Laurussia	68°N 132°W	undiff.	not reported	limestone, dolomite, chert	Patton and Matzko, 1959
Gondwana	34°N 73°E	e	pellets	dolomite, chert	Latif, 1972
PENNSYLVANIAN					
China	21°N 106°E (composite)	l	not reported	limestone	B.S.C., 1971, p. 163
China	11°N 104°E	l	cement	limestone	B.S.C., 1971, p. 165
Laurussia	32°N 97°W (composite)	e	nodules, oolite	shale, conglomerate, glauconitic clay	B.S.C., 1971, p. 23
Laurussia	38°N 95°W (composite)	m,l	nodules	bituminous shale	B.S.C., 1971, p. 24; Heckel, 1977
Laurussia	51°N 115°W	m	not reported	bituminous dolomite, sandstone, chert, pyrite	McGugan and Rapson, 1961
PERMIAN					
China	29°N 116°E	undiff.	nodules	sandy shale	Ch'ang Ta, 1959
Pangaea	27°S 116°E (composite)	e,m	not reported	shale, siltstone, limestone, greywacke	Howard, 1972; Geol. Surv. W. Australia, 1975

Appendix I (cont.)

Plate	Lat.-Long.	Ages	Form in which the phosphorite occurs	Associated lithologies	References
Pangaea	18°S 125°E (composite)	m,l	not reported	coal, shale, siltstone, ferruginous and calcareous sandstone, limestone	Howard, 1972; Geol. Surv. W. Australia, 1975
Pangaea	24°S 150°E (composite)	m	phosphatic sediments	sandstone, sandy limestone, siltstone	Brown et al., 1968; Howard, 1972
Pangaea	52°N 118°W	m	phosphatic sediments	sandstone, chert	Douglas, 1970
Pangaea	46°N 117°W (composite)	l	oolite, pellets, shell replacement	black shale, chert	B.S.C., 1971, p. 12; Sheldon, 1964; McKelvey et al., 1967
Pangaea	43°N 118°W (composite)	l	oolite, pellets, shell replacement	black shale, chert	B.S.C., 1971, p. 14, 17; Sheldon, 1964; McKelvey et al., 1967
Pangaea	43°N 112°W (composite)	l	oolite, pellets, shell replacement	black shale, chert	Sheldon, 1964; McKelvey et al., 1967
Pangaea	34°N 73°E	undiff.	lenses	dolomite, limestone, shale, sandstone, chert	Bhatti, 1977
Pangaea	64°N 58°E	undiff.	not reported	black shale, calcareous clay, chert	Chalyshev, 1968
Pangaea	57°N 133°W	undiff.	not reported	dolomite	Christie, 1980

PALEOZOIC BLACK SHALES IN RELATION TO CONTINENTAL MARGIN UPWELLING

Thomas J.M. Schopf

Department of Geophysical Sciences
University of Chicago
Chicago, Illinois 60637, U.S.A.

ABSTRACT

The geographic position of 26 better known Paleozoic black shales are plotted on appropriate paleogeographic maps. These occur in four major paleogeographic settings. A third to a half of the ancient black shales may be related to regions consistent with coastal upwelling as a factor in their origins: nine of the 26 (35%) occur in places where the inferred oceanographic setting is similar to "classical" upwelling of western margins of continents, seven of the 26 (28%) occur in east-west trending belts, perhaps analogous to the present setting with upwelling off northern South America. In contrast, six of the 26 (24%) occur along eastern sides of continents, perhaps analogous to modern basins off eastern Asia (such as the Sea of Japan) and 3 of the 26 (12%) occur in continental interiors.

INTRODUCTION

At various times from about 700 million to 225 million years ago, black shales formed on continental crust. Chemical analyses have been published for some of these shales. The rocks contain approximately 4% or more organic carbon by volume equivalent to approximately 2% or more organic carbon by weight (Schmoker, 1980), because organic matter is about half the density of the clay minerals composing the bulk of the rock. These rocks are not truly "black" on the Munsell system of colors. They best fit the medium gray to dark gray range (Hosterman and Whitlow, 1981), but to our eyes they look black and that has been the common term for them.

The origin of the deposits which gave rise to these rocks has been a recurrent problem for geologists for more than a century (reviewed by Potter, Maynard and Pryor, 1980). No readily apparent general modern analogue has been accepted for many black shales. The major black shale deposits of the geologic record are marine in origin; thus, most deposits are not directly analogous to brackish Black Sea type deposits of enclosed basins. In addition, a few of the ancient deposits are of far greater areal extent than any well-known modern organic-rich shale.

The purpose of the present paper is to inquire if there is a plausible relationship between black shale deposits of the Paleozoic geologic record, and inferred regions of continental margin upwelling. As far as I am aware, this is the first systematic attempt to relate these deposits to oceanographic conditions in a paleogeographic setting. The detailed paleogeographic setting of most of these deposits is known so provisionally, however, that it appears at this time to serve no useful purpose to try to be more specific with regard to inferences about depth, oxygen content of the water, etc., especially as these factors are very difficult to determine in paleoceanography (see chapters in Schopf, 1980). As in most papers of this type, the author has not been able to work directly on most of the deposits which are discussed. Also, I have emphasized the oceanographic setting rather than the paleotectonic setting, but both should be investigated in future work. Nevertheless, it is hoped that those who wish to consider black shales in the larger oceanographic setting may find the paper of interest.

Geographic Setting

I thought it would be of interest to see how black shales sort themselves out geographically. I have plotted the approximate position of 26 of the (at this time) best known Paleozoic marine deposits which, loosely speaking, fall under the rubric of black shale. Specifics of the deposits are given in Table 1. (Other Paleozoic black shales have existed, and some of these have been eroded away, and are today represented by a minor deposit. Yet other organic-rich shales are best known from oil company records, and these are certainly underrepresented here, however, there are more than enough black shales which are known for us to wonder how they were formed.)

The deposits are arranged in four groups according to geologic age, and their distribution is plotted on appropriate Ziegler-Scotese continental reconstructions of the Paleozoic era (C. Scotese, pers. comm.). As Ziegler and Scotese are the first to acknowledge, many aspects of these or any other Paleozoic reconstructions at present are unsettled. [For example, the well known "Atlantic" Cambrian and Ordovician fauna which strongly indicates a close biogeographic connection (Wilson, 1966; Bergström, 1973), is split apart in their Cambrian reconstruction]. Indeed, if there is a strong conflict between

Table 1. Deposits selected for discussion

Formation	Maximum Thickness (m); and Age*	Place	Comments	References
1. Glenogle Shale	600; E. Ord.	W. Canada; southern Cordilleran geosyncline	Graptolites	Douglas et al, 1970: 382, 391; Kummel, 1961:635.
2. Marathon, Ouachitas and Arbuckles shales	<100; E. Ord.	S. United States	Graptolites	Berry, 1960:36.
3. Deepkill Shale	<100; E. Ord.	New York	Graptolites	Berry, 1960:36.
4. Trout Brook, MacLean Brook and equivalents	400; M.-L. Camb.	Cape Breton Is.; New Brunswick	Trilobites	North, 1971:241; Poole et al, 1971:240.
5. Manuels River, Clarenville, Elliott Cove and equivalents	300; M.-L. Camb.; and E. Ord.	SE Newfoundland	Trilobites	North, 1971:237; Poole et al., 1971: 249.
6. Unnamed	50; E.-M. Camb.	Anabara Massif, Siberian Platform		Nalivkin, 1973:212.
7. Oakdale Fm. and equivalents	300; L. Ord.	NSW, Australia; Tasman Geosyncline		Packam et al., 1969: 82, 89, Fig. 3.9.
8. Goldie Shale	6000; L. Camb.-E. Ord.	Victoria, Australia; Tasman Geosyncline		Thomas, Spencer-Jones and Tattam, 1976:14.
9. Ibbett Bay	1000; E. Ord.-E. Sil.	Arctic Canada; Franklin Geosyncline		Thorsteinsson and Tozer, 1970:559.
10. Cape Phillips	1000; L. Ord.-Dev.	Arctic Canada; Franklin Geosyncline		Thorsteinsson and Tozer, 1970:559.
11. Unnamed	3500; Ord.-E. Sil.	Siberian Platform	Graptolites	Vinogradov et al., 1973:185.
12. Utica Shale and equivalents	300; M.-L. Ord.	E. North America	Graptolites	Sweet and Bergström, 1976:149.

Table 1. Continued

Formation	Maximum Thickness (m); and Age	Place	Comments	References
13. Kukruse horizon	3; M.-L. Ord.	Estonia	"Spores"	Nalivkin, 1973:66, 68, 164; Matveyev, 1974:12.
14. Goldwyer	700; M. Ord.	Canning Basin, NW Australia	Graptolites	McTavish and Legg, 1976:464.
15. Gisbornian, Eastonian, Bolindian equivalents	500 m; M.-L. Ord.	Victoria, Australia; Lachlan Geosyncline	Graptolites	Brown, Campbell and Crook, 1968:78.
16. Various	1000; E.-M. Ord.	New Zealand; Geosynclinal	Graptolites	Brown, Campbell and Crook, 1968:87.
17. Kayak Shale	200; E. Carb.	Brooks Range, Alaska	Graptolites	Bowsher and Dutro, 1957:6;
"Black Shale"	E.-M. Dev.		Graptolites	Gryc et al.,1967:712
18. Prong's Creek and equivalents;	800; M. Dev.	NWT, Canada, Cordilleran Geosyncline		Douglas et al.,1970: 400;
Besa River Shale	300; M. Dev.-E. Carb.			Norris, 1967:758; Douglas et al.,1970: 403; Pelzer, 1966.
19. Chainman Shale	1500; E. Carb.	W. North America		Summarized in Keroher et al, 1967: 706.
20. Chattanooga Shale Ohio Shale, and equivalents	200; L. Dev.-E. Carb.	E. North America		Duncan and Swanson, 1965:11; Conant and Swanson, 1961.
21. Unnamed units	100; L. Dev.-E. Carb.	N. South America	Chattanooga Equivalents	S. Barrett, personal communication.
22. Black Slates	50; M. Dev.	Thuringia, C. Germany	Graptolites	Grabe, Schlegel and Wucher, 1967:1285.

Table 1. Continued

Formation	Maximum Thickness (m); and Age	Place	Comments	References
23. Barzasski	50; M. Dev.	Kuznetsk Basin, Kazakhastan		Nalivkin, 1973:471
24. Bonaparte Beds	2000; L. Dev.-E Carb.	W. Australia		Playford et al., 1975:378
25. Phosphoria	7; Perm.	W. North America		Swanson, 1960:18
26. Iratí	<50; L. Perm.	S. Brazil-Uruguay		Kraemer, 1950:15; Padula, 1969:591

* Abbreviations used in table include E. (Early), M. (Middle), L. (Late), Camb. (Cambrian), Ord. (Ordovician), Sil. (Silurian), Dev. (Devonian), Carb. (Carboniferous), Perm. (Permian), Sh. (Shale), Fm. (Formation), C. (Central). Data on volumes of some of these "oil shales" are given by Hodgkins (1961), Duncan and Swanson (1965) and Imperial Mineral Resources Bureau (1924).

one's oceanographic intuition and the suggested reconstruction, there is a good chance that the intuition is correct. This should be taken into account in perusing these charts.

The four particular times I have chosen provide a framework for arranging the data. In terms of time dependence, the amount of organic carbon per unit time may vary by a factor of 2 over time intervals of 10 to 20 million years, or by a factor of 3 over 50 m.y. (Budyko and Ronov, 1979, Table 1; Ronov, 1976, Table 3). This is not a lot of variation by most geologic standards. Indeed, if Garrels and Lerman (1981) are correct, variation in the organic carbon reservoir through the Phanerozoic may be even less than these earlier estimates (see also Schopf, 1980:229).

On these reconstructions, continent-size pieces are supposed to reflect the sizes of plates at the time of the reconstruction. The plates are positioned by taking into account evidence from paleomagnetism, paleoclimatology, geology, and biogeography. The equator, 30° and 45° latitude lines are shown. The four major continental plates indicated are: (1) North America, (2) Europe and the Baltic Shield, (3) Siberian Shield, and (4) Gondwana with Africa, South America, Antarctica, Australia and India. Black shales are known from each of these four areas. Data on the black shale deposits are chiefly from articles, monographs and books on the geology of different countries (see Table 1).

In the first map (Fig. 1), the deposits are (1) Glenogle Shale from the southern Cordilleran geosyncline of western Canada, (2) shales of the Marathon, Ouachita and Arbuckle regions of southern United States, (3) the Deepkill Shale of eastern North America, (4) and (5) units of Cape Breton Island and Newfoundland, (6) unnamed units of the Siberian platform, and (7) the Oakdale and equivalents

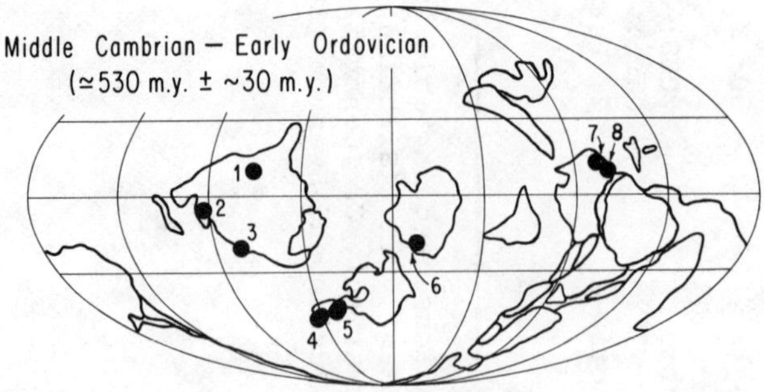

Fig. 1. Paleographic position of selected black shale deposits of the early Cambiran to early Ordovician. See Table 1 for localities 1-8. Map courtesy C.R. Scotese and A.M. Ziegler.

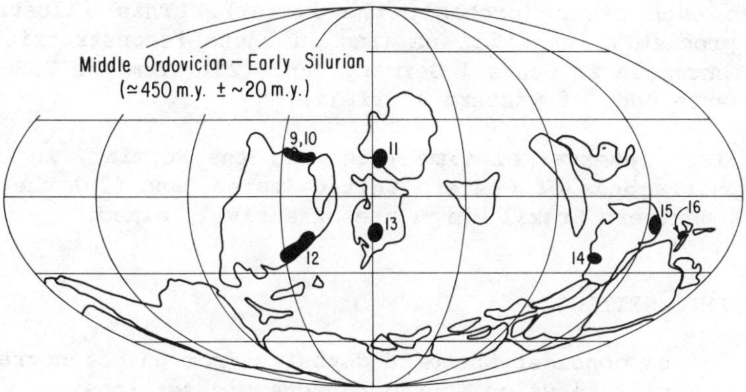

Fig. 2. Paleogeographic position of selected black shale deposits of the early Ordovician to Devonian. See Table 1 for localities 9-16. Map courtesy C.R. Scotese and A.M. Ziegler.

and (8) the Goldie Shale, both of eastern Australia and the Tasman geosyncline. The Goldie Shale is reported to have a thickness of 6000 meters of chiefly black shale, which represents a sediment thickness of two to nine times that, as the deposits initially are 50 to 90% water.

The second unit of time (Fig. 2) includes the Middle Ordovician to Early Silurian. The same four continental regions apply as before. The deposits are: (9) and (10) the Ibbett Bay and Cape Phillips of the Franklin Geosyncline of the present Arctic region; (11) unnamed units of the Siberian Platform; (12) the Utica Shale and its equivalents, which occur from the southern Appalachians to Quebec City along the Appalachian Geosyncline; (13) the Kukruse horizon of Estonia, one of the most heavily mined oil shales; (14) the Goldwyer of the Canning Basin, N.W. Australia; and (15) and (16) a variety of shales of Victoria, Australia, and of New Zealand, described as of geosynclinal origin.

The third time unit (Fig. 3) encompasses deposits chiefly of Middle Devonian to Early Carboniferous age. Beginning with this third time interval, land plants have made their appearance and the Baltic Shield has joined with North America. The deposits are (17) the Kayak Shale and other black shales of the Brooks Range, Alaska; (18) the Prong's Creek and Besa River Shale of the Cordilleran Geosyncline of the NWT, Canada; (19) the Chainman Shale (of later Mississippian age) of Nevada; (20) the Chattanooga Shale, and its equivalents, including (21) unnamed deposits in northern South America. The affinities of the South American fauna are said to be entirely Appalachian in aspect (S. Barrett, pers. comm.), and so, either that part of South America belongs to the north, or the positions of Gondwana and North America have to be adjusted to bring them much closer to each other (probably the latter). This illustrates the type of problem which still remains in these reconstructions. (22) is from Thuringia in central Germany, and (23) from the USSR; (24) is the Bonaparte Beds of western Australia.

The last interval of time (Fig. 4), the Permian, includes (25) the Phosphoria beds of western United States, and (26) the Iratí oil shales of southern Brazil which are extensively mined.

OCEANOGRAPHIC SETTING

Now let us consider these 26 deposits from an oceanographic perspective. I will focus on four oceanographic settings:

1. "classical" coastal upwelling, i.e., along western sides of continents;
2. basins of eastern sides of continents perhaps analogous to the Sea of Japan and other similarly situated basins;
3. continental interior situations, such as exist today in the Baltic and Black seas; and
4. east-west-trending basins, perhaps as along the northern coast of modern South America.

In the Middle Cambrian to Early Ordovician perspective (Fig. 1), the classical upwelling situation may have existed along the whole westward-facing margin of North America from the Marathon region in West Texas through the Ouachitas, Arbuckles and eastern New York in the Deepkill Shale, and possibly extending into Quebec. As modern analogues, one might think in terms of a possible geographical admixture of the California borderlands and the situation off Peru. The Siberian platform deposits (6) may also be of this classical type. In contrast, basin deposits along the eastern sides of continents, perhaps analogous to the Sea of Japan, might be responsible for deposits of the Baltic Shield (4,5). The Australian deposits (7,8) and the Glenogle Shale of Canada (1) are described as forming in geosynclinal troughs adjacent to a continental margin, presumably with east to west tropical circulation. Recall that all of the Cambrian to Early Ordovician deposits lack terrestrial organic matter, as land plants are not known until the Late Silurian or Early Devonian.

On the second map (Fig. 2), for the Middle Ordovician to Early Silurian, only the Australian Canning Basin deposits (14) seem likely to have been in a classical upwelling situation along western margins of a continent. Eastern margins include the Utica Shale (12) and the Australian (15) and New Zealand deposits (16). East-west-trending marginal deposits include the Ibbett Bay (9) and Cape Phillips (10) of the Franklinian Geosyncline, and the Siberian Platform (11) deposits. The thin but very organic-rich Kukruse horizon (13) may have been an internal basin. Similar to the earlier deposits, the Middle

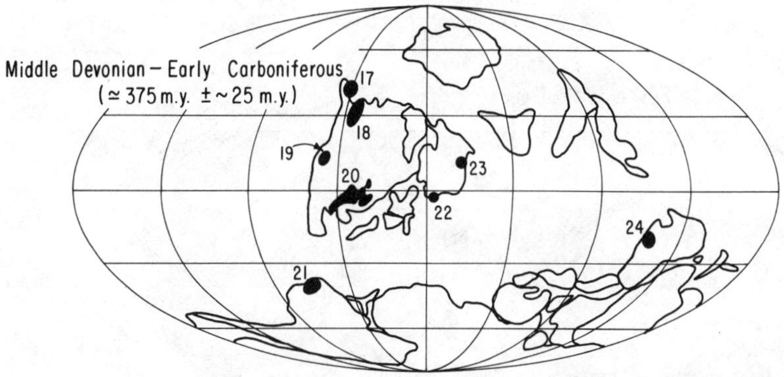

Fig. 3. Paleogeographic position of selected black shale deposits of the early Devonian to early Carboniferous. See Table 1 for localities 17-24. Map courtesy C.R. Scotese and A.M. Ziegler.

Ordovician to Early Silurian organic deposits lack terrestrial organic matter.

On the third map (Fig. 3), the Middle Devonian to Early Carboniferous, the Bonapart beds (24) of Australia, the Chainman Shale (19) of western North America, the Cordilleran Geosyncline deposits (18), and possibly the Alaskan (17) Brooks Range black shales seem to be along the western continental margin, in the right areas for classical upwelling to have contributed to the extensive organic production. Only the USSR deposits (23) are fairly certain to have been formed along the eastern margin of a continent. The black shales of Thuringia (22) may have been situated in an east-west-trending tropical belt. Lastly, the Chattanooga and coeval deposits (20,21) appear to have formed in a large interior seaway. This conclusion, which had been favored for the Chattanooga and similar-aged deposits by many previous students (e.g., Conant and Swanson, 1961; Lewis and Schwietering, 1971) is reflected in these present paleogeographic data. Marine sediments of tropical areas (such as the Chattanooga) are characterized in the modern world by much runoff and, especially near shore, may have a clearly recognizable component of terrestrial organic matter. Examples include sediments of the Niger Delta (Klingebiel, 1976), the inner shelf of the northern Gulf of Mexico (Hedges and Parker, 1976), and western United States (Peters, Sweeney and Kaplan, 1978), and in the Cariaco Trench (north of Venezuela) from a lower stand of sea level (Deuser, 1973). The hydrogen content of the organic matter in the central and eastern exposures of the Chattanooga is approximately 5% (moisture and ash-free basis; Breger and Brown, 1963). Values this low are closer to lignin (5.5) than to protein (6.5 to 7.3) (Emery, 1960:284). Therefore, organic matter in the Chattanooga Shale was believed to be "primarily of humic (terrestrial) origin" (Breger and Brown, 1962:224).

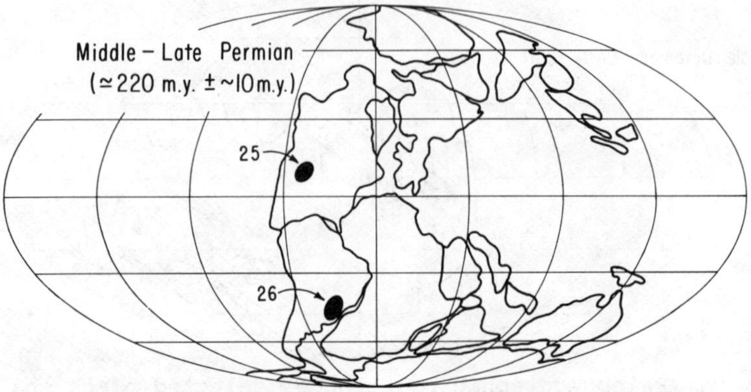

Fig. 4. Paleogeographic position of selected black shale deposits of the Permian. See Table 1 for localities 25-26. Map courtesy C.R. Scotese and A.M. Ziegler.

On the fourth and last map (Fig. 4), the Phosphoria (25) may represent a classical upwelling situation, as has been favored for more than 40 years by Kazakov (1937) and his later followers (reviewed by Sheldon, 1981). The Iratí shale (26), on the other hand, is pretty clearly a large interior basin, with the Baltic perhaps the best general modern analogue.

DISCUSSION

The most general conclusion further documents what many authors have understood for decades (e.g., Woolnough, 1937)--that black shales can form in different geologic settings. These deposits are found here to occur in four oceanographic settings as summarized in Table 2.

1. The first group would include those situations where the inferred oceanographic setting is similar to "classical" upwelling of western margins of continents. Among various aspects of ocean circulation, upwelling appears to be ready-made for geologic recognition because the strong signal of deposits is confined to a particular geographic setting. The marine sedimentological productivity signal can be differentiated from that related to river drainage because of the characteristic upwelling association with marine organic matter instead of with terrigenous organic matter. This first group of black shales has examples in each of the four time intervals; specifically, the Marathon to Deepkill deposits of the first time interval; the Canning Basin deposits of the second time interval; the Bonaparte beds, Chainman Formation and possibly the Alaskan and NWT geosynclinal deposits of the third time interval; and the Phosphoria of the fourth interval for a total of 9 of the 26 deposits mapped, or approximately 35%.

2. The second group of deposits would include those of the east-west-trending belts, perhaps analogous to the situation off northern South America. The deposits include the Glenogle of the Cordillerian Geosyncline and the Goldie Shale and Oakdale Shale of the Tasman Geosyncline of Australia of the first time interval; the Ibbett Bay and Cape Phillips Franklinian Geosyncline and Siberian platform deposits of the second; and the German Thuringian black slates of the third time interval. Seven of the 26 deposits are in this group, or approximately 28%.

3. The third group of black shales includes deposits along eastern sides of continents, perhaps oceanographically analogous to modern basins off eastern Asia (such as the Sea of Japan). In the first time interval these deposits include the Cape Breton and Newfoundland deposits; in the second, the Utica Shale of eastern North America and similar deposits of eastern Australia and New Zealand. The Kazakhastanian deposits are included in the third time interval. In total, 6 of 26 units appear to belong to this third group of black shales, or approximately 24%.

4. The fourth and last group of black shales includes three deposits (or 12%)--those of continental interiors. Here occur the Kukruse of the second time interval, the Chattanooga and related beds of the third time interval, and the Iratí of the fourth time interval. Among these three deposits, the Chattanooga is exceptional for its size (on the order of 10^6 km^2, Conant and Swanson, 1961), and for the fact that it was marine throughout nearly all of its formation. The other two deposits were, at least in part, of brackish origin (see Padula, 1969, on the Iratí).

The marine portion of the organic matter of black shales consists chiefly of phytoplankton, as delivered to the bottom by zooplankton fecal pellets (Porter and Robbins, 1981). W. Stürmer has discovered comparable Devonian zooplankton in his x-radiographs of Bundenbach black shales, and these forms may have been responsible for "conveyor belt" transport of phytoplankton to the bottom of Devonian seas.

The initial major factor in the accumulation of black shales appears to be rate of sedimentation. In modern sediments, sedimentation rate is closely correlated with burial of organic carbon over three orders of magnitude of rate of sedimentation (Heath, Moore and Dauphin, 1977; Müller and Suess, 1979; Suess and Müller, 1980). Preservation of shales with 4 to 10% organic carbon is consistent with organic production on the order of 10^3 $gC/m^2/y$, if sedimentation rate is on the order of 1 to 10 $cm/10^3/y$ (Suess and Müller, 1980, Fig. 2). The strong positive correlation between sedimentation rate and burial of organic carbon is also reflected in ancient marine sediments (Ibach, 1982).

Table 2. Paleozoic black shales arranged by inferred oceanographic and geographic setting, and by geologic age.

Oceanographic and Geographic Setting	Total	Geologic Age			
I. Western side of ocean	9	M. Cambrian to E. Ordovician Marathon etc. Deepkill Sh. Siberian platform	M. Ordovician to E. Silurian Canning Basin	M. Devonian to E. Carboniferous Bonaparte Beds Chainman Fm. (?)Brooks range deposits NWT Cordilleran geosyncline deposits	Permian Phosphoria
II. East-west trending	7	Glenogle Goldie Sh. Oakdale Fm.	Ibbett Bay Cape Phillips Siberian Platform	Thuringia	
III. Eastern side of ocean	6	Cape Breton Is. units Newfoundland units	Utica Sh. Gisbornian etc. New Zealand units	Kazakhakstan	
IV. Continental interior	4	---	Kukruse	Chattanooga Ohio Sh., northern S. America equivalents	Iratí
	26				

Burial is critical for the preservation of organic carbon because burial sets a limit on the extent of oxidation. Organic carbon is oxidized first by oxygen, then (if oxygen is depleted) by nitrogen compounds, then, if nitrate, nitrite and N_2O are depleted, by bacterial sulfate reduction (Richards and Redfield, 1954; Orr and Gaines, 1973; Cohen, 1978; Wilson, 1978). In the water column, sulfate is always in excess and, thus, organic carbon will continue to be oxidized, hence yielding no large increase in organic carbon in anoxic water columns (Richards, 1971). Once organic matter is buried (and hence removed from the influence of the water column), sulfate is in limited supply. The more rapid the sedimentation rate, the more rapidly organic matter is removed by oxidizing molecules.

Some black shales, however, are deposited with sedimentation rates of $\leqslant 1$ cm/10^3 years, and yet have organic carbon contents of several percent [e.g., the Ordovician Utica Shale (Churkin, Carter and Johnson, 1977) and the Devonian Chattanooga shale]. These high values of organic carbon are in excess of those which would be expected from this sedimentation rate under present-day conditions. This anomalous accumulation is usually attributed to one of two causes.

i. The "excess" organic carbon may not be excess because the time interval for deposition of individual beds has been artificially lengthened owing to a rate of sedimentation which is given as an <u>average</u> over long periods of time. In point of fact, sedimentation is episodic, and the rate of sedimentation is scale dependent; rate of sedimentation $\simeq$ 1/duration of period being sampled (Schindel, 1980). Thus, the true rate of sedimentation for individual shale layers is much higher than a time-averaged value for all shale layers. In the case of the Ordovician Utica shale, the paleogeographic setting is along the eastern margin of a continent. Neither the presumed wind system nor the geographic setting is consistent with upwelling. The source of the organic matter was marine (this is prior to the evolution of land plants). Possibly the "excess carbon" of this major black shale deposit is due to <u>episodic</u> high sedimentation rate owing to rapid burial by turbidites (Cisne, 1973).

ii. The "excess" organic carbon may be a true excess (after sedimentation rate has been factored out) caused by a higher than normal preservation: the organic matter is not oxidized in an anoxic environment. Indeed, a positive correlation between the occurrence of water low in oxygen and the occurrence of organic carbon in underlying sediments was specifically tested and confirmed by Richards and Redfield in 1954 along the northern margin of the Gulf of Mexico. This has subsequently been substantiated in many regions (Emery and Uchupi, 1972:303; Demaison and Moore, 1980), sometimes owing to the occurrence of more refractile terrestrial organic matter, as in the Black Sea sediments (Pelet and Debyser, 1977). This association of low oxygen and high organic carbon content is a common feature of continental margin sedimentation (as emphasized and much more fully discussed by many authors, e.g., Schlanger and Jenkyns, 1976; Arthur and Schlanger, 1979, and references therein) and of silled anoxic basins (33 of which are known today, as reviewed by Deuser, 1975). Whether this explanation applies to particular deposits must be judged after the effect of sedimentation rate is evaluated.

ACKNOWLEDGEMENTS

For helpful comments, information, or earlier review of a draft of this paper, I thank C.R. Scotese, H. Jenkyns, W. Deuser, H. Thierstein, M. Arthur, J.F. Read, D. Smith, and participants at the Coastal Upwelling Conference.

REFERENCES

Arthur, M.A. and Schlanger, S.O., 1979, Cretaceous oceanic anoxic events as causal factors in development of reef-reservoired giant oil fields, American Association of Petroleum Geologists, Bulletin, 63:870-885.

Bergström, S.M., 1973, Ordovician conodonts, in: "Atlas of Biogeography," A. Hallam, ed., Elsevier, Amsterdam, 47-58.

Berry, W.B.N., 1960, Graptolite faunas of the Marathon region, West Texas, University of Texas, Publication No. 6005, 179 pp.

Bowsher, A.L. and Dutro, J.T., Jr., 1957, The Paleozoic section in the Shainin Lake Area, Central Brooks Range, Alaska, U.S. Geological Survey Professional Paper, 303A:1-39.

Breger, I.A. and Brown, A., 1962, Kerogen in the Chattanooga Shale, Science, 137:221-224.

Breger, I.A. and Brown, A., 1963, Distribution and types of organic matter in a barred marine basin, Transactions New York Academy of Science, Series II, 25:741-755.

Brown, D.A., Campbell, K.S.W. and Crook, K.A.W., 1968, "The Geological Evolution of Australia and New Zealand," Pergamon Press, London, 409 pp.

Budyko, M.I. and Ronov, A.B., 1979, Chemical evolution of the atmosphere in the Phanerozoic, Geochemistry International, 16(3):1-9.

Churkin, M., Jr., Carter, C. and Johnson, B.R., 1977, Subdivision of Ordovician and Silurian time scale using accumulation rates of graptolitic shale, Geology, 5:452-456.

Cisne, J.L., 1973, Beecher's trilobite bed revisited: ecology of an Ordovician deepwater fauna, Postilla, 160, 25 pp.

Cohen, Y., 1978, Consumption of dissolved nitrous oxide in an anoxic basin, Saanich Inlet, British Columbia, Nature, 272:235-237.

Conant, L.C. and Swanson, V.E., 1961, Chattanooga Shale and related rocks of Central Tennessee and nearby areas, U.S. Geological Survey Professional Paper 357, 91 pp.

Demaison, G.J. and Moore, G.T., 1980, Anoxic environments and oil source bed genesis, Organic Geochemistry, 2:9-31.

Deuser, W.G., 1973, Cariaco Trench: oxidation of organic matter and residence time of anoxic water, Nature, 242:601-603.

Deuser, W.G., 1975, Reducing environments, in: "Chemical Oceanography," 2nd Edition, J.P. Riley and G. Skirrow, eds., Academic Press, London, 3:1-37.

Douglas, R.J.W., Gabrielse, H., Wheeler, J.O., Stott, D.F., and Belyea, H.R., 1970, Geology of western Canada, in: "Geology and Economic Minerals of Canada," R.J.W. Douglas, ed., Geological Survey of Canada, Economic Geology Report No. 1, Department of Energy, Mines and Resources, 366-488.

Duncan, D.C. and Swanson, V.E., 1965, Organic-rich shale of the United States and world land areas, U.S. Geological Survey Circular 523, 30 pp.

Emery, K.O., 1960, "The Sea off Southern California," John Wiley, New York, 366 pp.

Emery, K.O. and Uchupi, E., 1972, "Western North Atlantic Ocean: Topography, Rocks, Structure, Water, Life, and Sediment," American Association of Petroleum Geologists, Memoir, 17, 532 pp.

Garrels, R.M. and Lerman, A., 1981, Phanerozoic cycles of sedimentary carbon and sulfur, Proceedings of the National Academy of Sciences, 78:4652-4655.

Grabe, R., Schlegel, G. and Wucher, K., 1967, Environment and paleogeography of the Devonian in the area of the Berga Anticline, Thuringia, Germany, in: "International Symposium on the Devonian System," Vol. 2, D.H. Oswald, ed., Alberta Society of Petroleum Geologists, Calgary, 1283-1296.

Gryc, G., Dutro, J.T., Jr., Brosge, W.G., Tailleur, I.L., and Churkin, M., Jr., 1967, Devonian of Alaska, in: "International Symposium on the Devonian System," Vol. 1, D.H. Oswald, ed., Alberta Society of Petroleum Geologists, Calgary, 703-716.

Heath, G.R., Moore, T.C., Jr. and Dauphin, J.P., 1977, Organic carbon in deep-sea sediments, in: "The Fate of Fossil Fuel CO_2 in the Oceans," N.R. Anderson and A. Malahoff, eds., Plenum Press, New York, 605-625.

Hedges, J.I. and Parker, P.L., 1976, Land-derived organic matter in surface sediments from the Gulf of Mexico, Geochimica et Cosmochimica Acta, 40:1019-1029.

Hodgkins, J.A., 1961, "Soviet Power: Energy Resources, Production and Potentials," Prentice-Hall, Inc., Englewood Cliffs, 190 pp.

Hosterman, J.W. and Whitlow, S.I., 1981, Munsell color value as related to organic carbon in Devonian shale of Appalachian basin, American Association of Petroleum Geologists, Bulletin, 65:333-335.

Ibach, L.E.J., 1982, Relationship between sedimentation rate and total organic carbon content in ancient marine sediments, American Association of Petroleum Geologists, Bulletin, 66:170-188.

Imperial Mineral Resources Bureau, 1924, "Petroleum and Allied Products (1913-1919), The Mineral Industry of the British Empire and Foreign Countries.

Kazakov, A.V., 1937, The phosphorite facies and the genesis of phosphorites, in: "Geological Investigations of Agricultural Ores Transactions: Russian Scientific Institute Fertilizers and Insecto-Fungicides," No. 142, Publication of the 17th International Geological Congress, Leningrad, 95-113.

Keroher, G.C. et al., 1967, Lexicon of geologic names of the United States for 1936-1960, U.S. Geological Survey Bulletin 1200, Lexique Stratigraphique International, Vol. VII, Amérique du Nord, Fascicule 1, États-Unis, tome 1, A-F, 1448 pp.

Klingebiel, A., 1976, Sédiments et milieux sédimentaires dans le Golfe de Bénin, Bulletin du Centre de Recherches Pau-SNPA, 10: 129-148.

Kraemer, A.J., 1950, Oil shale in Brazil, Bureau of Mines Report of Investigations 4655, 36 pp.

Kummel, B., 1961, "History of the Earth," Second Edition, W.H. Freeman and Co., San Francisco, 707 pp.

Lewis, T.L. and Schwietering, J.F., 1971, Distribution of the Cleveland black shale in Ohio, Geological Society of America, Bulletin, 82:3477-3482.

Matveyev, A.K., editor, 1974, "Oil Shales Outside the Soviet Union," Deposits of Fossil Fuels, Vol. 4, G.K. Hall and Co., Boston.

McTavish, R.A. and Legg, D.P., 1976, The Ordovician of the Canning Basin, Western Australia, in: "The Ordovician System," M.G. Bassett, ed., University of Wales Press, Cardiff, Wales, 447-478.

Müller, P.J. and Suess, E., 1979, Productivity, sedimentation rate and sedimentary organic carbon content in the oceans. I. Organic carbon preservation, Deep-Sea Research, 26:1347-1362.

Nalivkin, D.V., 1973, "Geology of the U.S.S.R.," English translation by N. Rast; N. Rash and T.S. Westoll, eds., University of Toronto Press, Toronto, 855 pp.

Norris, A.W., 1967, Devonian of Northern Yukon/Territory and adjacent District of Mackenzie, in: "International Symposium on the Devonian System," D.H. Oswald, ed., Alberta Society of Petroleum Geologists, Calgary, 753-780.

North, F.K., 1971, The Cambrian of Canada and Alaska, in: "Cambrian of the New World," C.H. Holland, ed., Wiley-Interscience, London, 219-324.

Orr, W.L. and Gaines, A.G., Jr., 1973, Observations on rate of sulfate reduction and organic matter oxidation in the bottom waters of an estuarine basin: the upper basin of the Pettaquamscutt River (Rhode Island), Advances in Organic Geochemistry, 791-812.

Packam, G.H. et al., 1969, Ordovician System, in: "The Geology of New South Wales," G.H. Packam, ed., Journal of Geological Society of Australia, Vol. 16.

Padula, V.T., 1969, Oil shale of Permian Iratí Formation, Brazil, American Association of Petroleum Geologists, Bulletin, 53:591-602.

Pelet, R. and Debyser, Y., 1977, Organic geochemistry of Black Sea cores, Geochimica et Cosmochimica Acta, 41:1575-1586.
Pelzer, E.E., 1966, Mineralogy, geochemistry and stratigraphy of the Besa River Shale, British Columbia, Bulletin of Canadian Petroleum Geology, 14:273-321.
Peters, K.E., Sweeney, R.E. and Kaplan, I.R., 1978, Correlation of carbon and nitrogen stable isotope ratios in sedimentary organic matter, Limnology and Oceanography, 23:598-604.
Playford, P.E. et al., 1975, Phanerozoic, Geology of Western Australia, Western Australia Geological Survey Memoir, 2, 223-433.
Poole, W.H., Sanford, B.V., Williams, H., and Kelley, D.G., 1971, Geology of southeastern Canada, in: "Geology and Economic Minerals of Canada," R.J.W. Douglas, ed., Geological Survey of Canada, Economic Geology Report No. 1, Department of Energy, Mines and Resources, 228-304.
Porter, K.G. and Robbins, E.I., 1981, Zooplankton fecal pellets link fossil fuel and phosphate deposits, Science, 212:931-933.
Potter, P.E., Maynard, J.B. and Pryor, W.A., 1980, "Sedimentology of Shale," Springer-Verlag, New York, 310 pp.
Richards, F.A., 1971, Anoxic versus oxic environments, in: "Organic Matter in Natural Waters," D.W. Wood, ed., Institute of Marine Science Occasional Publication No. 1, University of Alaska, 399-411.
Richards, F.A. and Redfield, A.C., 1954, A correlation between the oxygen content of sea water and the organic content of marine sediments, Deep-Sea Research, 1:279-281.
Ronov, A.B., 1976, Volcanism, carbonate accumulation, life: the regularities of the global geochemistry of carbon, Geokyimiya, No. 8, 1252-1277 (in Russian).
Schindel, D.E., 1980, Microstratigraphic sampling and the limits of paleontologic resolution, Paleobiology, 6:408-426.
Schlanger, S.O. and Jenkyns, H.C., 1976, Cretaceous oceanic anoxic events: causes and consequences, Geologie en Mijnbouw, 55:179-184.
Schmoker, J.W., 1980, Organic content of Devonian shale in western Appalachian basin, American Association of Petroleum Geologists, Bulletin, 64:2156-2165.
Schopf, T.J.M., 1980, "Paleoceanography," Harvard University Press, Cambridge, 341 pp.
Sheldon, R.P., 1981, Ancient marine phosphorites, Annual Review of Earth and Planetary Sciences, 9:251-284.
Suess, E. and Müller, P.J., 1980, Productivity, sedimentation rate and sedimentary organic matter in the oceans, II. Elemental fractionation, in: "Biogéochimie de la matiére organique á l'interface eau-sédiment marin," Colloques Internationaux de Centre National de la Recherche Scientifique, No. 293, 17-26.
Swanson, V.E., 1960, Oil yield and uranium content of black shales, U.S. Geological Survey Professional Paper 365-A, 1-44.
Sweet, W.C., and Bergström, 1976, Conodont biostratigraphy of the Middle and Upper Ordovician of the United States midcontinent,

in: "The Ordovician System," M.G. Bassett, ed., University of Wales Press, Cardiff, 121-155.

Thomas, D.E., Spencer-Jones, D. and Tattam, C.M., 1976, Cambrian, in: "Geology of Victoria," J.G. Douglas and J.A. Ferguson, eds., Geological Society of Australia Special Publication No. 5, 11-24.

Thorsteinsson, R., and Tozer, E.T., 1970, Geology of the Arctic Archipelago, in: "Geology and Economic Minerals of Canada," R.J.W. Douglas, ed., Geological Survey of Canada, Economic Geology Report No. 1, Department of Energy, Mines and Resources, 548-590.

Vinogradov, V.A., et al., 1973, Main features of geologic structure and history of north-central Siberia, in: "Arctic Geology," M.G. Pitcher, ed., American Association of Petroleum Geologists, Memoir, 19, 181-193.

Wilson, J.T., 1966, Did the Atlantic close and then re-open? Nature, 211:676-681.

Wilson, T.R.S., 1978, Evidence for denitrification in aerobic pelagic sediments, Nature, 274:354-356.

Woolnough, W.G., 1937, Sedimentation in barred basins, and source rocks of oil, American Association of Petroleum Geologists, Bulletin, 31:1101-1157.

PARTICIPANTS

PARTICIPANTS

ABRANTES, Fatima Filomena, Servicos Geologicos de Portugal, Rua Academia das Ciencias, 19-2°, 1294 Lisboa Codex, Portugal

ALVEIRINHO DIAS, João Manuel, Servicos Geologicos de Portugal, Rua Academia das Ciencias, 19-2°, 1294 Lisboa Codex, Portugal

ARMENTROUT, John, Mobil Exploration and Production, P.O. Box 900, Dallas, Texas 75221, U.S.A.

BAIE, Lyle, Cities Service Company, P.O. Box 50408, Tulsa, Oklahoma 74110, U.S.A.

BARBER, Richard T., Marine Laboratory, Duke University, Beaufort, North Carolina 28516, U.S.A.

BRASSELL, Simon C., Organic Geochemistry Unit, School of Chemistry, University of Bristol, Bristol BS8 1TS, United Kingdom

BREMNER, J. Michael, Marine Geoscience Unit of the Geological Survey, University of Cape Town, Rondebosch 7700, South Africa

BRÖCKEL, Klaus von, Institute of Oceanography, University of Kiel, Düsternbrooker Weg 20, D-2300 Kiel 1, Federal Republic of Germany

BRONGERSMA-SANDERS, Margaretha, Houtlaan 3, 2334 CJ Leiden, The Netherlands

BRUMSACK, Hans J., Geochemisches Institut, Goldschmidtstr. 1, D-3400 Göttingen, Federal Republic of Germany

BURNETT, William C., Department of Oceanography, The Florida State University, Tallahassee, Florida 32306, U.S.A.

CALVERT, Stephen E., Department of Oceanography, The University of British Columbia, Vancouver, British Columbia, V6T 1W5, Canada

CODISPOTI, Louis A., Bigelow Laboratory for Ocean Sciences, McKown Point, W. Boothbay Harbor, Maine 04575, U.S.A.

COPELIN, Edward C., Union Oil Company of California, Science and Technology Division, 376 S. Valencia Avenue, P.O. Box 76, Brea, California 92621, U.S.A.

DENIS, Jérôme, Laboratoire de Géochimie des Eaux, Université de Paris 7, 2 Place Jussieu, F-75221 Paris Cedex 05, France

DEROO, Gérard, Institut Francais du Pétrole, 1 et 4, Avenue de Bois-Preau, B.P. 311, F-92506 Rueil-Malmaison Cedex, France

DIESTER-HAASS, Liselotte, Fachrichtung Geographie, Universität des Saarlandes, D-6600 Saarbrücken, Federal Republic of Germany

DOUGLAS, Robert G., Department of Geological Sciences, University of Southern California, Los Angeles, California 90007, U.S.A.

DUGDALE, Richard C., Department of Biological Sciences, University of Southern California, Los Angeles, California 90007, U.S.A.

DUNBAR, Robert B., Department of Geology, Rice University, Houston, Texas 77001, U.S.A.

DYER, Robin R., Robertson Research International, Ltd., 'Ty'N-Y-Coed' Llanrhos, Llandudno, Gwynedd, North Wales LL30 15A, United Kingdom

EGLINTON, Geoffrey, Organic Geochmistry Unit, School of Chemistry, University of Bristol, Contock's Close, Bristol BS8 1TS, United Kingdom

EINSELE, Gerhard, Institut und Museum für Geologie und Paläontologie, Universität Tübingen, Sigwartstrasse 10, D-7400 Tübingen, Federal Republic of Germany

FISCHER, Kathy, School of Oceanography, Oregon State University, Corvallis, Oregon 97331, U.S.A.

FIUZA, Armando, Oceanography Group, Department of Physics, University of Lisbon, Rua da Escola Politecnica 58, P-1200 Lisboa, Portugal

FLEET, Andrew, Geochemistry Branch, Exploration and Production Division, British Petroleum Research Center, Sudbury-on-Thames, Middlesex TW16 7LN, United Kingdom

FÜTTERER, Dieter, Alfred-Wegener-Institute for Polar Research, Columbus Center, D-2850 Bremerhaven, Federal Republic of Germany

GAGOSIAN, Robert B., Department of Chemistry, Woods Hole Oceanographic Institution, Woods Hole, Massachusetts 02543, U.S.A.

GANSSEN, Gerald, Geologisch-Paläontologisches Institut und Museum, Christian-Albrechts-Universität, Olshausenstrasse 40/60, D-2300 Kiel, Federal Republic of Germany

GARFIELD, Paula, Bigelow Laboratory for Ocean Sciences, McKown Point, W. Boothbay Harbor, Maine 04575, U.S.A.

GASPAR, Luis Caralho, Servicos Geologicos de Portugal, Rua Academia das Ciencias, 19-2°, 1294 Lisboa Codex, Portugal

GORSLINE, Donn S., Department of Geological Sciences, University of Southern California, Los Angeles, California 90007, U.S.A.

INGLE, James L., Department of Geology, Stanford University, Stanford, California 94305, U.S.A.

ITTEKKOT, Venu, Geologisch-Paläontologisches Institut und Museum, Universität Hamburg, Geomatikum, Bundesstrasse 55, D-2000 Hamburg, Federal Republic of Germany

JACOBS, Lucinda, Department of Oceanography, University of Washington, Seattle, Washington 98195, U.S.A.

JUILLET, Anne, Centre des Faibles Radioactivités, C.N.R.S., Place de l'église, F-91190 Gif-sur-Yvette, France

KELLER, George H., Oregon State University, School of Oceanography, Corvallis, Oregon 97331, U.S.A.

KEMPER, Edwin, Bundesanstalt für Geowissenschaften und Rohstoffe, Postfach 51 01 53, D-3000 Hannover 51, Federal Republic of Germany

KRISSEK, Lawrence A., Department of Geology, Ohio State University, Columbus, Ohio 43210, U.S.A.

LABRACHERIE, Monique, Département de Géologie et Oceanographie, Institut de Geologie du Bassin d'Aquitaine, Université de Bordeaux I, Avenue des Facultés, F-33405 Talence Cedex, France

MANGINI, Augusto, Institut für Umweltphysik, Universität Heidelberg Im Neuenheimer Feld 366, D-6900 Heidelberg, Federal Republic of Germany

MARTIN, John H., Moss Landing Marine Laboratories, P.O. Box 223, Moss Landing, California 95039, U.S.A.

MEYERS, Philip A., Department of Atmospheric and Oceanic Sciences, The University of Michigan, 2455 Haywood Avenue, Ann Arbor, Michigan 48109, U.S.A.

MITTELSTAEDT, Ekkehard, Deutsches Hydrographisches Institut, Bernhard-Nochtstrasse 78, Postfach 220, D-2000 Hamburg 3, Federal Republic of Germany

MOLINA-CRUZ, Adolfo, Instituto de Ciencias del Mar y Limnologia, Apartado Postal 70-305, 04510 Mexico, D.F., Mexico

MONTEIRO, Jose Hipolito, Servicos Geologicos de Portugal, Rua Academia das Ciencias, 19-2°, 1294 Lisboa Codex, Portugal

MOOERS, Christopher N.K., Naval Postgraduate School, Department of Oceanography, Monterey, California 93940, U.S.A.

MORRIS, Robert J., Institute of Oceanographic Sciences, Wormley, Godalming, Surrey GU8 5UB, United Kingdom

MÜLLER, German, Institut für Sedimentforschung, Universität, Heidelberg, Im Neuenheimer Feld 236, Postfach 10 30 20, D-6900 Heidelberg 1, Federal Republic of Germany

PARRISH, Judith Totman, U.S. Geological Survey, M.S. 940, Denver Federal Center, P.O. Box 25046, Denver, Colorado 80225, U.S.A.

PRELL, Warren L., Department of Geological Sciences, Brown University, Providence, Rhode Island 02912, U.S.A.

REIMERS, Clare E., Marine Biology Research Division, Scripps Institution of Oceanography, La Jolla, California 92093, U.S.A.

ROBINSON, Stephen W., U.S. Geological Survey, 345 Middlefield Road, Menlo Park, California 94025, U.S.A.

RULLKÖTTER, Jürgen, Institute of Petroleum and Organic Geochemistry, KFA Jülich G.m.b.H., P.O. Box 1913, D-5170 Jülich 1, Federal Republic of Germany

SARNTHEIN, Michael, Geologisch-Paläontologisches Institut und Museum, Christian-Albrechts-Universität, Olshausenstrasse 40/60, D-2300 Kiel, Federal Republic of Germany

SCHEIDEGGER, Kenneth F., School of Oceanography, Oregon State University, Corvallis, Oregon 97331, U.S.A.

SCHOPF, Thomas J.M., Geophysical Sciences Department, University of Chicago, 5734 S. Ellis Avenue, Chicago, Illinois 60637, U.S.A.

SCHRÖTER, Thomas, Brauerstr. 31, 1000 Berlin 45, Germany

SELLNER, Kevin G., Academy of Natural Sciences, Benedict Estuarine Research Laboratory, Benedict, Maryland 20612, U.S.A.

SHELDON, Richard P., U.S. Geologial Survey, Reston, Virginia 22092, U.S.A.

SHILLER, Alan M., Department of Earth and Planetary Sciences, Massachusetts Institute of Technology, Cambridge, Massachusetts 02139, U.S.A.

SIMONEIT, Bernd R.T., School of Oceanography, Oregon State University, Corvallis, Oregon 97331, U.S.A.

SMITH, Robert L., School of Oceanography, Oregon State University, Corvallis, Oregon 97331, U.S.A.

STARESINIC, Nick, Woods Hole Oceanographic Institution, Woods Hole, Massachusetts 02543, U.S.A.

STOFFERS, Peter, Institut für Sedimentforschung, Universität Heidelberg, Im Neuenheimer Feld 236, Postfach 10 30 20, D-6900 Heidelberg 1, Federal Republic of Germany

SUESS, Erwin, School of Oceanography, Oregon State University, Corvallis, Oregon 97331, U.S.A.

SUMMERHAYES, Colin P., BP Research Centre, Chertsey Road, Sunbury-on-Thames, Middlesex TW16 7LN, England, United Kingdom

THIEDE, Jörn, Institutt for geologi, Universitetet i Oslo, Postboks 1047, Blindern, Oslo 3, Norway

TRAGANZA, Eugene D., Naval Postgraduate School, Department of Oceanography, Monterey, California 92093, U.S.A.

VEEH, H. Herbert, School of Earth Sciences, The Flinders University of South Australia, Bedford Park, South Australia 5042, Australia

VORTISCH, Walter, Institut für Geologie und Paläontologie, Fachbereich Geowissenschaften, Phillips-Universität-Lahnberge, D-3550 Marburg/Lahn, Federal Republic of Germany

WASSMANN, Paul, Institutt for marin biologi, Universitetet i Bergen, N-5065 Blomsterdalen, Norway

WEFER, Gerold, Geologisch-Paläontologisches Institut und Museum, Christian-Albrechts-Universität, Olshausenstrasse 40/60, D-2300 Kiel, Federal Republic of Germany

WETZEL, Andreas, Institut und Museum für Geologie und Paläontologie, Universität Tübigen, Sigwartstrasse 10, D-7400 Tübingen, Federal Republic of Germany

WIEDMANN, Jost, Institut und Museum für Geologie und Paläontologie, Universität Tübigen, Sigwartstrasse 10, D-7400 Tübingen, Federal Republic of Germany

ZIMMERLE, Winfried, Deutsche Texaco AG, D-3109 Wietze/Celle, Federal Republic of Germany

INDEX